Introductory Incompressible Fluid Mechanics

This introduction to the mathematics of incompressible fluid mechanics and its applications keeps prerequisites to a minimum – only a background knowledge in multivariable calculus and differential equations is required. Part I covers inviscid fluid mechanics, guiding readers from the very basics of how to represent fluid flows through to the incompressible Euler equations and many real-world applications. Part II covers viscous fluid mechanics, from the stress/rate of strain relation to deriving the incompressible Navier–Stokes equations, through to Beltrami flows, the Reynolds number, Stokes flows, lubrication theory and boundary layers. Also included is a self-contained guide on the global existence of solutions to the incompressible Navier–Stokes equations. Students can test their understanding on 100 progressively structured exercises and look beyond the scope of the text with carefully selected mini-projects. Based on the authors' extensive teaching experience, this is a valuable resource for undergraduate and graduate students across mathematics, science and engineering.

Frank H. Berkshire is currently Principal Teaching Fellow in Dynamics in the Department of Mathematics at Imperial College London, where he has been a member of the Academic Staff since 1970 – latterly for 25 years as Director of Undergraduate Studies until formal 'retirement' in 2011. He has long-term experience of delivering lecture courses and projects, and has received awards for teaching excellence. He has promoted mathematics extensively in the UK and overseas, and is co-author with Tom Kibble of the textbook *Classical Mechanics* (1996, 2004). His research interests are in theoretical and practical dynamics, with wide application in, for example, waves, vortices, planetary motion, chaos, sport and gambling.

Simon J. A. Malham is Associate Professor of Mathematics at Heriot-Watt University. After obtaining his PhD from Imperial College London in 1993, he was a Visiting Assistant Professor at the University of Arizona (1993–6), a temporary lecturer at the University of Nottingham (1996–8) and Imperial College London (1998–2000), before joining Heriot-Watt as a lecturer in 2000. His research interests include Navier–Stokes regularity, the stability of nonlinear travelling fronts, computational spectral theory, the optimal simulation of stochastic differential equations including Lie group methods, Hopf algebra analysis and almost-exact methods, Grassmannian flows and the representation and integrability of non-commutative local and nonlocal nonlinear partial differential systems.

J. Trevor Stuart is Professor Emeritus in the Department of Mathematics at Imperial College London. After obtaining his PhD from Imperial in 1951, he worked at the National Physical Laboratory in the Aerodynamics Division from 1951 to 1966, before joining the Mathematics Department at Imperial in 1966. He retired in 1994. Professor Stuart was a Visiting Professor at MIT in 1956–7 and 1965–6, as well as at Brown

University in 1988–9. He became a Fellow of the Royal Society (FRS) in 1974 and was awarded the Otto Laporte Award from the American Physical Society in 1985, as well as the Senior Whitehead Prize from the London Mathematical Society in 1984. He holds honorary Doctor of Science degrees from Brown University and the University of East Anglia.

Introductory Incompressible Fluid Mechanics

FRANK H. BERKSHIRE
Imperial College of Science, Technology and Medicine, London

SIMON J. A. MALHAM
Heriot-Watt University, Edinburgh

J. TREVOR STUART
Imperial College of Science, Technology and Medicine, London

CAMBRIDGE
UNIVERSITY PRESS

CAMBRIDGE
UNIVERSITY PRESS

University Printing House, Cambridge CB2 8BS, United Kingdom

One Liberty Plaza, 20th Floor, New York, NY 10006, USA

477 Williamstown Road, Port Melbourne, VIC 3207, Australia

314–321, 3rd Floor, Plot 3, Splendor Forum, Jasola District Centre, New Delhi – 110025, India

103 Penang Road, #05–06/07, Visioncrest Commercial, Singapore 238467

Cambridge University Press is part of the University of Cambridge.

It furthers the University's mission by disseminating knowledge in the pursuit of education, learning, and research at the highest international levels of excellence.

www.cambridge.org
Information on this title: www.cambridge.org/highereducation/isbn/9781316513736
DOI: 10.1017/9781009075947

First published 2022

Printed in the United Kingdom by TJ Books Limited, Padstow, Cornwall 2022

A catalogue record for this publication is available from the British Library.

Library of Congress Cataloging-in-Publication Data
Names: Berkshire, F. H. (Frank H.), author. | Malham, Simon J. A., author. | Stuart, J. Trevor, author.
Title: Introductory incompressible fluid mechanics / Frank H. Berkshire, Imperial College of Science, Technology and Medicine, London, Simon J. A. Malham, Heriot-Watt University, Edinburgh, J. Trevor Stuart, Imperial College of Science, Technology and Medicine, London.
Description: Cambridge, United Kingdom ; New York, NY, USA : Cambridge University Press, 2022. | Includes bibliographical references and index.
Identifiers: LCCN 2021029734 | ISBN 9781316513736 (hardback)
Subjects: LCSH: Fluid mechanics.
Classification: LCC QA901 .B395 2022 | DDC 532/.0535–dc23
LC record available at https://lccn.loc.gov/2021029734

ISBN 978-1-316-51373-6 Hardback
ISBN 978-1-009-07470-4 Paperback

Additional resources for this publication at www.cambridge.org/berkshire-malham-stuart

To Rosie and Duncan, Charlie, Chloe and Emily

To Tamsin, Alexander, William and Amelia

To the late Christine Mary Stuart and Andrew, David and Katherine

Contents

Preface

The goal of this textbook is to introduce the mathematical theory of incompressible fluid mechanics and its applications. It grew out of lectures for a new course on the subject given by SJAM to final-year undergraduate as well as graduate students in autumn 2010. The basic material for the lectures was drawn from lecture notes on 'Ideal Fluid Mechanics' devised by Frank and lecture notes on 'Viscous Fluid Mechanics' devised by Trevor. In October 2015, David Tranah from Cambridge University Press suggested these lecture notes might be the seed for a textbook. After some delay, we all met in autumn 2018, and the plans and structure for the book were laid. Once writing was in full swing, the book grew and developed substantially, and slightly transformed its character. However, we hope the character of the book retains/includes the following essential features, it is accessible, comprehensive, mathematical, practical, engaging and useful. Let us briefly expand on these. We attempt to take interested students on a full journey. This journey starts with the very basic notion of how to represent the flow, carries the reader through analytical solutions and/or pragmatic approximations to practical flow scenarios, problems and applications, and ends at/on the issue of the global existence of solutions to the fluid equations themselves. The practical and mathematical go hand in hand, and a central theme of the book is the rigorous pursuit of the underlying mathematics and mathematical equations to obtain analytical solutions for the applied flow scenarios concerned. Another complementary and distinguishing theme is the plethora of extensive exercises we have included. These are styled progressively and aimed to complement the main theory and examples in the respective chapters.

We believe that Part I of the book (Inviscid Flow) is suitable for penultimate-year or final-year undergraduate students, while Part II (Viscous Flow) is suitable for final-year undergraduate and graduate students. This of course depends on mathematical background and knowledge. We include here mathematics, physics, science and engineering students, who are our target audience. We hope such students use this book as an introduction, springboard and long-term reference for their fluid mechanics knowledge and experience.

Many classical and contemporary classical textbooks have been invaluable in our preparation and exposition. Their influence is discernable throughout parts of the book. In particular, in general throughout, we relied on the classical textbooks by Batchelor, Currie, Kundu and Cohen, Lamb, Landau and Lifshitz, Lighthill, Panton, Paterson, Tritton and Van Dyke, as well as some classical, more contemporary textbooks by Childress, Chorin and Marsden, Majda and Bertozzi and Ockendon and Ockendon.

Other classical textbooks were invaluable to specific parts of the book and are cited at those junctures.

We would like to thank Darren Crowdy, Heiko Gimperlein, Robin Knops, Andrew Lacey, Marcel Oliver and Bernd Schroers for reading through the manuscript and their very helpful comments and suggestions. We are particularly grateful to Bernd Schroers for 'road-testing' the original notes and his invaluable suggestions for improvement. We also thank Daniel Coutand, Ioannis Stylianidis and Callum Thompson for their suggestions and input, as well as all the undergraduate/graduate students who pointed out typos along the way. We are also in debt to, and gratefully thank, the referees whose helpful constructive comments and suggestions significantly helped to improve the original raw manuscript. We would also like to thank Rachel Norridge, Anna Scriven and the whole production team at CUP for their invaluable help and the fantastic smooth job they made of the final production stages of the book. Lastly, many thanks to David Tranah of Cambridge University Press for seeing the potential, and then deftly and patiently guiding us through the whole writing and production process.
The Matlab files used to generate Figures 1.1, 1.2, 1.6, 2.6, 2.16, 3.14, 3.15, 3.16, 3.18, 4.4, 4.5, 5.6 and 6.4 are available on request from SJAM.

Introduction

The derivation of the equations of motion for an *ideal fluid* by Euler in 1755, and then for a *viscous fluid* by Navier (in 1822) and Stokes (in 1845), was a *tour-de-force* of eighteenth and nineteenth-century mathematics. These equations have been used to describe and explain so many physical phenomena around us in nature that currently billions of dollars of research grants in mathematics, science and engineering now revolve around them. They can be used to model the coupled atmospheric and ocean flow used by the meteorological office for weather prediction (generally incompressible flow) down to any application in chemical engineering you can think of, say to development of the thrusters on NASA's Apollo programme rockets (generally compressible flow). The incompressible *Navier–Stokes equations* are given by

$$\frac{\partial \boldsymbol{u}}{\partial t} + \boldsymbol{u} \cdot \nabla \boldsymbol{u} = \nu \nabla^2 \boldsymbol{u} - \frac{1}{\rho} \nabla p + \boldsymbol{f},$$

$$\nabla \cdot \boldsymbol{u} = 0,$$

where $\boldsymbol{u} = \boldsymbol{u}(\boldsymbol{x}, t)$ is a fluid velocity vector, $p = p(\boldsymbol{x}, t)$ is the pressure and $\boldsymbol{f} = \boldsymbol{f}(\boldsymbol{x}, t)$ is an external force field (per unit mass). The vector $\boldsymbol{x}$ records a position in space and t records the time elapsed. The constants ρ and ν are the mass density and kinematic viscosity, respectively. The frictional force due to stickiness of a fluid is represented by the term $\nu \nabla^2 \boldsymbol{u}$. An inviscid flow corresponds to the case $\nu = 0$, when the equations above are known as the incompressible *Euler equations* for an ideal flow. A viscous flow corresponds to $\nu > 0$ and the incompressible Navier–Stokes equations above. Herein we derive the incompressible Euler and Navier–Stokes equations and in the process learn about the subtleties of fluid mechanics and along the way see lots of interesting applications.

Part I of this book focuses on inviscid flow, while Part II focuses on viscous flow, though the two parts do naturally filter into and interact with each other. Our target audience is expected to have general *background mathematical knowledge* as follows, while further knowledge specific to particular chapters is outlined in the brief overview of the book just below. We expect the reader to have mastered some basic real analysis, multivariable differential and integral calculus, including knowledge of the gradient, divergence and curl operators, as well as path integration, Stokes' Theorem and the Divergence Theorem. Some knowledge of the solution methods for ordinary differential equations as well as basic methods of solution, such as separation of variables, for linear partial differential equations is also expected. Lastly, the chain rule for multivariable functions is so important we record it here. Given a scalar differentiable function $f = f(u, v, w)$ of three variables u, v and w, which are themselves scalar differentiable functions $u = u(x, y, z)$, $v = v(x, y, z)$ and $w = w(x, y, z)$ of three variables x, y and z, then we have:

$$\frac{\partial f}{\partial x} = \frac{\partial f}{\partial u}\frac{\partial u}{\partial x} + \frac{\partial f}{\partial v}\frac{\partial v}{\partial x} + \frac{\partial f}{\partial w}\frac{\partial w}{\partial x},$$

$$\frac{\partial f}{\partial y} = \frac{\partial f}{\partial u}\frac{\partial u}{\partial y} + \frac{\partial f}{\partial v}\frac{\partial v}{\partial y} + \frac{\partial f}{\partial w}\frac{\partial w}{\partial y},$$

$$\frac{\partial f}{\partial z} = \frac{\partial f}{\partial u}\frac{\partial u}{\partial z} + \frac{\partial f}{\partial v}\frac{\partial v}{\partial z} + \frac{\partial f}{\partial w}\frac{\partial w}{\partial z},$$

with the obvious extension to more variables if required. The matrix version of this identity $\nabla_{\boldsymbol{x}} f = (\nabla_{\boldsymbol{x}} \boldsymbol{u})^{\mathrm{T}} \nabla_{\boldsymbol{u}} f$ is often convenient, where $\boldsymbol{x} = (x, y, z)^{\mathrm{T}}$ and $\boldsymbol{u} = (u, v, w)^{\mathrm{T}}$ are vectors with the superscript 'T' denoting transpose, and ∇ represents the gradient operator.

There are seven chapters in this book, three in Part I and four in Part II. A brief *overview* is as follows. In Chapter 1 we discuss how to represent or record the fluid flow in terms of the local velocity field. We also consider the law of conservation of mass which locally, mathematically, is represented by the continuity equation. With this in hand we can quantify what the adjective 'incompressible' means in our title. We further quantify the transport of parcels of fluids and their acceleration. We round off the chapter by introducing the vorticity vector field and the rate of strain matrix. We use Newton's Second Law in Chapter 2, resolving the relevant forces and locally equating them to mass times acceleration, to derive the evolution equation for the fluid velocity field, the Euler equations for an inviscid fluid. The rest of this chapter explores many and varied applications of the Euler equations to real-world fluid flows from the Venturi tube through to channel flow and water waves. We consider a special class of inviscid fluid flow in Chapter 3, one where the flow is two-dimensional and the vorticity vector, which quantifies local fluid rotation, is everywhere zero. In other words, the fluid flow is irrotational. Remarkably, such a flow both applies to aid the basic understanding of dynamic lift of an aerofoil and thus flight, and can be represented by complex variables. Indeed, complex analysis plays a crucial role in the analysis of this important real-world application, and a basic knowledge is assumed. Further subtleties of this application requiring a viscous flow prescription are revisited in Chapter 6, discussed presently. Part II begins with Chapter 4, in which we again apply Newton's Second Law to derive the incompressible Navier–Stokes equations, only this time we include more realistic viscous or diffusive forces. Resolving the viscous force is a much more subtle affair and a large section is devoted to its careful derivation. Some background knowledge on eigenvalues and eigenvectors is assumed. The rest of the chapter is concerned with special classes of incompressible flows that have exact solutions and finishes by introducing the Reynolds number. Chapter 5 looks at flows which are characterised by small Reynolds numbers, which roughly translates to them being either akin to sticky flows, think of treacle or syrup, or flows restricted to a thin layer. The former flows can be approximated by the simpler 'Stokes flow', while the latter flows are relevant to lubrication theory, and together they constitute the two main sections of this chapter. The exercises in this chapter explore the consequences and limits of such approximations. Though there is no need for it to be a prerequisite, some asymptotic analysis is utilised in a few of these exercises. The practical theory of flight is reprised in Chapter 6, where the notion of boundary layers and the equations prescribing them are introduced. Boundary layer and/or flow separation is also discussed, though a full quantitative treatment is beyond the scope of this book. Finally, Chapter 7 covers the basic results concerning the well-posedness and regularity of the three-dimensional incompressible Navier–Stokes equations. A comprehensive analysis of the classical regularity results is presented and/or given as explicit, carefully staged exercises. Some basic knowledge of functional and Fourier analysis is assumed. See below for more details on this chapter. The full journey through all seven chapters spans a large section of applied mathematical knowledge. In practice, for most students and/or courses this will often not be required or expected, and certainly not in a single semester or maybe even two semesters.

A natural associated topic to remark on at this point is that of fluid 'turbulence'. For the sake of brevity we have not included any discussion on this herein, apart from naturally occuring allusions to the topic when we briefly mention the stochastically driven Navier–Stokes equations in Section 4.9(7) and turbulent boundary layers in Chapter 6. Such, possibly chaotic, features are relevant for weather prediction, see Section 2.10(7), and many problems in engineering and physics.

We consider a unique feature of the book to be the breadth and depth of the *exercises* provided at the end of each chapter. There are nearly one hundred pages of exercises. They are an integral part of the learning process and knowledge acquisition intended for the reader. They either consolidate knowledge or extend knowledge given in the main text of the chapter. All the exercises are staged, i.e. they are split into structured progressive parts which combine to establish an overall result or outcome. Each part generally states a partial goal/result to achieve. Thus, if the reader is temporarily stuck completing one part, they can still use the partial result in the following parts to try to establish/complete them. A large proportion of the exercises have been questions on past exam papers and some of the more advanced/extended exercises could certainly be used as components of a take-home exam or project. Indeed, at the end of most chapters we also present some *projects* that involve understanding some particular material of the chapter, completing the relevant exercises and using that as a basis to explore further the topic/concept concerned. At the end of most chapters is also a *notes* section, where further details, applications, extensions, questions or related material are explored. We also use these sections to cover, briefly, specific technical material beyond the scope of the main text in the chapter.

Another unique feature of the book is Chapter 7. The material therein on the regularity of solutions to the three-dimensional incompressible Navier–Stokes equations is not normally included in textbooks at this level. However, the importance of this material and question, especially as it constitutes one of the Millennium Prize Problems, impels us to include it. The goal is to make the material and results accessible, and to be as direct and self-contained as possible. To achieve this we have taken a slightly different approach to the standard one and have chosen to prove the existence of weak solutions via a Fourier-series representation for the solution. This means our setting is countably discrete and we only require more elementary concepts from real analysis, convergence, compactness and so forth. We have endeavoured to cover the classical results on weak and strong solutions as comprehensively, explicitly and succinctly as possible.

Lastly, let us remark on modelling and analytical solution strategies. Our focus herein is on incompressible flow scenarios. Liquid flows are in general incompressible, indeed low-Mach-number gas flows can be considered as such. Another distinguishing issue is whether it is sufficient to model the scenario by the inviscid flow Euler equations, or viscous flow Navier–Stokes equations, or both – such as in boundary-layer theory where in one region viscosity effects are important while in another viscous effects are negligible. The issue we wish to emphasise is this. Given a modelling scenario to which either the incompressible Euler or Navier–Stokes equations apply, the goal is to derive analytical solutions either to these equations themselves or to reduced versions of these equations under some underlying physical or geometric assumptions. Hence, for example, the flow may be particularly sticky or a slow/creeping flow corresponding to a low Reynolds number and to a very good approximation we can reduce the incompressible Navier–Stokes equations to the Stokes equation. Or the flow may take place in an asymptotically thin fluid layer and to a very good approximation we can reduce the incompressible Navier–Stokes equations to the Reynolds lubrication equation, and so forth. In all such cases we should utilise further properties. For example, to an accurate approximation is the flow scenario rotationally invariant about a given axis, i.e. axisymmetric? Is there any swirl? Is the flow steady/stationary? Any such symmetries can be utilised to simplify the model equations further to obtain a more tractable reduced system of differential equations. Once all natural symmetries have been taken into account, can we solve the resulting system of differential equations? To find solutions to such reduced equations it is often useful to make a guess at the general form of the solution flow, i.e. pose a solution ansatz. For example, the boundary conditions may indicate an overall solution form allowing us to reduce the approximate model equations further. Or prescribing such a specific flow ansatz in particular flow regions can also be helpful, for example, is the flow near the axis of rotation in an axisymmetric flow approximately a solid-body flow? In which case we might want to employ that fact to represent the flow in that region and couple it by continuity to the neighbouring flow further away from the axis of symmetry. Having obtained a solution, is it physically realistic? Does it represent what is observed? Does it predict flow behaviour not hitherto seen? For example, Moffatt vortices were predicted before they were observed! How can we utilise the knowledge we have gained to optimise processes, such as energy transmission from a wind turbine, or via the introduction of a control mechanism into the system, pushing the system towards a given desired state? As a simple example, the peloton of riders in the Tour de France will change its shape from a V-shaped headed mass in head winds to different-sized echelons depending on the direction of cross winds, in order to minimise energy expenditure due to air resistance. Besides the practical flow scenarios touched upon in this book, there are many more real-world flows the analysis and equations developed herein can be applied to and we should be on the lookout for. These days there are also many high-quality videos available online of all sorts of fluid-flow phenomena, to add to the classical *An Album of Fluid Motion* by Van Dyke. Chris Hadfield's experiments on the International Space Station also provide an out-of-this-world fluid-flow and surface-tension experience.

Part I

Inviscid Flow

1 Flow and Transport

1.1 Fluids and the Continuum Hypothesis

A material exhibits *flow* if shear forces, however small, lead to a deformation which is unbounded – we could use this as a definition of a *fluid*. A *solid* has a fixed shape, or at least a strong limitation on its deformation when force is applied to it. Within the category of 'fluids', we include liquids and gases. The main distinguishing feature between these two fluids is the notion of compressibility. Gases are usually compressible – as we know from everyday aerosols. Liquids are generally incompressible – a feature essential to all modern car brakes. However, some gas flows can also be incompressible, particularly at low speeds.

Fluids can be further subcategorised. There are *ideal* or *inviscid* fluids. In such fluids, the *only* internal force present is pressure, which acts so that fluid flows from a region of high pressure to one of low pressure. The equations for an ideal fluid have been applied to wing and aircraft design (as a limit of high Reynolds-number flow). However, fluids can exhibit internal frictional forces which model a 'stickiness' property of the fluid which involves energy loss – these are known as *viscous* fluids. Some fluids/material known as 'non-Newtonian or complex fluids' exhibit even stranger behaviour, their reaction to deformation may depend on: (i) past history (earlier deformations), for example some paints; (ii) temperature, for example some polymers or glass; (iii) the size of the deformation, for example some plastics or silly putty.

For any *real* fluid there are three natural length scales:

1. $L_{\text{molecular}}$, the molecular scale characterised by the mean-free-path distance of molecules between collisions;
2. L_{fluid}, the medium scale of a fluid parcel, the fluid droplet in the pipe or ocean flow;
3. L_{macro}, the macro-scale which is the scale of the fluid geometry, the scale of the container the fluid is in, whether a beaker or an ocean.

And, of course, we have the asymptotic inequalities

$$L_{\text{molecular}} \ll L_{\text{fluid}} \ll L_{\text{macro}}.$$

Continuum Hypothesis. We will assume that the properties of an elementary volume/parcel of fluid, however small, are the same as for the fluid as a whole – i.e. we suppose that the properties of the fluid at scale L_{fluid} propagate all the way down and through the molecular scale $L_{\text{molecular}}$. This is the *continuum assumption*. For everyday fluid mechanics engineering, this assumption is extremely accurate, see Chorin and Marsden (1990, p. 2).

1.2 Conservation Principles

Our derivation of the basic equations underlying the dynamics of fluids is based on three basic conservation principles:

1. *Conservation of mass*, mass is neither created or destroyed.
2. *Newton's Second Law/balance of momentum*, for a parcel of fluid the rate of change of momentum equals the force applied to it.
3. *Conservation of energy*, energy is neither created nor destroyed.

In turn these principles generate the:

1. *Continuity equation*, which governs how the density of the fluid evolves locally and thus indicates compressibility properties of the fluid.
2. *Navier–Stokes equations* of motion for a fluid, which indicate how the fluid moves around from regions of high pressure to those of low pressure and with the effects of viscosity.
3. *Equation of state*, which indicates the mechanism of energy exchange within the fluid.

1.3 Fluid Prescription

A crucial task is to decide how we wish to represent the fluid flow. We will use the Eulerian prescription as follows. Consider a fluid in a container, what information do we need in order to fully describe the 'state' of the fluid flow? Well, imagine that at every fixed spatial position in the fluid we placed a weather vane that could pivot three-dimensionally, i.e. at every fixed position there is a pointer that points in the direction (three-dimensional) the fluid is flowing at that position. Further, suppose the vane/pointer is also able to record the speed with which the fluid is flowing at that position. If we know the fluid-flow direction and speed at each spatial position, then we know the fluid-flow velocity vector at those positions. Of course that velocity vector could change with time. Thus at any given time the prescription of the velocity vector at every fixed spatial position in the fluid flow, along with some other concomitant fluid-related quantities such as the scalar pressure and mass density given at each spatial position, should be enough to describe the 'state' of the fluid. Given the velocity field at every position and time, we know the fluid flow and we can in principle determine fluid-particle trajectories under that flow. This is the focus of the next section.

1.4 Trajectories and Streamlines

Suppose that our fluid is contained within a region/domain $\mathcal{D} \subseteq \mathbb{R}^d$, where $d = 3$, and we use Cartesian coordinates $\boldsymbol{x} = (x, y, z)^{\mathrm{T}} \in \mathcal{D}$ to label points/positions in $\mathcal{D}$. Imagine a small fluid particle or a speck of dust moving in a fluid flow field prescribed by the *velocity field* $\boldsymbol{u} = \boldsymbol{u}(\boldsymbol{x}, t)$, where $\boldsymbol{u}$ has $d = 3$ components as follows: $\boldsymbol{u} = (u, v, w)^{\mathrm{T}}$. Suppose the position of the particle at time t is recorded by the variables $\boldsymbol{x}(t) = (x(t), y(t), z(t))^{\mathrm{T}}$, i.e. by the vector $\boldsymbol{x}(t) = x(t)\boldsymbol{i} + y(t)\boldsymbol{j} + z(t)\boldsymbol{k}$, where $\boldsymbol{i}$, $\boldsymbol{j}$ and $\boldsymbol{k}$ are the unit vectors in the respective coordinate directions x, y and z. We thus have the following equations for the velocity of the particle at time t at position $\boldsymbol{x}(t) = (x(t), y(t), z(t))^{\mathrm{T}}$:

$$\frac{\mathrm{d}}{\mathrm{d}t}x(t) = u(x(t), y(t), z(t), t),$$

$$\frac{\mathrm{d}}{\mathrm{d}t}y(t) = v(x(t), y(t), z(t), t),$$

$$\frac{\mathrm{d}}{\mathrm{d}t}z(t) = w(x(t), y(t), z(t), t).$$

Definition 1.1 (Particle path or trajectory) The *particle path* or *trajectory* of a fluid particle is the curve traced out by the particle as time progresses. If the particle starts at position $\boldsymbol{x}_0 = (x_0, y_0, z_0)^{\mathrm{T}}$ then its particle path $\boldsymbol{x} = \boldsymbol{x}(t)$ is the solution to the following system of differential equations (the same as those above but here in shorter vector notation) with initial conditions $\boldsymbol{x}(0) = \boldsymbol{x}_0$:

$$\frac{\mathrm{d}}{\mathrm{d}t}\boldsymbol{x}(t) = \boldsymbol{u}(\boldsymbol{x}(t), t).$$

Definition 1.2 (Streamline) A *streamline* is an integral curve of the velocity field $\boldsymbol{u} = \boldsymbol{u}(\boldsymbol{x}, t)$ for t fixed, i.e. it is a curve $\boldsymbol{x} = \boldsymbol{x}(s)$ parameterised by the variable s, that satisfies the following system of differential equations with t held constant and $\boldsymbol{x}(0) = \boldsymbol{x}_0$ at $s = 0$:

$$\frac{\mathrm{d}}{\mathrm{d}s}\boldsymbol{x}(s) = \boldsymbol{u}(\boldsymbol{x}(s), t).$$

Remark 1.3 (Stationary/steady flows) Flows for which $\partial\boldsymbol{u}/\partial t = \boldsymbol{0}$ are said to be *stationary/steady*. For such flows the velocity field $\boldsymbol{u}$ is time-independent, so $\boldsymbol{u} = \boldsymbol{u}(\boldsymbol{x})$ only, and trajectories and streamlines coincide.

Example 1.4 (Solid-body rotation flow) Suppose a velocity field $\boldsymbol{u} = (u, v, w)^{\mathrm{T}}$ depends on position $\boldsymbol{x}$ only and is given by

$$\begin{pmatrix} u \\ v \\ w \end{pmatrix} = \begin{pmatrix} -\Omega y \\ \Omega x \\ 0 \end{pmatrix}$$

for some non-zero constant $\Omega \in \mathbb{R}$. The particle path for a particle that starts at $\boldsymbol{x}_0 = (x_0, y_0, z_0)^{\mathrm{T}}$ is the integral curve of the system of differential equations

$$\begin{aligned}\frac{\mathrm{d}x}{\mathrm{d}t} &= -\Omega y,\\ \frac{\mathrm{d}y}{\mathrm{d}t} &= \Omega x,\\ \frac{\mathrm{d}z}{\mathrm{d}t} &= 0,\end{aligned}$$

with initial condition $\boldsymbol{x}(0) = \boldsymbol{x}_0$. This is a coupled pair of differential equations as the solution to the last equation is $z(t) = z_0$ for all $t \geqslant 0$. There are several methods for solving the pair of equations, one method is as follows. Differentiating the first equation with respect to t we find

$$\frac{\mathrm{d}^2x}{\mathrm{d}t^2} = -\Omega\frac{\mathrm{d}y}{\mathrm{d}t} \qquad \Leftrightarrow \qquad \frac{\mathrm{d}^2x}{\mathrm{d}t^2} = -\Omega^2 x.$$

In other words, we are required to solve the linear second-order differential equation for $x = x(t)$ shown. The general solution is

$$x(t) = A\cos(\Omega t) + B\sin(\Omega t),$$

where A and B are arbitrary constants. We now find $y = y(t)$ by substituting this solution for $x = x(t)$ into the first differential equation above as follows:

$$\begin{aligned}y(t) &= -\frac{1}{\Omega}\frac{\mathrm{d}x}{\mathrm{d}t}\\ &= -\frac{1}{\Omega}(-\Omega A\sin(\Omega t) + \Omega B\cos(\Omega t))\\ &= A\sin(\Omega t) - B\cos(\Omega t).\end{aligned}$$

Using that $x(0) = x_0$ and $y(0) = y_0$ we find that $A = x_0$ and $B = -y_0$, so the particle path of the particle that is initially at $\boldsymbol{x}_0 = (x_0, y_0, z_0)^{\mathrm{T}}$ is given by

$$\begin{aligned}x(t) &= x_0\cos(\Omega t) - y_0\sin(\Omega t),\\ y(t) &= x_0\sin(\Omega t) + y_0\cos(\Omega t),\\ z(t) &= z_0.\end{aligned}$$

This particle thus traces out a horizontal circular particle path at height $z = z_0$ of radius $(x_0^2 + y_0^2)^{1/2}$. Since this flow is stationary, streamlines coincide with particle paths for this flow. See Figure 1.1.

Example 1.5 (Two-dimensional oscillating flow) Consider the two-dimensional flow field $\boldsymbol{u} = (u, v)^{\mathrm{T}}$, which depends on the two-dimensional position $\boldsymbol{x} = (x, y)^{\mathrm{T}}$ vector and time $t \geqslant 0$, given by

$$\begin{pmatrix}u\\ v\end{pmatrix} = \begin{pmatrix}u_0\\ v_0\cos(kx - \alpha t)\end{pmatrix},$$

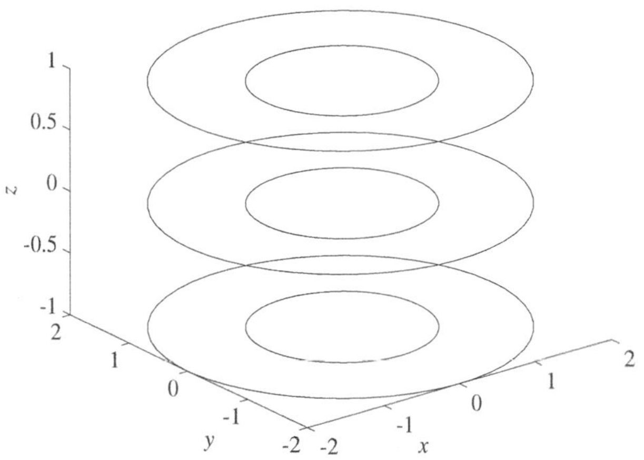

Figure 1.1 Six particle paths corresponding to the solid-body rotation flow in Example 1.4 are shown. Three correspond to the initial positions $x_0 = \sqrt{2}$, $y_0 = \sqrt{2}$ and z_0 equals either -1, 0 or 1. The other three particle paths correspond to $x_0 = \sqrt{2}/2$, $y_0 = \sqrt{2}/2$ and again z_0 equals either -1, 0 or 1. The particles trace out horizontal circular paths, with those tracing out paths, of larger radii travelling faster. All the particles complete one revolution in the same time, hence thc nomination as a solid-body rotation flow.

where u_0, v_0, k and α are constants. Let us find the particle path and streamline for the particle at $\boldsymbol{x}_0 = (x_0, y_0)^{\mathrm{T}} = (0, 0)^{\mathrm{T}}$ at $t = 0$. Starting with the *particle path*, we are required to solve the coupled pair of differential equations

$$\begin{aligned}\frac{\mathrm{d}x}{\mathrm{d}t} &= u_0,\\ \frac{\mathrm{d}y}{\mathrm{d}t} &= v_0 \cos(kx - \alpha t).\end{aligned}$$

We can solve the first differential equation, which tells us

$$x(t) = u_0 t,$$

where we used that $x(0) = 0$. We now substitute this expression for $x = x(t)$ into the second differential equation and integrate with respect to time using $y(0) = 0$ as follows:

$$\begin{aligned}&\frac{\mathrm{d}y}{\mathrm{d}t} = v_0 \cos((ku_0 - \alpha)t)\\ \Leftrightarrow\quad & y(t) = 0 + \int_0^t v_0 \cos((ku_0 - \alpha)\tau)\,\mathrm{d}\tau\\ \Leftrightarrow\quad & y(t) = \frac{v_0}{ku_0 - \alpha} \sin((ku_0 - \alpha)t).\end{aligned}$$

If we eliminate time t between the formulae for $x = x(t)$ and $y = y(t)$ we find that the trajectory through $(0, 0)^{\mathrm{T}}$ is

$$y = \frac{v_0}{ku_0 - \alpha} \sin\Big(\Big(k - \frac{\alpha}{u_0}\Big)x\Big).$$

To find the *streamline* through $(0,0)^{\mathrm{T}}$, we fix t and solve the pair of differential equations

$$\frac{\mathrm{d}x}{\mathrm{d}s} = u_0,$$
$$\frac{\mathrm{d}y}{\mathrm{d}s} = v_0 \cos(kx - \alpha t).$$

As above we can solve the first equation so that $x(s) = u_0 s$ using that $x(0) = 0$. We can substitute this into the second equation and integrate with respect to s, remembering that t is constant, to get

$$\begin{aligned} & \frac{\mathrm{d}y}{\mathrm{d}s} = v_0 \cos(k u_0 s - \alpha t) \\ \Leftrightarrow\quad & y(s) = 0 + \int_0^s v_0 \cos(k u_0 r - \alpha t)\,\mathrm{d}r \\ \Leftrightarrow\quad & y(s) = \frac{v_0}{k u_0}\big(\sin(k u_0 s - \alpha t) - \sin(-\alpha t)\big). \end{aligned}$$

If we eliminate the parameter s between $x = x(s)$ and $y = y(s)$ above, we find the equation for the streamline is

$$y = \frac{v_0}{k u_0}\big(\sin(kx - \alpha t) + \sin(\alpha t)\big).$$

The equation of the streamline through $(0,0)^{\mathrm{T}}$ at time $t = 0$ is thus given by

$$y = \frac{v_0}{k u_0}\sin(kx).$$

As the underlying flow is *not* stationary, as expected, the particle path and streamline through $(0,0)^{\mathrm{T}}$ at time $t = 0$ are distinguished; see Figure 1.2. Finally, let us examine

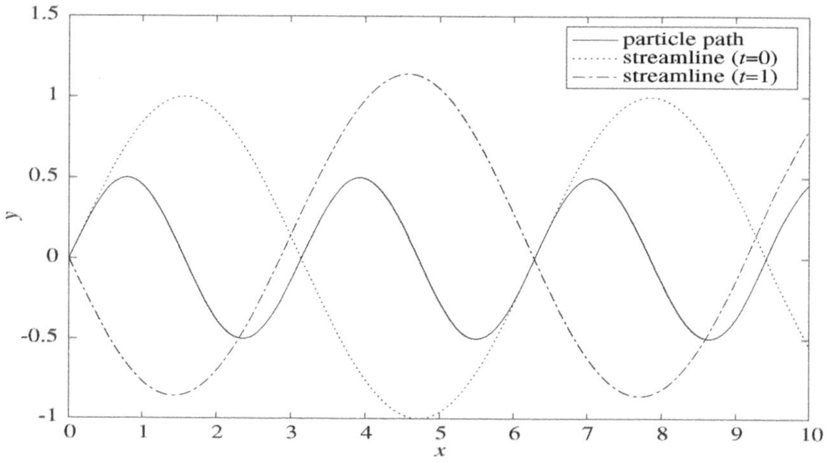

Figure 1.2 For the oscillatory flow in Example 1.5, we plot both the particle path associated with the particle starting from the origin (solid line) and the streamlines through the origin when time is instantaneously frozen, first at $t = 0$ (dotted line) and second at $t = 1$ (dash-dotted line). The other parameters were fixed as $k = 1$, $\alpha = 3$, $u_0 = 1$ and $v_0 = 1$.

two special limits for this flow. As $\alpha \to 0$ the flow becomes stationary and correspondingly the particle path and streamline coincide. As $k \to 0$ the flow is not stationary. In this limit the particle path through $(0,0)^{\mathrm{T}}$ is $y = (v_0/\alpha)\sin(\alpha x/u_0)$, i.e. it is sinusoidal, whereas the streamline is given by $x = u_0 s$ and $y = v_0 s$, which is a straight line through the origin.

Remark 1.6 (Streaklines) A streakline is the locus of all the fluid elements which at some time have passed through a particular point, say $(x_0, y_0, z_0)^{\mathrm{T}}$. We can obtain the equation for a streakline through $(x_0, y_0, z_0)^{\mathrm{T}}$ by solving the ordinary differential equations $(\mathrm{d}/\mathrm{d}t)\boldsymbol{x}(t) = \boldsymbol{u}(\boldsymbol{x}(t), t)$ assuming at $t = t_0$ we have $(x(t_0), y(t_0), z(t_0))^{\mathrm{T}} = (x_0, y_0, z_0)^{\mathrm{T}}$. Eliminating t_0 between the equations generates the streakline corresponding to $(x_0, y_0, z_0)^{\mathrm{T}}$. For example, ink dye injected at the point $(x_0, y_0, z_0)^{\mathrm{T}}$ in the flow will trace out a streakline.

1.5 Continuity Equation

Recall that we suppose our fluid is contained within a region/domain $\mathcal{D} \subseteq \mathbb{R}^d$. Here we assume $d = 3$, but everything we say is true for the collapsed two-dimensional case $d = 2$. Hence $\boldsymbol{x} = (x, y, z)^{\mathrm{T}} \in \mathcal{D}$ is a position/point in $\mathcal{D}$. At each time t we suppose that the fluid has a well-defined *mass density* $\rho = \rho(\boldsymbol{x}, t)$ at the point $\boldsymbol{x}$. Indeed, invoking the continuum hypothesis, at each time t we can compute the mass of fluid inside a small volume centred at $\boldsymbol{x}$, and then consider the ratio of that mass to the volume in the limit as the volume shrinks to zero around the point $\boldsymbol{x}$. The limiting ratio generates the mass density $\rho = \rho(\boldsymbol{x}, t)$. In addition, we note that each fluid particle traces out a well-defined path in the fluid, and its motion along that path is governed by the *velocity field* $\boldsymbol{u} = \boldsymbol{u}(\boldsymbol{x}, t)$ at position $\boldsymbol{x}$ at time t. Consider an arbitrary fixed subregion $\mathcal{V} \subseteq \mathcal{D}$; see Figure 1.3. The total mass of fluid contained inside the region $\mathcal{V}$ at time t is

$$\int_{\mathcal{V}} \rho(\boldsymbol{x}, t)\, \mathrm{d}V(\boldsymbol{x}),$$

where $\mathrm{d}V = \mathrm{d}V(\boldsymbol{x})$ is the volume measure in $\mathbb{R}^d$. Let us now consider the rate of change of mass inside $\mathcal{V}$. By the principle of conservation of mass, the rate of increase

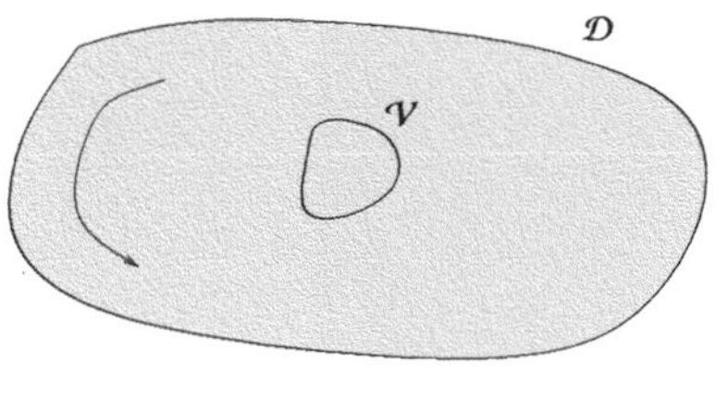

Figure 1.3 The fluid of mass density $\rho(\boldsymbol{x}, t)$ swirls around inside the container $\mathcal{D}$, while $\mathcal{V}$ is an arbitrary fixed subregion.

of the mass in $\mathcal{V}$ is given by the mass of fluid entering/leaving the boundary $\partial\mathcal{V}$ of $\mathcal{V}$ per unit time.

To compute the total mass of fluid entering/leaving the boundary $\partial\mathcal{V}$ per unit time, we consider a small area patch $\mathrm{d}S = \mathrm{d}S(\boldsymbol{x})$ on the boundary of $\partial\mathcal{V}$, which has unit outward normal $\boldsymbol{n}$. The total mass of fluid flowing out of $\mathcal{V}$ through the area patch $\mathrm{d}S = \mathrm{d}S(\boldsymbol{x})$ per unit time is (where '×' is just scalar multiplication)

$$\text{mass density} \times \text{fluid volume leaving per unit time}$$

which is, to leading order,

$$\rho(\boldsymbol{x},t) \times \boldsymbol{u}(\boldsymbol{x},t) \cdot \boldsymbol{n}(\boldsymbol{x})\,\mathrm{d}S(\boldsymbol{x}),$$

where, say, $\boldsymbol{x}$ is at the centre of the area patch $\mathrm{d}S$ on $\partial\mathcal{V}$. Note that to estimate the fluid volume leaving per unit time we have decomposed the fluid velocity at $\boldsymbol{x} \in \partial\mathcal{V}$, time t, into velocity components normal $(\boldsymbol{u} \cdot \boldsymbol{n})$ and tangent to the surface $\partial\mathcal{V}$ at that point. The velocity component tangent to the surface pushes fluid along the surface – no fluid enters or leaves $\mathcal{V}$ via this component. Hence we only retain the normal component – see Figure 1.4.

Returning to the principle of conservation of mass, this is now equivalent to the *integral form of the law of conservation of mass*, which is given by

$$\frac{\mathrm{d}}{\mathrm{d}t}\int_{\mathcal{V}} \rho\,\mathrm{d}V = -\int_{\partial\mathcal{V}} \rho\boldsymbol{u} \cdot \boldsymbol{n}\,\mathrm{d}S.$$

That the rate of change of the total mass in $\mathcal{V}$ equals the total rate of change of mass density in $\mathcal{V}$, and the divergence theorem, imply respectively

$$\frac{\mathrm{d}}{\mathrm{d}t}\int_{\mathcal{V}} \rho\,\mathrm{d}V = \int_{\mathcal{V}} \frac{\partial\rho}{\partial t}\,\mathrm{d}V \quad \text{and} \quad \int_{\partial\mathcal{V}} (\rho\boldsymbol{u}) \cdot \boldsymbol{n}\,\mathrm{d}S = \int_{\mathcal{V}} \nabla \cdot (\rho\boldsymbol{u})\,\mathrm{d}V.$$

Using these two relations, the law of conservation of mass is equivalent to

$$\int_{\mathcal{V}} \frac{\partial\rho}{\partial t}\,\mathrm{d}V = -\int_{\mathcal{V}} \nabla \cdot (\rho\boldsymbol{u})\,\mathrm{d}V \quad \Leftrightarrow \quad \int_{\mathcal{V}} \left(\frac{\partial\rho}{\partial t} + \nabla \cdot (\rho\boldsymbol{u})\right)\mathrm{d}V = 0.$$

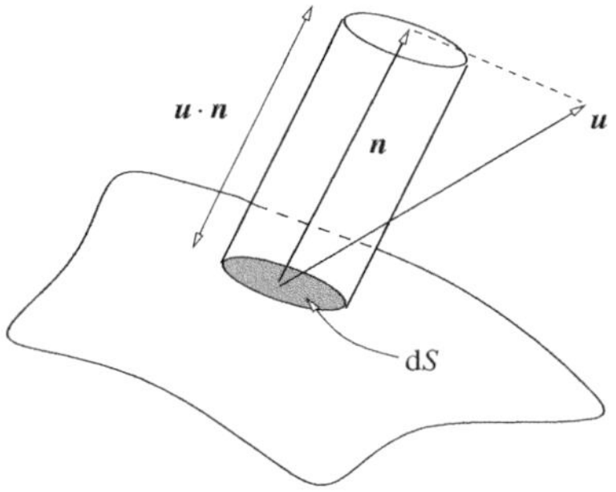

Figure 1.4 The total mass of fluid moving through the patch $\mathrm{d}S$ on the surface $\partial\mathcal{V}$ per unit time is given by the mass density $\rho(\boldsymbol{x},t)$ times the volume of the cylinder shown, which is $\boldsymbol{u} \cdot \boldsymbol{n}\,\mathrm{d}S$.

We now use that $\mathcal{V}$ is arbitrary to deduce the *differential form of the law of conservation of mass* or *continuity equation* applied pointwise, as follows.

Theorem 1.7 (Continuity equation) *Given a velocity field $\boldsymbol{u} = \boldsymbol{u}(\boldsymbol{x}, t)$, the mass density $\rho = \rho(\boldsymbol{x}, t)$ satisfies the first-order partial differential equation*

$$\frac{\partial \rho}{\partial t} + \nabla \cdot (\rho \boldsymbol{u}) = 0.$$

1.6 Reynolds Transport Theorem

Recall our image of a small fluid particle moving in a prescribed fluid velocity field $\boldsymbol{u} = \boldsymbol{u}(\boldsymbol{x}, t)$. The velocity of a particle at time t at position $\boldsymbol{x} = \boldsymbol{x}(t)$ is

$$\frac{\mathrm{d}}{\mathrm{d}t}\boldsymbol{x}(t) = \boldsymbol{u}(\boldsymbol{x}(t), t).$$

As the particle moves in the velocity field $\boldsymbol{u} = \boldsymbol{u}(\boldsymbol{x}, t)$, say from position $\boldsymbol{x} = \boldsymbol{x}(t)$ to a nearby position an instant in time later, two dynamical contributions change: (i) a small instant in time has elapsed and the velocity field $\boldsymbol{u}(\boldsymbol{x}, t)$, which depends on time, will have changed a little; (ii) the position of the particle has changed in that short time as it moved slightly, and the velocity field $\boldsymbol{u} = \boldsymbol{u}(\boldsymbol{x}, t)$, which depends on position, will be slightly different at the new position. Let us compute the *acceleration* of the particle to observe these two contributions explicitly. By using the chain rule we see that

$$\begin{aligned}
\frac{\mathrm{d}^2}{\mathrm{d}t^2}\boldsymbol{x}(t) &= \frac{\mathrm{d}}{\mathrm{d}t}\boldsymbol{u}(\boldsymbol{x}(t), t) \\
&= \frac{\partial \boldsymbol{u}}{\partial x}\frac{\mathrm{d}x}{\mathrm{d}t} + \frac{\partial \boldsymbol{u}}{\partial y}\frac{\mathrm{d}y}{\mathrm{d}t} + \frac{\partial \boldsymbol{u}}{\partial z}\frac{\mathrm{d}z}{\mathrm{d}t} + \frac{\partial \boldsymbol{u}}{\partial t}\frac{\mathrm{d}t}{\mathrm{d}t} \\
&= \left(\frac{\mathrm{d}x}{\mathrm{d}t}\frac{\partial}{\partial x} + \frac{\mathrm{d}y}{\mathrm{d}t}\frac{\partial}{\partial y} + \frac{\mathrm{d}z}{\mathrm{d}t}\frac{\partial}{\partial z}\right)\boldsymbol{u} + \frac{\partial \boldsymbol{u}}{\partial t} \\
&= (\boldsymbol{u} \cdot \nabla)\,\boldsymbol{u} + \frac{\partial \boldsymbol{u}}{\partial t}.
\end{aligned}$$

Indeed, for any function $F = F(\boldsymbol{x}(t), t)$, scalar- or vector-valued, the chain rule implies

$$\frac{\mathrm{d}}{\mathrm{d}t}F(\boldsymbol{x}(t), t) = \frac{\partial F}{\partial t} + (\boldsymbol{u} \cdot \nabla)F.$$

Definition 1.8 (Material derivative) Given a velocity field $\boldsymbol{u} = \boldsymbol{u}(\boldsymbol{x}, t)$ with components $\boldsymbol{u} = (u, v, w)^{\mathrm{T}}$, the partial differential operator $\boldsymbol{u} \cdot \nabla$ is

$$\boldsymbol{u} \cdot \nabla := u\frac{\partial}{\partial x} + v\frac{\partial}{\partial y} + w\frac{\partial}{\partial z}.$$

We thus define the *material derivative* following the fluid to be

$$\frac{\partial}{\partial t} + \boldsymbol{u} \cdot \nabla.$$

Note that, given $\boldsymbol{u}$, this is a first-order partial differential operator.

Notation 1.9 (Material derivative) The symbol $\mathrm{D}/\mathrm{D}t$ is often used as an abbreviation for $\partial/\partial t + \boldsymbol{u} \cdot \nabla$.

Suppose the region within which the fluid is moving is $\mathcal{D}$. For a fixed position $\boldsymbol{a} \in \mathcal{D}$ we denote by $\boldsymbol{x}(\boldsymbol{a}, t)$ the position of the particle at time t, which at time $t = 0$ was at $\boldsymbol{a}$.

Definition 1.10 (Flowmap) The map $\boldsymbol{a} \mapsto \boldsymbol{x}(\boldsymbol{a}, t)$ that advances each particle at position $\boldsymbol{a}$ at time $t = 0$ to its position $\boldsymbol{x} = \boldsymbol{x}(\boldsymbol{a}, t)$ at time $t \geqslant 0$ later is called the fluid *flowmap*. Note at time $t = 0$ for all $\boldsymbol{a} \in \mathcal{D}$ we have $\boldsymbol{x}(\boldsymbol{a}, 0) = \boldsymbol{a}$.

Remark 1.11 We generally assume throughout this book that the flowmap is sufficiently smooth and invertible for all our manipulations. We examine the validity of this assumption in detail in Chapter 7.

Suppose $\mathcal{V}$ is a subregion of $\mathcal{D}$ identified at time $t = 0$. As the fluid flow evolves, the fluid particles that originally made up $\mathcal{V}$ will subsequently fill out a volume $\mathcal{V}_t$ at time t. We think of $\mathcal{V}_t$ as the volume *moving with the fluid*. See Figure 1.5. The flowmap naturally advances $\mathcal{V}$ to $\mathcal{V}_t$, i.e. the flowmap naturally induces the mapping $\mathcal{V} \mapsto \mathcal{V}_t$. We will use the following Transport Theorem to deduce both the Euler and Cauchy equations of motion from the *primitive* integral form of the balance of momentum in Sections 2.2 and 4.2; see Chorin and Marsden (1990) and Majda and Bertozzi (2002).

Theorem 1.12 (Reynolds Transport Theorem) *For any function $F = F(\boldsymbol{x}, t)$ and volume $\mathcal{V}_t \subseteq \mathcal{D}$ moving with the fluid, we have*

$$\frac{\mathrm{d}}{\mathrm{d}t} \int_{\mathcal{V}_t} F \,\mathrm{d}V = \int_{\mathcal{V}_t} \left(\frac{\partial F}{\partial t} + \nabla \cdot (\boldsymbol{u}\, F) \right) \mathrm{d}V.$$

Proof To begin, we consider making a change of variables from $\boldsymbol{a}$ to $\boldsymbol{x} = \boldsymbol{x}(\boldsymbol{a}, t)$ for any fixed $t \geqslant 0$ as follows. Let $J = J(\boldsymbol{a}, t)$ denote the Jacobian for this transformation so the volume measures are related by $\mathrm{d}V(\boldsymbol{x}) = |J(\boldsymbol{a}, t)|\,\mathrm{d}V(\boldsymbol{a})$. If $\boldsymbol{a} = (a, b, c)^{\mathrm{T}}$ and $\boldsymbol{x} = (x, y, z)^{\mathrm{T}}$, then by definition the Jacobian is

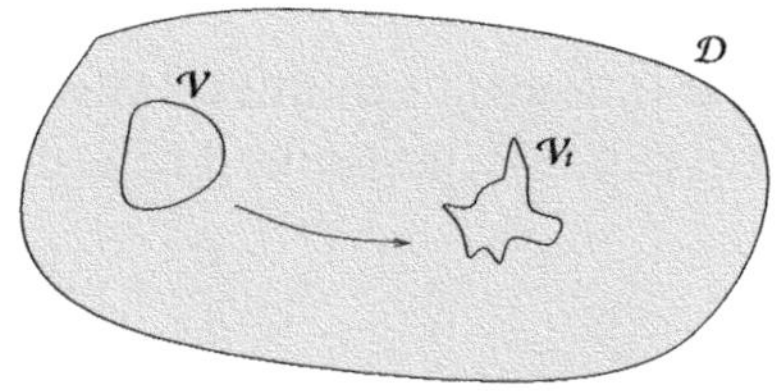

Figure 1.5 The fluid particles originally making up $\mathcal{V}$ are transported by the fluid flow continuously so they subsequently fill out the volume $\mathcal{V}_t$ at time t shown.

$$J(\boldsymbol{a},t) := \det\begin{pmatrix} \partial x/\partial a & \partial x/\partial b & \partial x/\partial c \\ \partial y/\partial a & \partial y/\partial b & \partial y/\partial c \\ \partial z/\partial a & \partial z/\partial b & \partial z/\partial c \end{pmatrix},$$

or more succinctly $J(\boldsymbol{a},t) := \det(\nabla_{\boldsymbol{a}}\, \boldsymbol{x}(\boldsymbol{a},t))$. Note for $\mathcal{V}_t$ we integrate over volume elements $\mathrm{d}V = \mathrm{d}V(\boldsymbol{x})$, i.e. with respect to the $\boldsymbol{x}$ coordinates, whereas for $\mathcal{V}$ we integrate over volume elements $\mathrm{d}V = \mathrm{d}V(\boldsymbol{a})$, i.e. with respect to the *fixed* coordinates $\boldsymbol{a}$. Hence by direct computation

$$\begin{aligned}
\frac{\mathrm{d}}{\mathrm{d}t}\int_{\mathcal{V}_t} F\,\mathrm{d}V &= \frac{\mathrm{d}}{\mathrm{d}t}\int_{\mathcal{V}_t} F(\boldsymbol{x},t)\,\mathrm{d}V(\boldsymbol{x}) \\
&= \frac{\mathrm{d}}{\mathrm{d}t}\int_{\mathcal{V}} F(\boldsymbol{x}(\boldsymbol{a},t),t)\,|J(\boldsymbol{a},t)|\,\mathrm{d}V(\boldsymbol{a}) \\
&= \int_{\mathcal{V}} \frac{\mathrm{d}}{\mathrm{d}t}\Big(F(\boldsymbol{x}(\boldsymbol{a},t),t)\,|J(\boldsymbol{a},t)|\Big)\,\mathrm{d}V(\boldsymbol{a}) \\
&= \int_{\mathcal{V}} \Big(\frac{\mathrm{d}}{\mathrm{d}t}\big(F(\boldsymbol{x}(\boldsymbol{a},t),t)\big)\,|J(\boldsymbol{a},t)| + F(\boldsymbol{x}(\boldsymbol{a},t),t)\,\frac{\mathrm{d}}{\mathrm{d}t}|J(\boldsymbol{a},t)|\Big)\,\mathrm{d}V(\boldsymbol{a}),
\end{aligned}$$

where in the last step we used the product rule. Let us examine the two terms under the integral on the right in turn. First, note by the chain rule we have

$$\begin{aligned}
\frac{\mathrm{d}}{\mathrm{d}t}\big(F(\boldsymbol{x}(\boldsymbol{a},t),t)\big) &= \frac{\partial F}{\partial x}\frac{\partial x}{\partial t} + \frac{\partial F}{\partial y}\frac{\partial y}{\partial t} + \frac{\partial F}{\partial z}\frac{\partial z}{\partial t} + \frac{\partial F}{\partial t}\frac{\partial t}{\partial t} \\
&= \frac{\partial F}{\partial t} + (\boldsymbol{u}\cdot\nabla_{\boldsymbol{x}})\,F.
\end{aligned}$$

Second, consider the term $(\mathrm{d}/\mathrm{d}t)|J(\boldsymbol{a},t)|$. In other words, let us consider how the Jacobian evolves in time. We know that a particle at position $\boldsymbol{x}(\boldsymbol{a},t)$, which started at $\boldsymbol{a}$ at time $t=0$, evolves according to

$$\frac{\mathrm{d}}{\mathrm{d}t}\boldsymbol{x}(\boldsymbol{a},t) = \boldsymbol{u}(\boldsymbol{x}(\boldsymbol{a},t),t).$$

Taking the gradient with respect to $\boldsymbol{a}$ of this relation, and swapping over the gradient and $\mathrm{d}/\mathrm{d}t$ operations on the left, we see that

$$\frac{\mathrm{d}}{\mathrm{d}t}\nabla_{\boldsymbol{a}}\,\boldsymbol{x}(\boldsymbol{a},t) = \nabla_{\boldsymbol{a}}\,\boldsymbol{u}(\boldsymbol{x}(\boldsymbol{a},t),t).$$

Using the chain rule we have $\nabla_{\boldsymbol{a}}\,\boldsymbol{u}(\boldsymbol{x}(\boldsymbol{a},t),t) = (\nabla_{\boldsymbol{x}}\,\boldsymbol{u}(\boldsymbol{x}(\boldsymbol{a},t),t))(\nabla_{\boldsymbol{a}}\,\boldsymbol{x}(\boldsymbol{a},t))$, or more succinctly $\nabla_{\boldsymbol{a}}\,\boldsymbol{u} = (\nabla_{\boldsymbol{x}}\,\boldsymbol{u})(\nabla_{\boldsymbol{a}}\,\boldsymbol{x})$. Here the product on the right is simply the matrix product of the two 3×3 matrices $\nabla_{\boldsymbol{x}}\,\boldsymbol{u}$ and $\nabla_{\boldsymbol{a}}\,\boldsymbol{x}$. Using this chain rule formula we see

$$\frac{\mathrm{d}}{\mathrm{d}t}\nabla_{\boldsymbol{a}}\,\boldsymbol{x} = (\nabla_{\boldsymbol{x}}\,\boldsymbol{u})\,\nabla_{\boldsymbol{a}}\,\boldsymbol{x}.$$

Liouville's Formula, see Lemma 1.13 below, implies that if $\nabla_{\boldsymbol{a}}\, \boldsymbol{x}$ evolves according to the system of equations above, then $\det \nabla_{\boldsymbol{a}}\, \boldsymbol{x}$ evolves according to

$$\frac{\mathrm{d}}{\mathrm{d}t} \det \nabla_{\boldsymbol{a}}\, \boldsymbol{x} = (\mathrm{tr}\,(\nabla_{\boldsymbol{x}}\, \boldsymbol{u}))\, \det \nabla_{\boldsymbol{a}}\, \boldsymbol{x},$$

where $\mathrm{tr}\,(\nabla_{\boldsymbol{x}}\, \boldsymbol{u})$ denotes the trace of the matrix $\nabla_{\boldsymbol{x}}\, \boldsymbol{u}$; this is equal to the sum of its diagonal elements. Since $\mathrm{tr}\,(\nabla_{\boldsymbol{x}}\, \boldsymbol{u}) \equiv \nabla_{\boldsymbol{x}} \cdot \boldsymbol{u}$ we have established that

$$\begin{aligned} & \frac{\mathrm{d}}{\mathrm{d}t} J(\boldsymbol{a},t) = (\nabla_{\boldsymbol{x}} \cdot \boldsymbol{u}(\boldsymbol{x}(\boldsymbol{a},t),t))\, J(\boldsymbol{a},t) \\ \Leftrightarrow \quad & J(\boldsymbol{a},t) = J(\boldsymbol{a},0)\, \exp\left(\int_0^t \nabla_{\boldsymbol{x}} \cdot \boldsymbol{u}(\boldsymbol{x}(\boldsymbol{a},\tau),\tau)\, \mathrm{d}\tau \right). \end{aligned}$$

Since $\boldsymbol{x}(\boldsymbol{a},0) = \boldsymbol{a}$ and thus $J(\boldsymbol{a},0) = 1$, we see that necessarily $J(\boldsymbol{a},t) > 0$ and we can replace $|J(\boldsymbol{a},t)|$ by $J(\boldsymbol{a},t)$. Substituting this result for the Jacobian as well as the first chain rule result above into our main calculation, we find

$$\begin{aligned} \frac{\mathrm{d}}{\mathrm{d}t} \int_{\mathcal{V}_t} F \,\mathrm{d}V &= \int_{\mathcal{V}} \left(\frac{\partial F}{\partial t} + (\boldsymbol{u} \cdot \nabla_{\boldsymbol{x}}) F + F\, \nabla_{\boldsymbol{x}} \cdot \boldsymbol{u} \right)(\boldsymbol{x}(\boldsymbol{a},t),t)\, J(\boldsymbol{a},t)\, \mathrm{d}V(\boldsymbol{a}) \\ &= \int_{\mathcal{V}} \left(\frac{\partial F}{\partial t} + \nabla_{\boldsymbol{x}} \cdot (\boldsymbol{u} F) \right)(\boldsymbol{x}(\boldsymbol{a},t),t)\, J(\boldsymbol{a},t)\, \mathrm{d}V(\boldsymbol{a}) \\ &= \int_{\mathcal{V}_t} \left(\frac{\partial F}{\partial t} + \nabla_{\boldsymbol{x}} \cdot (\boldsymbol{u} F) \right)(\boldsymbol{x},t)\, \mathrm{d}V(\boldsymbol{x}), \end{aligned}$$

where we used that $\nabla \cdot (\boldsymbol{u} F) \equiv (\nabla \cdot \boldsymbol{u})\, F + \boldsymbol{u} \cdot \nabla F$. The result is thus established. □

We used the following result in the proof above; it naturally generalises to any square matrices.

Lemma 1.13 (Liouville's Formula) *Suppose a* 3×3 *matrix* $X = X(t)$ *evolves according to the linear system of equations*

$$\frac{\mathrm{d}X}{\mathrm{d}t} = AX,$$

where $A = A(t)$ *is a given* 3×3 *matrix. Then* $\det X$ *evolves according to the equation*

$$\frac{\mathrm{d}}{\mathrm{d}t} \det X = (\mathrm{tr}\, A)\, \det X,$$

where tr *denotes the* trace *operator on matrices; the trace of a matrix is the sum of its diagonal elements.*

Proof Recall that the determinant of a matrix, where here we focus on the 3×3 case, is given by the formula

$$\det X = \sum_{\pi} X_{1,\pi(1)} X_{2,\pi(2)} X_{3,\pi(3)}.$$

The sum is over all permutations of $\{1, 2, 3\}$. By direct computation we have

$$\begin{aligned} \frac{\mathrm{d}}{\mathrm{d}t} \det X &= \sum_{\pi} \frac{\mathrm{d}}{\mathrm{d}t} \left(X_{1,\pi(1)} X_{2,\pi(2)} X_{3,\pi(3)} \right) \\ &= \sum_{\pi} \left(\left(\frac{\mathrm{d}}{\mathrm{d}t} X_{1,\pi(1)} \right) X_{2,\pi(2)} X_{3,\pi(3)} + X_{1,\pi(1)} \left(\frac{\mathrm{d}}{\mathrm{d}t} X_{2,\pi(2)} \right) X_{3,\pi(3)} \right. \end{aligned}$$

$$
\begin{aligned}
&+ X_{1,\pi(1)} X_{2,\pi(2)} \left(\frac{\mathrm{d}}{\mathrm{d}t} X_{3,\pi(3)} \right) \Big) \\
&= \sum_{\pi} \sum_{j=1}^{3} \big(A_{1j} X_{j,\pi(1)} X_{2,\pi(2)} X_{3,\pi(3)} + X_{1,\pi(1)} A_{2j} X_{j,\pi(2)} X_{3,\pi(3)} \\
&+ X_{1,\pi(1)} X_{2,\pi(2)} A_{3j} X_{j,\pi(3)} \big).
\end{aligned}
$$

Let us focus on the first sum shown on the right, the other terms can be treated similarly. We see that

$$
\begin{aligned}
\sum_{\pi} \sum_{j=1}^{3} A_{1j} X_{j,\pi(1)} X_{2,\pi(2)} X_{3,\pi(3)} &= \sum_{\pi} A_{11} X_{1,\pi(1)} X_{2,\pi(2)} X_{3,\pi(3)} \\
&+ \sum_{j\neq 1} \sum_{\pi} A_{1j} X_{j,\pi(1)} X_{2,\pi(2)} X_{3,\pi(3)}.
\end{aligned}
$$

Now further recall that the determinant of a matrix is zero if any two columns or rows are the same. Hence the second sum on the right above is zero; indeed, the individual terms for $j = 2$ and $j = 3$ are zero. Using the same argument on the other terms in our main calculation above, we find

$$
\frac{\mathrm{d}}{\mathrm{d}t} \det X = (A_{11} + A_{22} + A_{33}) \det X,
$$

which gives the result. □

A straightforward corollary of the Transport Theorem which we will often use in practice is as follows.

Corollary 1.14 (Transport Theorem with Density) *For any function $F = F(\boldsymbol{x}, t)$ and density $\rho = \rho(\boldsymbol{x}, t)$ satisfying the continuity equation, we have*

$$
\frac{\mathrm{d}}{\mathrm{d}t} \int_{\mathcal{V}_t} \rho F \, \mathrm{d}V = \int_{\mathcal{V}_t} \rho \left(\frac{\partial F}{\partial t} + (\boldsymbol{u} \cdot \nabla) F \right) \mathrm{d}V.
$$

Proof By direct computation, replacing F by ρF in the Transport Theorem and using the identity $\nabla \cdot (\boldsymbol{v} F) \equiv (\nabla \cdot \boldsymbol{v}) F + (\boldsymbol{v} \cdot \nabla) F$ with $\boldsymbol{v} = \rho \boldsymbol{u}$, we find

$$
\begin{aligned}
\frac{\partial}{\partial t}(\rho F) + \nabla \cdot (\boldsymbol{u} (\rho F)) &= \frac{\partial \rho}{\partial t} F + \rho \frac{\partial F}{\partial t} + \rho (\boldsymbol{u} \cdot \nabla) F + F \nabla \cdot (\rho \boldsymbol{u}) \\
&= F \left(\frac{\partial \rho}{\partial t} + \nabla \cdot (\rho \boldsymbol{u}) \right) + \rho \left(\frac{\partial F}{\partial t} + (\boldsymbol{u} \cdot \nabla) F \right) \\
&= \rho \left(\frac{\partial F}{\partial t} + (\boldsymbol{u} \cdot \nabla) F \right),
\end{aligned}
$$

where in the last step we used the continuity equation. □

1.7 Incompressible Flow

We now characterise a subclass of flows which are *incompressible*. We almost exclusively consider only incompressible flows in this book. The classic examples of incompressible flows are water, and the brake fluid in a car whose incompressibility properties are vital to the effective transmission of pedal pressure to brake-pad pressure. Herein we closely follow the presentation given in Chorin and Marsden (1990) and Majda and Bertozzi (2002). As in Section 1.6 above, given any subregion $\mathcal{V} \subseteq \mathcal{D}$ we use $\mathcal{V}_t \subseteq \mathcal{D}$ to denote the corresponding volume moving with the fluid and $J = J(\boldsymbol{a}, t)$ is the Jacobian determinant $J(\boldsymbol{a}, t) := \det(\nabla_{\boldsymbol{a}} \boldsymbol{x}(\boldsymbol{a}, t))$.

Definition 1.15 (Incompressible flow) A flow is said to be *incompressible* if for any subregion $\mathcal{V} \subseteq \mathcal{D}$, the volume of $\mathcal{V}_t$ is constant in time.

Lemma 1.16 (Equivalent incompressibility statements) *The following statements are equivalent, the:*

1. *fluid is incompressible;*
2. *Jacobian* $J \equiv 1$*;*
3. *velocity field* $\boldsymbol{u} = \boldsymbol{u}(\boldsymbol{x}, t)$ *is divergence-free, i.e.* $\nabla \cdot \boldsymbol{u} = 0$.

Proof For any subregion $\mathcal{V} \subseteq \mathcal{D}$ of the domain transported to $\mathcal{V}_t$ by the fluid flow, using the Transport Theorem 1.12 with $F \equiv 1$, we see that

$$\frac{\mathrm{d}}{\mathrm{d}t}\mathrm{vol}(\mathcal{V}_t) = \frac{\mathrm{d}}{\mathrm{d}t}\int_{\mathcal{V}_t} \mathrm{d}V = \int_{\mathcal{V}_t} (\nabla \cdot \boldsymbol{u})\, \mathrm{d}V.$$

Hence the fluid is incompressible if and only if $\nabla \cdot \boldsymbol{u} = 0$. Recall from the proof of the Transport Theorem that the Jacobian $J = J(\boldsymbol{a}, t)$ satisfies

$$J(\boldsymbol{a}, t) = J(\boldsymbol{a}, 0)\, \exp\left(\int_0^t \nabla_{\boldsymbol{x}} \cdot \boldsymbol{u}(\boldsymbol{x}(\boldsymbol{a}, \tau), \tau)\, \mathrm{d}\tau\right).$$

Since at time $t = 0$ we have $\boldsymbol{x}(\boldsymbol{a}, 0) \equiv \boldsymbol{a}$, we see that by definition $J(\boldsymbol{a}, 0) = 1$. Hence $J(\boldsymbol{a}, t) \equiv 1$ if and only if $\nabla \cdot \boldsymbol{u} = 0$. The result is thus established. □

The continuity equation and the identity $\nabla \cdot (\rho\, \boldsymbol{u}) \equiv (\boldsymbol{u} \cdot \nabla)\, \rho + \rho\, (\nabla \cdot \boldsymbol{u})$ imply

$$\frac{\partial \rho}{\partial t} + (\boldsymbol{u} \cdot \nabla)\, \rho + \rho\, (\nabla \cdot \boldsymbol{u}) = 0.$$

Hence, since $\rho > 0$, a flow is incompressible if and only if

$$\frac{\partial \rho}{\partial t} + (\boldsymbol{u} \cdot \nabla)\, \rho = 0,$$

i.e. the fluid density is constant following the fluid.

Definition 1.17 (Homogeneous fluid) A fluid is said to be *homogeneous* if its mass density ρ is constant in space.

The following is a direct result of this definition and the last observation.

Proposition 1.18 *If a fluid is homogeneous then it is incompressible if and only if ρ is constant in time.*

We can derive an explicit formula for the evolution of the mass density ρ. Setting $F \equiv 1$ in the Transport Theorem with Density Corollary 1.14 we find

$$\frac{\mathrm{d}}{\mathrm{d}t}\int_{\mathcal{V}_t} \rho \,\mathrm{d}V = 0 \quad \Leftrightarrow \quad \int_{\mathcal{V}_t} \rho(\boldsymbol{x},t)\,\mathrm{d}V(\boldsymbol{x}) = \int_{\mathcal{V}} \rho(\boldsymbol{a},0)\,\mathrm{d}V(\boldsymbol{a}).$$

Making a change of variables, this is equivalent to the statement

$$\int_{\mathcal{V}} \rho(\boldsymbol{x}(\boldsymbol{a},t),t)\; J(\boldsymbol{a},t)\,\mathrm{d}V(\boldsymbol{a}) = \int_{\mathcal{V}} \rho(\boldsymbol{a},0)\,\mathrm{d}V(\boldsymbol{a}).$$

Since $\mathcal{V}$ is arbitrary, we deduce $\rho(\boldsymbol{x}(\boldsymbol{a},t),t)\; J(\boldsymbol{a},t) = \rho(\boldsymbol{a},0)$, which prescribes the evolution of ρ. We now see if the flow is incompressible so $J(\boldsymbol{a},t) \equiv 1$, then

$$\rho(\boldsymbol{x}(\boldsymbol{a},t),t) = \rho(\boldsymbol{a},0).$$

Hence we conclude the following.

Corollary 1.19 *If an incompressible fluid is homogeneous at time $t = 0$ then it remains homogeneous thereafter.*

Remark 1.20 (Liquids and gases) Incompressibility is a modelling approximation, as quoting from Acheson, (1990, p. 7): 'All fluids are to some extent compressible. . . .' Liquids are generally essentially incompressible to a very high degree of accuracy. With these last two statements in mind, we observe that whales and dolphins communicate by sound, which in water travels at about 1500 $\mathrm{m\,s^{-1}}$, so water has compressibility. For a more detailed discussion, see Batchelor (1967, pp. 55–56). Importantly, quoting from Batchelor (1967, p. 75): 'A fluid is said to be *incompressible* when the density of an element of fluid is not affected by changes in the pressure. We shall see that pressure variations in some common flow fields are such a small fraction of the absolute pressure that even gases may behave as if they were almost completely incompressible.' In other words, as we shall see, quoting again from Acheson, (1990, p. 7): 'Air is, of course, highly compressible, but it can behave like an incompressible fluid if the flow speed is much smaller than the speed of sound. . . .'

1.8 Stream Functions

A *stream function* exists for a given field $\boldsymbol{u} = (u, v, w)^{\mathrm{T}}$ if $\boldsymbol{u}$ satisfies $\nabla \cdot \boldsymbol{u} = 0$ *and* the field possesses an additional symmetry that allows us to eliminate one coordinate. For example, a two-dimensional incompressible fluid flow $\boldsymbol{u} = \boldsymbol{u}(x, y, t)$ with $\boldsymbol{u} = (u, v)^{\mathrm{T}}$ satisfies

$$\nabla \cdot \boldsymbol{u} = 0 \qquad \Leftrightarrow \qquad \frac{\partial u}{\partial x} + \frac{\partial v}{\partial y} = 0$$

and has the symmetry that it is uniform with respect to z.

Proposition 1.21 (Stream function: Cartesian coordinates) *A two-dimensional field $\boldsymbol{u} = \boldsymbol{u}(x, y, t)$ with $\boldsymbol{u} = (u, v)^{\mathrm{T}}$ satisfies $\nabla \cdot \boldsymbol{u} = 0$ if and only if there exists a stream function $\psi = \psi(x, y, t)$ such that*

$$\frac{\partial \psi}{\partial y} = u(x, y, t) \qquad \textit{and} \qquad -\frac{\partial \psi}{\partial x} = v(x, y, t).$$

Remark 1.22 We emphasise the following points:

(i) The function ψ is called *Lagrange's stream function.*

(ii) For axisymmetric flows there is also the *Stokes stream function*; see Remark 5.9 in Section 5.1.

(iii) Fluid flows for which a stream function exists can thus be prescribed solely in terms of a single scalar function, i.e. the stream function itself.

(iv) We use stream functions extensively in Chapter 3 on two-dimensional irrotational flows, in two-dimensional incompressible flows in general, as well as in the theory of Stokes flows in Section 5.1.

(v) The 'if' statement of the proof, i.e. that if u and v are expressed in terms of the stream function then $\nabla \cdot \boldsymbol{u} = 0$, straightforwardly follows from the fact that for differentiable fields the order of mixed derivatives can be interchanged.

(vi) Strictly, the 'only if' statement in Proposition 1.21 only holds in flow regions that are simply connected (which ensures the stream function is not multi-valued). We examine this issue in more detail later in Section 2.8 in the context of establishing the existence of potential functions for irrotational flows; see in particular the discussion after Example 2.48. Also see Lemma 5.5.

(vii) A stream function is only defined up to an arbitrary additive constant – add any constant to ψ above and the two partial differential equations remain unchanged. Indeed, this is true for an arbitrary additive function of time.

(viii) Note for t fixed, or for stationary flows, streamlines are constant contour lines of ψ. To see this we observe that if $\boldsymbol{u}$ is given in terms of a stream function ψ, then $\nabla\psi \cdot \boldsymbol{u} = 0$ everywhere. Hence streamlines are orthogonal to $\nabla\psi$ and must therefore be aligned with contours of ψ; see Remark 2.3.

Definition 1.23 (Orthogonal derivative) Given a two-dimensional gradient field $\nabla \upsilon$ at $(x, y) \in \mathbb{R}^2$ associated with a scalar function $\upsilon = \upsilon(x, y)$, we define the *orthogonal derivative* $\nabla^{\perp}\upsilon$ at $(x, y) \in \mathbb{R}^2$ to be

$$\nabla^{\perp}\upsilon := \begin{pmatrix} \partial\upsilon/\partial y \\ -\partial\upsilon/\partial x \end{pmatrix}$$

which has the property $\nabla^{\perp}\upsilon \cdot \nabla\upsilon = 0$ at $(x, y) \in \mathbb{R}^2$.

Hence we can conveniently express the stream function relation in the form

$$\boldsymbol{u} = \nabla^{\perp}\psi.$$

There is an analogous result to Proposition 1.21 for plane polar coordinates.

Proposition 1.24 (Stream function: plane polar coordinates) *The two-dimensional field $\boldsymbol{u} = \boldsymbol{u}(r, \theta, t)$ with $\boldsymbol{u} = u_r\,\hat{\boldsymbol{r}} + u_\theta\,\hat{\boldsymbol{\theta}}$ satisfies $\nabla \cdot \boldsymbol{u} = 0$, i.e.*

$$\frac{1}{r}\frac{\partial}{\partial r}(r\,u_r) + \frac{1}{r}\frac{\partial u_\theta}{\partial \theta} = 0,$$

if and only if there exists a stream function $\psi = \psi(r, \theta, t)$ such that

$$\frac{1}{r}\frac{\partial \psi}{\partial \theta} = u_r(r, \theta, t) \qquad \text{and} \qquad -\frac{\partial \psi}{\partial r} = u_\theta(r, \theta, t).$$

Remark 1.25 In Section 5.1 on Stokes flows we explore stream functions, in particular the Stokes stream function, in detail for axisymmetric flows with no swirl in both cylindrical and spherical polar coordinates.

Example 1.26 (Strain flow) Suppose we are given the following two-dimensional strain flow $\boldsymbol{u} = u\boldsymbol{i} + v\boldsymbol{j}$ for some constant $\gamma \in \mathbb{R}$:

$$\begin{pmatrix} u \\ v \end{pmatrix} = \begin{pmatrix} \gamma x \\ -\gamma y \end{pmatrix}.$$

Note that $\nabla \cdot \boldsymbol{u} = 0$ so there exists a stream function $\psi = \psi(x, y)$ satisfying

$$\frac{\partial \psi}{\partial y} = \gamma x \qquad \text{and} \qquad -\frac{\partial \psi}{\partial x} = -\gamma y.$$

Integrating the first partial differential equation with respect to y, we find

$$\psi = \gamma x y + C(x),$$

where $C(x)$ is an arbitrary function of x. However, we know ψ must simultaneously satisfy the second partial differential equation above. Hence substituting this last relation into the second partial differential equation above we find

$$-\frac{\partial \psi}{\partial x} = -\gamma y \qquad \Rightarrow \qquad -\gamma y - C'(x) = -\gamma y.$$

We deduce $C'(x) = 0$ and therefore C is an arbitrary constant. Since a stream function is only defined up to an arbitrary constant, we take $C = 0$ for simplicity, and the stream function is given by

$$\psi = \gamma x y.$$

Now suppose we used plane polar coordinates instead. The corresponding flow $\boldsymbol{u} = u_r\hat{\boldsymbol{r}} + u_\theta\hat{\boldsymbol{\theta}}$ can be computed from $\boldsymbol{u} = u\boldsymbol{i} + v\boldsymbol{j}$ with $u = \gamma x$ and $v = -\gamma y$ as follows. We have the following identities relating $\boldsymbol{i}$ and $\boldsymbol{j}$ to $\hat{\boldsymbol{r}}$ and $\hat{\boldsymbol{\theta}}$:

$$\begin{aligned} \boldsymbol{i} &= \cos\theta\,\hat{\boldsymbol{r}} - \sin\theta\,\hat{\boldsymbol{\theta}}, \\ \boldsymbol{j} &= \sin\theta\,\hat{\boldsymbol{r}} + \cos\theta\,\hat{\boldsymbol{\theta}}. \end{aligned}$$

Hence using that $u = \gamma x = \gamma r\cos\theta$ and $v = -\gamma y = -\gamma r\sin\theta$, we find

$$\begin{aligned}\boldsymbol{u} &= \gamma r\cos\theta\,\boldsymbol{i} - \gamma r\sin\theta\,\boldsymbol{j}\\ &= \gamma r\cos\theta\,(\cos\theta\,\hat{\boldsymbol{r}} - \sin\theta\,\hat{\boldsymbol{\theta}}) - \gamma r\sin\theta\,(\sin\theta\,\hat{\boldsymbol{r}} + \cos\theta\,\hat{\boldsymbol{\theta}})\\ &= \gamma r(\cos^2\theta - \sin^2\theta)\,\hat{\boldsymbol{r}} - \gamma r(2\sin\theta\cos\theta)\,\hat{\boldsymbol{\theta}}\\ &= (\gamma r\cos 2\theta)\,\hat{\boldsymbol{r}} + (-\gamma r\sin 2\theta)\,\hat{\boldsymbol{\theta}}.\end{aligned}$$

Hence we deduce $u_r = \gamma r\cos 2\theta$ and $u_\theta = -\gamma r\sin 2\theta$. Note that $\nabla\cdot\boldsymbol{u} = 0$ using the polar coordinate form for $\nabla\cdot\boldsymbol{u}$ indicated above. Hence there exists a stream function $\psi = \psi(r,\theta)$ satisfying

$$\frac{1}{r}\frac{\partial\psi}{\partial\theta} = \gamma r\cos 2\theta \qquad \text{and} \qquad -\frac{\partial\psi}{\partial r} = -\gamma r\sin 2\theta.$$

As above, consider the first partial differential equation shown, and integrate with respect to θ to get

$$\psi = \tfrac{1}{2}\gamma r^2\sin 2\theta + C(r).$$

Substituting this into the second equation above reveals that $C'(r) = 0$, so that C is a constant. As discussed above, we can for convenience set $C = 0$ so that

$$\psi = \tfrac{1}{2}\gamma r^2\sin 2\theta.$$

Comparing this with its Cartesian equivalent above reveals they are the same. See Figure 1.6 for an example plot with $\gamma = 2$.

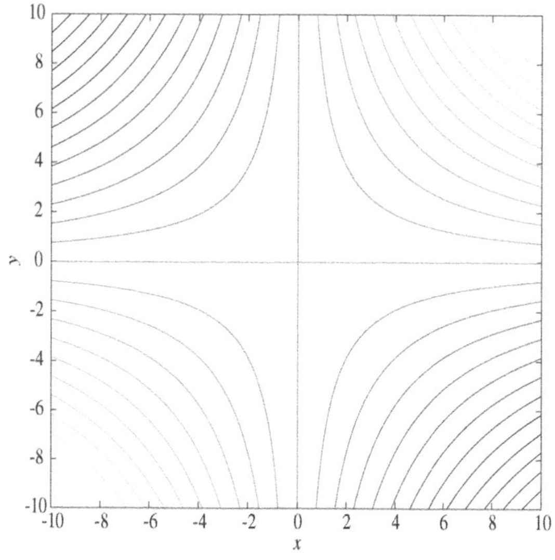

Figure 1.6 We plot the stream function $\psi = \gamma xy$ from Example 1.26 with the parameter $\gamma = 2$. The flow is directed along the streamlines, inwards towards the x-axis in the vertical direction and outwards away from the y-axis in the horizontal direction, consistent with the velocity field components $u = \gamma x$ and $v = -\gamma y$.

1.9 Rate of Strain and Vorticity

Consider a fluid flow in a region $\mathcal{D} \subseteq \mathbb{R}^3$. Suppose $\boldsymbol{x}$ and $\boldsymbol{x} + \boldsymbol{h}$ are two nearby points in the interior of $\mathcal{D}$. How is the velocity field at $\boldsymbol{x}$ related to that at $\boldsymbol{x} + \boldsymbol{h}$? By Taylor expansion, suppressing explicit t-dependence temporarily, we have

$$\boldsymbol{u}(\boldsymbol{x}+\boldsymbol{h}) = \boldsymbol{u}(\boldsymbol{x}) + (\nabla\boldsymbol{u}(\boldsymbol{x}))\,\boldsymbol{h} + O(|\boldsymbol{h}|^2),$$

where $(\nabla\boldsymbol{u})\,\boldsymbol{h}$ is simply matrix multiplication of the 3×3 matrix $\nabla\boldsymbol{u}$ by the 3×1 column vector $\boldsymbol{h}$. The expression $O(|\boldsymbol{h}|^2)$ denotes terms bounded by a constant times $|\boldsymbol{h}|^2$, as $|\boldsymbol{h}| \to 0$. Recall $\nabla\boldsymbol{u}$ is defined to be

$$\nabla\boldsymbol{u} := \begin{pmatrix} \partial u/\partial x & \partial u/\partial y & \partial u/\partial z \\ \partial v/\partial x & \partial v/\partial y & \partial v/\partial z \\ \partial w/\partial x & \partial w/\partial y & \partial w/\partial z \end{pmatrix}.$$

We call $\nabla\boldsymbol{u}$ the *velocity gradient matrix*. We can decompose the matrix $\nabla\boldsymbol{u}$ into its symmetric and antisymmetric parts as follows:

$$\nabla\boldsymbol{u} = \tfrac{1}{2}((\nabla\boldsymbol{u}) + (\nabla\boldsymbol{u})^{\mathrm{T}}) + \tfrac{1}{2}((\nabla\boldsymbol{u}) - (\nabla\boldsymbol{u})^{\mathrm{T}}).$$

Definition 1.27 (Rate of strain matrix) The symmetric part of $\nabla\boldsymbol{u}$ given by $D := \tfrac{1}{2}((\nabla\boldsymbol{u}) + (\nabla\boldsymbol{u})^{\mathrm{T}})$ is called the *rate of strain* matrix.

The antisymmetric part $R := \tfrac{1}{2}((\nabla\boldsymbol{u}) - (\nabla\boldsymbol{u})^{\mathrm{T}})$ of $\nabla\boldsymbol{u}$ has the explicit form

$$R = \tfrac{1}{2}\begin{pmatrix} 0 & \partial u/\partial y - \partial v/\partial x & \partial u/\partial z - \partial w/\partial x \\ \partial v/\partial x - \partial u/\partial y & 0 & \partial v/\partial z - \partial w/\partial y \\ \partial w/\partial x - \partial u/\partial z & \partial w/\partial y - \partial v/\partial z & 0 \end{pmatrix}.$$

Note if we set

$$\omega_1 := \frac{\partial w}{\partial y} - \frac{\partial v}{\partial z}, \quad \omega_2 := \frac{\partial u}{\partial z} - \frac{\partial w}{\partial x} \quad \text{and} \quad \omega_3 := \frac{\partial v}{\partial x} - \frac{\partial u}{\partial y},$$

then R is more simply expressed as

$$R = \tfrac{1}{2}\begin{pmatrix} 0 & -\omega_3 & \omega_2 \\ \omega_3 & 0 & -\omega_1 \\ -\omega_2 & \omega_1 & 0 \end{pmatrix}.$$

We now need the following lemma which is straightforwardly established by direct computation.

Lemma 1.28 (Rotation action) *Suppose Θ is an antisymmetric 3×3 matrix which we have expressed in the form*

$$\Theta := \begin{pmatrix} 0 & -\theta_3 & \theta_2 \\ \theta_3 & 0 & -\theta_1 \\ -\theta_2 & \theta_1 & 0 \end{pmatrix}.$$

If we set $\boldsymbol{\theta} := (\theta_1, \theta_2, \theta_3)^{\mathrm{T}}$, then for any vector $\boldsymbol{h} \in \mathbb{R}^3$ we have $\Theta\boldsymbol{h} \equiv \boldsymbol{\theta} \times \boldsymbol{h}$.

Using this lemma we observe that

$$Rh = \tfrac{1}{2}\omega \times h,$$

where $\omega = \omega(x)$ is the vector with three components ω_1, ω_2 and ω_3. At this point, we have thus established the following.

Proposition 1.29 *If x and $x + h$ are two nearby interior points of $\mathcal{D}$, then*

$$u(x+h) = u(x) + D(x)\,h + \tfrac{1}{2}\omega(x) \times h + O(|h|^2).$$

Since D is symmetric, there is an orthonormal basis e_1, e_2, e_3 in which D is diagonal, in other words if $X = [e_1, e_2, e_3]$ then

$$X^{-1}DX = \begin{pmatrix} d_1 & 0 & 0 \\ 0 & d_2 & 0 \\ 0 & 0 & d_3 \end{pmatrix}.$$

The vector field ω is equivalently given by $\omega = \nabla \times u$; we compute this below.

Definition 1.30 (Vorticity field) Associated to any differentiable velocity vector field u is the *vorticity* vector field given by

$$\omega := \nabla \times u.$$

The vorticity field ω encodes the magnitude of, and direction of, the axis about which, the fluid rotates, locally. It can be computed as follows:

$$\nabla \times u = \det \begin{pmatrix} i & j & k \\ \partial/\partial x & \partial/\partial y & \partial/\partial z \\ u & v & w \end{pmatrix} = \begin{pmatrix} \partial w/\partial y - \partial v/\partial z \\ \partial u/\partial z - \partial w/\partial x \\ \partial v/\partial x - \partial u/\partial y \end{pmatrix}.$$

Note the three components of $\nabla \times u$ shown on the right correspond to the respective three components ω_1, ω_2 and ω_3 preceding Lemma 1.28.

Now consider the motion of a fluid particle labelled by $x + h$, where x is fixed and h is small; for example, suppose that only a short time elapses. Then the velocity of the particle is given by

$$\begin{aligned} & \frac{\mathrm{d}}{\mathrm{d}t}(x+h) = u(x+h) \\ \Leftrightarrow\quad & \frac{\mathrm{d}x}{\mathrm{d}t} + \frac{\mathrm{d}h}{\mathrm{d}t} = u(x) + D(x)\,h + \tfrac{1}{2}\omega(x) \times h, \\ \Leftrightarrow\quad & \frac{\mathrm{d}h}{\mathrm{d}t} = D(x)\,h + \tfrac{1}{2}\omega(x) \times h, \end{aligned}$$

where we neglect terms $O(|h|^2)$. Let us consider the effects, separately, of each term shown on the right. Recall we suppressed explicit t-dependence above. We could have carried it explicitly all the way through so that we should have $D = D(x,t)$ and $\omega = \omega(x,t)$. However, as indicated, let us assume only a short time elapses and approximate $D(x,t) \approx D(x)$ and $\omega(x,t) \approx \omega(x)$, where $D(x)$ and $\omega(x)$ are the rate of strain matrix and vorticity, respectively, evaluated at the beginning of the time interval. Then, first isolating $D(\boldsymbol{x}) \cdot \boldsymbol{h}$, we have

$$\frac{\mathrm{d}\boldsymbol{h}}{\mathrm{d}t} = D(\boldsymbol{x})\,\boldsymbol{h}.$$

Making a local change of coordinates so that $\boldsymbol{h} = X\hat{\boldsymbol{h}}$, where $X = X(\boldsymbol{x})$ is the matrix of eigenvectors of $D = D(\boldsymbol{x})$ above, we get

$$\begin{aligned}
& & \frac{\mathrm{d}}{\mathrm{d}t}(X\hat{\boldsymbol{h}}) &= D(\boldsymbol{x})\,X\hat{\boldsymbol{h}} \\
&\Leftrightarrow & X\frac{\mathrm{d}\hat{\boldsymbol{h}}}{\mathrm{d}t} &= D(\boldsymbol{x})\,X\hat{\boldsymbol{h}} \\
&\Leftrightarrow & \frac{\mathrm{d}\hat{\boldsymbol{h}}}{\mathrm{d}t} &= X^{-1}D(\boldsymbol{x})\,X\hat{\boldsymbol{h}} \\
&\Leftrightarrow & \frac{\mathrm{d}}{\mathrm{d}t}\begin{pmatrix}\hat{h}_1\\ \hat{h}_2\\ \hat{h}_3\end{pmatrix} &= \begin{pmatrix} d_1 & 0 & 0\\ 0 & d_2 & 0\\ 0 & 0 & d_3\end{pmatrix}\begin{pmatrix}\hat{h}_1\\ \hat{h}_2\\ \hat{h}_3\end{pmatrix}.
\end{aligned}$$

Thus we have $\mathrm{d}\hat{h}_i/\mathrm{d}t = d_i\,\hat{h}_i$, for $i = 1, 2, 3$. We see we have pure expansion or contraction, depending on the sign of d_i, in each of the characteristic directions associated with the respective coordinates $\hat{h}_i$, $i = 1, 2, 3$. By direct computation using the product rule, the small linearised volume element $\hat{h}_1\hat{h}_2\hat{h}_3$ satisfies

$$\frac{\mathrm{d}}{\mathrm{d}t}(\hat{h}_1\hat{h}_2\hat{h}_3) = (d_1 + d_2 + d_3)(\hat{h}_1\hat{h}_2\hat{h}_3).$$

Recall $d_1 + d_2 + d_3 = \mathrm{tr}\,D = \nabla \cdot \boldsymbol{u}$. Now, second isolating $\frac{1}{2}\omega(\boldsymbol{x}) \times \boldsymbol{h}$, we have

$$\frac{\mathrm{d}\boldsymbol{h}}{\mathrm{d}t} = \tfrac{1}{2}\omega(\boldsymbol{x}) \times \boldsymbol{h}.$$

This is an equation well-known from classical mechanics. For brevity we suppress the $\boldsymbol{x}$-dependence in $\boldsymbol{\omega}$. Note the vector $\boldsymbol{\omega} \times \boldsymbol{h}$ is orthogonal to the plane prescribed by the vectors $\boldsymbol{\omega}$ and $\boldsymbol{h}$. Suppose for the moment $\boldsymbol{\omega}$ is aligned along $\boldsymbol{k}$ so $\boldsymbol{\omega} = \omega\,\boldsymbol{k}$, with $\omega = |\boldsymbol{\omega}|$. By direct computation $\boldsymbol{\omega} \times \boldsymbol{h} = (-\omega h_2)\,\boldsymbol{i} + (\omega h_1)\,\boldsymbol{j}$ and the differential equation above is in this case equivalent to

$$\frac{\mathrm{d}h_1}{\mathrm{d}t} = -\tfrac{1}{2}\omega h_2, \qquad \frac{\mathrm{d}h_2}{\mathrm{d}t} = \tfrac{1}{2}\omega h_1 \qquad \text{and} \qquad \frac{\mathrm{d}h_3}{\mathrm{d}t} = 0.$$

We have seen this system of equations before in Example 1.4, where we also derive the solution, which is given by

$$\begin{aligned}
h_1(t) &= h_1(0)\,\cos(\tfrac{1}{2}\omega t) - h_2(0)\,\sin(\tfrac{1}{2}\omega t),\\
h_2(t) &= h_1(0)\,\sin(\tfrac{1}{2}\omega t) + h_2(0)\,\cos(\tfrac{1}{2}\omega t),\\
h_3(t) &= h_3(0).
\end{aligned}$$

Hence $\boldsymbol{h} = \boldsymbol{h}(t)$ rotates about the vector $\boldsymbol{\omega} = \omega\,\boldsymbol{k}$ and we conclude that the term $\frac{1}{2}\boldsymbol{\omega}\times\boldsymbol{h}$ generates rotation in the fluid, with a rotation rate half the magnitude of the vorticity. See Exercise 1.8, where we derive the general solution to the differential equation for $\boldsymbol{h} = \boldsymbol{h}(t)$ generated by the vector field $\frac{1}{2}\boldsymbol{\omega} \times \boldsymbol{h}$ with $\boldsymbol{\omega}$ a constant vector.

Exercises

1.1 Trajectories and streamlines: Expanding jet flow. Find the trajectories and streamlines when

$$\begin{pmatrix} u \\ v \\ w \end{pmatrix} = \begin{pmatrix} xe^{2t-z} \\ ye^{2t-z} \\ 2e^{2t-z} \end{pmatrix}.$$

What is the track of the particle passing through $(1, 1, 0)^{\mathrm{T}}$ at time $t = 0$?

1.2 Trajectories and streamlines: Compressible flow. Suppose a velocity field $\boldsymbol{u}(\boldsymbol{x}, t) = (u, v, w)^{\mathrm{T}}$ is given for $t > -1$ by

$$u = \frac{x}{1+t}, \qquad v = \frac{y}{1+\frac{1}{2}t} \qquad \text{and} \qquad w = z.$$

Find the particle paths and streamlines for a particle starting at $(x_0, y_0, z_0)^{\mathrm{T}}$.

1.3 Trajectories and streamlines: Jet flow with swirl. Sketch streamlines for the flow field

$$\begin{pmatrix} u \\ v \\ w \end{pmatrix} = \alpha(t) \begin{pmatrix} x - y \\ x + y \\ 0 \end{pmatrix}.$$

Show the streamlines are exponential spirals. Here $\alpha = \alpha(t)$ is an arbitrary function of t. (*Hint:* Convert to cylindrical polar coordinates (r, θ, z) first. Note that in these coordinates the equations for trajectories are

$$\frac{\mathrm{d}r}{\mathrm{d}t} = u_r, \qquad r\frac{\mathrm{d}\theta}{\mathrm{d}t} = u_\theta \qquad \text{and} \qquad \frac{\mathrm{d}z}{\mathrm{d}t} = u_z,$$

where u_r, u_θ and u_z are the velocity components in the corresponding coordinate directions.)

1.4 Trajectories and streamlines: Steady channel flow. An incompressible fluid is in steady two-dimensional flow in the channel $-\infty < x < \infty$, $-\pi/2 < y < \pi/2$, with velocity

$$\boldsymbol{u} = \begin{pmatrix} 1 + x \sin y \\ \cos y \end{pmatrix}.$$

Find the equation of the streamlines and sketch them. Show that the flow has stagnation points at $(1, -\pi/2)$ and $(-1, \pi/2)$.

1.5 Stream functions: Channel shear flow. Consider the two-dimensional channel flow (with U a given constant)

$$\boldsymbol{u} = \begin{pmatrix} 0 \\ U(1 - x^2/a^2) \\ 0 \end{pmatrix},$$

between the two walls $x = \pm a$. Show that there is a *stream function* and find it. (*Hint:* A stream function ψ exists for a velocity field $\boldsymbol{u} = (u, v, w)^{\mathrm{T}}$ when $\nabla \cdot \boldsymbol{u} = 0$ and we have an additional symmetry. Here the symmetry is uniformity with respect to z.

You thus need to verify $u = \partial\psi/\partial y$ and $v = -\partial\psi/\partial x$ iff $\nabla \cdot \boldsymbol{u} = 0$. Then solve this system of equations to find ψ.)

Show that approximately 91% of the volume flux across $y = y_0$ for some constant y_0 flows through the central part of the channel $|x| \leqslant \frac{3}{4}a$.

1.6 Stream functions: Flow inside and around a disc. Calculate the stream function ψ for the flow field

$$\boldsymbol{u} = \left(U\cos\theta\left(1 - \frac{a^2}{r^2}\right)\right)\hat{\boldsymbol{r}} - \left(U\sin\theta\left(1 + \frac{a^2}{r^2}\right) - \frac{\gamma}{2\pi r}\right)\hat{\boldsymbol{\theta}},$$

in plane polar coordinates, where U, a, γ are constants.

1.7 Stream functions: Hamiltonian. Consider a two-dimensional incompressible flow field $\boldsymbol{u} = \boldsymbol{u}(x, y, t)$ in $\mathcal{D} \subseteq \mathbb{R}^2$ representable in terms of a stream function $\psi = \psi(x, y, t)$ via $\boldsymbol{u} = \nabla^{\perp}\psi$. Suppose $\boldsymbol{x} = \boldsymbol{x}(\boldsymbol{a}, t)$ represents the particle trajectory starting at $\boldsymbol{a} \in \mathcal{D}$ so that $\boldsymbol{x}(\boldsymbol{a}, 0) = \boldsymbol{a}$.

(a) Show the system of ordinary differential equations prescribing the particle path $\boldsymbol{x} = \boldsymbol{x}(\boldsymbol{a}, t)$ with components $x = x(\boldsymbol{a}, t)$ and $y = y(\boldsymbol{a}, t)$ is

$$\begin{aligned}\frac{\mathrm{d}}{\mathrm{d}t}(x(\boldsymbol{a}, t)) &= \frac{\partial\psi}{\partial y}(\boldsymbol{x}(\boldsymbol{a}, t), t),\\ \frac{\mathrm{d}}{\mathrm{d}t}(y(\boldsymbol{a}, t)) &= -\frac{\partial\psi}{\partial x}(\boldsymbol{x}(\boldsymbol{a}, t), t).\end{aligned}$$

(b) Explain why the system of equations in part (a) is Hamiltonian and that the Hamiltonian is given by ψ. What are the generalised coordinate and corresponding generalised momentum?

(c) Assuming the flow is stationary, explain why the Hamiltonian ψ is constant along particle paths and why these coincide with streamlines, i.e. level curves of ψ.

(d) Show at stagnation points $\boldsymbol{a}_*$ of the flow, where the velocity field is zero, we have $\boldsymbol{x}(\boldsymbol{a}_*, t) = \boldsymbol{a}_*$, and such points are stationary points of ψ.

(e) Describe the flow locally in the vicinity of a stagnation point, as described in part (d), when it is either a maximum or minimum of ψ or otherwise a saddle point of ψ.

1.8 Rate of strain and vorticity: Rotation equation. The goal of this exercise is to derive the general solution of the following ordinary differential equation for the vector $\boldsymbol{h} = \boldsymbol{h}(t)$:

$$\frac{\mathrm{d}\boldsymbol{h}}{\mathrm{d}t} = \tfrac{1}{2}\boldsymbol{\omega} \times \boldsymbol{h},$$

where $\boldsymbol{\omega} = (\omega_1, \omega_2, \omega_3)^{\mathrm{T}}$ is a given constant vector.

(a) To begin, we rescale time and set $\tau = \frac{1}{2}|\boldsymbol{\omega}|\, t$, and define the vector-valued function $\boldsymbol{v} = \boldsymbol{v}(\tau)$ by $\boldsymbol{v}(\tau) := \boldsymbol{h}(t)$. Use the chain rule to show

$$\frac{\mathrm{d}\boldsymbol{v}}{\mathrm{d}\tau} = \frac{2}{|\boldsymbol{\omega}|}\frac{\mathrm{d}\boldsymbol{h}}{\mathrm{d}t},$$

and hence deduce the equation prescribing the evolution of $\boldsymbol{v} = \boldsymbol{v}(\tau)$:

$$\frac{\mathrm{d}\boldsymbol{v}}{\mathrm{d}\tau} = \hat{\boldsymbol{\omega}} \times \boldsymbol{v},$$

where $\hat{\omega} := \omega/|\omega|$ is the unit vector in the direction ω.

(b) Reversing the step we made in Section 1.9, we know from Lemma 1.28 that we can rewrite the equation for $\boldsymbol{v} = \boldsymbol{v}(\tau)$ in part (a) in the form

$$\frac{\mathrm{d}\boldsymbol{v}}{\mathrm{d}\tau} = \hat{R}\boldsymbol{v},$$

where $\hat{R}$ is the 3×3 antisymmetric matrix shown in Lemma 1.28 populated with the components of $\hat{\omega} = (\hat{\omega}_1, \hat{\omega}_2, \hat{\omega}_3)^{\mathrm{T}}$ instead of the components of $\boldsymbol{\theta} = (\theta_1, \theta_2, \theta_3)^{\mathrm{T}}$. Show the solution to this linear system of ordinary differential equations with constant coefficient matrix $\hat{R}$ is

$$\boldsymbol{v}(\tau) = \exp(\tau\hat{R})\,\boldsymbol{v}_0,$$

where $\boldsymbol{v}(0) = \boldsymbol{v}_0$ is the initial condition. Here $\exp(\tau\hat{R})$ is the matrix exponential, which for square matrices $\hat{R}$ is defined by the exponential series

$$\exp(\tau\hat{R}) := I + \tau\hat{R} + \frac{1}{2!}(\tau\hat{R})^2 + \frac{1}{3!}(\tau\hat{R})^3 + \cdots .$$

(*Hint:* You can either differentiate the series to show it satisfies the differential equation or you can iterate the differential equation in time and use Picard iteration to generate the exponential series. In either case you can then invoke uniqueness to establish it is 'the' solution.)

(c) You may take as *given* Rodrigues' formula for the exponential of an antisymmetric matrix of the form $\hat{R}$, whose entries are populated from the unit vector $\hat{\omega}$, which states that

$$\exp(\tau\hat{R}) \equiv I + (\sin\tau)\,\hat{R} + (1 - \cos\tau)\,\hat{R}^2.$$

Using this formula and Lemma 1.28, show that the solution to the differential equation for $\boldsymbol{v} = \boldsymbol{v}(\tau)$ in part (a) can be expressed in the form

$$\boldsymbol{v}(\tau) = \boldsymbol{v}_0 + (\sin\tau)\,\hat{\omega} \times \boldsymbol{v}_0 + (1 - \cos\tau)\,\hat{\omega} \times (\hat{\omega} \times \boldsymbol{v}_0).$$

(d) By replacing τ by its formula in terms of t given in part (a) and recalling $\boldsymbol{v}(\tau) = \boldsymbol{h}(t)$, show the solution to the original ordinary differential equation for $\boldsymbol{h} = \boldsymbol{h}(t)$ is given by

$$\boldsymbol{h}(t) = \boldsymbol{h}_0 + \left(\frac{\sin(\frac{1}{2}|\omega|t)}{|\omega|}\right)\omega \times \boldsymbol{h}_0 + \left(\frac{1 - \cos(\frac{1}{2}|\omega|t)}{|\omega|^2}\right)\omega \times (\omega \times \boldsymbol{h}_0).$$

By substitution, verify this is indeed the solution. You will need the vector triple-product formula $\boldsymbol{a} \times (\boldsymbol{b} \times \boldsymbol{c}) \equiv (\boldsymbol{a} \cdot \boldsymbol{c})\,\boldsymbol{b} - (\boldsymbol{a} \cdot \boldsymbol{b})\,\boldsymbol{c}$.

(e) Using the vector triple-product formula from part (d), explain how the solution for $\boldsymbol{h} = \boldsymbol{h}(t)$ also given in part (d) corresponds to $\boldsymbol{h} = \boldsymbol{h}(t)$ rotating around the constant vector ω.

2 Ideal Fluid Flow

2.1 Internal Fluid Forces

Let us consider the forces that act on a small parcel of fluid in a fluid flow. There are two types:

1. *External or body forces*, these may be due to gravity or external electromagnetic fields. They exert a force per unit volume on the continuum.
2. *Surface or stress forces*, these are forces, molecular in origin, that are applied by the neighbouring fluid across the surface of the fluid parcel.

The surface or stress forces are normal stresses and tangential or shear stresses. In this chapter we only include the stress forces across the fluid parcels due to pressure differentials, representing a specific component of the normal stresses, and we entirely ignore the shear stresses. In other words, we leave out the stress components essentially resulting from molecular diffusion. This is what defines *ideal fluid flow*. In Part II of this book we discuss normal and shear stresses in detail in Section 4.3, prior to our derivation of the Navier–Stokes equations in Section 4.4. Here though, we proceed as follows.

The force per unit area exerted across a surface (imaginary in the fluid) is called the *stress*. Let $\mathrm{d}S$ be a small imaginary surface in the fluid centred on the point $\boldsymbol{x}$ with unit normal $\boldsymbol{n}$; see Figure 2.1. We assume the force $\mathrm{d}\boldsymbol{F}$ on side (2) by side (1) of $\mathrm{d}S$ in the fluid/material is given by

$$\mathrm{d}\boldsymbol{F} = -p\,\boldsymbol{n}\,\mathrm{d}S.$$

The scalar quantity $p = p(\boldsymbol{x}, t)$ represents the fluid *pressure*. It generates a force along the normal direction. In other words, we assume that the force $\mathrm{d}\boldsymbol{F}$ across the surface $\mathrm{d}S$ acts only in the normal direction $\boldsymbol{n}$ and its magnitude per unit area is given by the scalar $p = p(\boldsymbol{x}, t)$, usually positive. Essentially the origin of this force is due to neighbouring fluid parcels pushing (usually) any given parcel out of their path/way. Naturally the force due to the pressure acts from high to low pressure along the direction of greatest rate of decrease of the pressure field.

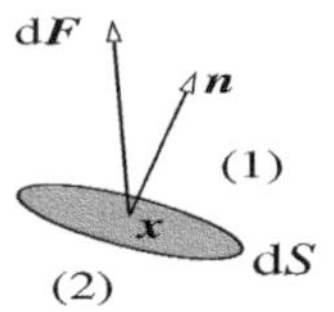

Figure 2.1 For an inviscid or ideal fluid, the force $\mathrm{d}\boldsymbol{F}$ on side (2) by side (1) of $\mathrm{d}S$ is given by $-p\,\boldsymbol{n}\,\mathrm{d}S$, where $p = p(\boldsymbol{x}, t)$ is the pressure field. In Chapter 4, where we consider viscous fluids, the force $\mathrm{d}\boldsymbol{F}$ is given more generally by $\boldsymbol{\Sigma}(\boldsymbol{n}, \boldsymbol{x}, t)\,\mathrm{d}S$, where $\boldsymbol{\Sigma} = \boldsymbol{\Sigma}(\boldsymbol{n}, \boldsymbol{x}, t)$ is the stress field.

2.2 Euler Equations of Fluid Motion

Consider an arbitrary subregion $\mathcal{V} \subseteq \mathcal{D}$ identified at $t = 0$. As the fluid flow evolves to time $t > 0$, let $\mathcal{V}_t \subseteq \mathcal{D}$ denote the volume of the fluid occupied by particles that originally made up $\mathcal{V}$. The total force exerted on the fluid inside $\mathcal{V}_t$ through the pressure exerted across its boundary $\partial\mathcal{V}_t$ is given by

$$-\int_{\partial\mathcal{V}_t} p\,\boldsymbol{n}\,\mathrm{d}S = -\int_{\mathcal{V}_t} \nabla p\,\mathrm{d}V.$$

This identity can be deduced from the Divergence Theorem as follows. If $\boldsymbol{v}$ is any fixed vector, then

$$\boldsymbol{v}\cdot\int_{\partial\mathcal{V}_t} p\,\boldsymbol{n}\,\mathrm{d}S = \int_{\partial\mathcal{V}_t} p\,\boldsymbol{v}\cdot\boldsymbol{n}\,\mathrm{d}S = \int_{\mathcal{V}_t} \nabla\cdot(p\,\boldsymbol{v})\,\mathrm{d}V = \boldsymbol{v}\cdot\int_{\mathcal{V}_t}\nabla p\,\mathrm{d}V,$$

where in the last step we used the identity $\nabla\cdot(p\,\boldsymbol{v}) = p(\nabla\cdot\boldsymbol{v}) + \boldsymbol{v}\cdot\nabla p$. Since $\boldsymbol{v}$ is fixed and arbitrary, the result follows.

If $\boldsymbol{f}$ is a body force (external force) per unit mass, which can depend on position and time, then the body force on the fluid inside $\mathcal{V}_t$ is

$$\int_{\mathcal{V}_t} \rho\,\boldsymbol{f}\,\mathrm{d}V,$$

where recall that $\rho = \rho(\boldsymbol{x}, t)$ is the mass density. Note the product of the mass density ρ, i.e. mass per unit volume, and the body force $\boldsymbol{f}$ per unit mass, gives the force per unit volume. Hence the expression $\rho\,\boldsymbol{f}\,\mathrm{d}V$ represents the total force acting on the infinitesimal volume element $\mathrm{d}V$. Thus on any parcel of fluid $\mathcal{V}_t$, the total force acting on it is

$$\int_{\mathcal{V}_t} (-\nabla p + \rho\,\boldsymbol{f})\,\mathrm{d}V.$$

Hence, using Newton's Second Law (force = rate of change of total momentum), we have

$$\frac{\mathrm{d}}{\mathrm{d}t}\int_{\mathcal{V}_t} \rho\,\boldsymbol{u}\,\mathrm{d}V = \int_{\mathcal{V}_t} (-\nabla p + \rho\,\boldsymbol{f})\,\mathrm{d}V.$$

Now we use the Transport Theorem with Density Corollary 1.14 with $F \equiv \boldsymbol{u}$ and that $\mathcal{V}$ and thus $\mathcal{V}_t$ are arbitrary. We see for each $\boldsymbol{x} \in \mathcal{D}$ and $t \geqslant 0$, the following relation must hold – the differential form of the *balance of momentum* in this case:

$$\rho\left(\frac{\partial \boldsymbol{u}}{\partial t} + (\boldsymbol{u} \cdot \nabla)\,\boldsymbol{u}\right) = -\nabla p + \rho \boldsymbol{f}.$$

We thus conclude the following.

Theorem 2.1 (Euler equations of motion) *For an* ideal fluid *for which we only include the pressure stress, the fluid flow for the velocity field* $\boldsymbol{u} = \boldsymbol{u}(\boldsymbol{x}, t)$ *is governed by the Euler equations of motion, given by*

$$\frac{\partial \boldsymbol{u}}{\partial t} + (\boldsymbol{u} \cdot \nabla)\,\boldsymbol{u} = -\frac{1}{\rho}\nabla p + \boldsymbol{f},$$

where $p = p(\boldsymbol{x}, t)$ *is the pressure field and* $\rho = \rho(\boldsymbol{x}, t)$ *is the fluid mass density.*

These were derived by Euler in 1755. They consist of three equations, one for each of the three components of velocity in $\boldsymbol{u} = (u, v, w)^{\mathrm{T}}$. However, we have five unknown variables, namely the three components of $\boldsymbol{u}$, the pressure p and mass density ρ. We need more equations to 'close the system'. Since we are concerned with homogeneous incompressible flows herein, we assume the flows we consider hereafter, unless otherwise specified, are indeed homogeneous and incompressible, so that

$$\nabla \cdot \boldsymbol{u} = 0,$$

everywhere in $\mathcal{D} \subseteq \mathbb{R}^3$, and the mass density ρ is constant in space and time. Hence, the mass density is given by its initial value. We now have four equations, the three equations making up the Euler equations of motion as well as the incompressibility condition $\nabla \cdot \boldsymbol{u} = 0$. And there are now only four unknowns, namely the three components of velocity $\boldsymbol{u}$ and the pressure field p. Thus we have a closed system of equations.

Now that we have partial differential equations that determine how the fluid flow evolves, we complement them with *boundary and initial conditions*. The initial condition is the velocity profile $\boldsymbol{u}_0 = \boldsymbol{u}_0(\boldsymbol{x})$ at time $t = 0$, i.e. $\boldsymbol{u}_0$ is the state in which the flow starts so that $\boldsymbol{u}(\boldsymbol{x}, 0) = \boldsymbol{u}_0(\boldsymbol{x})$. We naturally assume that $\nabla \cdot \boldsymbol{u}_0 = 0$ throughout the domain $\mathcal{D}$ of the flow. To have a *well-posed* evolutionary partial differential system for the evolution of the fluid flow, we also need to specify how the flow behaves near boundaries. Here a boundary could be a rigid boundary, for example the walls of the container the fluid is confined to or the surface of an obstacle in the fluid flow. Another example of a boundary is the free surface between two immiscible fluids – such as between seawater and air on the ocean surface. Here we will focus on rigid boundaries. For ideal fluid flow, i.e. one evolving according to the Euler equations, we simply need to specify that there is no net flow normal to the boundary – the fluid does not cross the boundary but can move tangentially to it. Mathematically, this means we specify that everywhere on the rigid boundary:

$$\boldsymbol{u} \cdot \boldsymbol{n} = 0,$$

where $\boldsymbol{n} = \boldsymbol{n}(\boldsymbol{x})$ is the normal to the boundary at the point $\boldsymbol{x}$ on the boundary.

Remark 2.2 We observe:

(i) That there are no tangential forces in an ideal fluid has some important consequences, quoting from Chorin and Marsden (1990, p. 5): '...there is no way for rotation to start in a fluid, nor, if there is any at the beginning, to stop... here we can detect trouble for ideal fluids because of the abundance of rotation in real fluids (near the oars of a rowboat, in tornadoes, etc.)'.

(ii) We do not need to specify data for the pressure field as it can be recovered from the velocity at any time $t \geqslant 0$ as follows. Taking the divergence of Euler's equations for a homogeneous incompressible flow, we observe

$$\nabla \cdot ((\boldsymbol{u} \cdot \nabla)\,\boldsymbol{u}) = -\Delta\left(\frac{p}{\rho}\right) + \nabla \cdot \boldsymbol{f} \quad \Leftrightarrow \quad \Delta p = -\rho\nabla \cdot ((\boldsymbol{u} \cdot \nabla)\,\boldsymbol{u}) + \rho\nabla \cdot \boldsymbol{f}.$$

Hence, given $\boldsymbol{u}$ and $\boldsymbol{f}$, solving this Poisson equation generates $p = p(\boldsymbol{x}, t)$.

Remark 2.3 (Force due to the pressure field) The gradient of any scalar field points in the direction of greatest rate of increase. Hence the force due to the pressure field '$-\nabla p/\rho$' points in the direction of greatest rate of decrease, from high pressure to low pressure, as one might intuitively expect. That the gradient of a function $f = f(x, y)$ does indeed point in the direction of greatest rate of increase is established as follows; see McCallum (1997, pp. 122–3). The argument easily generalises to higher dimensions. The directional derivative $\partial f/\partial \boldsymbol{\ell}$ of the function f at the point (x, y) in the direction of the unit vector $\boldsymbol{\ell} = \ell_x\,\boldsymbol{i} + \ell_y\,\boldsymbol{j}$, is defined as the limit as $h \to 0$ of the difference quotient $(f(x + h\ell_x, y + h\ell_y) - f(x, y))/h$. Taylor expansion reveals $\partial f/\partial \boldsymbol{\ell} \equiv \nabla f \cdot \boldsymbol{\ell}$. Consider the point (x, y) shown in Figure 2.2. Suppose the value of the contour of f through the point is $f = c$ shown. Consider the contours $f = c + \Delta c$ and $f = c - \Delta c$ shown, where $\Delta c > 0$. Let $\boldsymbol{n}$ denote the unit vector orthogonal to the contour at (x, y). Locally, as shown in Figure 2.2, the contours will be parallel anywhere where $\nabla f \neq 0$. Consider the difference ratio, of the function f at the slightly higher value contour $f = c + \Delta c$ minus that of $f = c$ at (x, y) (the fixed difference is Δc), divided by the distance in the plane from (x, y) to points on the contour $f = c + \Delta c$. If we go directly from (x, y) to the nearest point on the contour $c + \Delta c$, i.e. along $\boldsymbol{n}$, then the difference ratio will attain its largest value. Compare it, for example, to the other point on the contour $c + \Delta c$ shown in Figure 2.2, for which the distance between the respective points on the contour is longer. Now we can always decompose $\nabla f = (\nabla f)_{\boldsymbol{n}}\,\boldsymbol{n} + (\nabla f)_{\boldsymbol{t}}\,\boldsymbol{t}$, where $\boldsymbol{t}$ represents a unit vector orthogonal to $\boldsymbol{n}$ and tangent/parallel to the contour $f = c$ at (x, y). However, $(\nabla f)_{\boldsymbol{t}} = \partial f/\partial \boldsymbol{t} \equiv \nabla f \cdot \boldsymbol{t}$ must be zero as f does not change value along a contour. Hence, $\nabla f = (\partial f/\partial \boldsymbol{n})\,\boldsymbol{n}$ and does indeed point in the direction of greatest rate of increase of f.

We round off this section by making a connection to one of the earliest and most striking results in fluid mechanics: *Archimedes' Principle*. Consider a static fluid for which the velocity field $\boldsymbol{u} \equiv \boldsymbol{0}$. Suppose the body force per unit mass $\boldsymbol{f}$ is constant. In this case the Euler momentum equation becomes

$$\nabla p = \rho\,\boldsymbol{f}.$$

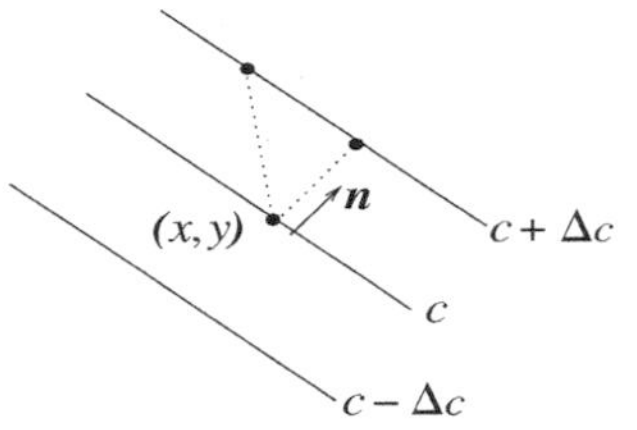

Figure 2.2 The gradient vector ∇f at (x, y) points in the direction of greatest rate of increase of f, along $\boldsymbol{n}$ which is orthogonal to the contour of f through (x, y). This is established in Remark 2.3.

We thus have a closed equation for the pressure, assuming we know $\boldsymbol{f}$ and ρ. The continuity equation from Theorem 1.7 in this case becomes

$$\frac{\partial \rho}{\partial t} = 0.$$

Hence in general in this context we can have $\rho = \rho(\boldsymbol{x})$. Now consider a submerged closed volume $\mathcal{V}$ with surface S and outward normal $\boldsymbol{n}$. The net force exterted by the fluid on the submerged body is given by

$$-\int_S p\,\boldsymbol{n}\,\mathrm{d}S = -\int_{\mathcal{V}} \nabla p\,\mathrm{d}V = -\int_{\mathcal{V}} \rho\,\boldsymbol{f}\,\mathrm{d}V = -\boldsymbol{f}\left(\int_{\mathcal{V}} \rho\,\mathrm{d}V\right).$$

Note the second factor in the final expression on the right is the mass of fluid displaced by the submerged body. Hence, for example, in Cartesian coordinates with $\boldsymbol{k}$ denoting the upwards vertical direction, the body force per unit mass is $\boldsymbol{f} = -g\boldsymbol{k}$, where g is the acceleration due to gravity, and the body experiences the Archimedean upthrust force given by

$$\text{upthrust force} = g \times (\text{mass of fluid displaced by the body}).$$

We observe that if $\mathcal{V}$ is occupied by some solid, then it will rise or fall depending on whether its mass is respectively less than or greater than that of the fluid it displaces. In the former case the upthrust exceeds the downward force due to gravity on the body given by the mass of the submerged body times g. While in the latter case the upthrust force is less then the force due to gravity. If the volume $\mathcal{V}$ is just occupied by the fluid itself, then this remains in equilibrium with the upthrust in balance with the weight of the fluid.

Problems involving Archimedes' Principle are not always trivial, as we see in the next examples and Exercises 2.2–2.7.

Example 2.4 (The stability of a floating boat) Consider a boat floating on water as shown in Figure 2.3. In the left panel, the centre of mass of the boat is G, and the centre of buoyancy, i.e. the centre of mass of displaced water, is H. As the boat is upright, both lie on the axis of symmetry of the boat. In the right panel, the boat is tilted. Now while the centre of mass of the boat G remains in the same position on the axis of symmetry of the boat, the centre of buoyancy of displaced water has moved to H' and lies off

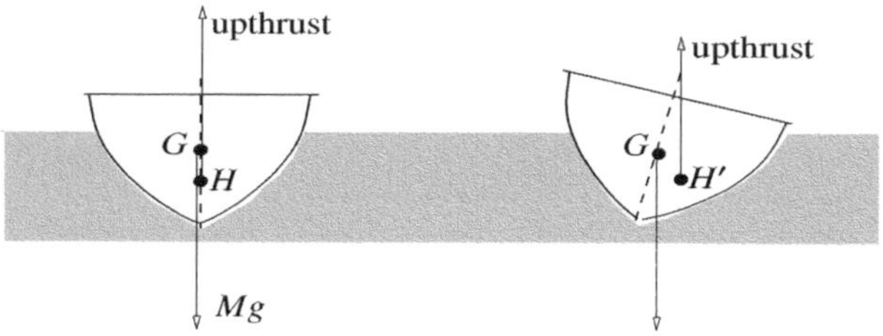

Figure 2.3 The stability of a floating boat. On the left, the centre of mass of the boat is G, while the centre of mass of displaced water is H. On the right, when the boat is tilted, the centre of buoyancy of displaced water is now H'.

the axis of symmetry of the boat. The upthrust in both cases matches the weight of the boat. However, when the boat is tilted, the thrust acts through the centre of buoyancy of a different submerged shape. The resulting couple, for the shape of hull shown, tends to restore the boat to the upright position so the boat is 'stable'. Sailing boats naturally experience strong surface lateral winds blowing into the sails, tending to tilt the boat potentially at a large angle. To prevent such sailing boats tipping over, they are built/equipped with a keel that lowers the centre of mass of the boat along the axis of symmetry. This increases the magnitude of couple resulting from the displacement of the centre of buoyancy to H' when the boat is tilted. This is because, with the boat tilted, and G lower on the axis of symmetry, the horizontal distance between G and H', which contributes to the restoring couple as a linear factor, is greater. The keel also acts as an underwater foil to minimise the lateral motion of the boat under sail; see Chapter 3.

Example 2.5 (Floating beam) A simpler version of the last example. The stability of configuration for a beam of square cross-section (see Figure 2.4) depends on the relative density of the beam material and the liquid in which it is floating. The configuration on the right is stable for e.g. softwood; the orientation on the left is stable for a very light material (balsa) or a very hard material (teak). However, this state of affairs also depends on the aspect ratio in the case of a beam of rectangular cross-section and its density distribution.

Example 2.6 (Uniform density) When the density ρ is uniform and the body force per unit mass is $\boldsymbol{f} = -g\boldsymbol{k}$, then the equations $\nabla p = \rho \boldsymbol{f}$ for the pressure imply $\partial p/\partial x = 0$, $\partial p/\partial y = 0$ and thus

$$\frac{\mathrm{d}p}{\mathrm{d}z} = -\rho g \qquad \Leftrightarrow \qquad p = p_0 - \rho g z,$$

where z measures the depth below the surface, $z = 0$ where the pressure is p_0. Since $z \leqslant 0$, we observe the pressure increases with depth z. One atmosphere of pressure is $101{,}325\,\mathrm{kg}\,(\mathrm{m\,s}^2)^{-1}$, and the density of water is naturally $1000\,\mathrm{kg\,m}^{-3}$, since a

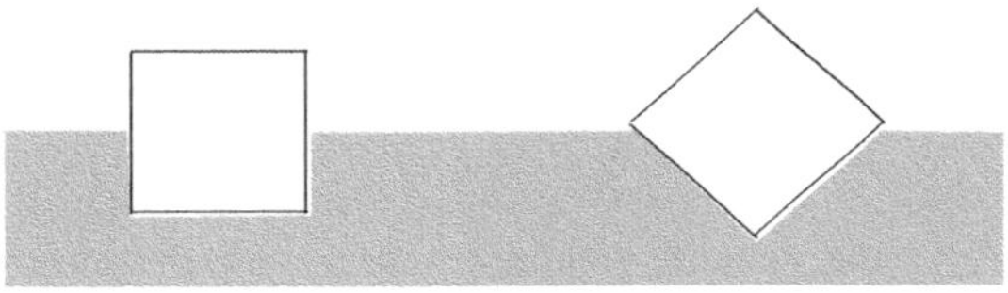

Figure 2.4 Floating beam. A beam of square cross-section floats on water. The orientation on the right is preferred for softwood, and that on the left for balsa wood or teak, although less or more water would of course be displaced respectively at equilibrium in these cases!

$10 \times 10 \times 10\ \text{cm}^3$ volume of water corresponds to a kilogram mass. Hence a pressure difference of one atmosphere at any depth corresponds to a vertical distance of

$$z = -\frac{101,325}{\rho g}\,\frac{\text{kg}\,(\text{m}\,\text{s}^2)^{-1}}{\text{kg}\,\text{m}^{-3} \times \text{m}\,\text{s}^{-2}} = -\frac{101.325}{9.81}\,\text{m} \approx -10\,\text{m},$$

i.e. the pressure increases by one atmosphere for every 10 m of depth.

2.3 Vorticity Formulation

Recall from Section 1.9 that the *vorticity field* of a flow with velocity field $\boldsymbol{u}$ is defined as $\boldsymbol{\omega} := \nabla \times \boldsymbol{u}$, which records the magnitude of, and direction of the axis about which, the fluid rotates locally. Using the Euler equations for a homogeneous incompressible fluid, we can in fact derive a closed system of equations governing the evolution of vorticity as follows.

Corollary 2.7 (Evolution of vorticity) *If the velocity field $\boldsymbol{u} = \boldsymbol{u}(\boldsymbol{x}, t)$ and pressure field $p = p(\boldsymbol{x}, t)$ are solutions to the Euler equations for a homogeneous incompressible fluid with a driving body force $\boldsymbol{f}$, then the vorticity field $\boldsymbol{\omega} = \nabla \times \boldsymbol{u}$ satisfies the closed system of equations*

$$\frac{\partial \boldsymbol{\omega}}{\partial t} + (\boldsymbol{u} \cdot \nabla)\,\boldsymbol{\omega} = D\boldsymbol{\omega} + \nabla \times \boldsymbol{f},$$

where the velocity field $\boldsymbol{u}$ is determined from the vorticity field $\boldsymbol{\omega}$ via

$$\Delta \boldsymbol{u} = -\nabla \times \boldsymbol{\omega}.$$

In the above, $D := \frac{1}{2}((\nabla \boldsymbol{u}) + (\nabla \boldsymbol{u})^{\mathrm{T}})$ is the rate of strain matrix.

Proof Using the identity $(\boldsymbol{u} \cdot \nabla)\,\boldsymbol{u} \equiv \frac{1}{2}\nabla(|\boldsymbol{u}|^2) - \boldsymbol{u} \times (\nabla \times \boldsymbol{u})$, we can equivalently represent the Euler equations in the form (ρ is the constant mass density)

$$\frac{\partial \boldsymbol{u}}{\partial t} + \tfrac{1}{2}\nabla(|\boldsymbol{u}|^2) - \boldsymbol{u} \times \boldsymbol{\omega} = -\nabla\left(\frac{p}{\rho}\right) + \boldsymbol{f}.$$

If we take the curl of this equation and use the identity

$$\nabla \times (\boldsymbol{u} \times \boldsymbol{\omega}) \equiv \boldsymbol{u}\,(\nabla \cdot \boldsymbol{\omega}) - \boldsymbol{\omega}\,(\nabla \cdot \boldsymbol{u}) + (\boldsymbol{\omega} \cdot \nabla)\,\boldsymbol{u} - (\boldsymbol{u} \cdot \nabla)\,\boldsymbol{\omega},$$

noting that $\nabla \cdot \boldsymbol{u} = 0$ and $\nabla \cdot \boldsymbol{\omega} = \nabla \cdot (\nabla \times \boldsymbol{u}) \equiv 0$, we get

$$\frac{\partial \boldsymbol{\omega}}{\partial t} + (\boldsymbol{u} \cdot \nabla)\,\boldsymbol{\omega} = (\boldsymbol{\omega} \cdot \nabla)\,\boldsymbol{u} + \nabla \times \boldsymbol{f}.$$

We recover the velocity field $\boldsymbol{u}$ from the vorticity $\boldsymbol{\omega}$ using the identity $\nabla \times (\nabla \times \boldsymbol{u}) = \nabla(\nabla \cdot \boldsymbol{u}) - \Delta \boldsymbol{u}$. This implies $\Delta \boldsymbol{u} = -\nabla \times \boldsymbol{\omega}$ as stated, closing the system of partial differential equations for $\boldsymbol{\omega}$ and $\boldsymbol{u}$. That we can replace the 'vortex stretching' term $(\boldsymbol{\omega} \cdot \nabla)\,\boldsymbol{u}$ in the evolution equation for the vorticity by $D\boldsymbol{\omega}$, where D is the 3×3 rate of strain matrix, follows if we decompose the velocity gradient matrix $\nabla \boldsymbol{u}$ into its symmetric D and antisymmetric R parts: $\nabla \boldsymbol{u} = D + R$. We then observe that $R\boldsymbol{\omega} \equiv \boldsymbol{0}$ as follows. From the Rotation Action Lemma 1.28, we have $R\boldsymbol{h} = \frac{1}{2}\boldsymbol{\omega} \times \boldsymbol{h}$ for any vector $\boldsymbol{h} \in \mathbb{R}^3$. Hence when $\boldsymbol{h} = \boldsymbol{\omega}$ we observe $R\boldsymbol{\omega} \equiv \boldsymbol{0}$. Thus we have $(\boldsymbol{\omega} \cdot \nabla)\,\boldsymbol{u} \equiv (\nabla \boldsymbol{u})\boldsymbol{\omega} = D\boldsymbol{\omega} + R\boldsymbol{\omega} = D\boldsymbol{\omega}$. □

Can we recover the Euler equations for a homogeneous incompressible fluid from the vorticity formulation given in Corollary 2.7? The answer is in principle 'yes', and certainly when the domain of the flow is $\mathcal{D} = \mathbb{R}^3$, which we assume here.

Corollary 2.8 (Incompressible Euler equations from the vorticity formulation) *Given initial data in the form $\boldsymbol{\omega}_0 = \nabla \times \boldsymbol{u}_0$ in $\mathbb{R}^3$, assume $\boldsymbol{\omega}$ and $\boldsymbol{u}$ satisfy the vorticity formulation given in Corollary 2.7, i.e. the field $\boldsymbol{\omega}$ satisfies the evolution equation shown therein and the field $\boldsymbol{u}$ is given by $\Delta \boldsymbol{u} = -\nabla \times \boldsymbol{\omega}$. Then the field $\boldsymbol{u} = \boldsymbol{u}(\boldsymbol{x}, t)$ satisfies $\nabla \cdot \boldsymbol{u} = 0$, and there exists a scalar field $p = p(\boldsymbol{x}, t)$ such that $\boldsymbol{u}$ and p satisfy Euler's equations of motion.*

Proof Assume $\boldsymbol{\omega}$ and $\boldsymbol{u}$ satisfy the vorticity formulation. Taking the divergence of the equation $\Delta \boldsymbol{u} = -\nabla \times \boldsymbol{\omega}$ reveals $\Delta(\nabla \cdot \boldsymbol{u}) = 0$. The only square-integrable solution to Laplace's equation in $\mathbb{R}^3$ is the zero solution. Hence we conclude $\nabla \cdot \boldsymbol{u} = 0$. Now consider the evolution equation for $\boldsymbol{\omega}$ given in Corollary 2.7, expressed with the vortex stretching term $(\boldsymbol{\omega} \cdot \nabla)\boldsymbol{u}$ instead of $D\boldsymbol{\omega}$. Using the incompressibility condition $\nabla \cdot \boldsymbol{u} = 0$ and the identity for $\nabla \times (\boldsymbol{u} \times \boldsymbol{\omega})$ given in the proof of Corollary 2.7, we find

$$\frac{\partial \boldsymbol{\omega}}{\partial t} + \boldsymbol{u}(\nabla \cdot \boldsymbol{\omega}) = \nabla \times (\boldsymbol{u} \times \boldsymbol{\omega}) + \nabla \times \boldsymbol{f}.$$

Taking the divergence of this evolution equation and subsequently using the identity $\nabla \cdot (\boldsymbol{u}(\nabla \cdot \boldsymbol{\omega})) \equiv (\nabla \cdot \boldsymbol{u})(\nabla \cdot \boldsymbol{\omega}) + (\boldsymbol{u} \cdot \nabla)(\nabla \cdot \boldsymbol{\omega})$, we find

$$\frac{\partial}{\partial t}(\nabla \cdot \boldsymbol{\omega}) + (\boldsymbol{u} \cdot \nabla)(\nabla \cdot \boldsymbol{\omega}) = 0.$$

Now consider particle trajectories $\boldsymbol{x} = \boldsymbol{x}(\boldsymbol{a}, t)$ starting at $\boldsymbol{a}$, so $\boldsymbol{x}(\boldsymbol{a}, 0) = \boldsymbol{a}$, and satisfying $\boldsymbol{x}'(\boldsymbol{a}, t) = \boldsymbol{u}(\boldsymbol{x}(\boldsymbol{a}, t), t)$, where the prime denotes 'd/dt'. By the chain rule we deduce, evaluating $\nabla \cdot \boldsymbol{\omega}$ along the particle paths $\boldsymbol{x} = \boldsymbol{x}(\boldsymbol{a}, t)$, that

$$\frac{\mathrm{d}}{\mathrm{d}t}\Big((\nabla \cdot \boldsymbol{\omega})(\boldsymbol{x}(\boldsymbol{a}, t), t)\Big) = 0 \quad \Leftrightarrow \quad (\nabla \cdot \boldsymbol{\omega})(\boldsymbol{x}(\boldsymbol{a}, t), t) = (\nabla \cdot \boldsymbol{\omega}_0)(\boldsymbol{a}).$$

Since the initial data has the form $\boldsymbol{\omega}_0 = \nabla \times \boldsymbol{u}_0$ whose divergence is zero, we conclude $\nabla \cdot \boldsymbol{\omega} = 0$ throughout the flow. With the knowledge $\nabla \cdot \boldsymbol{u} = 0$ and $\nabla \cdot \boldsymbol{\omega} = 0$, we can reverse the sequence of steps in the proof of Corollary 2.7 that generated the equation for the evolution of the vorticity with the 'vortex stretching' term in the form $(\boldsymbol{\omega} \cdot \nabla)\, \boldsymbol{u}$ to deduce $\nabla \times (\partial \boldsymbol{u}/\partial t + (\boldsymbol{u} \cdot \nabla)\, \boldsymbol{u} - \boldsymbol{f}) = \boldsymbol{0}$. Hence the argument of the curl operator in this last equation equals the gradient of some function which we set to be $p(\boldsymbol{x}, t)/\rho$, where $\rho > 0$ is a constant. □

Remark 2.9 (Two dimensions: vorticity formulation) The vorticity formulation above has a special character in two dimensions; here we assume the domain of flow is $\mathcal{D} = \mathbb{R}^2$. We can view such a two-dimensional flow as a subflow of a three-dimensional flow for which the velocity field is independent of z and $\boldsymbol{u} = (u, v, 0)^{\mathrm{T}}$, i.e. the velocity field component in the direction $\boldsymbol{k}$ is zero: $w = 0$. Using these two facts we see from the definition of the vorticity field $\boldsymbol{\omega}$ as the curl of $\boldsymbol{u}$, we have $\boldsymbol{\omega} = (0, 0, \omega)^{\mathrm{T}}$ where $\omega = \partial v/\partial x - \partial u/\partial y$. Note in two dimensions the 'vortex stretching' term $(\boldsymbol{\omega} \cdot \nabla)\, \boldsymbol{u}$ is identically zero:

$$(\boldsymbol{\omega} \cdot \nabla)\, \boldsymbol{u} = \omega \frac{\partial}{\partial z} \begin{pmatrix} u \\ v \\ 0 \end{pmatrix} \equiv \boldsymbol{0}.$$

Also the advection term $(\boldsymbol{u} \cdot \nabla)\, \boldsymbol{\omega}$ simplifies so that its only non-zero component is the third component which is given by $(u(\partial/\partial x) + v(\partial/\partial y))\, \omega$. With this all in hand now suppose $\boldsymbol{u} = (u, v)^{\mathrm{T}}$ only. Then the vorticity formulation in Corollary 2.7 simplifies to the following form for the scalar vorticity field ω:

$$\begin{aligned} \frac{\partial \omega}{\partial t} + \boldsymbol{u} \cdot (\nabla \omega) &= 0, \\ \Delta \boldsymbol{u} + \nabla^{\perp} \omega &= 0. \end{aligned}$$

The second condition above is computed from the full three-dimensional condition prescribing how to recover the velocity field from the vorticity field in the vorticity formulation in Corollary 2.7. Indeed, the result follows by directly substituting $\boldsymbol{\omega} = (0, 0, \omega)^{\mathrm{T}}$ and recalling from Definition 1.23 in Section 1.8 that the orthogonal derivative $\nabla^{\perp}$ is defined so that $\nabla^{\perp} \upsilon := (\partial \upsilon/\partial y, -\partial \upsilon/\partial x)^{\mathrm{T}}$ for any scalar field $\upsilon = \upsilon(x, y, t)$.

In the next sections we explore simple examples of homogeneous incompressible Euler flows. We have already seen many example flows in Section 1.4 on trajectories and streamlines (e.g. solid-body rotation) and Section 1.8 on stream functions (e.g. two-dimensional strain flow), as well as in the exercises at the end of Chapter 1 (e.g. a combination of solid-body rotation and strain flow in Exercise 1.3). In Section 2.4 we consider flows where the velocity field is linear in the spatial variable $\boldsymbol{x} \in \mathbb{R}^3$; the flows we have just highlighted are example flows in this category. In Section 2.5 we consider axisymmetric flows. The flows we consider in these sections will often reappear hereafter throughout subsequent chapters. In Section 2.6 we consider Bernoulli's Theorem on the conservation of a local energy density before going on to consider notions of circulation, irrotational and potential flows, and then water waves.

2.4 Linear Flows

The first category of solution flows we consider are characterised by the property that the velocity field is linear in the spatial coordinates $\boldsymbol{x} \in \mathbb{R}^3$ with a time-dependent coefficient matrix. As a consequence, the vorticity field is time-dependent only and satisfies a linear system of ordinary differential equations, and the highest-degree term in the pressure field is naturally quadratic in $\boldsymbol{x} \in \mathbb{R}^3$. Further, since the velocity field is linear in $\boldsymbol{x} \in \mathbb{R}^3$, these flows are exact solutions to both the incompressible Euler and Navier–Stokes equations. The latter equations, which we see in Part II, contain the additional term $\Delta \boldsymbol{u}$, i.e. the Laplacian of the velocity field. Since here the velocity field is linear, this term is identically zero. The main result for exact-solution linear velocity flow fields for the homogeneous incompressible Euler or Navier–Stokes equations is as follows. It is a variant of Prop. 1.5 in Majda and Bertozzi (2002); see Exercise 2.9. A proof is given at the end of this section, after exploring some examples. The mass density ρ is constant and we assume the body force is conservative. This means it has the representation $\boldsymbol{f} = -\nabla\Phi$, for some potential function Φ; see Section 2.8 for more details.

Proposition 2.10 (Linear flow) *Let $V = V(t)$ be a real 3×3 time-dependent matrix with $\operatorname{tr} V = 0$ and with the symmetric–antisymmetric decomposition*

$$V = D + R,$$

where $D = D(t)$ and $R = R(t)$ are the real 3×3 symmetric and antisymmetric matrices given by $D := \frac{1}{2}(V + V^{\mathrm{T}})$ and $R := \frac{1}{2}(V - V^{\mathrm{T}})$, respectively. Suppose we represent the antisymmetric matrix $R = R(t)$ in the form

$$R := \frac{1}{2}\begin{pmatrix} 0 & -\omega_3 & \omega_2 \\ \omega_3 & 0 & -\omega_1 \\ -\omega_2 & \omega_1 & 0 \end{pmatrix},$$

and set $\boldsymbol{\omega} = \boldsymbol{\omega}(t)$ to be the vector of components $\boldsymbol{\omega} = (\omega_1, \omega_2, \omega_3)^T$. Then the velocity field $\boldsymbol{u} = \boldsymbol{u}(\boldsymbol{x}, t)$ and the pressure field $p = p(\boldsymbol{x}, t)$ given by

$$\begin{aligned} \boldsymbol{u}(\boldsymbol{x}, t) &= V(t)\boldsymbol{x}, \\ p(\boldsymbol{x}, t) &= -\tfrac{1}{2}\rho\boldsymbol{x}^{\mathrm{T}}\left(D'(t) + D^2(t) + R^2(t)\right)\boldsymbol{x} - \rho\Phi, \end{aligned}$$

are an exact solution to the incompressible Euler and Navier–Stokes equations in $\mathbb{R}^3$ driven by a conservative body force $\boldsymbol{f} = -\nabla\Phi$, for some potential function Φ. Further, necessarily, the vorticity field $\nabla \times \boldsymbol{u}$ is given by $\boldsymbol{\omega} = \boldsymbol{\omega}(t)$ and satisfies the system of ordinary differential equations $\boldsymbol{\omega}'(t) = D(t)\,\boldsymbol{\omega}(t)$.

Remark 2.11 Naturally $\operatorname{tr} V = 0$ if and only if $\nabla \cdot \boldsymbol{u} = 0$. Further, we observe:

(i) The pressure includes a term quadratic in $\boldsymbol{x} \in \mathbb{R}^3$, so these solutions only have meaningful interpretations locally.

(ii) The conservative body force suggests in such cases we define a modified pressure $p' := p/\rho + \Phi$ and subsequently proceed as if there is no body force.

Recall from the Rotation Action Lemma 1.28, if Θ is an antisymmetric 3×3 matrix of the form

$$\Theta := \begin{pmatrix} 0 & -\theta_3 & \theta_2 \\ \theta_3 & 0 & -\theta_1 \\ -\theta_2 & \theta_1 & 0 \end{pmatrix},$$

then for any vector $\boldsymbol{h} \in \mathbb{R}^3$ we have $\Theta \boldsymbol{h} \equiv \boldsymbol{\theta} \times \boldsymbol{h}$, where $\boldsymbol{\theta} := (\theta_1, \theta_2, \theta_3)^{\mathrm{T}}$. Hence using that $V = D + R$, and from the representation for R given in Proposition 2.10 that $R\boldsymbol{x} = \frac{1}{2}\boldsymbol{\omega}(t) \times \boldsymbol{x}$ for any $\boldsymbol{x} \in \mathbb{R}^3$, the velocity field $\boldsymbol{u}$ for such linear flows can be represented in the form

$$\boldsymbol{u}(\boldsymbol{x}, t) = D(t)\boldsymbol{x} + \tfrac{1}{2}\boldsymbol{\omega}(t) \times \boldsymbol{x}.$$

From the proposition note, once $D = D(t)$ is given, the evolution of $\boldsymbol{\omega} = \boldsymbol{\omega}(t)$ is prescribed by its initial state $\boldsymbol{\omega}_0$ and the ordinary differential system $\boldsymbol{\omega}' = D\boldsymbol{\omega}$.

Example 2.12 (Jet flow) This is an example of a three-dimensional strain flow. Suppose the initial vorticity $\boldsymbol{\omega}_0 = \boldsymbol{0}$, i.e. the flow is initially irrotational everywhere in $\mathbb{R}^3$. Further suppose matrix $D = \mathrm{diag}\{d_1, d_2, d_3\}$ is constant and diagonal with $d_1 + d_2 + d_3 = 0$ so that $\mathrm{tr}\, D = 0$, and thus the flow is incompressible. Hence the evolution equation $\boldsymbol{\omega}' = D\boldsymbol{\omega}$ for $\boldsymbol{\omega} = \boldsymbol{\omega}(t)$ implies $\boldsymbol{\omega}(t) = \boldsymbol{0}$ for all $t \geqslant 0$. The velocity field $\boldsymbol{u}$ is thus given by $\boldsymbol{u} = D\boldsymbol{x}$, or more explicitly

$$\boldsymbol{u}(\boldsymbol{x}, t) = \begin{pmatrix} d_1\, x \\ d_2\, y \\ d_3\, z \end{pmatrix}.$$

Consider a particle initially at $\boldsymbol{x}_0 = (x_0, y_0, z_0)^{\mathrm{T}}$. Its trajectory is given by the solution to the system of ordinary differential equations $\mathrm{d}\boldsymbol{x}/\mathrm{d}t = \boldsymbol{u}(\boldsymbol{x}, t)$, where now $\boldsymbol{x} = \boldsymbol{x}(t)$ with $\boldsymbol{x} = (x, y, z)^{\mathrm{T}}$ and $\boldsymbol{u}$ has the simple form stated above. The ordinary differential equations for $x = x(t)$, $y = y(t)$ and $z = z(t)$ decouple and can be solved individually to reveal: $x(t) = \mathrm{e}^{d_1 t} x_0$, $y(t) = \mathrm{e}^{d_2 t} y_0$ and $z(t) = \mathrm{e}^{d_3 t} z_0$. If $d_1 < 0$ and $d_2 < 0$, then necessarily $d_3 > 0$ and the flow resembles two jets streaming in opposite directions away from the $z = 0$ plane.

Example 2.13 (Strain flow) The strain flow we considered in Example 1.26 in Section 1.8 is an example linear flow. Suppose again the initial vorticity $\boldsymbol{\omega}_0 = \boldsymbol{0}$ and $D = \mathrm{diag}\{d_1, d_2, 0\}$ is a constant diagonal matrix with $d_1 + d_2 = 0$, so $\mathrm{tr}\, D = 0$ holds. Then as in the last example, the flow is irrotational with $\boldsymbol{\omega}(t) = \boldsymbol{0}$ for all $t \geqslant 0$. The velocity field $\boldsymbol{u}$ is thus given by $\boldsymbol{u} = D\boldsymbol{x}$, i.e. by

$$\boldsymbol{u}(\boldsymbol{x}, t) = \begin{pmatrix} d_1\, x \\ d_2\, y \\ 0 \end{pmatrix}.$$

Similar arguments to those in the last example imply the particle path for the particle at $\boldsymbol{x}_0 = (x_0, y_0, z_0)^{\mathrm{T}}$ at time $t = 0$ is given by $x(t) = \mathrm{e}^{d_1 t} x_0$, $y(t) = \mathrm{e}^{d_2 t} y_0$ and $z(t) = z_0$.

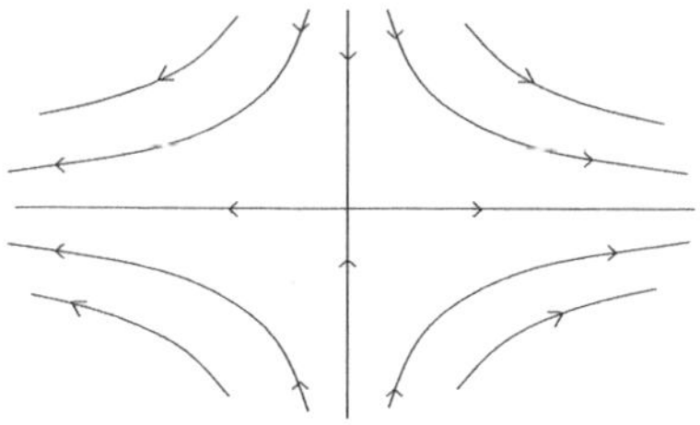

Figure 2.5 Strain flow example with $d_1 > 0$. Also see Figure 1.6.

Since $d_2 = -d_1$, the flow forms a strain flow as shown in Figure 2.5 – neighbouring particles are pushed together in one direction while being pulled apart in the other orthogonal direction. In the figure, $d_1 > 0$.

Example 2.14 (Solid-body rotation) The solid-body rotation flow we considered in Example 1.4 in Section 1.4 is also a linear flow. Here we suppose the initial vorticity is $\boldsymbol{\omega}_0 = (0, 0, 2\Omega)^{\mathrm{T}}$, where $\Omega \in \mathbb{R}$ is a non-zero constant. Further we assume $D = O$, the 3×3 zero matrix. Hence the evolution equation $\boldsymbol{\omega}' = D\boldsymbol{\omega}$ for $\boldsymbol{\omega} = \boldsymbol{\omega}(t)$ implies $\boldsymbol{\omega}(t) = \boldsymbol{\omega}_0$, with $\boldsymbol{\omega}_0 = (0, 0, 2\Omega)^{\mathrm{T}}$ for all $t \geqslant 0$. The velocity field $\boldsymbol{u}$ is thus given by $\boldsymbol{u}(\boldsymbol{x}, t) = \frac{1}{2}\boldsymbol{\omega} \times \boldsymbol{x}$, or more explicitly

$$\boldsymbol{u}(\boldsymbol{x}, t) = \begin{pmatrix} -\Omega y \\ \Omega x \\ 0 \end{pmatrix}.$$

In Example 1.4 we showed the particle trajectory for the particle starting at $\boldsymbol{x}_0 = (x_0, y_0, z_0)^{\mathrm{T}}$ at time $t = 0$ is given by $x(t) = x_0 \cos \Omega t - y_0 \sin \Omega t$, $y(t) = x_0 \sin \Omega t + y_0 \cos \Omega t$ and $z(t) = z_0$. These are circular trajectories at height $z = z_0$, and the flow resembles a solid-body rotation; see Figure 1.1.

Example 2.15 (Jet flow with swirl) We combine a strain flow together with a time-dependent solid-body rotation. Suppose the initial vorticity $\boldsymbol{\omega}_0 = (0, 0, \omega_0)^{\mathrm{T}}$ and $D = \mathrm{diag}\{d_1, d_2, d_3\}$ is a constant diagonal matrix with $d_1 + d_2 + d_3 = 0$ to ensure incompressibility. The evolution equation $\boldsymbol{\omega}' = D\boldsymbol{\omega}$ for $\boldsymbol{\omega} = \boldsymbol{\omega}(t)$ with $\boldsymbol{\omega} = (\omega_x, \omega_y, \omega_z)$ decouples into individual equations for the vorticity components: $\omega_x' = d_1\omega_x$, $\omega_y' = d_2\omega_y$ and $\omega_z' = d_3\omega_z$, with respective initial data being zero for the first two components and ω_0 for ω_z. Consequently, the only non-zero component of vorticity is $\omega_z = \omega_z(t)$, where $\omega_z(t) = \omega_0 \mathrm{e}^{d_3 t}$. The velocity field $\boldsymbol{u}$ is given by $\boldsymbol{u} = D\boldsymbol{x} + \frac{1}{2}\boldsymbol{\omega} \times \boldsymbol{x}$ and is given explicitly by

$$\boldsymbol{u}(\boldsymbol{x}, t) = \begin{pmatrix} d_1\, x - \frac{1}{2}\omega_z(t) y \\ d_2\, y + \frac{1}{2}\omega_z(t) x \\ d_3\, z \end{pmatrix}.$$

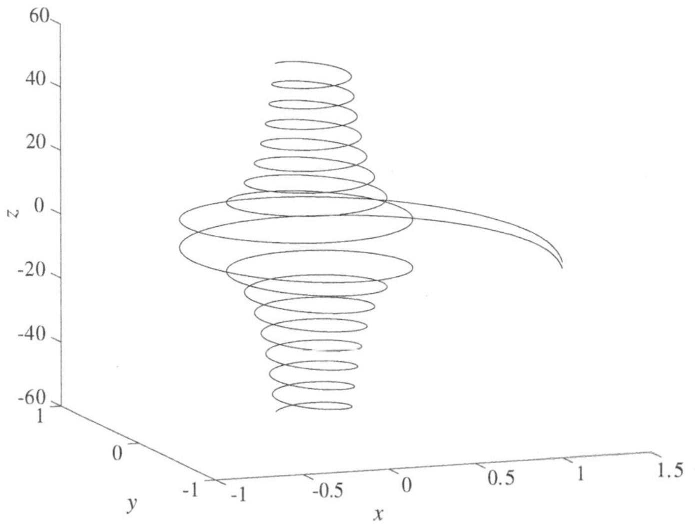

Figure 2.6 Jet flow with swirl example. The trajectories of two particles starting from $(1, -1, 1)^{\mathrm{T}}$ and $(1, -1, -1)^{\mathrm{T}}$ are shown. We set $d_1 = -0.2$ and $d_2 = -0.3$, so $d_3 = 0.5$. We also set $\omega_0 = 1$. The fluid particles rotate around and move closer to the z-axis whilst moving further from the $z = 0$ plane.

The particle trajectory for a particle at $\boldsymbol{x}_0 = (x_0, y_0, z_0)^{\mathrm{T}}$ at $t = 0$ can be described as follows. We see that $z(t) = z_0 \mathrm{e}^{d_3 t}$, while $x = x(t)$ and $y = y(t)$ satisfy the coupled system of ordinary differential equations

$$\frac{\mathrm{d}}{\mathrm{d}t}\begin{pmatrix} x \\ y \end{pmatrix} = \begin{pmatrix} d_1 & -\frac{1}{2}\omega_z(t) \\ \frac{1}{2}\omega_z(t) & d_2 \end{pmatrix}\begin{pmatrix} x \\ y \end{pmatrix}.$$

If $d_1 < 0$ and $d_2 < 0$, so $d_3 > 0$, the particles spiral around the z-axis with decreasing radius and increasing angular velocity $\frac{1}{2}\omega_z(t)$. The flow resembles a rotating jet flow; see Figure 2.6. We saw a variant of this flow in Exercise 1.3.

We now proceed to prove Proposition 2.10. Before we do so, it is helpful to establish some straightforward calculus identities, summarised as follows.

Lemma 2.16 *Suppose $S \in \mathbb{R}^{3\times 3}$ is a symmetric matrix while $\boldsymbol{\theta} \in \mathbb{R}^3$. Both S and $\boldsymbol{\theta}$ are assumed to be independent of $\boldsymbol{x} \in \mathbb{R}^3$. Then we observe the following calculus identities:*

(i) $\nabla(S\boldsymbol{x}) \equiv S$.
(ii) $\nabla \times (S\boldsymbol{x}) \equiv \boldsymbol{0}$.
(iii) $\nabla \times (\boldsymbol{\theta} \times \boldsymbol{x}) \equiv 2\boldsymbol{\theta}$.
(iv) $\nabla(\boldsymbol{x}^{\mathrm{T}} S\boldsymbol{x}) \equiv 2S\boldsymbol{x}$.

Result (i) is true for any matrix, not just symmetric ones.

Proof Proceeding in turn for (i), recall that by convention the gradient of a vector is a matrix where the different partial derivatives with respect to x, y and z are distributed across the columns of the matrix and the components/rows of the vector, on which the gradient and partial derivatives act, keep their row positions. Hence for the vector $\boldsymbol{Sx} \in \mathbb{R}^3$ we observe that

$$\nabla(S\boldsymbol{x}) = \nabla \begin{pmatrix} S_{11}x + S_{12}y + S_{13}z \\ S_{21}x + S_{22}y + S_{23}z \\ S_{31}x + S_{32}y + S_{33}z \end{pmatrix} = \begin{pmatrix} S_{11} & S_{12} & S_{13} \\ S_{21} & S_{22} & S_{23} \\ S_{31} & S_{32} & S_{33} \end{pmatrix},$$

where in the final step the columns 1, 2 and 3 are generated by taking the partial derivatives of the rows of $S\boldsymbol{x}$ with respect to x, y and z, respectively. Result (ii) follows by directly computing the curl of the vector $S\boldsymbol{x}$, which generates the vector $(S_{32} - S_{23}, S_{13} - S_{31}, S_{21} - S_{12})^{\mathrm{T}}$, which is the zero vector when S is symmetric. To establish (iii) we need identity #11 from the Appendix (Section A.1) which for any two vectors $\boldsymbol{u}, \boldsymbol{v} \in \mathbb{R}^d$ states

$$\nabla \times (\boldsymbol{u} \times \boldsymbol{v}) = \boldsymbol{u}\,(\nabla \cdot \boldsymbol{v}) - \boldsymbol{v}\,(\nabla \cdot \boldsymbol{u}) + (\boldsymbol{v} \cdot \nabla)\boldsymbol{u} - (\boldsymbol{u} \cdot \nabla)\boldsymbol{v}.$$

If we set $\boldsymbol{u} = \boldsymbol{\theta}$ and $\boldsymbol{v} = \boldsymbol{x}$, then this identity implies

$$\nabla \times (\boldsymbol{\theta} \times \boldsymbol{x}) = \boldsymbol{\theta}\,(\nabla \cdot \boldsymbol{x}) - (\boldsymbol{\theta} \cdot \nabla)\boldsymbol{x} = 3\boldsymbol{\theta} - \boldsymbol{\theta} = 2\boldsymbol{\theta}.$$

For (iv) we need the following product rule identity, which holds for any two differentiable $\mathbb{R}^3$-valued functions $\boldsymbol{f} = \boldsymbol{f}(\boldsymbol{x})$ and $\boldsymbol{g} = \boldsymbol{g}(\boldsymbol{x})$:

$$\nabla(\boldsymbol{f}^{\mathrm{T}} S\boldsymbol{g}) \equiv (\nabla\boldsymbol{f})^{\mathrm{T}} S\boldsymbol{g} + (\nabla\boldsymbol{g})^{\mathrm{T}} S^{\mathrm{T}}\boldsymbol{f}.$$

Setting $\boldsymbol{f} = \boldsymbol{g} = \boldsymbol{x}$ and noting that $\nabla\boldsymbol{x} = I$, the 3×3 identity matrix, we observe since S is symmetric: $\nabla(\boldsymbol{x}^{\mathrm{T}} S\boldsymbol{x}) = 2(\nabla\boldsymbol{x})^{\mathrm{T}} S\boldsymbol{x} = 2S\boldsymbol{x}$. □

Remark 2.17 Identities such as that for $\nabla(\boldsymbol{f}^{\mathrm{T}} S\boldsymbol{g})$ given in the proof of the lemma above are often most easily established using index notation directly. Hence, for example, to prove this identity let us identify the three Cartesian coordinates parameterising $\mathbb{R}^3$ as $\boldsymbol{x} := (x_1, x_2, x_3)^{\mathrm{T}}$ and the components of $\boldsymbol{f}$ and $\boldsymbol{g}$, respectively, as f_i and g_i, $i = 1, 2, 3$. Then we observe the ith component of the gradient of the scalar $\boldsymbol{f}^{\mathrm{T}} S\boldsymbol{g}$ is computed using the product rule as

$$[\nabla(\boldsymbol{f}^{\mathrm{T}} S\boldsymbol{g})]_i = \frac{\partial}{\partial x_i}\left(\sum_{j,k=1}^{3} f_j\, S_{jk}\, g_k\right) = \sum_{j,k=1}^{3}\left(\left(\frac{\partial f_j}{\partial x_i}\right) S_{jk}\, g_k + f_j\, S_{jk}\left(\frac{\partial g_k}{\partial x_i}\right)\right).$$

Now it is a question of interpreting the right-hand side back in terms of the appropriate matrix products. Indeed the right-hand side above equals

$$\sum_{j,k=1}^{3} (\nabla\boldsymbol{f})_{ji}\, S_{jk}\, g_k + \sum_{j,k=1}^{3} f_j\, S_{jk}\, (\nabla\boldsymbol{g})_{ki} = [(\nabla\boldsymbol{f})^{\mathrm{T}} S\boldsymbol{g}]_i + [(\boldsymbol{f}^{\mathrm{T}} S(\nabla\boldsymbol{g}))^{\mathrm{T}}]_i,$$

which then gives (after performing the overall transpose) the corresponding identity.

Proof of Proposition 2.10 First we substitute the linear form $\boldsymbol{u}(\boldsymbol{x}, t) = V(t)\boldsymbol{x}$ into the incompressible Euler/Navier–Stokes equations. Using that $(\boldsymbol{u} \cdot \nabla)\,\boldsymbol{u} \equiv (\nabla\boldsymbol{u})\boldsymbol{u}$, that

$\nabla(V\boldsymbol{x}) \equiv V$ from Lemma 2.16(i) for general matrices and that $\nu\Delta(V(t)\boldsymbol{x}) \equiv \boldsymbol{0}$, where $\nu > 0$ is the constant viscosity, direct substitution implies

$$\begin{aligned}\nabla\left(\frac{p}{\rho} + \Phi\right) &= -\left(\frac{\partial \boldsymbol{u}}{\partial t} + (\nabla \boldsymbol{u})\boldsymbol{u} - \nu\Delta\boldsymbol{u}\right)\\ &= -(V' + V^2)\,\boldsymbol{x}\\ &= -(D' + D^2 + R^2)\,\boldsymbol{x} - (R' + DR + RD)\,\boldsymbol{x}\\ &= -S\boldsymbol{x} - \boldsymbol{\theta} \times \boldsymbol{x},\end{aligned}$$

where $S = S(t)$ is the symmetric matrix $S := D' + D^2 + R^2$ and if $\Theta = \Theta(t)$ is the antisymmetric matrix $\Theta := R' + DR + RD$, then $\boldsymbol{\theta}$ is the vector of the corresponding off-diagonal components of Θ using the representation convention of the Rotation Action Lemma 1.28. Second, taking the curl of the expression for $\nabla(p/\rho + \Phi)$ above and using Lemma 2.16(ii) and (iii), we deduce $\boldsymbol{\theta} \equiv \boldsymbol{0}$. Hence we deduce $\nabla(p/\rho + \Phi) = -S\boldsymbol{x}$ and using Lemma 2.16(iv) that $p(\boldsymbol{x}, t) = -\frac{1}{2}\rho\boldsymbol{x}^{\mathrm{T}}S(t)\boldsymbol{x} - \rho\Phi$, which is the stated form for the pressure field. Third, as already observed, using that $V = D + R$ and $R\boldsymbol{x} = \frac{1}{2}\omega(t) \times \boldsymbol{x}$ for any $\boldsymbol{x} \in \mathbb{R}^3$ from Lemma 1.28, the velocity field $\boldsymbol{u}$ can be represented in the form

$$\boldsymbol{u}(\boldsymbol{x}, t) = D(t)\boldsymbol{x} + \tfrac{1}{2}\omega(t) \times \boldsymbol{x}.$$

Taking the curl of this form, observing $\nabla \times (D(t)\boldsymbol{x}) \equiv \boldsymbol{0}$ using Lemma 2.16(ii) and $\nabla \times (\omega \times \boldsymbol{x}) = 2\omega$ using Lemma 2.16(iii), we deduce $\nabla \times \boldsymbol{u} \equiv \omega(t)$. Fourth, since $\boldsymbol{\theta} = \boldsymbol{0}$ we deduce $\Theta = O$, where O is the 3×3 zero matrix. Hence we deduce $R' + DR + RD = O$. Since $\operatorname{tr} V = 0$, corresponding to incompressibility, and $\operatorname{tr} R = 0$, we deduce $\operatorname{tr} D = 0$. By direct enumeration, using that D is symmetric with $D_{11} + D_{22} + D_{33} = 0$, then extracting as a vector the off-diagonal elements of $DR + RD$ using the standard representation identification, we generate $-D\omega$. Since the same procedure generates ω' from R', that $\omega = \omega(t)$ satisfies the given system of ordinary differential equations, follows. □

2.5 Axisymmetric Flows

The next category of solution flows we consider are characterised by the property that they are axisymmetric. This means in *cylindrical coordinates* (r, θ, z) the fluid flow, i.e. the velocity field, is independent of the azimuthal angle θ. The corresponding velocity flow field component u_θ is known as the *swirl* velocity. We consider axisymmetric flow scenario examples both with no swirl, and with non-zero swirl. There is an equivalent characterisation in spherical polar coordinates which we see examples of later; for the examples here, we use cylindrical polar coordinates. Consider the Euler equations for a homogeneous incompressible fluid in cylindrical polar coordinates with the velocity field $\boldsymbol{u} = u_r\hat{\boldsymbol{r}} + u_\theta\hat{\boldsymbol{\theta}} + u_z\hat{\boldsymbol{z}}$. Using the form of the equations given in the Appendix (Section A.2) with $\nu = 0$, the Euler equations for a homogeneous incompressible fluid are

$$\frac{\partial u_r}{\partial t} + (\boldsymbol{u} \cdot \nabla)u_r - \frac{u_\theta^2}{r} = -\frac{1}{\rho}\frac{\partial p}{\partial r} + f_r,$$

$$\frac{\partial u_\theta}{\partial t} + (\boldsymbol{u} \cdot \nabla)u_\theta + \frac{u_r u_\theta}{r} = -\frac{1}{\rho r}\frac{\partial p}{\partial \theta} + f_\theta,$$
$$\frac{\partial u_z}{\partial t} + (\boldsymbol{u} \cdot \nabla)u_z = -\frac{1}{\rho}\frac{\partial p}{\partial z} + f_z,$$

where $p = p(r, \theta, z, t)$ is the pressure, ρ is the uniform constant density and $\boldsymbol{f} = f_r\hat{\boldsymbol{r}} + f_\theta\hat{\boldsymbol{\theta}} + f_z\hat{\boldsymbol{z}}$ is the body force per unit mass. Here we also have

$$\boldsymbol{u} \cdot \nabla = u_r\frac{\partial}{\partial r} + \frac{u_\theta}{r}\frac{\partial}{\partial \theta} + u_z\frac{\partial}{\partial z}.$$

The incompressibility condition $\nabla \cdot \boldsymbol{u} = 0$ in cylindrical coordinates is given by

$$\frac{1}{r}\frac{\partial(ru_r)}{\partial r} + \frac{1}{r}\frac{\partial u_\theta}{\partial \theta} + \frac{\partial u_z}{\partial z} = 0.$$

Under the assumption of axisymmetry, so no explicit dependence of $\boldsymbol{u}$ on θ, we have $\boldsymbol{u}(r, z, t) = u_r(r, z, t)\hat{\boldsymbol{r}} + u_\theta(r, z, t)\hat{\boldsymbol{\theta}} + u_z(r, z, t)\hat{\boldsymbol{z}}$ and all the terms with partial derivatives with respect to θ of u_r, u_θ or u_z in the equations above are zero.

We now consider a classical example of an axisymmetric flow below (hurricane) and many more in the exercises (Couette flow, bath drain flows, coffee stirred in a mug, Hill's spherical vortex). Recall our comments on modelling and analytical solution strategies in the Introduction.

Example 2.18 (Hurricane flow) We make some sensible simplifying assumptions to reduce the system of equations above to a set of partial differential equations we might be able to solve analytically. We assume the flow is:

(i) Homogeneous and thus has constant density ρ.

(ii) Axisymmetric, so partial derivatives of $\boldsymbol{u}$ with respect to θ are zero.

(iii) Steady, so partial derivatives of the velocity field components with respect to time are zero.

(iv) Acted upon by the external force of gravity only, so $f_r = f_\theta = 0$ and $f_z = -g$, since $\boldsymbol{f}$ is the body force per unit mass.

(v) Governed by the *swirl* velocity field u_θ only, i.e. $u_r = u_z = 0$ everywhere.

As a consequence of these assumptions, the fluid (air) particles move in horizontal circles – see Figure 2.23 later. Combining all these modelling assumptions reduces the incompressible Euler equations above to

$$\frac{u_\theta^2}{r} = \frac{1}{\rho}\frac{\partial p}{\partial r}, \qquad 0 = -\frac{1}{\rho r}\frac{\partial p}{\partial \theta} \qquad \text{and} \qquad 0 = -\frac{1}{\rho}\frac{\partial p}{\partial z} - g.$$

The incompressibility condition is satisfied trivially. The second equation tells us that the pressure p is independent of θ, as we might have already suspected. Hereafter we assume $p = p(r, z)$ and focus on the first and third equations.

The question now is, can we find solution functions $u_\theta = u_\theta(r, z)$ and $p = p(r, z)$ to the first and third partial differential equations? To help us in this direction we make some further assumptions. As a simple approximation to a real hurricane flow we assume the core performs a solid-body rotation, so the swirl velocity u_θ is proportional to r, whilst outside the core, the swirl velocity u_θ decays slowly with r. Indeed, if the core has radius a, then we assume

$$u_\theta = \begin{cases} \Omega r, & r \leqslant a, \\ \Omega a^{3/2}/r^{1/2}, & r > a. \end{cases}$$

The swirl flow we assume for $r \leqslant a$ represents solid-body rotation in the core region. Recall we already know this is an exact solution to the incompressible Euler equations as it is an example linear flow from Section 2.4; see Examples 2.14 and 1.4. To see this note $u_x = -\Omega y$, $u_y = \Omega x$ and $u_z = 0$ imply $\boldsymbol{u} = \Omega(-y\boldsymbol{i} + x\boldsymbol{j}) = \Omega r(-\sin\theta\,\boldsymbol{i} + \cos\theta\,\boldsymbol{j}) \equiv \Omega r\,\hat{\boldsymbol{\theta}}$. The flow we assume for $r > a$ represents the slow decay of the swirl velocity field. Note the flow field u_θ prescribed above is continuous. With $u_\theta = u_\theta(r)$ assumed to have this form, the question now is, can we find a corresponding pressure field $p = p(r, z)$ that supports this velocity flow field so that the first and third partial differential equations above are satisfied? For $r \leqslant a$, as just mentioned, we know that indeed such a pressure field exists and that it should depend quadratically and symmetrically (in this case) on x and y, and thus should depend on r^2. The case for $r > a$ is not so immediately obvious. For completeness, we look at both cases in detail.

Assume $r \leqslant a$. Using that $u_\theta = \Omega r$ in the first equation we see that

$$\frac{\partial p}{\partial r} = \rho\Omega^2 r \qquad \Leftrightarrow \qquad p(r, z) = \tfrac{1}{2}\rho\Omega^2 r^2 + C(z),$$

where $C(z)$ is an arbitrary function of z. If we then substitute this into the third equation, above we see that

$$-\frac{1}{\rho}\frac{\partial p}{\partial z} = -g \qquad \Leftrightarrow \qquad C'(z) = -\rho g,$$

and hence $C(z) = -\rho g z + C_0$ where C_0 is an arbitrary constant. Thus we now deduce that the pressure function is given by

$$p(r, z) = \tfrac{1}{2}\rho\Omega^2 r^2 - \rho g z + C_0.$$

What is the shape of an isobaric (constant pressure) surface for $r \leqslant a$? Consider a fixed pressure value P_* say. Then we see that for $r \leqslant a$,

$$P_* = \tfrac{1}{2}\rho\Omega^2 r^2 - \rho g z + C_0 \qquad \Leftrightarrow \qquad z = (\Omega^2/2g)\,r^2 - (C_0 - P_*)/\rho g,$$

so an isobaric surface is a paraboloid of revolution; see Figure 2.7.

For $r > a$, we use a completely analogous argument using $u_\theta = \Omega a^{3/2}/r^{1/2}$. Indeed, let us consider the slightly more general case $u_\theta = \Omega a^{\alpha+1}/r^\alpha$ for any $\alpha \in \mathbb{R}$. Then solving the first and third partial differential equations for $p = p(r, z)$ as we did for the case $r \leqslant a$, we find

$$p(r, z) = -\frac{\rho\Omega^2 a^2}{2\alpha}\left(\frac{a}{r}\right)^{2\alpha} - \rho g z + K_0,$$

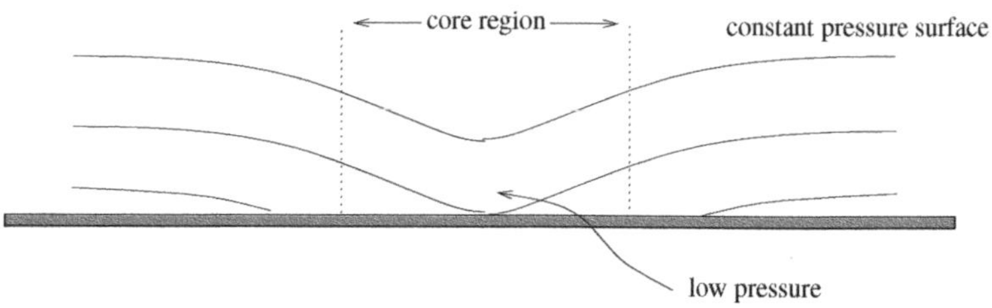

Figure 2.7 Simple model of a hurricane from Example 2.18. Isobaric, i.e. constant pressure, surfaces dip in the middle due to the huge swirl flow. Since in a still atmosphere pressure increases the closer to the ground one is, the dip in the isobaric surface means there is effectively low pressure at ground level in the central region of the hurricane. In the very middle, the eye of the storm, there is also very little airflow/wind. Hurricanes typically have diameters of approximately 500 km and last on a timescale of the order of one week.

where K_0 is an arbitrary constant. Since the pressure must be continuous at $r = a$, we substitute $r = a$ into the expression for the pressure here for $r > a$ and the expression for the pressure for $r \leqslant a$, and equate the two. This gives

$$\tfrac{1}{2}\rho\Omega^2 a^2 - \rho g z + C_0 = -\frac{\rho\Omega^2 a^2}{2\alpha}\left(\frac{a}{r}\right)^{2\alpha} - \rho g z + K_0 \quad \Leftrightarrow \quad K_0 = C_0 + \tfrac{1}{2}\rho\Omega^2 a^2\left(\frac{\alpha+1}{\alpha}\right).$$

Hence the pressure for $r > a$ is given by

$$p(r,z) = -\tfrac{1}{2}\rho\Omega^2 a^2 \frac{1}{\alpha}\left(\frac{a}{r}\right)^{2\alpha} - \rho g z + \tfrac{1}{2}\rho\Omega^2 a^2\left(\frac{\alpha+1}{\alpha}\right) + C_0.$$

If we set $p(r,z) = P_*$, the same fixed pressure prescribing the isobaric surface for $r \leqslant a$, we see that for $r > a$ the isobaric surface is given by

$$z = \tfrac{1}{2}\Omega^2 a^2 \frac{1}{\alpha g}\left(\alpha + 1 - \left(\frac{a}{r}\right)^{2\alpha}\right) - \frac{(C_0 - P_*)}{\rho g}.$$

See Figure 2.7. Since the pressure is only ever globally defined up to an additive constant, we can take $C_0 = 0$.

Remark 2.19 In the last example, once we assumed $u_\theta = u_\theta(r)$, we can solve the first and third partial differential equations to establish

$$p(r,z) = \rho\int_0^r \frac{u_\theta^2(r')}{r'}\,\mathrm{d}r' - \rho g z + C_0,$$

for some arbitrary constant C_0.

Remark 2.20 (Tornadoes, waterspouts and dust devils) The analysis above is also the basis of simple models for tornadoes, which have typical diameters from 100 m to 1 km and can last a few hours. Swirl velocity speeds can reach up to 640 km h^{-1}. Waterspouts are similar but occur on stretches of water such as lakes or out to sea.

In both cases the stark depression of the pressure surfaces means a violent pressure change as the tornado passes by. The extremely low pressure in the middle of the tornado at ground level creates an extreme suction effect – large enough to shift very large particles such as cars! Dust devils are smaller, with typically diameters of a few metres and lasting a few minutes.

In the following sections and Chapter 3, we see many more examples of homogeneous incompressible Euler flows. In addition, further examples can be found in Sections 4.6 and 4.7.

2.6 Bernoulli's Theorem

Theorem 2.21 (Bernoulli) *Consider an ideal homogeneous incompressible stationary flow with a conservative body force $\boldsymbol{f} = -\nabla\Phi$, where Φ is the potential function. Then the energy density per unit mass H is conserved along streamlines, where*

$$H := \tfrac{1}{2}|\boldsymbol{u}|^2 + \frac{p}{\rho} + \Phi.$$

Proof We need the following identity from Section A.1 of the Appendix (#7 with $\boldsymbol{v} = \boldsymbol{u}$):

$$\tfrac{1}{2}\nabla(|\boldsymbol{u}|^2) = (\boldsymbol{u}\cdot\nabla)\,\boldsymbol{u} + \boldsymbol{u}\times(\nabla\times\boldsymbol{u}).$$

Since the flow is stationary, Euler's equation of motion for an ideal fluid implies

$$(\boldsymbol{u}\cdot\nabla)\,\boldsymbol{u} = -\nabla\left(\frac{p}{\rho}\right) - \nabla\Phi.$$

Using the identity above, we see

$$\begin{aligned}
&\tfrac{1}{2}\nabla(|\boldsymbol{u}|^2) - \boldsymbol{u}\times(\nabla\times\boldsymbol{u}) = -\nabla\left(\frac{p}{\rho}\right) - \nabla\Phi \\
\Leftrightarrow\quad & \nabla\left(\tfrac{1}{2}|\boldsymbol{u}|^2 + \frac{p}{\rho} + \Phi\right) = \boldsymbol{u}\times(\nabla\times\boldsymbol{u}) \\
\Leftrightarrow\quad & \nabla H = \boldsymbol{u}\times(\nabla\times\boldsymbol{u}),
\end{aligned}$$

using the definition for H. Let $\boldsymbol{x} = \boldsymbol{x}(s)$ be a streamline that satisfies

$$\frac{\mathrm{d}}{\mathrm{d}s}\boldsymbol{x}(s) = \boldsymbol{u}(\boldsymbol{x}(s)).$$

By the Fundamental Theorem of Calculus and the chain rule, we have

$$\frac{\mathrm{d}}{\mathrm{d}s}H(\boldsymbol{x}(s)) = \big(\nabla H(\boldsymbol{x}(s))\big)\cdot\frac{\mathrm{d}}{\mathrm{d}s}\boldsymbol{x}(s).$$

Hence for any s_1 and s_2:

$$H(\boldsymbol{x}(s_2)) - H(\boldsymbol{x}(s_1)) = \int_{H(\boldsymbol{x}(s_1))}^{H(\boldsymbol{x}(s_2))} \mathrm{d}H(\boldsymbol{x}(s))$$

$$
\begin{aligned}
&= \int_{s_1}^{s_2} \frac{\mathrm{d}H}{\mathrm{d}s}(\boldsymbol{x}(s))\,\mathrm{d}s \\
&= \int_{s_1}^{s_2} \big(\nabla H(\boldsymbol{x}(s))\big) \cdot \frac{\mathrm{d}}{\mathrm{d}s}\boldsymbol{x}(s)\,\mathrm{d}s \\
&= \int_{s_1}^{s_2} (\boldsymbol{u} \times (\nabla \times \boldsymbol{u})) \cdot \boldsymbol{u}(\boldsymbol{x}(s))\,\mathrm{d}s \\
&= 0,
\end{aligned}
$$

where we used that $(\boldsymbol{u} \times \boldsymbol{a}) \cdot \boldsymbol{u} \equiv 0$ for any vector $\boldsymbol{a}$ (since $\boldsymbol{u} \times \boldsymbol{a}$ is orthogonal to $\boldsymbol{u}$). As s_1 and s_2 are arbitrary, we deduce that H does not change along streamlines. □

Remark 2.22 Note that:

(i) We assume a conservative body force, i.e. the body force can be expressed in the form $\boldsymbol{f} = -\nabla\Phi$, for some potential function Φ. For more details on conservative vector fields, see Section 2.8.

(ii) The quantity ρH has the units of an energy density per unit volume. Since ρ is constant here, we can also interpret Bernoulli's Theorem as saying that energy density per unit volume is constant along streamlines.

(iii) The incompressibility assumption for Bernoulli's Theorem can be relaxed as Bernoulli's Theorem applies for isentropic flows (Chorin and Marsden 1990, p. 14). See Section 2.10 later.

(iv) There is an unsteady version of Bernoulli's Theorem for irrotational flows; see Exercise 2.21.

Remark 2.23 (Asymptotic notation) We introduced the asymptotic notation 'O' at the beginning of Section 1.9. In the next example and hereafter we often use the additional asymptotic notation '$\ll$' (or '$\gg$') and '$\sim$'. Suppose the two functions $f = f(x)$ and $g = g(x)$ have well-defined limits as $x \to x_0$. We write $f \ll g$ if $f/g \to 0$ as $x \to x_0$. We write $f \sim g$ if $f/g \to 1$ as $x \to x_0$. The limit $x \to x_0$ is sometimes omitted if it is obvious from the context.

Example 2.24 (Torricelli's Theorem 1643) Consider the problem of an oil drum full of water that has a small hole punctured in it near the bottom. The problem is to determine the velocity of the fluid jetting out of the hole at the bottom and how that varies with the amount of water left in the tank – the setup is shown in Figure 2.8. We assume the hole has a small cross-sectional area α. Suppose that the cross-sectional area of the drum, and therefore of the free surface (water surface) at $z = 0$, is A. We naturally assume $\alpha \ll A$. Since the rate at which the amount of water is dropping inside the drum must equal the rate at which water is leaving the drum through the punctured hole, we have

$$
\left(-\frac{\mathrm{d}h}{\mathrm{d}t}\right) \cdot A = U \cdot \alpha \qquad \Leftrightarrow \qquad \left(-\frac{\mathrm{d}h}{\mathrm{d}t}\right) = \left(\frac{\alpha}{A}\right) \cdot U.
$$

We observe that $\alpha \ll A$, i.e. $\alpha/A \ll 1$, and hence we can deduce

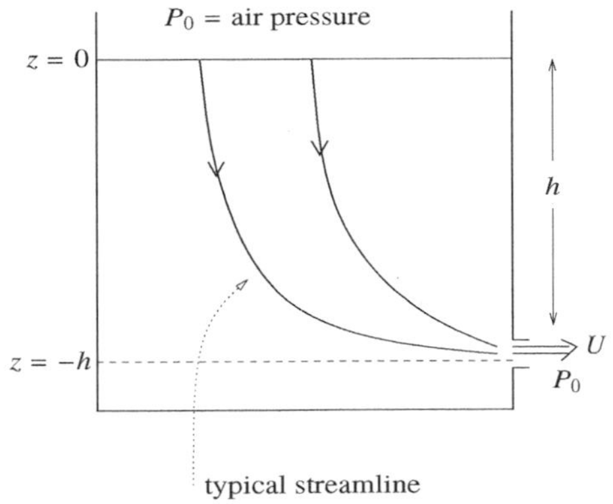

Figure 2.8 Torricelli problem. The pressure at the top surface and outside the puncture hole is atmospheric pressure P_0. Suppose the height of water above the puncture is h. The goal is to determine how the velocity of water U out of the puncture hole varies with h.

$$\frac{1}{U^2}\left(\frac{\mathrm{d}h}{\mathrm{d}t}\right)^2 = \left(\frac{\alpha}{A}\right)^2 \ll 1.$$

Since the flow is quasi-stationary, incompressible as it is water, and there is conservative body force due to gravity, we apply Bernoulli's Theorem for one of the typical streamlines shown in Figure 2.8. This implies that the quantity H is the same at the free surface and at the puncture hole outlet, hence

$$\tfrac{1}{2}\left(\frac{\mathrm{d}h}{\mathrm{d}t}\right)^2 + \frac{P_0}{\rho} = \tfrac{1}{2}U^2 + \frac{P_0}{\rho} - gh.$$

Thus, cancelling the P_0/ρ terms we can deduce

$$\begin{aligned} gh &= \tfrac{1}{2}U^2 - \tfrac{1}{2}\left(\frac{\mathrm{d}h}{\mathrm{d}t}\right)^2 \\ &= \tfrac{1}{2}U^2\left(1 - \frac{1}{U^2}\left(\frac{\mathrm{d}h}{\mathrm{d}t}\right)^2\right) \\ &= \tfrac{1}{2}U^2\left(1 - \left(\frac{\alpha}{A}\right)^2\right) \\ &\sim \tfrac{1}{2}U^2 \end{aligned}$$

for $\alpha/A \ll 1$ with an error of order $(\alpha/A)^2$. Thus, in the asymptotic limit we have $gh = \frac{1}{2}U^2$ and we deduce

$$U = \sqrt{2gh}.$$

Remark 2.25 We observe:

(i) The pressure inside the container at the puncture hole level is $P_0 + \rho g h$. The difference between this and the atmospheric pressure P_0 outside accelerates the water through the puncture hole. A similar application of Bernoulli's Theorem demonstrates the normal action of a siphon – relating the exit speed of drainage to the vertical

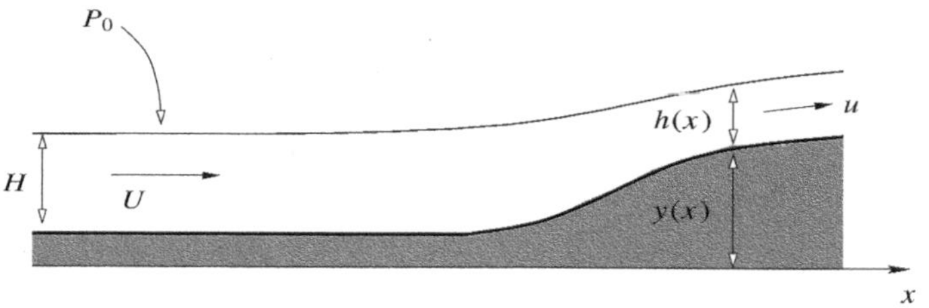

Figure 2.9 Channel flow problem. A steady flow of water, uniform in cross-section, flows over a gently undulating bed of height $y = y(x)$ as shown. The depth of the flow is given by $h = h(x)$. Upstream the flow is characterised by flow velocity U and depth H.

difference in height between that of the inlet reservoir and the (lower) exit from a U-shaped tube.

(ii) We can also interpret the asymptotic result above in the Torricelli problem as follows. Consider a fluid particle that starts at the top surface with zero initial velocity just as the hole is punctured. Then its velocity and height at time $t > 0$, under free fall due to the force of gravity, are $v = gt$ and $h = \frac{1}{2}gt^2$. Hence we deduce $t = \sqrt{2h/g}$, which substituted into $v = gt$ gives $v = \sqrt{2gh}$. In other words, the velocity with which the fluid particle leaves the puncture hole matches that it has gained (corresponding to the kinetic energy it has gained and thus the potential energy it has lost) due to free fall under gravity.

Example 2.26 (Channel flow, Froude number) Consider the problem of a steady flow of water in a channel over a gently undulating bed – see Figure 2.9. We assume that the flow is shallow and uniform in cross-section. Upstream the flow is characterised by flow velocity U and depth H. The flow then impinges on a gently undulating bed of height $y = y(x)$ as shown in Figure 2.9, where x measures distance downstream. The depth of the flow is given by $h = h(x)$, whilst the fluid velocity at that point is $u = u(x)$, which is uniform over the depth throughout. Reiterating, our assumptions are thus $|\mathrm{d}y/\mathrm{d}x| \ll 1$, i.e. the bed is gently undulating, and $|\mathrm{d}h/\mathrm{d}x| \ll 1$, i.e. variations in depth are small. The continuity equation (incompressibility here) implies that for all x the volume flux per unit breadth is fixed, and thus

$$uh = UH.$$

Then Euler's equations for a steady flow imply Bernoulli's Theorem which we apply to the surface streamline, for which the pressure is constant and equal to atmospheric pressure P_0. Hence for all x we have, cancelling P_0 on both sides:

$$\tfrac{1}{2}U^2 + gH = \tfrac{1}{2}u^2 + g(y + h).$$

Substituting for $u = u(x)$ from the incompressibility condition above, and rearranging, Bernoulli's Theorem implies that for all x we have the constraint

$$y = \frac{U^2}{2g} + H - h - \frac{(UH)^2}{2gh^2}.$$

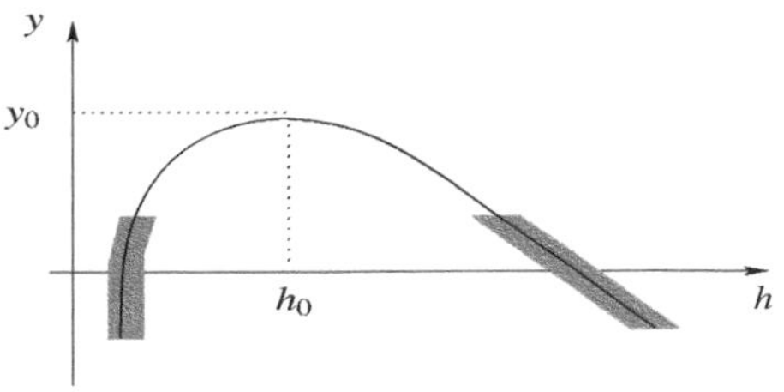

Figure 2.10 Channel flow problem. The flow depth $h = h(x)$ and undulation height $y = y(x)$ are related as shown, from Bernoulli's Theorem. Note that y has a maximum value y_0 at height $h_0 = H\,\mathrm{F}^{2/3}$, where $\mathrm{F} = U/\sqrt{gH}$ is the Froude number.

We can think of this as a parametric equation (implicitly) relating the fluid depth $h = h(x)$ to the undulation height $y = y(x)$, where the parameter x runs from $x = -\infty$ far upstream to $x = +\infty$ far downstream. We plot this relation, y as a function of h, in Figure 2.10. Note that y has a unique global maximum y_0 coinciding with the local maximum and given by

$$\frac{\mathrm{d}y}{\mathrm{d}h} = 0 \quad \Leftrightarrow \quad -1 + \frac{(UH)^2}{gh^3} = 0 \quad \Leftrightarrow \quad h = \frac{(UH)^{2/3}}{g^{1/3}}.$$

We set $h_0 := (UH)^{2/3}/g^{1/3}$, i.e. the height at which y attains its maximum, and

$$\mathrm{F} := U/\sqrt{gH},$$

where F is known as the *Froude number*. It is a dimensionless function of the upstream conditions and represents the ratio of the oncoming fluid speed to the wave (signal) speed in fluid depth H; see Section 2.9. Note $h_0 = H\,\mathrm{F}^{2/3}$.

Note that when $y = y(x)$ attains its maximum value at $h_0 = (UH)^{2/3}/g^{1/3}$, then

$$y = \frac{U^2}{2g} + H - \frac{(UH)^{2/3}}{g^{1/3}} - \frac{(UH)^{2/3}}{2g^{1/3}} = H\,(1 + \tfrac{1}{2}\mathrm{F}^2 - \tfrac{3}{2}\mathrm{F}^{2/3}).$$

We set $y_0 := H\,(1 + \frac{1}{2}\mathrm{F}^2 - \frac{3}{2}\mathrm{F}^{2/3})$ to be this maximum value. This puts a bound on the height of the bed undulation that is compatible with the upstream conditions. In Figure 2.11 we plot the normalised maximum permissible height y_0/H the undulation is allowed to attain as a function of the Froude number F. Note that two different values of the Froude number F give the same maximum permissible undulation height y_0: one with $\mathrm{F} < 1$ for which U is slower compared with $\sqrt{gH}$, and one with $\mathrm{F} > 1$ for which it is faster.

Let us now consider an actual given undulation $y = y(x)$. Suppose that it attains an actual maximum value $y_{\max}$. There are three cases to consider. In turn we shall consider $y_{\max} < y_0$, $y_{\max} = y_0$ and $y_{\max} > y_0$.

In the first case, $y_{\max} < y_0$, as x varies from $x = -\infty$ to $x = +\infty$, the undulation height $y = y(x)$ varies and is such that $y(x) \leqslant y_{\max}$. Consider Figure 2.10, which plots the constraint relationship between y and h resulting from Bernoulli's Theorem. Since $y(x) \leqslant y_{\max}$ as x varies from $-\infty$ to $+\infty$, the values of (h, y) are restricted to part of the branches

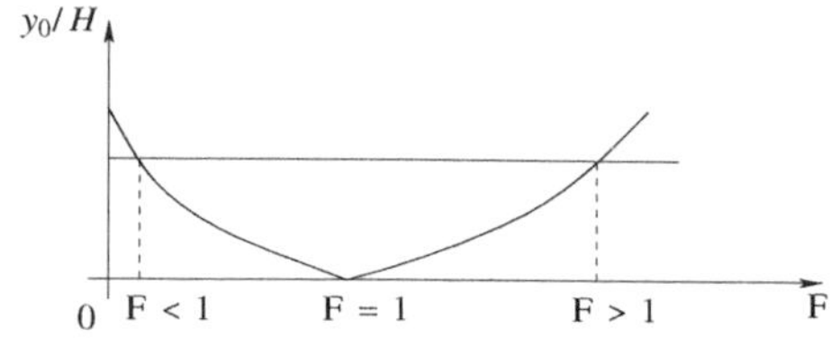

Figure 2.11 Channel flow problem. Two different values of the Froude number F give the same maximum permissible undulation height y_0. Note that we actually plot the normalised maximum possible height y_0/H on the ordinate axis.

of the graph either side of the global maximum (h_0, y_0). In the figure these parts of the branches are the locale of the shaded sections shown. Note the derivative $\mathrm{d}y/\mathrm{d}h = 1/(\mathrm{d}h/\mathrm{d}y)$ has the same fixed (and opposite) sign in each of the branches. In the branch for which $h < h_0$ we observe $\mathrm{d}y/\mathrm{d}h > 0$, while in the branch for which $h > h_0$ we observe $\mathrm{d}y/\mathrm{d}h < 0$. Indeed from the constraint condition, as we have seen, we have

$$\frac{\mathrm{d}y}{\mathrm{d}h} = -\left(1 - \frac{(UH)^2}{gh^3}\right).$$

Using the incompressibility condition to substitute for UH, we observe

$$\frac{\mathrm{d}y}{\mathrm{d}h} = -\left(1 - \frac{u^2}{gh}\right).$$

We can think of $u/\sqrt{gh}$ as a local Froude number. In any case, we are on one branch or the other, and in either case the sign of $\mathrm{d}y/\mathrm{d}h$ is fixed. This means, using the expression for $\mathrm{d}y/\mathrm{d}h$ we just derived, for any flow realisation the sign of $1 - u^2/gh$ is also fixed. When $x = -\infty$ this quantity has the value $1 - U^2/gH$. Hence the sign of $1 - U^2/gH$ determines the sign of $1 - u^2/gh$.

If $\mathrm{F} < 1$ then $U^2/gH = \mathrm{F}^2 < 1$ and therefore for all x we must have $u^2/gh < 1$. We also deduce we must be on the branch for which $h > h_0$ as $\mathrm{d}y/\mathrm{d}h$ is negative. The flow is said to be *subcritical* throughout and we see

$$\frac{\mathrm{d}h}{\mathrm{d}y} = \left(\frac{\mathrm{d}y}{\mathrm{d}h}\right)^{-1} = -\left(1 - \frac{u^2}{gh}\right)^{-1} < -1 \qquad \Rightarrow \qquad \frac{\mathrm{d}}{\mathrm{d}y}(h + y) < 0.$$

Hence, as the bed height y increases, the fluid depth h decreases and vice versa.

On the other hand, if $\mathrm{F} > 1$ then $U^2/gH > 1$ and thus $u^2/gh > 1$. We must be on the branch for which $h < h_0$ as $\mathrm{d}y/\mathrm{d}h$ is positive. The flow is said to be *supercritical* throughout and we have

$$\frac{\mathrm{d}h}{\mathrm{d}y} = -\left(1 - \frac{u^2}{gh}\right)^{-1} > 0 \qquad \Rightarrow \qquad \frac{\mathrm{d}}{\mathrm{d}y}(h + y) > 1.$$

Hence in this case, as the bed height y increases, the fluid depth h increases and vice versa. Both cases, $\mathrm{F} < 1$ and $\mathrm{F} > 1$, are illustrated in Figure 2.12.

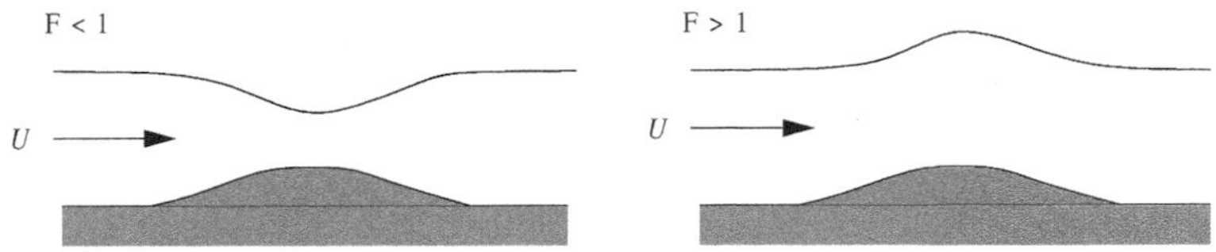

Figure 2.12 Channel flow problem. For the case $y_{max} < y_0$, when F < 1, as the bed height y increases, the fluid depth h decreases and vice versa. Hence we see a depression in the fluid surface above a bump in the bed. On the other hand, when F > 1, as the bed height y increases, the fluid depth h increases and vice versa. Hence we see an elevation in the fluid surface above a bump in the bed. The case F < 1 might correspond to river flow in a plain (relatively slow), while F > 1 would correspond to mountain streams (relatively fast).

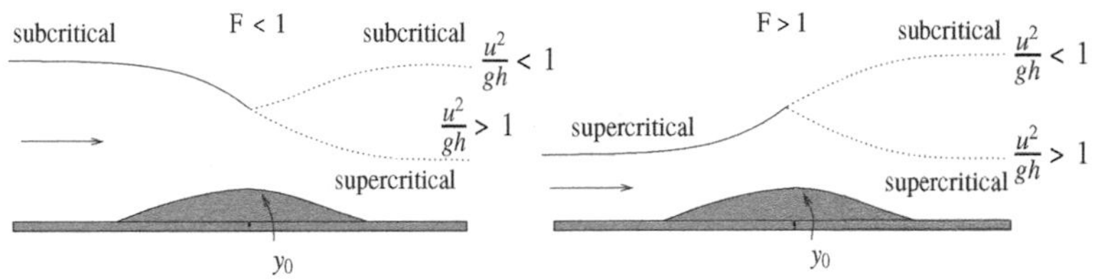

Figure 2.13 Channel flow problem. For the case $y_{max} = y_0$, when F < 1, upstream the fluid depth h decreases as the bed height y increases. When F > 1, upstream as the bed height y increases, the fluid depth h increases. At $y = y_0$ the flow becomes critical and there are two possibilities downstream. The supercritical downstream flow occurs in a real flow due to energy dissipation (the effect of viscosity).

In the second case, $y_{max} = y_0$, the free surface falls (F < 1) or rises (F > 1) as long as $y < y_0$. Then at $y = y_0$ the flow is locally critical as $u^2/gh = 1$ and $\mathrm{d}h/\mathrm{d}y$ is undefined. However, $\mathrm{d}h/\mathrm{d}x$ is normally bounded and there are two possible flows downstream; see Figure 2.13. This is revealed using a local expansion; see Exercise 2.20. For a real flow, the supercritical downstream flow occurs as a result of energy dissipation through internal friction, i.e. due to viscosity.

In the third case, $y_{max} > y_0$, the undulation height is larger than the maximum permissible height y_0 compatible with the upstream conditions. Indeed, $\mathrm{d}h/\mathrm{d}y$ becomes infinite at $y = y_0$ and in practice our steady model breaks down. Consider a surface ideal-fluid particle that will initially have total energy density per unit mass $\frac{1}{2}U^2 + gH$, i.e. its kinetic plus potential energy densities. As this particle rises and exchanges its kinetic energy for potential energy, the maximum height it can reach, when it will have zero velocity, is $Y := H + U^2/2g$. Qualitatively, if F < 1 the flow tries to modify the upstream conditions so the new U', H' make the flow just critical at y_{max}. If $y_{max} < Y$ the flow may then traverse the obstacle, but if $y_{max} > Y$ it cannot, and the flow is reflected. See the top two panels in Figure 2.14. If F > 1 there will be a jump in the surface height at $y = y_0$ where $\mathrm{d}h/\mathrm{d}y$ is infinite; see the bottom two panels in Figure 2.14. The jump region corresponds to a bore, i.e. a foaming dissipative region. If $y_{max} < Y$, the flow may traverse the obstacle, whilst if $y_{max} > Y$ there will be total reflection.

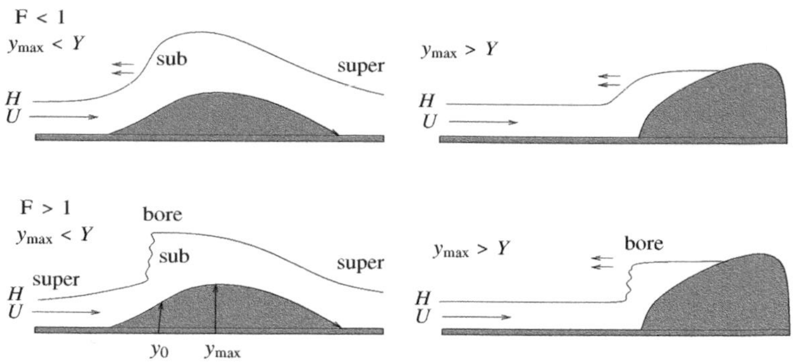

Figure 2.14 Channel flow problem. For the case $y_{\text{max}} > y_0$ our steady model breaks down. The height $Y := H + U^2/2g$ is the maximum height a fluid particle can reach, when it has zero velocity. The flow tries to modify the upstream conditions so that a new U', H' make the flow just critical at y_{max}. The result may be either of the flows in the left panels if $y_{\text{max}} < Y$. If $y_{\text{max}} > Y$ there must a total reflection of the flow as shown in the two right panels.

Remark 2.27 Example 2.26 above is a subtle case of channel flow with observable examples in e.g. rivers and streams.

2.7 Kelvin's Circulation Theorem

We turn our attention to important results centred on the persistence of circulation and vortex lines and tubes in inviscid flow. Our presentation herein closely follows that in Chorin and Marsden (1990, Sec. 1.2).

Definition 2.28 (Circulation) Let C be a simple closed contour in the fluid at time $t = 0$. Suppose C is carried by the flow to the closed contour C_t at time t. The *circulation* around such a material contour C_t is defined to be the line integral

$$\oint_{C_t} \boldsymbol{u} \cdot \mathrm{d}\boldsymbol{x}.$$

Theorem 2.29 (Kelvin's Circulation Theorem, 1869) *For an ideal, homogeneous, incompressible flow with a conservative body force, the circulation for any closed contour C_t is constant in time.*

Proof First, we establish a variant of the Transport Theorem for closed loops of fluid particles C_t as follows. Let $\boldsymbol{a} = \boldsymbol{a}(s)$ with $0 \leqslant s < 1$ denote a parameterisation of the closed loop C at time $t = 0$. A parameterisation of the evolved closed loop C_t is given by $\boldsymbol{x} = \boldsymbol{x}(\boldsymbol{a}(s), t)$ with $0 \leqslant s < 1$. The chain rule implies

$$\oint_{C_t} \boldsymbol{u} \cdot \mathrm{d}\boldsymbol{x} = \int_0^1 \boldsymbol{u}(\boldsymbol{x}(\boldsymbol{a}(s), t), t) \cdot \frac{\partial \boldsymbol{x}}{\partial s}(\boldsymbol{a}(s), t)\, \mathrm{d}s.$$

If we take the time derivative of this quantity, then we must use the product rule for the integrand with the dot product in the term on the right. When the time derivative falls on the second term in the integrand, we have

$$\begin{aligned}
\int_0^1 \boldsymbol{u}(\boldsymbol{x}(\boldsymbol{a}(s),t),t) &\cdot \frac{\mathrm{d}}{\mathrm{d}t}\left(\frac{\partial \boldsymbol{x}}{\partial s}(\boldsymbol{a}(s),t)\right)\mathrm{d}s \\
&= \int_0^1 \boldsymbol{u}(\boldsymbol{x}(\boldsymbol{a}(s),t),t) \cdot \frac{\partial}{\partial s}\left(\frac{\mathrm{d}\boldsymbol{x}}{\mathrm{d}t}(\boldsymbol{a}(s),t)\right)\mathrm{d}s \\
&= \int_0^1 \boldsymbol{u}(\boldsymbol{x}(\boldsymbol{a}(s),t),t) \cdot \frac{\partial}{\partial s}\big(\boldsymbol{u}(\boldsymbol{x}(\boldsymbol{a}(s),t),t)\big)\,\mathrm{d}s \\
&= \tfrac{1}{2}\int_0^1 \frac{\partial}{\partial s}\big(|\boldsymbol{u}|^2(\boldsymbol{x}(\boldsymbol{a}(s),t),t)\big)\,\mathrm{d}s,
\end{aligned}$$

which equals zero as C_t is closed. Hence the only non-zero contribution to the time derivative of the circulation comes from the case when the time derivative falls on the first term in the integrand and we observe that

$$\begin{aligned}
\frac{\mathrm{d}}{\mathrm{d}t}\oint_{C_t} \boldsymbol{u}\cdot \mathrm{d}\boldsymbol{x} &= \int_0^1 \left(\frac{\partial \boldsymbol{u}}{\partial t} + (\boldsymbol{u}\cdot\nabla)\,\boldsymbol{u}\right)(\boldsymbol{x}(\boldsymbol{a}(s),t),t)\cdot\frac{\partial \boldsymbol{x}}{\partial s}(\boldsymbol{a}(s),t)\,\mathrm{d}s \\
&= \int_{C_t}\left(\frac{\partial \boldsymbol{u}}{\partial t} + (\boldsymbol{u}\cdot\nabla)\,\boldsymbol{u}\right)\cdot \mathrm{d}\boldsymbol{x}.
\end{aligned}$$

Second, using this transport result and the Euler equations with a conservative body force of the form $\boldsymbol{f} = -\nabla\Phi$ for some potential function Φ, we observe

$$\frac{\mathrm{d}}{\mathrm{d}t}\oint_{C_t} \boldsymbol{u}\cdot\mathrm{d}\boldsymbol{x} = \oint_{C_t}\left(-\nabla\left(\frac{p}{\rho}\right) - \nabla\Phi\right)\cdot\mathrm{d}\boldsymbol{x} = -\oint_{C_t}\nabla\left(\frac{p}{\rho}+\Phi\right)\cdot\mathrm{d}\boldsymbol{x} = 0,$$

since C_t is closed, thus establishing the required result. □

Remark 2.30 The theorem does not require that the domain $\mathcal{D}$ of the fluid be simply connected, see Definition 2.31 just below, however C_t must be a continuous deformation of C. The homogeneity and incompressibility assumptions are also not essential. The body force is assumed to be conservative. For more details on conservative vector fields, see Section 2.8.

Definition 2.31 (Simply connected domains) A domain $\mathcal{D}$ is *simply connected* if any continuous closed curve in $\mathcal{D}$ can be continuously deformed to a point without leaving $\mathcal{D}$.

Remark 2.32 The exterior of a solid sphere is simply connected whilst the exterior to a disc in $\mathbb{R}^2$ is not simply connected.

Stokes' Theorem provides an equivalent definition for the circulation as

$$\oint_{C_t} \boldsymbol{u}\cdot\mathrm{d}\boldsymbol{x} = \iint_{S_t}(\nabla\times\boldsymbol{u})\cdot\boldsymbol{n}\,\mathrm{d}S = \iint_{S_t}\boldsymbol{\omega}\cdot\boldsymbol{n}\,\mathrm{d}S,$$

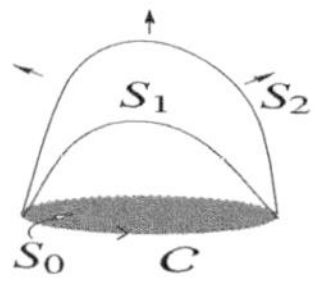

Figure 2.15 Stokes' Theorem tells us that the circulation around the closed contour C equals the flux of vorticity through any surface whose perimeter is C. For example, here the flux of vorticity through S_0, S_1 or S_2 is the same.

where $\mathcal{S}_t$ is any suitably smooth surface with suitably smooth perimeter C_t; see Figure 2.15. Naturally, to make sense of the surface $\mathcal{S}_t$ spanning the perimeter C_t when using Stokes' Theorem to give an equivalent definition of circulation means that we must restrict ourselves to simply connected domains $\mathcal{D}$. In this instance, we deduce that the circulation around C_t is equivalent to the flux of vorticity through the surface $\mathcal{S}_t$ with perimeter C_t. We conclude the following.

Corollary 2.33 (Flux of vorticity) *For an ideal, homogeneous, incompressible flow without external forces in a simply connected domain $\mathcal{D}$, the flux of vorticity across a surface moving with the fluid is constant in time.*

Definition 2.34 (Vortex lines, sheets and tubes) A *vortex line* is an integral curve/trajectory of the vorticity field $\boldsymbol{\omega}$, i.e. with t fixed they satisfy $(\mathrm{d}/\mathrm{d}s)\boldsymbol{x}(s) = \boldsymbol{\omega}(\boldsymbol{x}(s), t)$. Equivalently, they are continuous curves that are everywhere tangent to the local vorticity $\boldsymbol{\omega}$. A *vortex sheet* is a surface $\mathcal{E}_t$ that is tangent to the vorticity $\boldsymbol{\omega}$ at every point. A *vortex tube* consists of a two-dimensional surface $\mathcal{S}_t$ that is nowhere tangent to the vorticity field $\boldsymbol{\omega}$, with vortex lines drawn through each point of the bounding perimeter C_t of $\mathcal{S}_t$. The vortex lines are extended as far as possible in each direction. As emphasised in Chorin and Marsden (1990, p. 26), assuming $\boldsymbol{\omega} \neq 0$, we assume $\mathcal{S}_t$ is diffeomorphic to a disc $\mathcal{B}$, i.e. $\mathcal{S}_t$ can be smoothly bijectively transformed to a disc $\mathcal{B}$. Hence the resulting tube is diffeomorphic to $\mathcal{B} \times \mathbb{R}$.

Proposition 2.35 (Evolution of vortex lines and sheets) *For an ideal, homogeneous, incompressible flow, consider a line $\mathcal{L}$ or surface $\mathcal{E}$ that is, respectively, a vortex line or sheet at time $t = 0$. Then the corresponding evolved line $\mathcal{L}_t$ or surface $\mathcal{E}_t$ under the flow remains a vortex line or sheet.*

Proof It is sufficient to prove the result for a vortex sheet. Let $\boldsymbol{n} = \boldsymbol{n}(\boldsymbol{a})$ be the normal to $\mathcal{E}$ at every point $\boldsymbol{a} \in \mathcal{E}$ at $t = 0$. Since we assume $\mathcal{E}$ is a vortex sheet, $\boldsymbol{n} \cdot \boldsymbol{\omega}_0 = 0$ everywhere on $\mathcal{E}$. The flux of $\boldsymbol{\omega}_0$ across any portion $\mathcal{E}' \subseteq \mathcal{E}$ is thus zero. By Kelvin's Circulation Theorem, the flux of $\boldsymbol{\omega}$ across $\mathcal{E}'_t$ is also zero. Since $\mathcal{E}'$ and thus $\mathcal{E}'_t$ are arbitrary, $\boldsymbol{n} \cdot \boldsymbol{\omega} = 0$ everywhere on $\mathcal{E}_t$, where $\boldsymbol{n} = \boldsymbol{n}(\boldsymbol{x}, t)$ is now the normal at every $\boldsymbol{x} \in \mathcal{E}_t$. Hence $\mathcal{E}_t$ is a vortex sheet. □

We conclude this section by proving Helmholtz's Theorem for the 'strength' of a vortex tube.

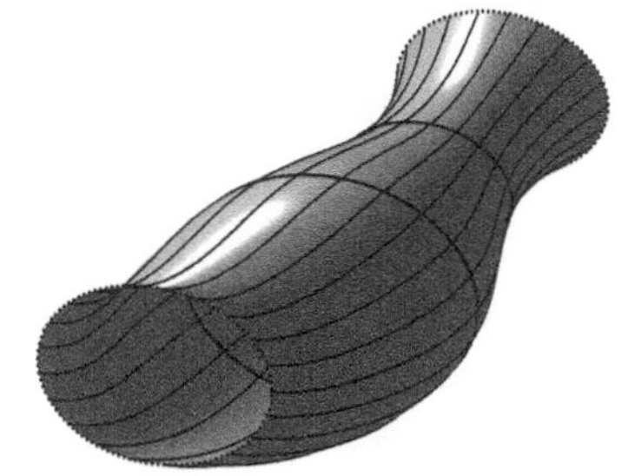

Figure 2.16 The strength of the vortex tube is given by the circulation around any curve that encircles the tube once. For example, the circulation around each of the two solid curves encircling the tube shown is the same.

Theorem 2.36 (Helmholtz's Theorem, 1858) *Consider an ideal, homogeneous, incompressible flow. If C_1 and C_2 are any two closed curves encircling a vortex tube once, with the same orientation, anywhere along its length, then the circulation around each is one and the same. Further, the circulation is constant in time as the tube moves with the fluid.*

Remark 2.37 (Vortex tube strength) We define this circulation to be the *strength* of the vortex tube. It is independent of the precise cross-sectional area $\mathcal{S}_t$ and/or the precise circuit C_t around the vortex tube taken; see Figure 2.16. Hence stretching of vortex lines/tubes leads to intensification of vorticity – and local rotation.

Proof Assume C_1 and C_2 are two closed loops as stated. For simplicity, assume they do not intersect. Assume $\mathcal{S}_1$ and $\mathcal{S}_2$ are two smooth spanning surfaces nowhere tangent to ω with respective perimeters C_1 and C_2. Note that the lateral surface $\mathcal{S}$ of the vortex tube, which by definition is everywhere tangent to ω, is an example vortex sheet. Let $\mathcal{V}$ denote the volume enclosed by the vortex tube between C_1 and C_2. Then by the Divergence and Stokes' Theorems,

$$\begin{aligned}
0 &= \int_{\mathcal{V}} \nabla \cdot \boldsymbol{\omega} \, \mathrm{d}V \\
&= \int_{\mathcal{S}_1} \boldsymbol{\omega} \cdot \boldsymbol{n} \, \mathrm{d}S + \int_{\mathcal{S}_2} \boldsymbol{\omega} \cdot \boldsymbol{n} \, \mathrm{d}S + \int_{\mathcal{S}} \boldsymbol{\omega} \cdot \boldsymbol{n} \, \mathrm{d}S \\
&= \oint_{C_1} \boldsymbol{u} \cdot \mathrm{d}\boldsymbol{x} - \oint_{C_2} \boldsymbol{u} \cdot \mathrm{d}\boldsymbol{x},
\end{aligned}$$

where we have used that the orientation of C_2 and the vorticity ω across $\mathcal{S}_2$ are opposite those for C_1 and $\mathcal{S}_1$. We also used that $\boldsymbol{\omega} \cdot \boldsymbol{n}$ is zero everywhere on $\mathcal{S}$ by definition. The first conclusion follows. That the circulation is constant in time follows from Kelvin's Circulation Theorem. □

2.8 Irrotational and Potential Flow

Many flows have extensive regions where the vorticity is zero; some have zero vorticity everywhere.

Definition 2.38 (Irrotational regions and flows) Suppose there exists a fluid flow in the domain $\mathcal{D} \subseteq \mathbb{R}^d$. A region $\mathcal{V} \subseteq \mathcal{D}$ is said to be an *irrotational* region of the flow when $\nabla \times \boldsymbol{u} = \boldsymbol{0}$ everywhere in $\mathcal{V}$, i.e. the vorticity is zero everywhere in the region. If the vorticity is zero everywhere in $\mathcal{D}$, then the flow is said to be *irrotational*.

Theorem 2.39 (The persistence of irrotational flow) *Consider an ideal, homogeneous, incompressible flow with a conservative body force. Suppose initially a simply connected region $\mathcal{V} \subseteq \mathcal{D}$ is irrotational. Then this region, carried by the flow to the region $\mathcal{V}_t$ at time t, remains irrotational.*

Proof By assumption, at time $t = 0$ the vorticity ω is zero everywhere in $\mathcal{V}$. Suppose at some time t later the vorticity ω in the region $\mathcal{V}_t$ was not zero everywhere in $\mathcal{V}_t$. In this case, by Stokes' Theorem,

$$\oint_{C_t} \boldsymbol{u} \cdot \mathrm{d}\boldsymbol{x} = \iint_{S_t} \omega \cdot \boldsymbol{n} \,\mathrm{d}S.$$

We can find a closed contour C_t contained in $\mathcal{V}_t$ for which the circulation is non-zero. However, by Kelvin's Circulation Theorem 2.29, this is not possible as C_t must have evolved from a closed contour C contained in $\mathcal{V}$ around which the circulation is zero. Hence the vorticity in $\mathcal{V}_t$ must be zero everywhere. □

Remark 2.40 In two dimensions the persistence of irrotational regions is more immediately obvious. This is because the vorticity formulation of the incompressible Euler equations in two dimensions in Remark 2.9 expresses the fact that the scalar vorticity in this case is simply transported by the flow.

Definition 2.41 (Conservative fields and potential functions) A given field $\boldsymbol{u} : \mathcal{D} \to \mathbb{R}^d$ is said to be *conservative* if and only if there exists a scalar function ϕ on $\mathcal{D}$, known as the *potential*, such that

$$\boldsymbol{u} = \nabla\phi.$$

Suppose we are given a conservative vector field $\boldsymbol{u} = \nabla\phi$ in $\mathcal{D}$. Then by the Fundamental Theorem of Calculus, for any smooth contour $C(\boldsymbol{x}_1, \boldsymbol{x}_2) \subset \mathcal{D}$ starting and then ending respectively at any $\boldsymbol{x}_1, \boldsymbol{x}_2 \in \mathcal{D}$, the line integral

$$\int_{C(\boldsymbol{x}_1,\boldsymbol{x}_2)} \boldsymbol{u} \cdot \mathrm{d}\boldsymbol{x} = \phi(\boldsymbol{x}_2) - \phi(\boldsymbol{x}_1).$$

Remark 2.42 (Path independence property) That the line integral in $\mathcal{D}$ is independent of the path of $C(\boldsymbol{x}_1, \boldsymbol{x}_2) \subset \mathcal{D}$ between $\boldsymbol{x}_1$ and $\boldsymbol{x}_2$ in $\mathcal{D}$ is known as the *path independence property*.

Lemma 2.43 *The path independence property for all $\boldsymbol{x}_1, \boldsymbol{x}_2 \in \mathcal{D}$ is equivalent to the condition that, for all simple closed curves $C \subset \mathcal{D}$, the circulation*

$$\oint_C \boldsymbol{u} \cdot \mathrm{d}\boldsymbol{x} = 0.$$

Proof The forward statement follows by setting $\boldsymbol{x}_2 = \boldsymbol{x}_1$. The reverse statement is established as follows. Given any $\boldsymbol{x}_1, \boldsymbol{x}_2 \in \mathcal{D}$, consider two smooth distinct contours $C(\boldsymbol{x}_1, \boldsymbol{x}_2)$ and $C'(\boldsymbol{x}_1, \boldsymbol{x}_2)$, both in $\mathcal{D}$, joining $\boldsymbol{x}_1$ and $\boldsymbol{x}_2$. Since the contour $C(\boldsymbol{x}_1, \boldsymbol{x}_2) - C'(\boldsymbol{x}_1, \boldsymbol{x}_2)$ is a closed loop and the circulation around any such loop is zero, we conclude that the line integrals along $C(\boldsymbol{x}_1, \boldsymbol{x}_2)$ and $C'(\boldsymbol{x}_1, \boldsymbol{x}_2)$ are the same. Since both contours are arbitrary, the reverse statement holds. □

The reverse statement to that directly after Definition 2.41 above also holds.

Lemma 2.44 *The path independence property for all $\boldsymbol{x}_1, \boldsymbol{x}_2 \in \mathcal{D}$ implies $\boldsymbol{u}$ is a conservative vector field in $\mathcal{D}$.*

Proof Fix the point $\boldsymbol{x}_0 \in \mathcal{D}$ and for any $\boldsymbol{x} \in \mathcal{D}$ define

$$\phi(\boldsymbol{x}) := \int_{C(\boldsymbol{x}_0, \boldsymbol{x})} \boldsymbol{u} \cdot \mathrm{d}\boldsymbol{x},$$

along any smooth contour $C(\boldsymbol{x}_0, \boldsymbol{x})$. This is well-defined using the path independence property. Taking the gradient of ϕ establishes $\boldsymbol{u} = \nabla\phi$. □

We now make the connection between irrotational flows and conservative flows. Since $\nabla \times (\nabla\phi) \equiv \boldsymbol{0}$, every conservative vector field $\boldsymbol{u}$ is irrotational. However the converse statement only holds when $\mathcal{D}$ is simply connected. Recall our discussion preceding Corollary 2.33.

Proposition 2.45 *In simply connected domains $\mathcal{D}$, irrotational flows are conservative.*

Proof For simply connected domains we can apply Stokes' Theorem: for an arbitrary closed loop C, which is the perimeter of a smooth surface $\mathcal{S}$, we have

$$\oint_C \boldsymbol{u} \cdot \mathrm{d}\boldsymbol{x} = \iint_{\mathcal{S}} (\nabla \times \boldsymbol{u}) \cdot \boldsymbol{n} \, \mathrm{d}S.$$

Since the flow is irrotational, the term on the right is zero and so the circulation around an arbitrary closed loop C is zero. Hence $\boldsymbol{u}$ is conservative in $\mathcal{D}$. □

If the fluid is also incompressible, then ϕ is *harmonic* since $\nabla \cdot \boldsymbol{u} = 0$ implies

$$\Delta\phi = 0.$$

Hence, for such situations, we only need to solve Laplace's equation $\Delta\phi = 0$ subject to certain boundary conditions. For example, for an ideal flow, the normal velocity $\boldsymbol{u} \cdot \boldsymbol{n} = \nabla\phi \cdot \boldsymbol{n} = \partial\phi/\partial n$ is given on the boundary, and this would constitute a Neumann problem for Laplace's equation, for which there is a unique solution (up to an arbitrary additive constant). Indeed we observe, for simply connected domains $\mathcal{D}$, such incompressible fluid flows $\boldsymbol{u}$ in $\mathcal{D}$ that are irrotational, and thus conservative, trivially satisfy the vorticity formulation of the Euler equations of motion, see Corollary 2.7. And from that corollary we deduce that there exists a scalar field $p = p(\boldsymbol{x})$ such that $\boldsymbol{u} = \nabla\phi$ satisfies

$$\frac{\partial \boldsymbol{u}}{\partial t} + (\boldsymbol{u} \cdot \nabla)\,\boldsymbol{u} = -\frac{1}{\rho}\nabla p. \tag{2.1}$$

Using the identity $(\boldsymbol{u} \cdot \nabla)\,\boldsymbol{u} = \frac{1}{2}\nabla(|\boldsymbol{u}|^2) - \boldsymbol{u} \times (\nabla \times \boldsymbol{u})$, we see $\partial\phi/\partial t + \frac{1}{2}|\boldsymbol{u}|^2 + p/\rho + \Phi$ is constant in space, where Φ is the body force potential. Hence solutions $\phi = \phi(\boldsymbol{x}, t)$ of the Laplace equation generate solutions $\boldsymbol{u} = \nabla\phi$ of the incompressible, irrotational Euler equations of motion and vice versa.

Example 2.46 (Linear flow) Consider the two-dimensional flow field in $\mathbb{R}^2$ given by $\boldsymbol{u} = (\gamma x, -\gamma y)^{\mathrm{T}}$ where $\gamma \in \mathbb{R}$ is a constant. This flow is irrotational everywhere. Hence there exists a flow potential which is in fact given by $\phi = \frac{1}{2}\gamma(x^2 - y^2)$. Since $\nabla \cdot \boldsymbol{u} = 0$ as well, we have $\Delta\phi = 0$. Further, since this flow is two-dimensional, there exists a stream function $\psi = \gamma xy$.

Remark 2.47 We see many two-dimensional irrotational flows in Chapter 3.

For non-simply connected domains, when a flow is irrotational, the circulation around a closed loop may be non-zero. More precisely the situation is as follows. Generally, for non-simply connected domains $\mathcal{D}$ when a flow is irrotational, if a closed loop C_1 can be continuously deformed to the closed loop C_2 without leaving $\mathcal{D}$, then the circulation around either contour is the same. In three dimensions, when the flow is around a bounded obstacle, the circulation is zero by Stokes' Theorem. In two dimensions, though the circulations around the closed loops C_1 and C_2 must be the same, they need not be zero. Let us establish the circulations are the same. The two closed loops, which may contain a component of the complement of $\mathcal{D}$ in the interior to both of them, also enclose a bounded subregion $\mathcal{S}$ of $\mathbb{R}^2$ between them; see Figure 2.17. By Stokes' Theorem, since $\boldsymbol{u}$ is irrotational in $\mathcal{D}$, we have

$$0 = \iint_{\mathcal{S}} \boldsymbol{\omega} \cdot \boldsymbol{n}\, \mathrm{d}S = \oint_{C_1} \boldsymbol{u} \cdot \mathrm{d}\boldsymbol{x} - \oint_{C_2} \boldsymbol{u} \cdot \mathrm{d}\boldsymbol{x}.$$

Note we have assumed the setup in Figure 2.17, but the argument is easily adapted to the case when C_1 and C_2 might intersect. Hence we deduce that the circulations around C_1 and C_2 must be the same. By Kelvin's Circulation Theorem, when moving with the flow or indeed otherwise, the circulation around any such continuously deformable closed loops must be constant. Hence, for potential flows, from Chorin and Marsden (1990): 'the circulation around an obstacle in the plane is well-defined and is constant in time'.

Example 2.48 (Line vortex) Consider the flow field given in cylindrical polar coordinates by $\boldsymbol{u} = (\kappa/r)\,\hat{\boldsymbol{\theta}}$, where $\kappa > 0$ is a constant. The velocity components u_r and u_z are zero. This is a line vortex – the idealisation of a thin vortex tube. Direct computation shows that $\nabla \times \boldsymbol{u} = \boldsymbol{0}$ everywhere *except* at $r = 0$, where $\nabla \times \boldsymbol{u}$ is undefined. The region $r > 0$ is not simply connected. For any closed circuit C, the circulation is

$$\oint_C \boldsymbol{u} \cdot \mathrm{d}\boldsymbol{x} = \oint_C \frac{\kappa}{r}\hat{\boldsymbol{\theta}} \cdot (\hat{\boldsymbol{\theta}}\, r\, \mathrm{d}\theta + \hat{\boldsymbol{r}}\, \mathrm{d}r) = \oint_C \kappa\, \mathrm{d}\theta = 2\pi\kappa N,$$

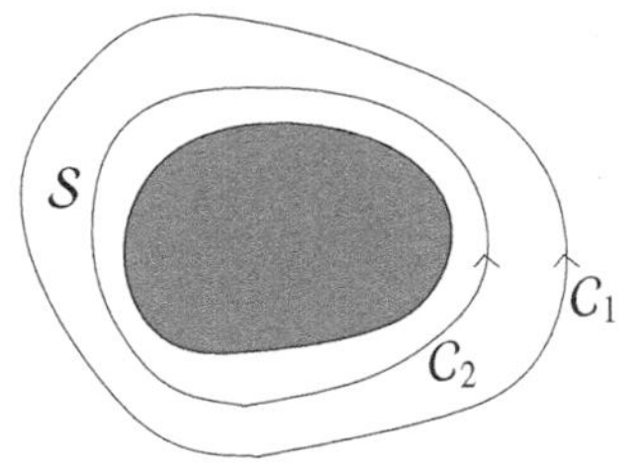

Figure 2.17 The circulation around the closed loops C_1 and C_2 which can be continuously deformed, one to the other, must be the same if the flow is a potential flow in $\mathcal{D}$, the exterior to the obstacle shown.

where N is the number of times the closed curve C winds round the origin $r = 0$. Naturally, the circulation will be zero for all circuits reducible continuously to a point without breaking the vortex, i.e. without passing through $r = 0$.

The line vortex example above illustrates the problem with establishing a potential when the domain is not simply connected. Suppose we fix the point $\boldsymbol{x}_0$ and define the potential for any point $\boldsymbol{x}$ as we did in the proof of Lemma 2.44:

$$\phi(\boldsymbol{x}) := \int_{C(\boldsymbol{x}_0,\boldsymbol{x})} \boldsymbol{u} \cdot \mathrm{d}\boldsymbol{x},$$

for any smooth contour $C(\boldsymbol{x}_0, \boldsymbol{x})$. The value of the integral on the right will be the same for all contours $C(\boldsymbol{x}_0, \boldsymbol{x})$ that do not wind round the line vortex at the origin. However, for contours $C(\boldsymbol{x}_0, \boldsymbol{x})$ that do wind round the origin, the value of ϕ will shift by $2\pi\kappa$ times the winding number of the contour. This *multi-valuedness* of the potential ϕ becomes even more complex for more complicated non-simply connected domains, though as Saffman (1992, p. 18) points out: 'The velocity potential in a multiply-connected region can be made single valued by the artifice of introducing barriers, or diaphragms, which are connected open surfaces with edges on the boundaries drawn across the "holes" so that the region is simply connected. The velocity potential may then be discontinuous, that is, have a finite jump, across the barrier. But the velocity field is continuous and the value of the jump is therefore the same over the barrier. It is also constant in time... '.

We now prove an important result for three-dimensional ideal potential flows known as D'Alembert's Theorem or D'Alembert's Paradox. The underlying conundrum is as follows. Consider a uniform flow into which we place an obstacle. We would naturally expect that the obstacle represents an obstruction to the fluid flow and the flow would exert a force on the obstacle, which if strong enough, might dislodge it and subsequently carry it downstream. However, for an ideal three-dimensional flow, as we are just about to prove, there is no net force exerted on such an obstacle placed in the midst of a uniform flow. Consider the setup shown in Figure 2.18. We assume that the flow around the sphere shown is a steady potential flow. We suppose further that the flow is axisymmetric. By this we mean the following. Use spherical polar coordinates to represent the flow with the south–north pole axis passing through the centre of the sphere and aligned with the uniform flow U at infinity; see Figure 2.18. Then the flow is

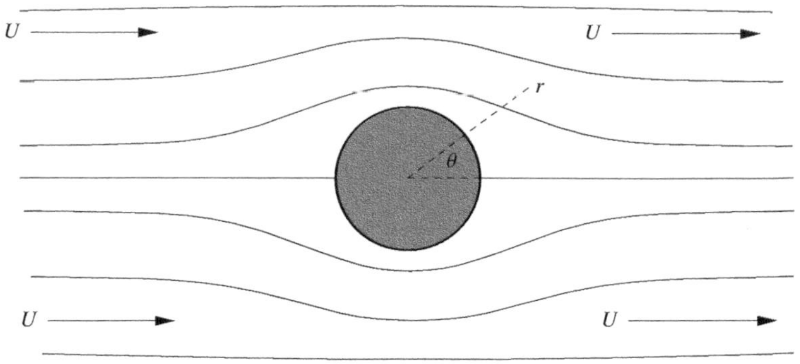

Figure 2.18 Consider an ideal steady, incompressible, irrotational and axisymmetric flow past a sphere as shown. The net force exerted on the sphere (obstacle) in the flow is zero. This is D'Alembert's Theorem.

axisymmetric if it is independent of the azimuthal angle φ of the spherical coordinates (r, θ, φ). Further, we also assume no swirl so $u_\varphi = 0$.

Theorem 2.49 (D'Alembert) *Consider a uniform ideal flow into which is placed a sphere. Assume the resulting flow diverting around the sphere is a steady potential flow that is axisymmetric, as outlined above, and there is no swirl. Then the flow does* not *exert either a lateral or drag force on the sphere.*

Proof We assume the sphere has radius a, as shown in Figure 2.18. Note that the domain exterior to the sphere is simply connected. Since the flow is a potential flow, we seek a potential function ϕ such that $\Delta\phi = 0$. In spherical polar coordinates this is equivalent to

$$\frac{1}{r^2}\left(\frac{\partial}{\partial r}\left(r^2\frac{\partial\phi}{\partial r}\right) + \frac{1}{\sin\theta}\frac{\partial}{\partial\theta}\left(\sin\theta\frac{\partial\phi}{\partial\theta}\right)\right) = 0.$$

The general solution to Laplace's equation is well known by the method of separation of variables for linear partial differential equations; see e.g. King *et al.* (2003) for more details. In the case of spherical polar coordinates and axisymmetry, the general solution is given by

$$\phi(r,\theta) = \sum_{n=0}^{\infty}\left(A_n r^n + \frac{B_n}{r^{n+1}}\right)P_n(\cos\theta),$$

where P_n are the Legendre polynomials, with $P_1(x) = x$. The coefficients A_n and B_n are constants, most of which, as we shall see presently, are necessarily zero in this case. For our problem we have two sets of boundary data. First, as $r \to \infty$ in any direction, the flow field is uniform and given by $\boldsymbol{u} = U\hat{z}$. Here U is the given constant background flow speed and $\hat{z}$ is the unit vector aligned along the z-axis, which constitutes the coordinate line through the south and north poles, and in that direction. We need to express $U\hat{z}$ in spherical polar coordinates. Since θ is the co-latitude angle with respect to the $\hat{z}$-axis, the projection of $U\hat{z}$ onto the $\hat{\boldsymbol{r}}$ and $\hat{\boldsymbol{\theta}}$ directions reveals

$U\hat{z} = U\cos\theta\,\hat{r} - U\sin\theta\,\hat{\theta}$. Hence in terms of the potential function $\phi = \phi(r,\theta)$, we have $\partial\phi/\partial r = U\cos\theta$ as well as $r^{-1}(\partial\phi/\partial\theta) = -U\sin\theta$. Thus as $r \to \infty$ we have

$$\phi \sim Ur\cos\theta.$$

Second, on the sphere $r = a$ we have the no-normal flow condition $\nabla\phi \cdot \boldsymbol{n} = 0$, which since $\boldsymbol{n} \equiv \hat{r}$ on the sphere surface, becomes

$$\frac{\partial\phi}{\partial r} = 0.$$

Using the first boundary condition for $r \to \infty$, we see all the A_n must be zero except $A_1 = U$. Using the second boundary condition on $r = a$, we see all the B_n must be zero except for $B_1 = \frac{1}{2}Ua^3$. Hence the unique potential for this flow is

$$\phi = U(r + a^3/2r^2)\cos\theta.$$

In spherical polar coordinates, the velocity $\boldsymbol{u} = \nabla\phi$ is $\boldsymbol{u} = u_r\hat{r} + u_\theta\hat{\theta}$, where

$$u_r = U(1 - a^3/r^3)\cos\theta \qquad \text{and} \qquad u_\theta = -U(1 + a^3/2r^3)\sin\theta.$$

Since the flow is ideal and stationary, Bernoulli's Theorem applies and so along a typical streamline $\frac{1}{2}|\boldsymbol{u}|^2 + P/\rho$ is constant. Indeed, since the conditions at infinity are uniform, the pressure P_∞ and velocity field U are the same everywhere there. This means for any streamline and in fact everywhere for $r \geqslant a$:

$$\frac{1}{2}|\boldsymbol{u}|^2 + P/\rho = \frac{1}{2}U^2 + P_\infty/\rho.$$

Rearranging and using our expression for the velocity field above, we have

$$\frac{P - P_\infty}{\rho} = \tfrac{1}{2}U^2\left(1 - (1 - a^3/r^3)^2\cos^2\theta - (1 + a^3/2r^3)^2\sin^2\theta\right).$$

On the sphere $r = a$ we see that

$$P = P_\infty + \tfrac{1}{2}\rho U^2\left(1 - \tfrac{9}{4}\sin^2\theta\right).$$

Note that on the sphere, the pressure is invariant under reflection $\theta \to \pi - \theta$. Recall from Section 2.1 that, in terms of the pressure, the force $\mathrm{d}\boldsymbol{F}$ on one side of an element of surface $\mathrm{d}S$ from the other is $\mathrm{d}\boldsymbol{F} = -p\boldsymbol{n}\,\mathrm{d}S$. Hence the total force on the sphere due to the pressure at $r = a$ is given by

$$\boldsymbol{F} = -\int_{\text{sphere}} p\boldsymbol{n}\,\mathrm{d}S = -\int_0^{2\pi}\int_0^{\pi} p(\theta)\,\hat{r}(\theta,\varphi)\,a^2\sin\theta\,\mathrm{d}\theta\,\mathrm{d}\varphi,$$

where we used that on the sphere $r = a$ we have $\boldsymbol{n} = \hat{r}(\theta,\varphi)$ and, at a given position (θ,φ) on the sphere, an element of the surface is given by $(a\sin\theta\,\mathrm{d}\varphi)(a\,\mathrm{d}\theta)$. Note that $\hat{r} = \sin\theta\,\cos\varphi\,\boldsymbol{i} + \sin\theta\,\sin\varphi\,\boldsymbol{j} + \cos\theta\,\boldsymbol{k}$, where $\boldsymbol{k} \equiv \hat{z}$, so $\boldsymbol{i}$ and $\boldsymbol{j}$ are fixed orthogonal, and thus lateral, directions. Since $p = p(\theta)$ only, if we substitute this expression for $\hat{r}$ into the integral for the total force above, then the coefficients of $\boldsymbol{i}$ and $\boldsymbol{j}$ will contain the respective factors consisting of the integral of $\cos\varphi$ and $\sin\varphi$ over $[0, 2\pi]$, which are zero. Hence there is no lateral force on the sphere, which we should have guessed already from the axisymmetry and no-swirl assumptions. The remaining component of $\boldsymbol{F}$ is the $\boldsymbol{k}$ component whose coefficient is given by

$$
\begin{aligned}
&-2\pi a^2 \int_0^\pi p(\theta)\,\cos\theta\,\sin\theta\,\mathrm{d}\theta \\
&= -2\pi a^2\Big((P_\infty + \tfrac{1}{2}\rho U^2)\int_0^\pi \cos\theta\,\sin\theta\,\mathrm{d}\theta - \tfrac{9}{8}\rho U^2\int_0^\pi \sin^3\theta\,\cos\theta\,\mathrm{d}\theta\Big) \\
&= -2\pi a^2\Big((P_\infty + \tfrac{1}{2}\rho U^2)\big[\tfrac{1}{2}\sin^2\theta\big]_{\theta=0}^{\theta=2\pi} - \tfrac{9}{8}\rho U^2\big[\tfrac{1}{4}\sin^4\theta\big]_{\theta=0}^{\theta=2\pi}\Big).
\end{aligned}
$$

We observe that both terms are zero. □

Hence the fluid exerts no net force on the sphere. There is no drag or lateral force. This result, in principle, applies to any shape of obstacle in such a flow. The analogous two-dimensional situation is more intricate, and we investigate such flows in detail in Chapter 3. Naturally this is only a paradox in that the fluid has been assumed to be ideal, with viscosity having been neglected; we return to this matter in Chapters 4–6.

2.9 Water Waves

Ocean waves are a permanent bountiful source of natural wonder, realised in many shapes and forms across many magnitudes of scales. These scales range from Rossby waves on the scale of planet earth, both in the ocean and upper atmosphere (the jetstream), to waves in rocky pools, or changing the context, across the surface of tea in a cup generated by blowing on it to cool it down more quickly. There are periodic waves such as those that travel across the Atlantic Ocean and solitary waves such as tsunamis and storm surges. We provide herein a brief introduction to the theory of *linear* water waves, in particular ones with small amplitude compared to the fluid depth as well as small amplitude compared to their wavelength.

We consider the following setup. We suppose a homogeneous incompressible irrotational fluid to be contained within $(x, y) \in \mathbb{R}^2$, which represent horizontal coordinates. In the vertical direction, when the fluid is completely at rest, it is completely contained within $-h \leqslant z \leqslant 0$, where the constant mean fluid depth is h. In this rest state the top boundary is $z = 0$, which we refer to hereafter as the average surface height or average free surface level. However, now suppose this fluid is disturbed for example by a driving wind across the top surface – we do not provide the theory of how this mechanism generates a surface disturbance here. We denote the height of the disturbed free surface to be $z = \zeta(x, y, t)$, which parameterises the deflection of the surface height from $z = 0$. The free surface $z = \zeta(x, y, t)$ is characterised by the fact that along it the pressure is constant and equal to the atmospheric pressure; see Figure 2.19.

Given the constant mean fluid depth is h, we assume herein that the maximum amplitude $\zeta_{\max}$ of any fluid surface disturbances will be small compared to h and small compared to the wavelength λ of any waves formed. In other words, we assume the asymptotic inequalities

$$\zeta_{\max} \ll h \qquad \text{and} \qquad \zeta_{\max} \ll \lambda.$$

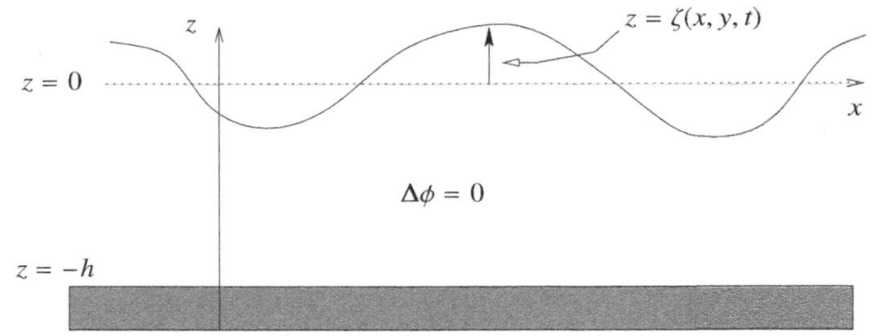

Figure 2.19 A homogeneous incompressible irrotational fluid is contained within the region $(x, y, z) \in \mathbb{R}^2 \times [-h, \zeta]$, where h is the constant mean fluid depth and $z = \zeta(x, y, t)$ is the height of the free surface above $z = 0$. The free surface is at constant atmospheric pressure and $z = 0$ is the 'average' surface height.

Since we assume that the fluid between the bottom and the free surface is irrotational, there exists a potential function ϕ such that the fluid velocity $\boldsymbol{u} = \nabla\phi$ everywhere in $(x, y, z) \in \mathbb{R}^2 \times [-h, \zeta]$. That the fluid is incompressible then implies

$$\Delta\phi = 0,$$

throughout this region. Further, as we know from the discussion succeeding Proposition 2.45 as well as from Exercise 2.21, we can choose the potential function ϕ such that the energy density per unit mass

$$H := \frac{\partial\phi}{\partial t} + \tfrac{1}{2}|\boldsymbol{u}|^2 + \frac{p}{\rho} + gz$$

is constant throughout the flow, where ρ is the constant fluid density and g is the acceleration due to gravity. Hence in particular it is equal to its value at the free surface where the pressure is atmospheric pressure, say p_0. Further, since any disturbances/waves are small amplitude, we neglect the quadratic term $\frac{1}{2}|\boldsymbol{u}|^2$. Absorbing the constants H and p_0/ρ into the potential ϕ, we deduce that we have the dynamic Bernoulli condition at $z = \zeta(x, y, t)$:

$$\frac{\partial\phi}{\partial t} + g\zeta = 0.$$

In addition we also have the following kinematic condition for the vertical velocity at $z = \zeta(x, y, t)$:

$$\frac{\partial\phi}{\partial z} = \frac{\partial\zeta}{\partial t} + (\nabla\phi) \cdot \nabla\zeta.$$

Again, since any disturbances/waves are small amplitude, we neglect the quadratic term $(\nabla\phi) \cdot \nabla\zeta$ so that the kinematic condition at $z = \zeta(x, y, t)$ becomes:

$$\frac{\partial\phi}{\partial z} = \frac{\partial\zeta}{\partial t}.$$

Since ζ is small we can Taylor expand the first term $(\partial\phi/\partial t)(x, y, \zeta, t)$ in the Bernoulli condition about $\zeta = 0$ and approximate it by $(\partial\phi/\partial t)(x, y, 0, t)$. We can perform the same operation for the kinematic condition and thus obtain the boundary conditions at $z = 0$

$$\frac{\partial \phi}{\partial t} + g\zeta = 0 \qquad \text{and} \qquad \frac{\partial \phi}{\partial z} = \frac{\partial \zeta}{\partial t}.$$

Lastly, at $z = -h$ we also have the no-normal flow boundary condition

$$\frac{\partial \phi}{\partial z} = 0.$$

Our goal is thus to solve Laplace's equation $\Delta\phi = 0$ subject to the boundary conditions at $z = 0$ and $z = -h$ above on the fixed domain $\mathbb{R}^2 \times [-h, 0]$. For these small-amplitude waves the approximation restriction to a fixed domain is crucial.

Remark 2.50 We observe:

(i) We can combine the boundary conditions at $z = 0$ and stipulate that we require the following single boundary condition: $\partial^2\phi/\partial t^2 + g\partial\phi/\partial z = 0$.

(ii) Given ϕ, we can always recover the surface elevation ζ from the Bernoulli boundary condition $\zeta(x, y, t) = -g^{-1}(\partial\phi/\partial t)(x, y, 0, t)$.

How do solutions to Laplace's equation with such boundary conditions generate waves? Let us look for solutions to Laplace's equation of the separated form $\phi(x, y, z, t) = \varphi(x, y, t) f(z)$. This generates the usual separation of variables condition:

$$\frac{f''(z)}{f(z)} = -\frac{\Delta_{x,y}\varphi}{\varphi} = k^2, \qquad \text{where} \qquad \Delta_{x,y} := \frac{\partial^2}{\partial x^2} + \frac{\partial^2}{\partial y^2},$$

where without loss of generality $k \in \mathbb{C}$ is a constant. However, since we are looking for real solutions and assume $\varphi(x, y, t)$ and $f(z)$ are both real, we thus expect k^2 itself to be real. The general solution for f is $f(z) = Ae^{kz} + Be^{-kz}$, for some constants $A, B \in \mathbb{C}$. The boundary condition $\partial\phi/\partial z = 0$ at $z = -h$ corresponds to $f'(-h) = 0$. Using this boundary condition generates the following solution form for f, namely:

$$f(z) = \tilde{A}\cosh(k(z + h)),$$

where $\tilde{A} \in \mathbb{C}$ is a rescaling of the constant A above. With this form for f, the Bernoulli boundary condition at $z = 0$ becomes

$$\zeta = -\frac{1}{g}\tilde{A}\cosh(kh)\frac{\partial \varphi}{\partial t}.$$

Applying the two-dimensional Laplace operator $\Delta_{x,y}$ to this boundary condition and utilising that $\Delta_{x,y}\varphi = -k^2\varphi$ implies ζ must satisfy the constraint

$$\Delta_{x,y}\zeta = -k^2\zeta.$$

If we combine the Bernoulli boundary condition above with the kinematic condition at $z = 0$ which has the form

$$\frac{\partial \zeta}{\partial t} = k\tilde{A}\sinh(kh)\varphi,$$

then we see that $\zeta = \zeta(x, y, t)$ satisfies

$$\frac{\partial^2 \zeta}{\partial t^2} = -kg \tanh(kh)\, \zeta.$$

Combined with the constraint on ζ just above, this implies that $\zeta = \zeta(x, y, t)$ satisfies the evolutionary linear partial differential equation

$$\frac{\partial^2 \zeta}{\partial t^2} = \frac{g}{k} \tanh(kh)\, \Delta_{x,y} \zeta.$$

We now seek planar wave solutions $\zeta = \zeta(x, y, t)$ to this equation of the form

$$\zeta(x, y, t) = A e^{\mathrm{i}(\boldsymbol{\nu} \cdot \boldsymbol{x} - \omega t)},$$

where the constant $\boldsymbol{\nu} = (\nu_x, \nu_y)^{\mathrm{T}} \in \mathbb{R}^2$ is the wavenumber vector, the variable vector $\boldsymbol{x} = (x, y)^{\mathrm{T}}$ and $\omega \in \mathbb{R}$ is the frequency. The arbitrary constant $A \in \mathbb{C}$ is related to the amplitude of the waves, and naturally we expect $A = A(\boldsymbol{\nu})$.

Definition 2.51 (Wavelength and period) For such planar waves characterised by wavenumber $\boldsymbol{\nu}$, their *wavelength* $\lambda := 2\pi/|\boldsymbol{\nu}|$ and their *period* $T := 2\pi/\omega$.

Remark 2.52 The planar wave ansatz given is complex-valued. We take it as *understood* that the physical fluid height above the free surface $z = 0$ is the real part of the complex-valued planar wave form.

Note that the constraint $\Delta_{x,y} \zeta = -k^2 \zeta$ on ζ implies

$$|\boldsymbol{\nu}|^2 \equiv k^2 \geqslant 0.$$

Hence substituting the planar wave ansatz for ζ above into the linear evolution equation for ζ above we find

$$\omega^2 = kg \tanh(kh).$$

This is the *dispersion relation* for the planar waves on the surface. Indeed, the phase speed of the waves, which is defined by $c_{\mathrm{p}} := \omega/|\boldsymbol{\nu}| \equiv \lambda/T$, is given by

$$c_{\mathrm{p}}^2 = \frac{g}{k} \tanh(kh).$$

Thus waves characterised by different wavenumbers $k = |\boldsymbol{\nu}|$ travel at different speeds, i.e. we have dispersion. Let us examine some special example cases. Recall that we always assume $\zeta_{\max} \ll h$ and $\zeta_{\max} \ll \lambda$.

Example 2.53 (Deep-water waves, $h \gg \lambda$) When the depth is asymptotically large compared to the wavelength, so $h \gg \lambda$ or equivalently $kh \gg 1$, since $\lambda = 2\pi/|\boldsymbol{\nu}|$, then $\tanh(kh) \sim 1$ and we observe

$$\omega^2 \sim gk \qquad \text{and} \qquad c_{\mathrm{p}}^2 \sim \frac{g}{k} \equiv \frac{g\lambda}{2\pi}.$$

These are deep-water waves. They have the characterisation that longer-wavelength waves travel faster.

Example 2.54 (Shallow-water long waves, $h \ll \lambda$) When the depth is asymptotically small compared to the wavelength, though still asymptotically large compared to the wave amplitude, so $h \ll \lambda$ or equivalently $kh \ll 1$, then $\tanh(kh) \sim kh$ at leading order, and we observe

$$\omega^2 \sim ghk^2 \qquad \text{and} \qquad c_{\mathrm{p}}^2 \sim gh.$$

These are long waves on shallow water. They have the characterisation that the wave speed is independent of the wavelength.

Remark 2.55 We make the following observations:

(i) The evolutionary linear partial differential equation satisfied by ζ, i.e. $\partial^2\zeta/\partial t^2 = (g/k)\tanh(kh)\Delta_{x,y}\zeta$, is dispersive. It is hyperbolic in the shallow-water long wave limit when it becomes $\partial^2\zeta/\partial t^2 = (gh)\Delta_{x,y}\zeta$.

(ii) The motion of the fluid particles is *local*: fluid particles move on elliptical paths with the focal distance independent of the depth. At the bottom the motion is oscillatory between the ends of the major axis; see Exercise 2.24.

(iii) Modern professional surfers famously scan the world's weather reports looking for storms, anticipating the arrival of the perfect waves in the perfect locations and jetting off to meet them. The realisation of such events can be further optimised/timed by studying the wavelengths of the waves impinging the coasts of target locations. For example, the 'first sign of an Atlantic storm is usually provided by the arrival at the west coast of Britain of very long waves of rather small amplitude (swell), which have been generated by the storm' (Paterson, 1983, p.294). Such very long waves may have a period of roughly 30 s, a wavelength of roughly 1430 m and speed near 47 m s^{-1}. They thus 'might be expected to travel about 4000 km in a day, and hence arrive well before the storm, which can take several days to cross the Atlantic'.

The energy transported by such surface waves does not necessarily travel at the same speed as the wave crests, indeed it travels at the so-called group velocity.

Definition 2.56 (Group velocity) Given a dispersion relation $\omega = \omega(\boldsymbol{\nu})$ for planar waves, we define the *group velocity* by $\boldsymbol{c}_{\mathrm{g}} := \nabla_{\boldsymbol{\nu}}\,\omega$.

We motivate this definition as follows. In a surface disturbance that generates planar waves there may be many waves of different frequencies present that, by superposition, constructively and destructively interfere with each other. We focus on the example of a wave packet as shown in Figure 2.20. Let us represent such a localised general disturbance via the Fourier integral

$$\zeta(\boldsymbol{x},t) = \int_{\mathbb{R}^2} A(\boldsymbol{\nu})\,\mathrm{e}^{\mathrm{i}(\boldsymbol{\nu}\cdot\boldsymbol{x}-\omega(\boldsymbol{\nu})t)}\,\mathrm{d}\boldsymbol{\nu}.$$

Here $\boldsymbol{x} = (x,y)^{\mathrm{T}}$ and, as previously, it is understood that the physical fluid height above the free surface $z = 0$ is the real part of the integral expression shown. Suppose the disturbance of concern is a single wave packet as shown in Figure 2.20 with the following characteristics. We assume for the wave packet: the amplitude varies slowly

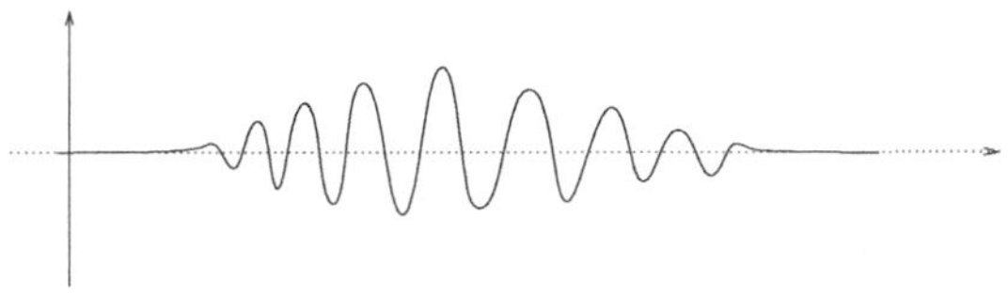

Figure 2.20 An example wave packet, i.e. a localised travelling disturbance.

with $\boldsymbol{x}$; it contains waves of nearly constant wavenumber $\boldsymbol{\nu}_0$ and $|\boldsymbol{\nu}_0|$ is large enough that it contains a large number of crests. In particular, we suppose that the amplitude $A = A(\boldsymbol{\nu})$ is vanishingly small except when $\boldsymbol{\nu}$ is close to $\boldsymbol{\nu}_0$. For $\boldsymbol{\nu}$ close to $\boldsymbol{\nu}_0$ we have

$$\omega(\boldsymbol{\nu}) \approx \omega(\boldsymbol{\nu}_0) + (\boldsymbol{\nu} - \boldsymbol{\nu}_0) \cdot ((\nabla\omega)(\boldsymbol{\nu}_0)).$$

Note the linear term on the right contains the group velocity $\boldsymbol{c}_{\mathrm{g}} = (\nabla\omega)(\boldsymbol{\nu}_0)$ associated with this wave packet of characteristic wavenumber $\boldsymbol{\nu}_0$. Given the character of $A = A(\boldsymbol{\nu})$ described above for our slowly modulated wave packet, we assume that the main contribution to the Fourier integral above comes from around $\boldsymbol{\nu} = \boldsymbol{\nu}_0$ and thus we replace $\omega = \omega(\boldsymbol{\nu})$ by the approximation shown, so

$$\zeta(\boldsymbol{x}, t) \approx \mathrm{e}^{\mathrm{i}(\boldsymbol{\nu}_0\cdot\boldsymbol{x} - \omega(\boldsymbol{\nu}_0)t)} \int_{\mathbb{R}^2} A(\boldsymbol{\nu})\, \mathrm{e}^{\mathrm{i}(\boldsymbol{\nu}-\boldsymbol{\nu}_0)\cdot(\boldsymbol{x}-\boldsymbol{c}_{\mathrm{g}}t)}\, \mathrm{d}\boldsymbol{\nu}.$$

The first factor in this expression represents a pure plane wave associated with the wavenumber $\boldsymbol{\nu}_0$, while the second factor is a function of $\boldsymbol{x} - \boldsymbol{c}_{\mathrm{g}}t$ and thus represents an envelope that travels with velocity $\boldsymbol{c}_{\mathrm{g}}$ so that the whole wave packet translates across the (x, y)-plane with the group velocity $\boldsymbol{c}_{\mathrm{g}} = \boldsymbol{c}_{\mathrm{g}}(\boldsymbol{\nu}_0)$.

Example 2.57 (Group velocity: deep water) Recall the dispersion relation in the asymptotic 'deep-water' limit $h \gg \lambda$ from Example 2.53. In the limit it has the form $\omega = (gk)^{1/2}$, where $k = |\boldsymbol{\nu}|$. Recall also that the phase speed of the waves in the limit is $c_{\mathrm{p}} = (g/k)^{1/2}$. The group velocity for these waves is given by

$$\boldsymbol{c}_{\mathrm{g}} = \nabla_{\boldsymbol{\nu}}\,\omega = \frac{\mathrm{d}\omega}{\mathrm{d}k}(\nabla_{\boldsymbol{\nu}}\,k) = \frac{\mathrm{d}\omega}{\mathrm{d}k}\frac{1}{k}\begin{pmatrix}\nu_x\\ \nu_y\end{pmatrix} \quad \Rightarrow \quad |\boldsymbol{c}_{\mathrm{g}}| = \frac{1}{2}\left(\frac{g}{k}\right)^{1/2}.$$

In this case the group velocity is half the phase speed. Hence for a wave packet the crests within the packet travel twice as fast as the packet as a whole. Hence the waves/crests within the packet move forwards through the packet, disappearing at the front with new crests appearing at the back. This might be observed in ripples created by a pebble splashing on the surface of a pond.

Example 2.58 (Group velocity: shallow water) In the asymptotic 'shallow-water long wave' limit $h \ll \lambda$ from Example 2.54, the dispersion relation and phase speed are $\omega = (gh)^{1/2}k$ and $c_{\mathrm{p}} = (gh)^{1/2}$, respectively. Now, the magnitude of the group velocity $|\boldsymbol{c}_{\mathrm{g}}| = |\mathrm{d}\omega/\mathrm{d}k| = (gh)^{1/2}$ matches the phase speed.

The group velocity is also the velocity of *energy transport*. In the case of a slowly modulated wave packet, as we considered above, this is straightforwardly verified. The energy associated with the localised disturbance moves as a whole with the packet, which moves with the group velocity. Further, Exercise 2.27 shows for deep-water plane waves, of wavenumber $k = |\boldsymbol{v}|$ and frequency $\omega = (gk)^{1/2}$, that energy is transported at the group speed $|\boldsymbol{c}_{\mathrm{g}}|$ rather than the phase/crest speed c_{p}. That generally for dispersive waves, energy is transported at the group velocity is proved in Lighthill (1978).

Remark 2.59 In light of Example 2.57, where we found that for deep-water waves the group velocity is half that of the phase or crest speed, we should revise our estimate in Remark 2.55(iii) for the speed at which a storm will cross the Atlantic to 2000 km per day.

2.10 Notes

Some further details and examples are as follows.

(1) *Astrophysics.* Exercises 2.5–2.7 on the structure of liquid planets/stars and polytropes extend the hydrostatic material and Archimedes' Principle at the end of Section 2.2. There are, however, many further dynamic applications of fluid mechanics to astrophysics, including: planetary atmospheres, Jupiter's red spot, the solar wind, flow/convection patterns in the sun, sun spots, rotating stars, pulsars, astrophysical jets, accretion disks, galactic structure, and so forth. See e.g. Guilliot and Gautier (2015) and indeed the treatise in which Guillot and Gautier is contained as a whole.

(2) *Bernoulli's Theorem.* There are many natural phenomena, additional to those above or in the exercises below, for which Bernoulli's Theorem provides an explanation. Some examples are:

(a) The Froude number plays an important role in flows over mountain ranges, i.e. *mountain meteorology*. See e.g. Baines (1987), Berkshire and Warren (1970), Durran (2015), Hunt *et al.* (1997), Scorer (1978) or Smith (2015). There are many websites with examples and simulations. For example, see the US National Weather Service website. It also plays a role in tidal flow in fjords, see Stashchuk *et al.* (2007).

(b) Lotus introduced a *Formula One* car with 'ground effects' for the 1977 season with substantially improved cornering. These 'ground effects' combined inverted aerofoils (see Chapter 3) under the car with side skirts to keep the flow channelled under the car and reduce pressure on the underside, thus creating a suction effect with the ground and consequently greater traction/grip. A mandatory flat underside and a ban on side skirts were introduced after the 1979 season.

(c) Set a powerful enough hairdryer going with a circular nozzle and point it upwards. Place a *ping-pong* ball in the middle of the flow just above the nozzle and watch the ball jiggle there within a contained region. You can even turn the flow at a slight angle to the vertical and again the ball remains. The constrained flow just above the middle of the nozzle is faster and at a lower pressure compared to the flow just above the perimeter of the nozzle.

(d) The 'peloton' in the *Tour de France* is the main group of riders, usually a hundred or more of them on flat stages. The following phenomenon can be observed whenever the peloton meets a constriction in the road. As riders reach the constriction they appear to be cycling slowly, but as they pass into the constriction they appear to accelerate up to another level speed. This is Bernoulli's Theorem at both a physical and psychological level. Nearing the constriction the individual riders jostle for position to find their path into the constricted region and the pressure is thus high. Once into the constriction the 'pressure' is off once they've accelerated faster to maintain the overall throughput of riders along the road.

(3) *Stability of Hill's spherical vortex.* Exercises 2.14 and 2.15 explore Hill's spherical vortex; see Hill (1894). Moffatt and Moore (1978) show that it is unstable to small axisymmetric disturbances; also see Pozrikidis (1986).

(4) *Vortex rings.* The study of vortex rings goes back to Kelvin (1867) and Hicks (1899). Fraenkel (1970) and Norbury (1972) considered their existence, and Norbury (1973) computed a one-parameter family of rings with infinitely thin vortex rings as one extreme member and Hill's spherical vortex as the other extreme. Also see Fraenkel (1972), Benjamin (1975), Widnall *et al.* (1977) and Akhmetov (2009). More recently their stability has been studied by Protas (2019), see citations therein. There are also many spectacular videos online of colliding vortex rings, leapfrogging vortex rings and smoke rings rising from volcanoes. See Mavroyiakoumou and Berkshire (2020) for the similar case of leapfrogging vortex line pairs.

(5) *Sound waves.* These are a compressible phenomenon. Introductory accounts can be found in, for example, Acheson, (1990), Lighthill (1978) or Paterson (1983).

(6) *Geophysical fluid dynamics.* In studies of atmospheric and ocean flow, the Coriolis force due to the rotation of the earth must be considered. This is a force that is linear in the velocity field but acting orthogonally to it. Reduced models have been derived that account for, and demonstrate, various meteorological and oceanographic effects, as well as to aid computation. Two reduced models are notable: the quasi-geostrophic equations and semi-geostrophic equations, respectively derived in Pedlosky (1979, Ch. 3 & Ch. 8). The former have been used to model ocean flow while the related surface quasi-geostrophic equations have been used to study finite-time singularities, see Held *et al.* (1995). The latter have been used in meteorology to study frontogenesis, i.e. the creation and evolution of weather fronts. See e.g. Hoskins (1982).

(7) *Probabilistic and numerical weather prediction.* A major application of geophysical fluid mechanics from (6) above is that of weather prediction. To achieve good-quality forecasts over suitable forecast windows, coupled equations for the fluid flow of both the atmosphere and ocean are augmented by equations for temperature, humidty, and so forth. Accurate predictions over timescales up to a week are possible by blending the models, which contain uncertainty, with satellite, weather station and weather buoy data. The goal is to minimise the uncertainty by incorporating the observation data using techniques from data assimilation, for example the ensemble Kalman filter. For a comprehensive introduction, see Reich and Cotter (2015). As an extension of this, the Norwegian Meteorogical Institute had an extremely accurate 1 to 2-day forecast simulation of the dust cloud density from the Icalandic Eyjafjallajökull eruption in April 2010.

(8) *Incompressible Euler equation solution as a geodesic flow.* The incompressible Euler equations of motion can also be derived as a geodesic submanifold flow. The submanifold is the group of measure-preserving diffeomorphisms; see Arnold and Keshin (1998).

(9) *Regularity of the incompressible Euler equations.* Given smooth initial data with finite kinetic energy to the three-dimensional ($d = 3$) homogeneous incompressible Euler equations, do smooth solutions blow up in finite time? This is an open problem. They do not blow up in two dimensions ($d = 2$); this is because the vortex stretching term '$(\boldsymbol{\omega} \cdot \nabla)\boldsymbol{u}$' or equivalently '$D\boldsymbol{\omega}$' in the vorticity formulation of the Euler equations (see Section 2.3) is identically zero and so $\boldsymbol{\omega}(\boldsymbol{x}(\boldsymbol{a}, t), t) = \boldsymbol{\omega}_0(\boldsymbol{a})$ in this case. In other words the vorticity is transported. In three dimensions the presence of the vorticity stretching term means that the equivalent solution formula is $\boldsymbol{\omega}(\boldsymbol{x}(\boldsymbol{a}, t), t) = (\nabla_{\boldsymbol{a}}\boldsymbol{x}(\boldsymbol{a}, t))\,\boldsymbol{\omega}_0(\boldsymbol{a})$; see Exercise 2.8. In other words, quoting from Constantin (2007), 'the three-dimensional vorticity is carried as well, but its magnitude is amplified or diminished by the gradient of the flow map'. A thorough review of the regularity of the incompressible Euler equations can be found in Constantin (2007), to where the interested reader is referred. However, two classical results are particularly notable. They provide a connection to the results we develop in Chapter 7 on the regularity of the incompressible Navier–Stokes equations, so we briefly mention them here. First, Beale *et al.* (1984) provided a criterion for blow-up as follows. Consider the time integral of the maximum magnitude of the vorticity:

$$\int_0^T \|\boldsymbol{\omega}(t)\|_{L^\infty}\,\mathrm{d}t.$$

If this integral is finite and the initial vorticity $\boldsymbol{\omega}_0$ and all its spatial partial derivatives up to order three are square-integrable, then the solution remains smooth on the time interval $[0, T]$. If the integral becomes infinite, then there is finite-time blow-up. Second, Constantin *et al.* (1996) provided a criterion on the direction of the vorticity field that rules out blow-up. See Chapter 7, where we discuss the analogous result for the three-dimensional incompressible Navier–Stokes equations in some detail. For recent developments on regularity for the three-dimensional incompressible Euler equations, see e.g. Ambrosio and Figalli (2008, 2010), De Lellis and Székelyhidi (2014), Buckmaster and Vicol (2019) and Besse (2020).

(10) *Kinetic energy and incompressibility.* An ideal flow for which all the energy is kinetic, is necessarily incompressible. To see this, assume from conservation of mass that we have the continuity equation

$\partial\rho/\partial t + \nabla\cdot(\rho\boldsymbol{u}) = 0$. Further assume from balance of momentum that we have Euler's equation of motion $\partial\boldsymbol{u}/\partial t + (\boldsymbol{u}\cdot\nabla)\boldsymbol{u} = -\nabla p/\rho + \boldsymbol{f}$. In three-dimensional space when $d = 3$, we have four equations, but five unknowns – namely $\boldsymbol{u}$, p and ρ. We cannot specify the fluid motion completely without specifying one more condition. We define the *kinetic energy* of the fluid in the region $\mathcal{V}_t \subseteq \mathcal{D}$ to be $\frac{1}{2}\int_{\mathcal{V}_t}\rho\,|\boldsymbol{u}|^2\,\mathrm{d}V$. By direct computation using the Transport Theorem, we find

$$\frac{\mathrm{d}}{\mathrm{d}t}\left(\frac{1}{2}\int_{\mathcal{V}_t}\rho\,|\boldsymbol{u}|^2\,\mathrm{d}V\right) = \int_{\mathcal{V}_t}\boldsymbol{u}\cdot\left(\rho\left(\frac{\partial\boldsymbol{u}}{\partial t} + \boldsymbol{u}\cdot\nabla\boldsymbol{u}\right)\right)\mathrm{d}V.$$

Here we assume *all* the energy is kinetic. The principal of conservation of energy states, quoting from Chorin and Marsden (1990, p. 13): 'the rate of change of kinetic energy in a portion of fluid equals the rate at which the pressure and body forces do work'. In other words, we have

$$\frac{\mathrm{d}}{\mathrm{d}t}\left(\frac{1}{2}\int_{\mathcal{V}_t}\rho\,|\boldsymbol{u}|^2\,\mathrm{d}V\right) = -\int_{\partial\mathcal{V}_t}p\,\boldsymbol{u}\cdot\boldsymbol{n}\,\mathrm{d}S + \int_{\mathcal{V}_t}\rho\,\boldsymbol{u}\cdot\boldsymbol{f}\,\mathrm{d}V.$$

We compare this with our expression above for the rate of change of the kinetic energy after inserting Euler's equation of motion, revealing $\int_{\mathcal{V}_t}(\nabla\cdot\boldsymbol{u})\,p\,\mathrm{d}V = 0$. Since $\mathcal{V}_t$ is arbitrary we see that the assumption that all the energy is kinetic implies $\nabla\cdot\boldsymbol{u} = 0$. Thus the remaining conservation law, conservation of energy, implies $\nabla\cdot\boldsymbol{u} = 0$, i.e. an ideal flow is incompressible.

(11) *Isentropic flows.* A compressible flow is *isentropic* if there is a function π, the *enthalpy*, such that $\nabla\pi = \nabla p/\rho$. The equations for an isentropic flow are thus $\partial\boldsymbol{u}/\partial t + \boldsymbol{u}\cdot\nabla\boldsymbol{u} = -\nabla\pi + \boldsymbol{f}$ and $\partial\rho/\partial t + \nabla\cdot(\rho\boldsymbol{u}) = 0$ in $\mathcal{D}$, and matching normal velocities assumed at any boundary. For compressible ideal gas flow, the pressure is often proportional to ρ^γ, for some constant $\gamma \geqslant 1$, i.e. $p = C\rho^\gamma$ for some constant C. This is a special case of an isentropic flow and closes the system. The relation $p = C\rho^\gamma$ is an example *equation of state*; see Chorin and Marsden (1990, p. 14).

2.11 Projects

Some suggested extended projects from this chapter are as follows.

(1) Hill's spherical vortex, vortex rings and their stability. Complete Exercises 2.14 and 2.15 and then investigate the phenomena and literature indicated in Section 2.10, parts (3) and (4).

(2) Bernoulli's Theorem and applications. Complete Exercises 2.16 to 2.19, and 2.21. Then investigate further natural phenomena explained by Bernoulli's Theorem, e.g. see Section 2.10, part (2): (b), (c) and (d).

(3) The Froude number and its applications. Complete Exercise 2.20 and then investigate further applications to mountain meteorology, see Section 2.10, part (2)(a), as well as weirs, dams, hydraulic jumps, bores, fjords and so forth.

(4) The Rayleigh–Taylor and Kelvin–Helmholtz instabilities. Complete Exercises 2.24 to 2.32. Then investigate the difference between capillary and gravity waves – the difference between a raindrop hitting water or a large stone (Acheson, 1990) – as well as further examples of the Kelvin–Helmholtz instability in nature such as in the atmosphere or stratified water flow.

Exercises

2.1 Euler's equations: Solid body rotation. Consider an ideal homogeneous incompressible fluid that is contained in an infinite fixed right-circular cylinder, with symmetry axis the z-axis. Suppose that the fluid moves under the influence of a body force field $\boldsymbol{f} = (\alpha x + \beta y, \gamma x + \delta y, 0)^{\mathrm{T}}$ per unit mass, where x and y are orthogonal planar coordinates, orthogonal to and centred on the z-axis. In this body force α, β, γ and δ are constants. Suppose that the velocity field in the Cartesian coordinate directions x, y and z are u, v and w, respectively. (*Hint*: It is easiest to use Cartesian coordinates throughout this question.)

(a) Use Euler's equations of motion to show that the solid-body rotation ansatz about the z-axis for which $u = -\Omega y$, $v = \Omega x$ and $w = 0$, where $\Omega = \Omega(t)$ only, satisfies the system of equations

$$-\frac{\mathrm{d}\Omega}{\mathrm{d}t}y - \Omega^2 x = -\frac{1}{\rho}\frac{\partial p}{\partial x} + \alpha x + \beta y,$$
$$\frac{\mathrm{d}\Omega}{\mathrm{d}t}x - \Omega^2 y = -\frac{1}{\rho}\frac{\partial p}{\partial y} + \gamma x + \delta y,$$

where $p = p(x, y, t)$ is the pressure field and ρ is the constant density.

(b) For the flow given by the solid-body rotation ansatz in part (a), for which $w = 0$ and u and v only depend on x, y and t, explain why there is only one non-zero component of vorticity, namely $\omega = -\partial u/\partial y + \partial v/\partial x$. Show for the solid-body rotation in part (a), this vorticity is $\omega = 2\Omega$.

(c) By considering the curl of the equations in part (a), i.e. compute $\partial/\partial x$ of the second equation minus $\partial/\partial y$ of the first equation, show that

$$\frac{\mathrm{d}\Omega}{\mathrm{d}t} = \tfrac{1}{2}(\gamma - \beta).$$

(d) Use part (c) to show that the solid-body rotation in part (a) is a solution to Euler's equations with $p = p_0 + \frac{1}{2}\rho((\Omega^2 + \alpha)x^2 + (\beta + \gamma)xy + (\Omega^2 + \delta)y^2)$, where p_0 is a constant.

2.2 Euler's equations (statics): Isothermal gas. The ideal gas law states that for an ideal gas its pressure p, temperature T and density ρ are related according to the formula

$$p = \rho RT,$$

where R is the gas constant. For this exercise, consider an isothermal gas such as air kept at constant temperature T. Assume a static scenario so the velocity field of the fluid $\boldsymbol{u} \equiv 0$. Hence the Euler momentum equation reduces to $\nabla p = \rho \boldsymbol{f}$. Using Cartesian coordinates, assume $\boldsymbol{f} = -g\boldsymbol{k}$, with $\boldsymbol{k}$ the unit vector in the vertical upwards direction. Here g is the acceleration due to gravity.

(a) Explain why in this situation we know that the pressure $p = p(z)$ only and satisfies the ordinary differential equation

$$\frac{\mathrm{d}p}{\mathrm{d}z} = -\rho g.$$

(b) Using the ideal gas law and part (a), show

$$p = p_0 \mathrm{e}^{-z/H} \qquad \text{and} \qquad \rho = \rho_0 \mathrm{e}^{-z/H},$$

where p_0 and ρ_0 are the pressure and density at $z = 0$ and the constant H is given by

$$H = \frac{RT}{g} = \frac{p_0}{\rho_0 g}.$$

(c) The constant H from part (b) is known as the *atmospheric scale height*. Using the units for pressure, density and acceleration, verify that indeed H has units of length. Then assuming $g = 9.81\,\mathrm{m\,s^{-2}}$, the temperature of the gas is $T = 294\,\mathrm{K}$, where we measure the temperature in degrees Kelvin with absolute zero at $0\,\mathrm{K}$ or $-273^{\circ}\mathrm{C}$, and using $R = 287\,\mathrm{m^2\,s^{-2}\,K^{-1}}$ for air, show $H \approx 8.6\,\mathrm{km}$.

(d) From part (b) we know $p_0 = H\rho_0 g$ or equivalently $p_0 = \rho_0 RT$. Take this to be the pressure at sea level $z = 0$. Using that the density of dry air at sea level ρ_0 is $1.2\,\mathrm{kg\,m^{-3}}$ and the values for R and T quoted in part (c), compute the atmospheric pressure – the weight of the column of air above a square metre – at sea level in units $\mathrm{kg\,(m\,s^2)^{-1}}$. Compare your answer to the value for this quantity quoted in Example 2.6.

2.3 Euler's equations (statics): Leaking rusty metal can. Consider a metal can of height h and cross-sectional area A floating vertically in water of density ρ as shown in Figure 2.21 (left panel). Suppose the can is a depth k underwater. Suppose rust generates a small hole in the base and water leaks in slowly so the temperature remains constant, i.e. the process is isothermal. The question is whether an equilibrium floating state for the punctured metal can exists as shown in Figure 2.21 (right panel). Let M denote the combined mass of the metal can and the air it contains.

(a) If p_0 is the atmospheric pressure, explain why the pressure in the water at depth z is $p_0 + \rho g z$.

(b) By equating forces before the leak, i.e. in the situation in Figure 2.21 (left panel), and using Archimedes' Principle from the end of Section 2.2, explain why $Mg = \rho g k A$, where the right-hand side is the weight of the water displaced by the can.

(c) As the water leaks into the punctured can, the air inside compresses isothermally so from the ideal gas law in Exercise 2.2, we know pV is constant where p is the pressure of the air inside and V is the volume of that air. Use this fact to show that in the final equilibrium position shown in Figure 2.21 (right panel), if s is the height of the water inside the can, then the pressure of the air pocket inside the can p_1 is given by

$$p_1 = \frac{p_0 h}{h - s}.$$

(d) Leakage of water into the can stops if the pressure just inside and outside the hole equalises. Explain why this statement is equivalent to the relation $p_1 + \rho g s = p_0 + \rho g \ell$, where ℓ is the depth the can is underwater with the leak. Further show, by equating forces and using the relation in part (b), that in this scenario we have $k + s = \ell$.

(e) Note that we can express $p_0 = \rho g \hat{H}$, where $\hat{H}$ is the depth of a column of water of density ρ, of unit cross-section, that generates one atmosphere of pressure. Using this relation, combine all the results of parts (c) and (d) to show

$$s = \frac{hk}{\hat{H} + k}.$$

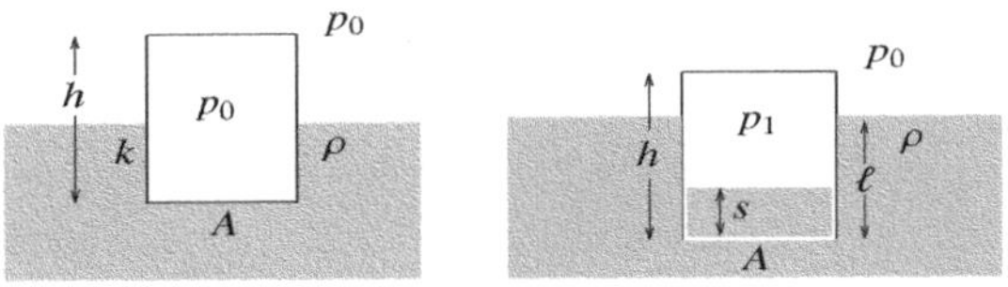

Figure 2.21 A floating closed rusty metal can develops a small hole in the base into which water leaks slowly, i.e. isothermally.

(f) Explain why, for the leaked equilibrium position to exist, we must have $\ell < h$. Explain why this condition holds provided $k/h \leqslant 1$ is not close to 1. What happens if k/h is close to 1?
(*Hint:* Note $\hat{H} \approx 10\,\text{m}$; see Example 2.6.)

2.4 Euler's equations (statics): Adiabatic gas. Consider a well-mixed adiabatic gas, i.e. no heat is being added/removed to/from the system, for which the pressure is proportional to a power of the density as follows:

$$p \propto \rho^{\gamma}.$$

The exponent γ is given by the ratio C_p/C_V of the specific heat of the gas at constant pressure C_p to the specific heat of the gas at constant volume C_V. Typically $\gamma \in [1, 2)$ and for air at 'ordinary' temperatures $\gamma = 1.4$. Further note, the ideal gas constant $R = C_p - C_V$. Recall the ideal gas law from Exercise 2.2.

(a) Let p_0, ρ_0 and T_0 denote the pressure, density and temperature, respectively, at a certain height z_0. Use the ideal gas law to show

$$\frac{T}{T_0} = \left(\frac{\rho}{\rho_0}\right)^{\gamma-1} = \left(\frac{p}{p_0}\right)^{1-1/\gamma}.$$

(b) By taking the logarithm of the relation between T and p in part (a) and differentiating with respect to z, show that the temperature gradient satisfies the relation

$$\frac{1}{T}\frac{\mathrm{d}T}{\mathrm{d}z} = \left(1 - \frac{1}{\gamma}\right)\frac{1}{p}\frac{\mathrm{d}p}{\mathrm{d}z}.$$

(c) Using the same arguments as those leading to the result in Exercise 2.2(a), explain how we know $\mathrm{d}p/\mathrm{d}z = -\rho g$. Using this result and the ideal gas law, show

$$\frac{\mathrm{d}T}{\mathrm{d}z} = -\left(1 - \frac{1}{\gamma}\right)\frac{g}{R} = -\frac{g}{C_p},$$

i.e. the temperature gradient is constant. It is also known as the *adiabatic lapse rate* which is important in meteorological convection.

(d) The specific heat C_p of dry air at 273 K at sea level is $C_p = 1003\,\text{m}^2\,\text{s}^{-2}\,\text{K}^{-1}$. Use this to show the adiabatic lapse rate is approximately $10\,\text{K}\,\text{km}^{-1}$.

(e) From part (c) we deduce that the temperature decays linearly with height z. Find the dependence of the pressure p and density ρ on height z as follows.

(i) Recall from part (c) that we know $\mathrm{d}p/\mathrm{d}z = -\rho g$. Substitute the relation between the pressure p and density ρ, e.g. from part (a), into this equation for the pressure gradient and integrate, to show

$$\rho = \rho_0 \left(1 - \frac{\rho_0 g}{p_0}\left(1 - \frac{1}{\gamma}\right) z\right)^{1/(\gamma-1)}.$$

(ii) Use the relation between the pressure p and density ρ from part (a) to deduce the corresponding dependence of the pressure p on height z.

(f) Suppose p_0 and ρ_0 are the pressure and density at height $z_0 = 0$. Show that the height z_* at which the density falls to zero is given by

$$z_* = \frac{p_0}{\rho_0 g}\left(\frac{\gamma}{\gamma - 1}\right).$$

(g) Explain why the mass of a column of gas with unit cross-section between the heights $z = 0$ and $z = z_*$ is given by

$$\int_0^{z_*} \rho(z)\,\mathrm{d}z.$$

Using part (f) show that this mass equals p_0/g and is thus independent of γ.

2.5 Euler's equations (statics): Liquid planet/star. We consider a liquid planet or star and assume spherical symmetry so that in spherical polar coordinates there is variation in the radial distance r only.

(a) Let $M = M(r)$ denote the mass of material within the radius r. Explain why

$$M(r) = 4\pi \int_0^r \tilde{r}^2 \rho(\tilde{r})\,\mathrm{d}\tilde{r}.$$

(b) The gravitational field $\boldsymbol{g} = \boldsymbol{g}(r)$ at radius r is given by

$$\boldsymbol{g}(r) = -\frac{GM(r)}{r^2}\hat{\boldsymbol{r}},$$

where G is the universal gravitational constant. For this static scenario for which $\boldsymbol{u} = \boldsymbol{0}$, explain why we know:

(i) The pressure field p satisfies $\nabla p = \rho \boldsymbol{g}$, where $\rho = \rho(r)$ is the density at radius r.

(ii) As a result of the spherical symmetry, we have

$$\frac{\mathrm{d}p}{\mathrm{d}r} = -\frac{G\rho(r)\,M(r)}{r^2}.$$

(c) Now suppose the density is constant and the liquid sphere forming the planet or star has radius a, i.e. $\rho(r) = \rho_0$, a constant, for $r \leqslant a$ and $\rho(r) = 0$ for $r > a$. Suppose $\boldsymbol{g}(r) = g(r)\,\hat{\boldsymbol{r}}$. Using parts (a) and (b), show

$$g(r) = \begin{cases} -\frac{4}{3}\pi\rho_0 Gr, & r \leqslant a, \\ -\frac{4}{3}\pi\rho_0 Ga^3/r^2, & r > a, \end{cases}$$

and

$$p(r) = \begin{cases} \frac{2}{3}\pi\rho_0^2 G(a^2 - r^2), & r \leqslant a, \\ 0, & r > a. \end{cases}$$

(d) Sketch $g = g(r)$ as a function of $r \geqslant 0$. Where is $g = g(r)$ greatest in magnitude and what is its maximum value? Where is the pressure $p = p(r)$ greatest and what is its value?

2.6 Euler's equations (statics): Non-uniform liquid planet. Consider a spherically symmetric liquid planet, call it planet Q, which is in 'hydrostatic' equilibrium and whose density $\rho = \rho(r)$ is a function of radial distance r, given by

$$\rho = \begin{cases} \rho_0(1 - \alpha(r/R)), & r \leqslant R, \\ 0, & r > R, \end{cases}$$

where $\rho_0 > 0$, $R > 0$ and $0 \leqslant \alpha < 1$ are constants.

(a) As in Exercise 2.5(a), let $M = M(r)$ denote the mass of material with radius r. Show that

$$M = \begin{cases} 4\pi\rho_0 r^3\left(\frac{1}{3} - \frac{\alpha r}{4R}\right), & r \leqslant R, \\ 4\pi\rho_0 R^3\left(\frac{1}{3} - \frac{1}{4}\alpha\right), & r > R. \end{cases}$$

(b) The gravitational field $\boldsymbol{g} = \boldsymbol{g}(r)$ is given by

$$\boldsymbol{g}(r) = -\frac{GM(r)}{r^2}\,\hat{\boldsymbol{r}},$$

where G is the universal gravitational constant. Use the 'hydrostatic' approximation from the end of Section 2.2 to explain why the pressure field p satisfies $\nabla p = \rho\,\boldsymbol{g}$, where $\rho = \rho(r)$ is the density at radius r. Use the spherically symmetric assumption to deduce

$$\frac{\mathrm{d}p}{\mathrm{d}r} = -\frac{G\rho(r)\,M(r)}{r^2}.$$

(c) Use parts (a) and (b) to show that for $r \leqslant R$, we have

$$\frac{\mathrm{d}p}{\mathrm{d}r} = -4\pi G\rho_0^2\, r\left(\frac{1}{3} - \frac{\alpha r}{4R}\right)\left(1 - \frac{\alpha r}{R}\right).$$

Hence, taking the pressure at the surface $r = R$ to be zero, integrate this last relation to show that the pressure at the centre of the planet is given by

$$p(0) = \pi G\rho_0^2 R^2\left(\tfrac{2}{3} - \tfrac{7}{9}\alpha + \tfrac{1}{4}\alpha^2\right).$$

(d) Now consider another planet with the same mass and radius as planet Q, but whose density $\overline{\rho}$ is uniform throughout. Find the pressure $\overline{p}(0)$ at the centre of this planet as follows.

(i) Compute the mass of the uniform planet in terms of $\overline{\rho}$ and equate that to $M(R)$ from part (a) above, to show

$$\overline{\rho} = \rho_0\left(1 - \tfrac{3}{4}\alpha\right).$$

(ii) Replace α by zero in the formula for $p(0)$ from part (c) as well as ρ_0 by $\overline{\rho}$. Why does this give a valid formula for $\overline{p}(0)$, the pressure at the centre of the uniform planet?

(iii) Use part (i) just above to show $\overline{p}(0) = \frac{2}{3}\pi G\rho_0^2 R^2(1 - \frac{3}{4}\alpha)^2$, and consequently show that $p(0) > \overline{p}(0)$ when $0 < \alpha < 1$. (This is an instance of a more general result of Chandrasekhar and Milne.)

2.7 Euler's equations (statics): Polytropes. There is a class of 'static' models of stellar structure in which radiation pressure is neglected and pressure depends on density in a simple way. These are called *polytropes*. The pressure dependence on the density is given by the *polytropic law*

$$p \propto \rho^{1+1/n},$$

where n is called the *polytropic index*. Following the arguments from Exercise 2.5 (a) and (b), note that the pressure satisfies

$$\frac{\mathrm{d}p}{\mathrm{d}r} = -\frac{G\rho(r)\,M(r)}{r^2},$$

where $M = M(r)$ is the mass of material within the radius r, an expression for which is given in Exercise 2.5(a).

(a) Differentiate the expression for $M = M(r)$ from Exercise 2.5(a). Then differentiate the equation just above for the pressure and insert your answer for $\mathrm{d}M/\mathrm{d}r$ to show that the pressure p satisfies the second-order ordinary differential equation

$$\frac{1}{r^2}\frac{\mathrm{d}}{\mathrm{d}r}\left(\frac{r^2}{\rho}\frac{\mathrm{d}p}{\mathrm{d}r}\right) = -4\pi G\rho.$$

(b) Suppose p_0, ρ_0 and T_0 represent a fixed reference pressure, density and temperature state. Using the polytropic law and ideal gas law (see Exercise 2.2), show

$$\frac{p}{p_0} = \left(\frac{\rho}{\rho_0}\right)^{1+1/n} \qquad \text{and} \qquad \frac{p}{p_0} = \frac{\rho T}{\rho_0 T_0}.$$

(c) We define the dimensionless variables ξ and $\Theta = \Theta(\xi)$ as follows:

$$\xi := \frac{r}{\alpha} \qquad \text{and} \qquad \Theta := \frac{T}{T_0},$$

where the scalar factor α, which has the dimensions of length, is to be determined presently.

(i) Using part (b), show

$$\frac{T}{T_0} = \left(\frac{\rho}{\rho_0}\right)^{1/n} \quad \text{and therefore} \quad \frac{p}{p_0} = \left(\frac{T}{T_0}\right)^{n+1} = \Theta^{n+1}.$$

(ii) By substituting the expression for the pressure p in terms of Θ from the relation in part (i) just above and making the change of variables from r to ξ, show that $\Theta = \Theta(\xi)$ satisfies the equation

$$\frac{1}{\xi^2}\frac{\mathrm{d}}{\mathrm{d}\xi}\left(\xi^2\frac{\mathrm{d}\Theta}{\mathrm{d}\xi}\right) = -\left(\frac{4\pi G\rho_0^2\alpha^2}{(n+1)p_0}\right)\Theta^n.$$

Explain why it is sensible to set

$$\alpha := \left(\frac{4\pi G\rho_0^2}{(n+1)p_0}\right)^{-1/2},$$

and confirm this form for α has the dimensions of length. This second-order ordinary differential equation is known as the *Lane–Emden equation* of index n, after Lane (1870) and Emden (1907). Different stellar materials are associated with different values for n.

(d) Solutions $\Theta = \Theta(\xi)$ to the Lane–Emden equation from part (c)(ii) are usually required to satisfy the natural boundary conditions at $\xi = 0$:

$$\Theta(0) = 1 \qquad \text{and} \qquad \Theta'(0) = 0.$$

Further, since T is absolute temperature measured in Kelvin, we require $\Theta > 0$. For the different values of $n = 0$, $n = 1$ and $n = 5$ establish the following exact solutions to the Lane–Emden equation satisfying the boundary conditions stated, by the methods outlined:

(i) $n = 0$. Integrate the Lane–Emden equation to show that $\Theta(\xi) = 1 - \frac{1}{6}\xi^2$ is a solution for $0 \leqslant \xi \leqslant \sqrt{6}$ or equivalently $0 \leqslant r \leqslant \sqrt{6}(p_0/4\pi\rho_0^2 G)^{1/2}$ (this corresponds to the case in Example 2.5).

(ii) $n = 1$. By substitution, show that $\Theta(\xi) = (\sin\xi)/\xi$ is a solution for $0 \leqslant \xi \leqslant \pi$ (this case is due to Ritter).

(iii) $n = 5$. By substitution, show that $\Theta(\xi) = \left(1 + \frac{1}{3}\xi^2\right)^{-1/2}$ is a solution for $0 \leqslant \xi \leqslant \infty$. In this case the model is infinite – it is of infinite extent – but the total mass is finite (this case is due to Schuster and Emden).

(*Note:* In summary, when $n < 5$ the model is finite with Θ vanishing at a finite radius while when $n > 5$ the model is infinite with infinite mass. The cases $n = 1.5$ and $n = 2.5$, respectively correspond to monatomic and diatomic gases. Any significant departures of such a pressure field from very close to hydrostatic equilibrium would have catastrophic consequences.)

2.8 Vorticity formulation: Vorticity formula. Assuming we know the Lagrangian paths of an ideal, homogeneous, incompressible flow, or equivalently we have knowledge of $\boldsymbol{u} = \boldsymbol{u}(\boldsymbol{x}, t)$, then the solution to the vorticity formulation of the Euler equations

$$\frac{\partial \omega}{\partial t} + (\boldsymbol{u} \cdot \nabla)\,\omega = (\omega \cdot \nabla)\,\boldsymbol{u}$$

along the Lagrangian paths $\boldsymbol{x} = \boldsymbol{x}(\boldsymbol{a}, t)$ is given by

$$\omega(\boldsymbol{x}(\boldsymbol{a}, t), t) = (\nabla_{\boldsymbol{a}}\boldsymbol{x}(\boldsymbol{a}, t))\,\omega_0(\boldsymbol{a}).$$

Our goal here is to establish this result by direct substitution.

(a) Recall for a particle emerging from $\boldsymbol{a} \in \mathbb{R}^3$ that the Lagrangian particle path or trajectory $\boldsymbol{x} = \boldsymbol{x}(\boldsymbol{a}, t)$ is the integral curve of the equation

$$\frac{\mathrm{d}}{\mathrm{d}t}\boldsymbol{x}(\boldsymbol{a}, t) = \boldsymbol{u}(\boldsymbol{x}(\boldsymbol{a}, t), t),$$

with $\boldsymbol{x}(\boldsymbol{a}, 0) = \boldsymbol{a}$. Using the chain rule, show that along particle paths the vorticity field $\omega = \omega(\boldsymbol{x}(\boldsymbol{a}, t), t)$ satisfies the ordinary differential equation

$$\frac{\mathrm{d}}{\mathrm{d}t}\omega(\boldsymbol{x}(\boldsymbol{a}, t), t) = ((\omega \cdot \nabla)\, \boldsymbol{u})(\boldsymbol{x}(\boldsymbol{a}, t), t).$$

(b) Taking the time derivative of the solution formula above for ω along $\boldsymbol{x} = \boldsymbol{x}(\boldsymbol{a}, t)$, and assuming all fields are sufficiently smooth so that the order of derivatives can be swapped, show that

$$\frac{\mathrm{d}}{\mathrm{d}t}\omega(\boldsymbol{x}(\boldsymbol{a}, t), t) = \frac{\mathrm{d}}{\mathrm{d}t}\Big((\nabla_{\boldsymbol{a}}\boldsymbol{x}(\boldsymbol{a}, t))\, \omega_0(\boldsymbol{a})\Big) = \nabla_{\boldsymbol{a}}\boldsymbol{u}(\boldsymbol{x}(\boldsymbol{a}, t), t)\, \omega_0(\boldsymbol{a}).$$

(c) Given the result in part (b), and noting the form of the vorticity formulation in part (a), our goal now is to show that $\nabla_{\boldsymbol{a}}\boldsymbol{u}(\boldsymbol{x}(\boldsymbol{a}, t), t)\, \omega_0(\boldsymbol{a})$ equals $(\omega \cdot \nabla \boldsymbol{u})(\boldsymbol{x}(\boldsymbol{a}, t), t)$. To this end:

  (i) Use the chain rule on $\nabla_{\boldsymbol{a}}\boldsymbol{u}$ to show

  $$\nabla_{\boldsymbol{a}}\boldsymbol{u} = (\nabla_{\boldsymbol{x}}\boldsymbol{u})\,(\nabla_{\boldsymbol{a}}\boldsymbol{x}),$$

  where the product on the right between the matrices $\nabla_{\boldsymbol{x}}\boldsymbol{u}$ and $\nabla_{\boldsymbol{a}}\boldsymbol{x}$ is just the usual matrix product.

  (ii) Use the result from part (i) just above and the solution formula to show the term on the right in part (b) is given by

  $$(\nabla_{\boldsymbol{a}}\boldsymbol{u})\, \omega_0 = (\nabla_{\boldsymbol{x}}\boldsymbol{u})\, \omega.$$

  (iii) To establish the overall result, show that $(\nabla \boldsymbol{u})\, \omega = (\omega \cdot \nabla)\, \boldsymbol{u}$.

  (*Note:* See Section 2.10, part (9).)

2.9 Linear flows: Alternative. Proposition 2.10 in Section 2.4 is a modified form of Proposition 1.5 in Majda and Bertozzi (2002). Their proposition statement is as follows. Let $D = D(t)$ be a real symmetric 3×3 matrix with $\operatorname{tr} D = 0$. Suppose the vorticity $\omega = \omega(t) \in \mathbb{R}^3$ is the solution of the ordinary differential system of equations $\omega'(t) = D(t)\, \omega(t)$, for some initial data $\omega(0) = \omega_0$. Suppose the antisymmetric matrix R and the vector ω are related by the formula $R\boldsymbol{h} = \frac{1}{2}\omega \times \boldsymbol{h}$. Then the velocity and pressure fields given by

$$\boldsymbol{u}(\boldsymbol{x}, t) = \tfrac{1}{2}\omega(t) \times \boldsymbol{x} + D(t)\, \boldsymbol{x},$$
$$p(\boldsymbol{x}, t) = -\tfrac{1}{2}\Big((D'(t) + D^2(t) + R^2(t))\, \boldsymbol{x}\Big) \cdot \boldsymbol{x},$$

are exact solutions to the incompressible Euler and Navier–Stokes equations in $\mathbb{R}^3$. Prove this result via the following steps.

(a) The velocity gradient matrix $\nabla \boldsymbol{u}$ can be decomposed into a direct sum of its symmetric and antisymmetric parts, which are the 3×3 matrices $D := \frac{1}{2}((\nabla \boldsymbol{u}) + (\nabla \boldsymbol{u})^{\mathrm{T}})$ and $R := \frac{1}{2}((\nabla \boldsymbol{u}) - (\nabla \boldsymbol{u})^{\mathrm{T}})$. Using the homogeneous (no body force) Navier–Stokes equations, show that $\nabla \boldsymbol{u}$ evolves according to the partial differential system

$$\frac{\partial}{\partial t}(\nabla \boldsymbol{u}) + (\boldsymbol{u} \cdot \nabla)(\nabla \boldsymbol{u}) + (\nabla \boldsymbol{u})^2 = \nu\, \Delta(\nabla \boldsymbol{u}) - \nabla\nabla p,$$

where $\nu > 0$ is the viscosity; see Chapter 4 and in particular Remark 4.13(ii). Note here that $(\nabla \boldsymbol{u})^2 = (\nabla \boldsymbol{u})(\nabla \boldsymbol{u})$ is simply matrix multiplication.

(b) Show by direct computation that $(\nabla \boldsymbol{u})^2 = (D + R)^2 = (D^2 + R^2) + (DR + RD)$, where the first term on the right is symmetric and the second is antisymmetric. Hence deduce that we can decompose the evolution of $\nabla \boldsymbol{u}$ into the coupled evolution of its symmetric and antisymmetric parts

$$\frac{\partial D}{\partial t} + (\boldsymbol{u} \cdot \nabla)\, D + D^2 + R^2 = \nu\, \Delta D - \nabla\nabla p,$$
$$\frac{\partial R}{\partial t} + (\boldsymbol{u} \cdot \nabla)\, R + DR + RD = \nu\, \Delta R.$$

(c) By directly computing the evolution for the three components of $\boldsymbol{\omega} = (\omega_1, \omega_2, \omega_3)^{\mathrm{T}}$ from the second system of equations in (b), show that the vorticity satisfies the system of equations

$$\frac{\partial \boldsymbol{\omega}}{\partial t} + (\boldsymbol{u} \cdot \nabla)\, \boldsymbol{\omega} = \nu\, \Delta \boldsymbol{\omega} + D\boldsymbol{\omega}.$$

(*Note:* (i) We could have derived this more directly as in Corollaries 2.7 and 4.15. (ii) That the off-diagonal elements of $DR + RD$ are equal to $-D\boldsymbol{\omega}$ using the standard representation can be demonstrated by direct enumeration, see Proposition 2.10.)

(d) Thus far we have not utilised the form for the velocity stated above. Now assume $\boldsymbol{u}(\boldsymbol{x}, t) = \frac{1}{2}\boldsymbol{\omega}(t) \times \boldsymbol{x} + D(t)\, \boldsymbol{x}$. Show that $\nabla \times \boldsymbol{u} = \boldsymbol{\omega}(t)$, independent of $\boldsymbol{x}$, and substitute this into the evolution equation for $\boldsymbol{\omega}$ above to reveal that the system of ordinary differential equations $\boldsymbol{\omega}' = D\boldsymbol{\omega}$ do indeed govern the evolution of $\boldsymbol{\omega} = \boldsymbol{\omega}(t)$.

(e) Hence show that the first system of equations from (b) for $D = D(t)$ are independent of $\boldsymbol{x}$ and reduce to the system of differential equations $D' + D^2 + R^2 = -\nabla\nabla p$. Noting that $R = R(t)$ only as $\boldsymbol{\omega} = \boldsymbol{\omega}(t)$, deduce that $\nabla\nabla p$ must be a function of t only. Hence deduce that $p = p(\boldsymbol{x}, t)$ can only quadratically depend on $\boldsymbol{x}$ and must have the form shown.

2.10 Axisymmetric flows: Couette flow

(a) Using the Euler equations for an ideal homogeneous incompressible flow in cylindrical polar coordinates (see Section A.2 in the Appendix) show, with z the coordinate in the vertical upward direction, for a *stationary* flow which is independent of θ with $u_r = u_z = 0$ that we have

$$\frac{u_\theta^2}{r} = \frac{1}{\rho}\frac{\partial p}{\partial r} \qquad \text{and} \qquad 0 = \frac{1}{\rho}\frac{\partial p}{\partial z} + g,$$

where $p = p(r, z)$ is the pressure, ρ is the constant uniform mass density and g is the acceleration due to gravity (this is the body force per unit mass). Verify that any such flow is indeed incompressible.

(b) Suppose an incompressible fluid resides in the region between two concentric cylinders of radii R_1 and R_2, where $R_1 < R_2$. Suppose the velocity field in cylindrical coordinates $\boldsymbol{u} = u_r\hat{\boldsymbol{r}} + u_\theta\hat{\boldsymbol{\theta}} + u_z\hat{\boldsymbol{z}}$ of the fluid flow between the concentric cylinders is given by

$$u_r = 0, \qquad u_z = 0 \qquad \text{and} \qquad u_\theta = \frac{A}{r} + Br,$$

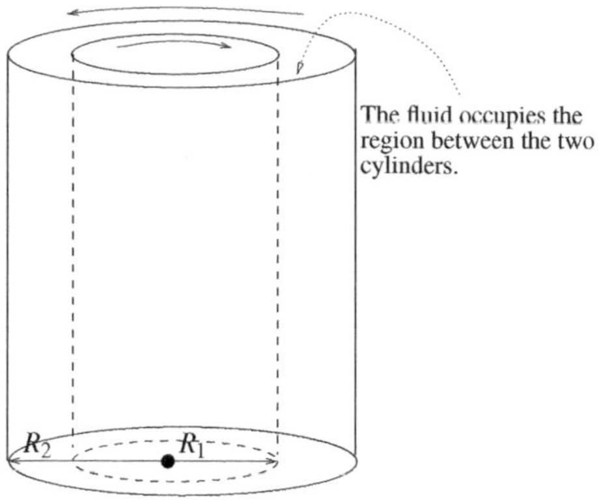
The fluid occupies the region between the two cylinders.
R_2
R_1

Figure 2.22 Couette flow between two concentric cylinders of radii $R_1 < R_2$.

where

$$A = -\frac{R_1^2 R_2^2(\omega_2 - \omega_1)}{R_2^2 - R_1^2} \qquad \text{and} \qquad B = -\frac{R_1^2\omega_1 - R_2^2\omega_2}{R_2^2 - R_1^2}.$$

This is known as a *Couette flow* – see Figure 2.22. Note that we ignore gravity and henceforth assume $g = 0$. Show

(i) The velocity field $\boldsymbol{u} = u_r\hat{\boldsymbol{r}} + u_\theta\hat{\boldsymbol{\theta}} + u_z\hat{\boldsymbol{z}}$ is a stationary solution of Euler's equations of motion for an ideal fluid with density $\rho \equiv 1$. (*Hint:* You need to find a pressure field p that is consistent with the velocity field given. Indeed, the pressure field should be $p = -A^2/2r^2 + 2AB \log r + B^2r^2/2 + C$ for some arbitrary constant C.)

(ii) The angular velocity of the flow (i.e. the quantity u_θ/r) is ω_1 on the cylinder $r = R_1$ and ω_2 on the cylinder $r = R_2$.

(iii) The vorticity field $\boldsymbol{\omega} = \nabla \times \boldsymbol{u} = 2B\hat{\boldsymbol{z}}$. (*Hint:* You need to use curl in cylindrical polar coordinates; see Section A.2 in the Appendix.)

2.11 Axisymmetric flows: Sink or bath drain flow. As we have all observed when water runs out of a bath or sink, the free surface of the water directly over the drain hole has a depression in it – see Figure 2.23. The question is, what is the form/shape of this free surface depression? The essential idea is that we know the pressure at the free surface is uniform; it is atmospheric pressure, say P_0, in the core region.

(a) Using the Euler equations for an ideal homogeneous incompressible flow in cylindrical polar coordinates, show, with z the coordinate in the vertical upward direction, for a *stationary* flow which is independent of θ with $u_r = u_z = 0$ that we have

$$\frac{u_\theta^2}{r} = \frac{1}{\rho}\frac{\partial p}{\partial r} \qquad \text{and} \qquad 0 = \frac{1}{\rho}\frac{\partial p}{\partial z} + g,$$

where $p = p(r, z)$ is the pressure, ρ is the constant uniform mass density and g is the acceleration due to gravity (i.e. the body force per unit mass). Verify that such a flow is incompressible.

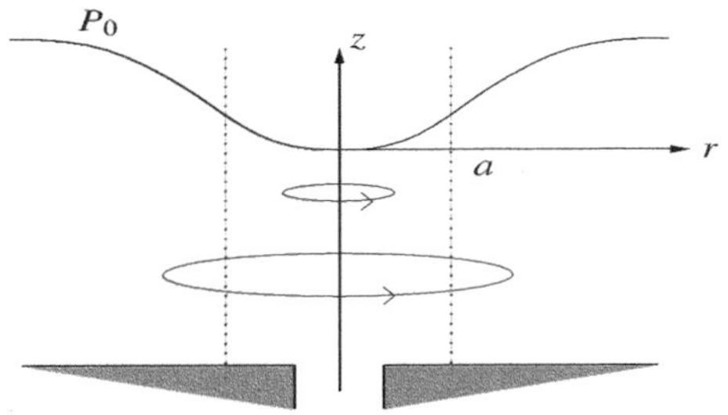

Figure 2.23 Water draining from a bath.

(b) In a *simple* model for water in a bath or sink slowly draining through a narrow plughole, each fluid particle traverses a horizontal circle whose centre is the fixed vertical axis. The angular velocity at a distance r from the axis is assumed to be (Ω and $a > 0$ are known constants)

$$u_\theta = \begin{cases} \Omega r, & \text{for } 0 \leqslant r \leqslant a, \\ \Omega a^2/r, & \text{for } r > a. \end{cases}$$

This is a *Rankine vortex* flow, in which non-zero vorticity and local rotation is confined to the core region $r \leqslant a$.

(i) Consider the flow given above in the inner region $0 \leqslant r \leqslant a$. Using the equations in part (a) above, show that the pressure in this region is given by (with C_0 an arbitrary constant)

$$p(r, z) = \tfrac{1}{2}\rho\Omega^2 r^2 - \rho g z + C_0.$$

(ii) Using that at the free surface of the water the pressure is constant atmospheric pressure P_0, show that the free surface for $r \leqslant a$ is given by

$$z = \frac{\Omega^2 r^2}{2g} - \frac{(C_0 - P_0)}{\rho g}.$$

(Hence the depression in the free surface for $r \leqslant a$ is a *parabolic surface* of revolution. Note, since the pressure is only ever globally defined up to an additive constant, we can take $C_0 = 0$ or $C_0 = P_0$.)

(iii) Now consider the flow in the outer region $r > a$. Using the equations in part (a) above, show that the pressure in this region is given by (with K_0 an arbitrary constant)

$$p(r, z) = -\frac{\rho\Omega^2 a^4}{2r^2} - \rho g z + K_0.$$

(iv) Using that the pressure must be continuous at $r = a$, show that $K_0 = \rho\Omega^2 a^2 + C_0$, and hence deduce that the pressure for $r > a$ is given by

$$p(r, z) = -\frac{\rho\Omega^2 a^4}{2r^2} - \rho g z + \rho\Omega^2 a^2 + C_0.$$

(v) Again using that the pressure at the free surface is $p(r,z) = P_0$, show that for $r > a$ the free surface is given by (taking $C_0 = P_0$)

$$z = -\frac{\Omega^2 a^4}{2gr^2} + \frac{\Omega^2 a^2}{g}.$$

(Note that the flow we assume for $r > a$ represents two-dimensional irrotational flow generated by a point source at the origin; see Section 2.8 and Chapter 3 for more details.)

2.12 Axisymmetric flows: No initial swirl. Consider a general homogeneous incompressible axisymmetric flow in cylindrical polar coordinates. Recall that axisymmetry implies $\boldsymbol{u} = \boldsymbol{u}(r,z,t)$ only. (Compare with the two-and-a-half-dimensional flows in Exercise 4.13.) Assume no body force. Our goal is to demonstrate from the Euler equations that if $u_\theta(r,z,0) = 0$ throughout the flow, i.e. initially there is no swirl, then $u_\theta(r,z,t) = 0$ throughout the flow for all times $t > 0$, i.e. there is no swirl thereafter.

(a) First use the Euler equation for u_θ in cylindrical polar coordinates to show that $\partial^2 p/\partial\theta^2 \equiv 0$. From this we deduce that p is linear in θ. Since we require p to be periodic in θ, we deduce that p is independent of θ.

(b) Consider an arbitrary fluid particle (characteristic) trajectory

$$\frac{\mathrm{d}}{\mathrm{d}t}r(t) = u_r(r(t),z(t),t) \qquad \text{and} \qquad \frac{\mathrm{d}}{\mathrm{d}t}z(t) = u_z(r(t),z(t),t).$$

From the Euler equation for $u_\theta = u_\theta(r(t),z(t),t)$, show that

$$\frac{\mathrm{d}}{\mathrm{d}t}u_\theta = \frac{1}{r}u_r u_\theta$$

along the characteristics, where of course $r = r(t)$ and $u_r = u_r(r(t),z(t),t)$.

(c) Show that the solution u_θ to the ordinary differential equation in part (b) has the form

$$u_\theta(r(t),z(t),t) = \exp\left(\int_0^t \frac{1}{r(\tau)}u_r(r(\tau),z(\tau),\tau)\,\mathrm{d}\tau\right) u_\theta(r_0,z_0,0).$$

Hence conclude the result if $u_\theta(r_0,z_0,0) = 0$ throughout the flow.

2.13 Axisymmetric flows: No swirl. Consider a general homogeneous incompressible axisymmetric flow with no swirl in cylindrical polar coordinates. Axisymmetry implies $\boldsymbol{u} = \boldsymbol{u}(r,z,t)$ only. Assume no body force. From the exercise just above, we know that if initially there is no swirl, then $u_\theta(r,z,t) = 0$ throughout the flow for all times $t > 0$.

(a) First show that for any vector $\boldsymbol{h}$ and scalar f satisfying

$$\left(\frac{\partial}{\partial t} + (\boldsymbol{u}\cdot\nabla)\right)\boldsymbol{h} = (\boldsymbol{h}\cdot\nabla)\,\boldsymbol{u} \qquad \text{and} \qquad \left(\frac{\partial}{\partial t} + (\boldsymbol{u}\cdot\nabla)\right)f = 0,$$

then

$$\left(\frac{\partial}{\partial t} + (\boldsymbol{u}\cdot\nabla)\right)(\nabla f\cdot\boldsymbol{h}) = 0.$$

(b) Since there is no swirl, i.e. $u_\theta \equiv 0$, we have $\boldsymbol{u} = u_r\hat{\boldsymbol{r}} + u_z\hat{\boldsymbol{z}}$. Hence show that $(\partial/\partial t + (\boldsymbol{u}\cdot\nabla))\,\theta = 0$, where $\boldsymbol{u}\cdot\nabla = u_r\partial/\partial r + r^{-1}u_\theta\partial/\partial\theta + u_z\partial/\partial z$.

(c) The only non-zero component of vorticity is $\omega_\theta := \partial u_r/\partial z - \partial u_z/\partial r$, which satisfies the vorticity equation $\partial\omega/\partial t + (\boldsymbol{u}\cdot\nabla)\,\omega = (\omega\cdot\nabla)\,\boldsymbol{u}$. Use part (b) to show that the result from part (a) with $\boldsymbol{h} = \omega$ and $f = \theta$ implies

$$\left(\frac{\partial}{\partial t} + u_r\frac{\partial}{\partial r} + u_z\frac{\partial}{\partial z}\right)\left(\frac{\omega_\theta}{r}\right) = 0.$$

(d) The flow is incompressible, i.e. $r^{-1}\partial(ru_r)/\partial r + \partial u_z/\partial z = 0$. Verify that u_r and u_z can be represented in terms of the Stokes stream function Ψ as follows (see Remark 5.9):

$$u_r = -\frac{1}{r}\frac{\partial\Psi}{\partial z} \quad \text{and} \quad u_z = \frac{1}{r}\frac{\partial\Psi}{\partial r}.$$

(e) Use the calculations in Section 5.1 on Stokes flow to show

$$\omega_\theta = -\frac{1}{r}\mathrm{D}^2_{\mathrm{cyl}}\Psi,$$

where

$$\mathrm{D}^2_{\mathrm{cyl}} := \frac{\partial^2}{\partial z^2} + r\frac{\partial}{\partial r}\left(\frac{1}{r}\frac{\partial}{\partial r}\right).$$

We deduce that we have a closed equation for the evolution of ω_θ in part (c) as we can in principle invert the relation above to find Ψ in terms of ω_θ and thus use part (d) to compute u_r and u_z in terms of ω_θ. For further results in this direction, with swirl, see Majda and Bertozzi (2002, pp. 64–9).

2.14 Axisymmetric flows: Spherical polar coordinates. Consider an axisymmetric flow with no swirl in *spherical polar coordinates*. This means that the velocity field is explicitly independent of the azimuthal angle φ. In other words, $u_r = u_r(r,\theta,t)$ and $u_\theta = u_\theta(r,\theta,t)$ only. In spherical polar coordinates no swirl corresponds to $u_\varphi \equiv 0$. Hence the velocity field is given by $\boldsymbol{u} = u_r\,\hat{\boldsymbol{r}} + u_\theta\,\hat{\boldsymbol{\theta}}$ only. Assume the flow domain to be $\mathbb{R}^3$. The vorticity formulation of a three-dimensional Euler flow for a homogeneous incompressible fluid (assuming no body force) is (with $\Delta\boldsymbol{u} = -\nabla\times\omega$)

$$\frac{\partial\omega}{\partial t} + (\boldsymbol{u}\cdot\nabla)\,\omega = (\omega\cdot\nabla)\,\boldsymbol{u}.$$

(a) Compute the vorticity field $\omega = \nabla\times\boldsymbol{u}$ in spherical polar coordinates and show that $\omega = \omega_\varphi\,\hat{\boldsymbol{\varphi}}$ only, where $\omega_\varphi = r^{-1}((\partial/\partial r)(ru_\theta) - \partial u_r/\partial\theta)$.

(b) Use the following coordinate identities:

$$\begin{aligned}
\hat{\boldsymbol{r}} &\equiv \sin\theta\cos\varphi\,\boldsymbol{i} + \sin\theta\sin\varphi\,\boldsymbol{j} + \cos\theta\,\boldsymbol{k},\\
\hat{\boldsymbol{\varphi}} &\equiv -\sin\varphi\,\boldsymbol{i} + \cos\varphi\,\boldsymbol{j},\\
\hat{\boldsymbol{\theta}} &\equiv \cos\theta\cos\varphi\,\boldsymbol{i} + \cos\theta\sin\varphi\,\boldsymbol{j} - \sin\theta\,\boldsymbol{k},
\end{aligned}$$

to show that $\partial\hat{\boldsymbol{r}}/\partial\varphi \equiv \sin\theta\,\hat{\boldsymbol{\varphi}}$ and $\partial\hat{\boldsymbol{\theta}}/\partial\varphi \equiv \cos\theta\,\hat{\boldsymbol{\varphi}}$.

(c) Using all the identities in part (b) and the formulations from Section A.3 in the Appendix for $(\boldsymbol{u}\cdot\nabla)\,\omega$ and $(\omega\cdot\nabla)\,\boldsymbol{u}$ in spherical polars, show that

$$(\boldsymbol{u}\cdot\nabla)\,\omega = \left(u_r\frac{\partial}{\partial r} + \frac{u_\theta}{r}\frac{\partial}{\partial\theta}\right)(\omega_\varphi\,\hat{\boldsymbol{\varphi}}) = \left(u_r\frac{\partial\omega_\varphi}{\partial r} + \frac{u_\theta}{r}\frac{\partial\omega_\varphi}{\partial\theta}\right)\hat{\boldsymbol{\varphi}}$$

and

$$(\boldsymbol{\omega}\cdot\nabla)\,\boldsymbol{u} = \frac{\omega_\varphi}{r\sin\theta}\frac{\partial}{\partial\varphi}(u_r\,\hat{\boldsymbol{r}} + u_\theta\,\hat{\boldsymbol{\theta}}) = \frac{\omega_\varphi}{r\sin\theta}(u_r\sin\theta + u_\theta\cos\theta)\,\hat{\boldsymbol{\varphi}}.$$

(d) Substituting both quantities shown in part (c) into the vorticity formulation of Euler's equations preceding part (a) above, show that the vorticity formulation in this case is equivalent to

$$\frac{\partial\omega_\varphi}{\partial t} + u_r\left(\frac{\partial\omega_\varphi}{\partial r} - \frac{\omega_\varphi}{r}\right) + \frac{u_\theta}{r}\left(\frac{\partial\omega_\varphi}{\partial\theta} - \omega_\varphi\frac{\cos\theta}{\sin\theta}\right) = 0.$$

Further, show that this evolution equation for ω_φ is equivalent to the condition

$$\left(\frac{\partial}{\partial t} + (\boldsymbol{u}\cdot\nabla)\right)\left(\frac{\omega_\varphi}{r\sin\theta}\right) = 0.$$

(*Hint*: It is easier to start with the condition above and work backwards.)

2.15 Axisymmetric flows: Hill's spherical vortex. Hill's spherical vortex is a non-trivial steady vortex flow localised to a sphere with irrotational flow outside the sphere. It is easily expressed in either cylindrical or spherical polar coordinates; here we choose the latter. The whole flow is axisymmetric and has no swirl. For this context, Exercise 2.14 shows how the Euler equations for a homogeneous incompressible fluid flow are equivalent to the evolution equation for the only non-zero component of the vorticity field ω_φ shown in part (d) of the exercise. Indeed, the stationary version of the equation is $(\boldsymbol{u}\cdot\nabla)(\omega_\varphi/(r\sin\theta)) = 0$. For Hill's spherical vortex we prescribe the vorticity field ω_φ to be

$$\omega_\varphi = \begin{cases} \Omega r\sin\theta, & r < a, \\ 0, & r > a, \end{cases}$$

where $a > 0$ and $\Omega \in \mathbb{R}$ are constants. This explicit form for ω_φ satisfies the stationary partial differential equation for ω_φ just above, and thus represents an ideal incompressible flow. Note that the vorticity jumps at the boundary $r = a$, however a solution velocity field that is continuous everywhere, and in particular at $r = a$, does exist, as we are about to demonstrate. The solution for $r > a$, where the flow is irrotational, can be found in Section 2.8 in the proof of Theorem 2.49, under the assumption that the same boundary conditions in the far field and at $r = a$ apply, which indeed we assume; see Figures 2.24 and 2.18. Note the positive $\theta = 0$ axis direction, i.e. the $\boldsymbol{k}$ or $\hat{\boldsymbol{z}}$-direction, is aligned as shown in Figure 2.18.

(a) Recall from the proof of Theorem 2.49 that the boundary conditions are $\boldsymbol{u} \sim U\boldsymbol{k}$ as $r \to \infty$ and $\boldsymbol{u}\cdot\hat{\boldsymbol{r}} = 0$ on $r = a$ corresponding to no normal flow. The latter condition is equivalent to $u_r = 0$ on $r = a$. Using the coordinate identities from Exercise 2.14(b) for $\hat{\boldsymbol{r}}$ and $\hat{\boldsymbol{\theta}}$, show that the far-field boundary conditions as $r \to \infty$ are equivalent to the conditions

$$u_r\cos\theta - u_\theta\sin\theta \sim U \qquad \text{and} \qquad u_r\sin\theta + u_\theta\cos\theta \sim 0.$$

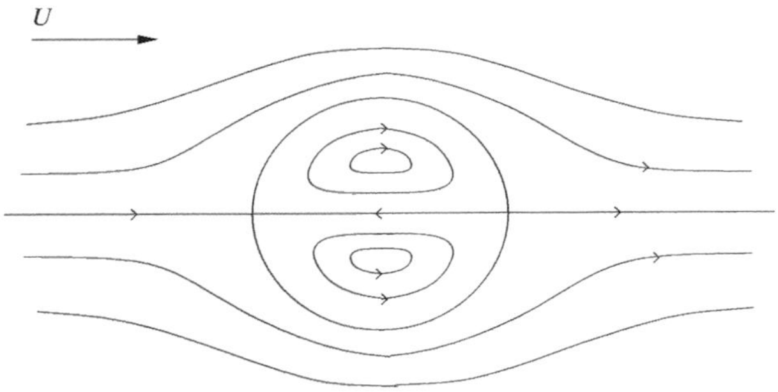

Figure 2.24 Hill's spherical vortex. This is a steady vortex flow localised to a sphere. The whole flow is axisymmetric with no swirl. In the far field the velocity field is uniform, of speed U. Outside the sphere the flow is irrotational and identical to that for D'Alembert's Theorem 2.49. Inside the sphere the vorticity $\omega_\varphi = \Omega r \sin\theta$ is non-trivial, however an explicit formula for Stokes stream function can be found. See Exercise 2.15.

(b) The solution velocity field for $r > a$ satisfying the given boundary conditions derived in the proof of Theorem 2.49 is given by

$$u_r = U(1 - a^3/r^3)\cos\theta \qquad \text{and} \qquad u_\theta = -U(1 + a^3/2r^3)\sin\theta.$$

The component u_r satisfies the boundary condition at $r = a$. Verify that this velocity field also satisfies the far field conditions derived in part (a).

(c) In spherical polar coordinates, with axisymmetry and no swirl, the solution flow can be represented by a stream function $\Psi = \Psi(r, \theta)$ via

$$u_r = \frac{1}{r^2 \sin\theta}\frac{\partial\Psi}{\partial\theta} \qquad \text{and} \qquad u_\theta = -\frac{1}{r\sin\theta}\frac{\partial\Psi}{\partial r}.$$

Solve these equations to show that the stream function Ψ for $r > a$ is

$$\Psi = \tfrac{1}{2}U(r^2 - a^3/r)\sin^2\theta.$$

Show that the boundary condition on $r = a$ becomes $\partial\Psi/\partial\theta = 0$ on $r = a$. Note that the slip velocity u_θ along $r = a$ is $u_\theta(a, \theta) = -\frac{3}{2}U\sin\theta$.

(d) Let us now consider the solution flow field, indeed the stream function $\Psi = \Psi(r, \theta)$, for $r < a$. From the arguments preceding Proposition 5.8 in Section 5.1, we know that the vorticity component can be expressed in terms of the stream function Ψ via $\omega_\varphi = -(r\sin\theta)^{-1}\mathrm{D}^2_{\mathrm{sph}}\Psi$, where $\mathrm{D}^2_{\mathrm{sph}}$ is the second-order partial differential operator given by

$$\mathrm{D}^2_{\mathrm{sph}} := \frac{\partial^2}{\partial r^2} + \frac{\sin\theta}{r}\frac{\partial}{\partial\theta}\left(\frac{1}{r\sin\theta}\frac{\partial}{\partial\theta}\right).$$

Since for $r < a$ we know $\omega_\varphi = \Omega r\sin\theta$, we see that Ψ must satisfy

$$\frac{\partial^2\Psi}{\partial r^2} + \frac{\sin\theta}{r}\frac{\partial}{\partial\theta}\left(\frac{1}{r\sin\theta}\frac{\partial\Psi}{\partial\theta}\right) = -\Omega(r\sin\theta)^2,$$

subject to the conditions on $r = a$: $\partial\Psi/\partial\theta = 0$ and $\partial\Psi/\partial r = \frac{3}{2}Ua\sin^2\theta$, to match the slip velocity. By looking for a solution of the form $\Psi(r,\theta) = F(r)\sin^2\theta$ for some function $F = F(r)$, show that F satisfies the following second-order ordinary differential equation (of Cauchy–Euler type):

$$r^2F'' - 2F = -\Omega r^4,$$

together with the conditions $F(a) = 0$ and $F'(a) = \frac{3}{2}Ua$.

(e) By looking for homogeneous solutions of the form r^α for some constant $\alpha \in \mathbb{R}$ and then a particular solution of the form cr^4 for some constant c, show that the general solution to the equation for F in part (d) is $F(r) = Ar^2 + Br^{-1} - \Omega r^4/10$, where A and B are arbitrary constants. For a bounded solution set $B = 0$. Then, using the condition $F(a) = 0$ and the slip velocity condition $F'(a) = \frac{3}{2}Ua$, show that $A = \Omega a^2/10$ and $\Omega = -15U/2a^2$. Hence deduce the solution $\Psi = \Psi(r,\theta)$ for $r < a$ is

$$\Psi = \tfrac{1}{10}\Omega r^2(a^2 - r^2)\sin^2\theta \equiv -\tfrac{3}{4}Ur^2\left(1 - \frac{r^2}{a^2}\right)\sin^2\theta.$$

Note that the boundary conditions ensure u_r and u_θ are continuous.

2.16 Bernoulli's Theorem: Venturi tube. Consider the Venturi tube of narrow cross-section and constriction shown in Figure 2.25. Assume that the ideal fluid flow through the constriction is homogeneous, incompressible and steady. The flow in the wider section of cross-sectional area A_1 has velocity u_1 and pressure p_1, while that in the narrower section of cross-sectional area A_2 has velocity u_2 and pressure p_2. Separately within the uniform wider and narrower sections, we assume the velocity and pressure are uniform themselves. In order to focus on the dynamic pressure changes, ignore gravity.

(a) Why does the relation $A_1u_1 = A_2u_2$ hold? Why is the flow faster in the narrower region of the tube compared to the wider region?

(b) Use Bernoulli's Theorem to show that

$$\tfrac{1}{2}u_1^2 + \frac{p_1}{\rho_0} = \tfrac{1}{2}u_2^2 + \frac{p_2}{\rho_0},$$

where ρ_0 is the constant uniform density of the fluid.

(c) Using the results in parts (a) and (b), compare the pressure in the narrow and wide regions of the tube.

(d) Give a practical application where the principles of the Venturi tube are used or might be useful.

(*Note:* A similar application to the measure of flow speed is the Pitot tube which involves a right-angled tube introduced into a flow so that the pressure difference can be ascertained between a point within the main flow and a stagnation point. This gives a direct measure of the main flow speed. Originally designed to measure the speed of the River Seine in Paris, these devices are commonly used as a measure of the air speeds of aircraft; when these have been rendered inaccurate, due e.g. to icing, then loss of the aircraft can follow – as happened with the Air France 447 disaster in 2009.)

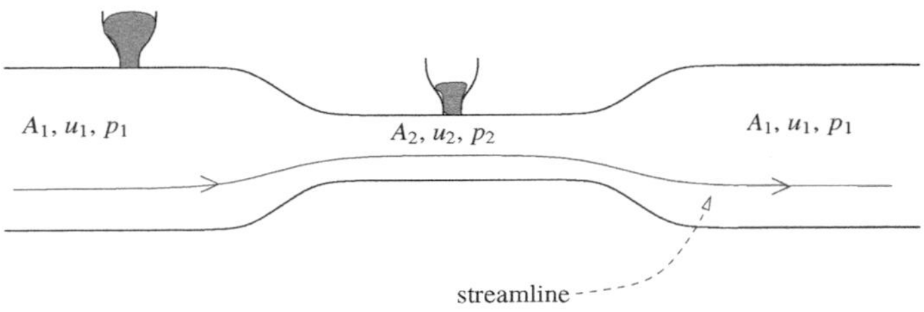

Figure 2.25 Venturi tube. The flow in the wider section of cross-sectional area A_1 has velocity u_1 and pressure p_1, while that in the narrower section of cross-sectional area A_2 has velocity u_2 and pressure p_2. Separately within the uniform wider and narrower sections, we assume the velocity and pressure are uniform. The two appendages shown are pressure sensors.

2.17 Bernoulli's Theorem: Clepsydra or water clock. A clepsydra has the form of a surface of revolution containing water and the level of the free surface of the water falls at a *constant* rate, as the water flows out through a small hole in the base. The basic setup is shown in Figure 2.26.

(a) Apply Bernoulli's Theorem to one of the typical streamlines shown in Figure 2.26 to show that

$$\tfrac{1}{2}\left(\frac{\mathrm{d}z}{\mathrm{d}t}\right)^2 = \tfrac{1}{2}U^2 - gz$$

where z is the height of the free surface above the small hole in the base, U is the velocity of the water coming out of the small hole and g is the acceleration due to gravity.

(b) If S is the cross-sectional area of the hole in the bottom, and A is the cross-sectional area of the free surface, explain why we must have

$$-A\frac{\mathrm{d}z}{\mathrm{d}t} = SU.$$

(c) Assuming that $S \ll A$, combine parts (a) and (b) to explain why we can deduce $U \sim \sqrt{2gz}$.

(d) Now combine the results from (b) and (c) above, to show that the shape of the container that guarantees that the free surface of the water drops at a constant rate must have the form $z = Cr^4$ in cylindrical polars, where C is a constant.

2.18 Bernoulli's Theorem: Tap flow. Consider water flowing from the nozzle of a tap of narrow circular cross-section of radius a; see Figure 2.27. We assume the water is emerging from the tap at a constant speed U uniformly across the cross-section. As is commonly observed, provided the flow is steady and not too fast, as the water leaves the tap nozzle, the cross-section of the flow remains circular but becomes narrower the further the distance from the nozzle. The question is, what is the formula for the radius r of the cross-section of the flow with distance z from the nozzle? Let $\rho > 0$ denote the uniform density of the water. Further, let u denote the vertical speed of the flow at distance z from the nozzle, which we assume to be uniform across the flow cross-section. Lastly we use P_0 to denote atmospheric pressure. We remark that this

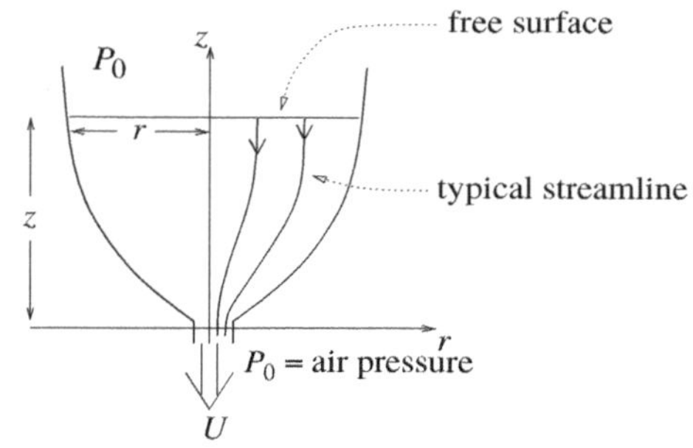

Figure 2.26 Clepsydra (water clock).

setup is simple, and a more careful analysis of the flow just as it emerges from the nozzle is ideally required. However, we can obtain a rough formula as follows.

(a) By applying Bernoulli's Theorem to a typical streamline along the surface of the flow, show that

$$\tfrac{1}{2}U^2 + \frac{P_0}{\rho} = \tfrac{1}{2}u^2 + \frac{P_0}{\rho} + gz,$$

and hence deduce that

$$\tfrac{1}{2}u^2 = \tfrac{1}{2}U^2 - gz.$$

(b) Explain why we know that $a^2U = r^2u$. Using this together with part (a), show that

$$r = a\left(1 - \frac{2gz}{U^2}\right)^{-1/4}.$$

2.19 Bernoulli's Theorem: Coffee in a mug. A coffee mug in the form of a right-circular cylinder (diameter $2a$, height h), closed at one end, is initially filled to a depth $d > \frac{1}{2}h$ with static inviscid coffee. Suppose the coffee is then made to rotate inside the mug – see Figure 2.28.

(a) Using the Euler equations for an ideal homogeneous incompressible flow in cylindrical polar coordinates, show for a flow which is independent of θ with $u_r = u_z = 0$ that the Euler equations reduce to

$$\frac{u_\theta^2}{r} = \frac{1}{\rho}\frac{\partial p}{\partial r} \qquad \text{and} \qquad 0 = -\frac{1}{\rho}\frac{\partial p}{\partial z} + g,$$

where $p = p(r, z)$ is the pressure, ρ is the constant uniform fluid density and g is the acceleration due to gravity (assume this to be the body force per unit mass). Verify that any such flow is indeed incompressible.

(b) Assume that the coffee in the mug is rotating as a solid body with constant angular velocity Ω so that the velocity component u_θ at a distance r from the axis of symmetry for $0 \leqslant r \leqslant a$ is

$$u_\theta = \Omega r.$$

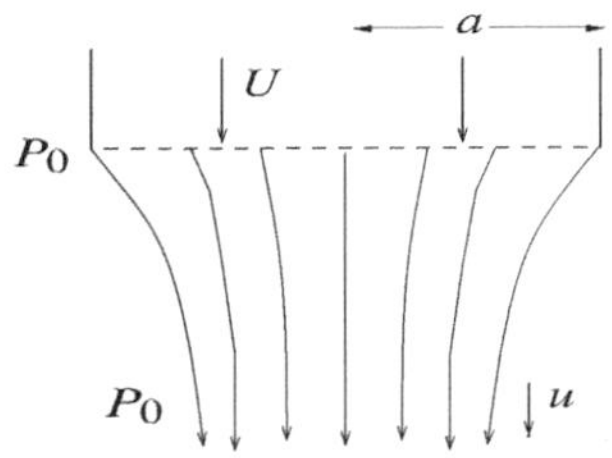

Figure 2.27 Water flowing out of a tap of circular cross-section.

Use the equations in part (a) to show that the pressure in this region is

$$p = \tfrac{1}{2}\rho\Omega^2 r^2 - g\rho z + C,$$

where C is an arbitrary constant. At the free surface between the coffee and air, the pressure is constant and equal to the atmospheric pressure P_0. Use this to show that the shape of the free surface has the form

$$z = \frac{\Omega^2}{2g} r^2 + \frac{C - P_0}{\rho g}.$$

(c) We are free to choose $C = P_0$ in the equation of the free surface, so it is described by $z = \Omega^2 r^2/2g$. This is equivalent to choosing the origin of our cylindrical coordinates to be the centre of the dip in the free surface. Suppose this origin is a distance z_0 from the bottom of the mug.

(i) Explain why the total volume of coffee is $\pi a^2 d$. Then, by using incompressibility, explain why the following constraint must be satisfied:

$$\pi a^2 z_0 + \int_0^a \frac{\Omega^2 r^2}{2g} \cdot 2\pi r \, dr = \pi a^2 d.$$

(ii) By computing the integral in the constraint in part (i), show some coffee will be spilled out of the mug if $\Omega^2 > 4g(h - d)/a^2$. Explain why this formula does not apply when the mug is initially less than half full.

2.20 Bernoulli's Theorem: Channel flow, Froude number. Recall the scenario of the steady channel flow over a gently undulating bed given in Example 2.26. Consider the case when the maximum permissible height y_0 compatible with the upstream conditions and the actual maximum height y_{max} of the undulation are exactly equal, i.e. $y_{max} = y_0$. The goal of this exercise is to show that the flow becomes locally critical immediately above y_{max} and, by a local expansion about that position, to show that there are subcritical and supercritical flows downstream consistent with the continuity and Bernoulli equations. Friction in a real flow leads to the latter being preferred. See Figure 2.29. Recall the general theory from Example 2.26. We saw there that in the case when the actual maximum value of the river bed undulation y_{max} does not reach the maximum possible bound y_0, i.e. $y_{max} < y_0$, when the Froude number $F < 1$, then $(d/dy)(h + y) < 0$, while when $F > 1$ then $(d/dy)(h + y) > 0$. In the case when $y_{max} = y_0$, these last two properties still hold upstream until y reaches it maximum $y_{max} = y_0$ – all the arguments in the text for the case $y_{max} < y_0$ apply until

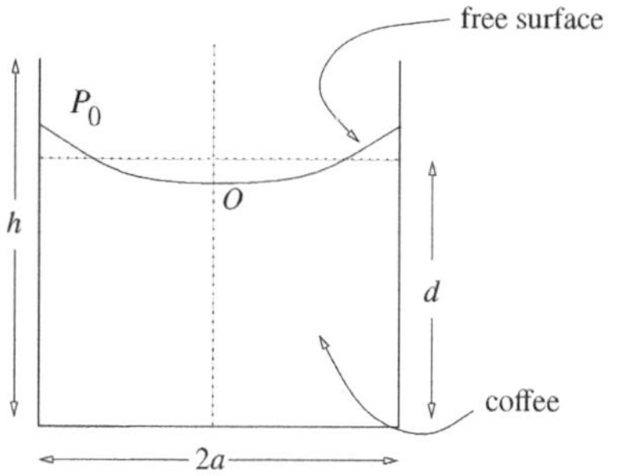

Figure 2.28 Coffee mug. We consider a mug of coffee of diameter $2a$ and height h, which is initially filled with coffee to a depth d. The coffee is then made to rotate about the axis of symmetry of the mug. The free surface between the coffee and the air takes up the characteristic shape shown, dipping down towards the middle (axis of symmetry). The goal is to specify the shape of the free surface.

y actually reaches its maximum. Recall the upper bound y_0 occurred when $h = h_0$, where $h_0 := (UH)^{2/3}/g^{1/3}$.

(a) Using the incompressibility condition $UH = uh$, so that at the maximum $y = y_0$ we have $UH = uh_0$, and the characterisation for h_0 just above, show that $u^2/(gh_0) = 1$ at $y = y_0$. Hence, since $\mathrm{d}h/\mathrm{d}y = -(1 - u^2/gh)^{-1}$, deduce that as $y \to y_0$ then $\mathrm{d}h/\mathrm{d}y \to \infty$.

(b) We know $\mathrm{d}y/\mathrm{d}x = 0$ at $y_{\max} = y_0$ so the question now is: what happens to the free surface at this point, in particular what is $\mathrm{d}h/\mathrm{d}x$? To answer this we need to expand locally and look at higher-order terms. Suppose $x = 0$ at the maximum value for $y = y(x)$ so $y(0) = y_{\max} = y_0$. Assume expansions for $y = y(x)$ and $h = h(x)$ about $x = 0$ as follows:

$$y = y_0 - Kx^2 + \cdots \quad \text{and} \quad h = h_0(1 + a_1x + a_2x^2 + \cdots),$$

where K, a_1 and a_2 are all constants. You can ignore terms of degree x^3 and higher. Note that $K > 0$, why is this?

(i) Use the Binomial expansion for $(1 + X)^{-2}$ with $X := a_1x + a_2x^2 + \cdots$ to expand h^{-2} and show that $h^{-2} = h_0^{-2}(1 - 2a_1x + (3a_1^2 - 2a_2)x^2 + \cdots)$.

(ii) Substitute the expansions above into the Bernoulli constraint $y = (U^2/2g) + H - h - ((UH)^2/(2gh^2))$. By equating the coefficients of powers of x show that the relation between the coefficients at x^0 reproduces the definition for the maximum value y_0, the relation between the coefficients at x^1 generates no new information, while the relation between the coefficients at x^2 implies $a_1 = \pm\sqrt{2K/3h_0}$.

(iii) Using the Binomial expansion around $x = 0$, show that

$$\frac{u^2}{gh} = \frac{(UH)^2}{gh_0^3}(1 - 3a_1x + \cdots).$$

Hence deduce that, locally, the flow is subcritical so $\mathrm{F} < 1$ if $a_1x > 0$, and supercritical so $\mathrm{F} > 1$ if $a_1x < 0$; see Figure 2.29. Friction forces the supercritical solution downstream in each case.

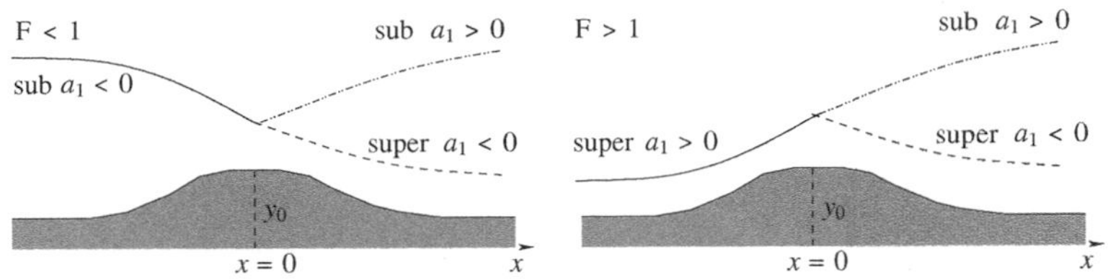

Figure 2.29 The cases for the two incident (upstream) Froude numbers are shown. The flow is subcritical if $a_1 x > 0$ and supercritical if $a_1 x < 0$. For a real flow, friction will force the supercritical solution downstream in each case. For example, when F > 1 upstream, the gradient of $h = h(x)$ will be discontinuous at $x = 0$. Also see Figure 2.12.

(*Note:* For flow over a broad-crested weir the upstream flow is subcritical, the flux Q adjusts so that the flow is locally critical immediately above the crest, and the original water level is not reattained in the supercritical flow downstream. Then we can show that $Q = (2/3)^{3/2}(gD^3)^{1/2}$, where D is the height difference between that original water level and the weir crest.)

2.21 Bernoulli's Theorem: Irrotational unsteady flow. Consider the Euler equations of motion for an ideal homogeneous incompressible fluid, with $\boldsymbol{u} = \boldsymbol{u}(\boldsymbol{x}, t)$ denoting the fluid velocity at position $\boldsymbol{x}$ and time t, ρ the uniform constant density, $p = p(\boldsymbol{x}, t)$ the pressure, and $\boldsymbol{f}$ denoting the body force per unit mass. Suppose that the flow is unsteady, but irrotational, i.e. we know that $\nabla \times \boldsymbol{u} = \boldsymbol{0}$ throughout the flow. This means that we know there exists a scalar potential function $\varphi = \varphi(\boldsymbol{x}, t)$ such that $\boldsymbol{u} = \nabla\varphi$. Also suppose that the body force is conservative so that $\boldsymbol{f} = -\nabla\Phi$ for some potential function $\Phi = \Phi(\boldsymbol{x}, t)$.

(a) Using the identity $\boldsymbol{u} \cdot \nabla \boldsymbol{u} = \frac{1}{2}\nabla(|\boldsymbol{u}|^2) - \boldsymbol{u} \times (\nabla \times \boldsymbol{u})$, show from Euler's equations of motion that the Bernoulli quantity

$$H := \frac{\partial \varphi}{\partial t} + \tfrac{1}{2}|\boldsymbol{u}|^2 + \frac{p}{\rho} + \Phi$$

satisfies $\nabla H = \boldsymbol{0}$ throughout the flow.

(b) From part (a) above we can deduce that H can only be a function of t throughout the flow, say $H = f(t)$ for some function f. By setting

$$V := \varphi - \int_0^t f(\tau)\,\mathrm{d}\tau,$$

show that the following Bernoulli quantity is constant throughout the flow:

$$\frac{\partial V}{\partial t} + \tfrac{1}{2}|\boldsymbol{u}|^2 + \frac{p}{\rho} + \Phi.$$

2.22 Bernoulli's Theorem: Vorticity, streamlines with gravity. A homogeneous incompressible inviscid fluid, under the influence of gravity, has the velocity field $\boldsymbol{u} = 2\alpha y\,\boldsymbol{i} - \alpha x\,\boldsymbol{j}$ with the z-axis, i.e. $\boldsymbol{k}$-direction, vertically upwards. Here α is constant. Naturally the density ρ is also constant. Verify that $\boldsymbol{u}$ satisfies the governing equations and find the pressure p. Show that the Bernoulli function $H = p/\rho + \frac{1}{2}|\boldsymbol{u}|^2 + \Phi$ is constant on streamlines and vortex lines, where Φ is the gravitational potential.

2.23 Kelvin's Circulation Theorem: Vorticity and streamlines. An inviscid incompressible fluid of uniform density ρ is in steady two-dimensional horizontal motion. Show that the Euler equations are equivalent to

$$\frac{\partial H}{\partial x} = v\omega \qquad \text{and} \qquad \frac{\partial H}{\partial y} = -u\omega,$$

where $H = p/\rho + \frac{1}{2}(u^2 + v^2)$ with p the dynamical pressure, $(u, v)^{\mathrm{T}}$ is the velocity field and ω is the vorticity. Deduce that ω is constant along streamlines and that this is in accord with Kelvin's Circulation Theorem.

2.24 Water waves: Particle paths. Recall the model equations for small-amplitude water waves we derived in Section 2.9. For an incompressible irrotational fluid (water), the velocity field potential $\phi = \phi(x, y, z, t)$ satisfies Laplace's equation $\Delta\phi = 0$ in the region $\mathbb{R}^2 \times [-h, 0]$, where $z = -h$ is the fluid bottom boundary. The boundary condition at $z = -h$ is $\partial\phi/\partial z = 0$. The free surface $z = \zeta(x, y, t)$ is constrained so the following two conditions are satisfied at $z = 0$ (respectively the Bernoulli and kinematic conditions):

$$\frac{\partial\phi}{\partial t} + g\zeta = 0 \qquad \text{and} \qquad \frac{\partial\phi}{\partial z} = \frac{\partial\zeta}{\partial t}.$$

Our goal is to describe the particle paths for planar waves in this context. Hence we look for solutions to this model of the form

$$\phi(x, z, t) = \varphi(x, t)\cosh(k(z + h)),$$

with a plane wave ansatz for the surface elevation:

$$\zeta(x, t) = \mathrm{Re}\big\{A\,\mathrm{e}^{\mathrm{i}(kx-\omega t)}\big\},$$

where $k \in \mathbb{R}$ is the wave-number, $\omega \in \mathbb{R}$ is the frequency and $A \in \mathbb{C}$ is a constant. Note that we assume without loss of generality the plane waves are aligned with the x-direction, and hereafter ignore all y-dependence.

(a) Use the second kinematic boundary condition above at $z = 0$ to show that $k\sinh(kh)\,\varphi(x, t) = \mathrm{Re}\{(-\mathrm{i}\omega)A\,\mathrm{e}^{\mathrm{i}(kx-\omega t)}\}$ and so

$$\phi(x, z, t) = \frac{\cosh(k(z + h))}{k\sinh(kh)}\,\mathrm{Re}\big\{(-\mathrm{i}\omega)A\,\mathrm{e}^{\mathrm{i}(kx-\omega t)}\big\}.$$

(b) Set $A = \alpha\mathrm{e}^{\mathrm{i}\beta}$, where $\alpha = |A|$ and $\beta = \arg(A)$. Explain why we can deduce that the velocity field components u and w, respectively in the x- and z-directions, are given by

$$u = \omega\alpha\,\frac{\cosh(k(z + h))}{\sinh(kh)}\,\cos(kx - \omega t + \beta)$$

and

$$w = \omega\alpha\,\frac{\sinh(k(z + h))}{\sinh(kh)}\,\sin(kx - \omega t + \beta).$$

(c) Assume a fluid particle has position $x(t) = \overline{x} + X(t)$ and $z(t) = \overline{z} + Z(t)$ where X and Z are small displacements from its mean position $(\overline{x}, \overline{z})$. Approximate the velocity of the particle in part (b) by replacing the positions x and z by their mean values $\overline{x}$ and $\overline{z}$, and hence show that

$$X(t) - X_0 = -\alpha \frac{\cosh(k(\overline{z} + h))}{\sinh(kh)} \sin(k\overline{x} - \omega t + \beta)$$

and

$$Z(t) - Z_0 = \alpha \frac{\sinh(k(\overline{z} + h))}{\sinh(kh)} \cos(k\overline{x} - \omega t + \beta).$$

(d) The equations for X and Z in part (c) are the parametric equations for an ellipse with semi-axes given by the factors $\alpha \cosh(k(\overline{z} + h))/\sinh(kh)$ and $\alpha \sinh(k(\overline{z} + h))/\sinh(kh)$. The particle motion is thus *local*.

(i) Show that as $z \to -h$ the vertical amplitude tends to zero and thus fluid particles at the bottom oscillate horizontally. This can be observed when snorkelling in wavy conditions close to shore!

(ii) What happens in the deep-water case when $kh \to \infty$?

2.25 Water waves: Group velocity. In Section 2.9, given a dispersion relation $\omega = \omega(k)$ for planar waves of wavenumber k, we defined the group velocity as the derivative of the dispersion relation, i.e. $c_g := d\omega/dk$. Note here that we restrict ourselves to planar waves travelling in, say, the x-direction. Suppose the surface elevation $\zeta = \zeta(x, t)$ is composed of two such planar waves of almost equal wavenumbers k and k' and thus corresponding frequencies ω and ω'. Thus assume we have

$$\zeta = A\cos(kx - \omega t) + A\cos(k'x - \omega' t).$$

The phase speed of the individual waves is approximately the same and given by $c_p := \omega/k \approx \omega'/k'$.

(a) Using the standard trigonometric identities for the cosine of the sum of two angles and so forth, show that

$$\zeta = 2A\cos\left(\tfrac{1}{2}(k + k')x - \tfrac{1}{2}(\omega + \omega')t\right)\cos\left(\tfrac{1}{2}(k - k')x - \tfrac{1}{2}(\omega - \omega')t\right).$$

(b) Observe from the form for ζ in part (a) that ζ is the product of a long wave of wavelength $2\pi/(\frac{1}{2}(k - k'))$ and a short wave of wavelength $2\pi/(\frac{1}{2}(k + k'))$. Explain why the speed of the modulating wave, with the long wavelength, which may not equal c_p in general, is given by

$$\frac{\omega - \omega'}{k - k'}.$$

(c) With reference to Figure 2.20, the modulating wave describes the dynamics of the envelope of the wave packet shown. And indeed in this case there are similar wave packets periodically adjacent to that shown in each direction. That the modulation speed is different to the phase speed means that waves move through the packet, in one end and out the other, from and into adjacent packets. In the limit $k' \to k$, the speed of the modulating wave approaches $d\omega/dk$, agreeing with our definition for, and representing the group velocity $c_g = d\omega/dk$. For deep-water waves show that $c_g = \frac{1}{2}c_p$ and explain the consequences of this effect for any disturbances on the water surface.

2.26 Water waves: Plane wave energy density. We assume a similar context to that in Exercise 2.24, except that we restrict ourselves here to plane waves in deep water so $\omega^2 = gk$ and also the velocity potential in Section 2.9 has the form $\phi(x, z, t) = \varphi(x, t)\mathrm{e}^{kz}$ with $k > 0$. In particular we assume the planar waves travel in the x-direction with the following plane wave ansatz for the surface elevation of wavenumber $k \in \mathbb{R}$:

$$\zeta(x, t) = \mathrm{Re}\big\{A\,\mathrm{e}^{\mathrm{i}(kx-\omega t)}\big\},$$

where $\omega \in \mathbb{R}$ is the frequency and $A \in \mathbb{C}$ is a constant.

(a) Use the kinematic boundary condition at $z = 0$ to show that

$$k\,\varphi(x, t) = \mathrm{Re}\big\{(-\mathrm{i}\omega)A\,\mathrm{e}^{\mathrm{i}(kx-\omega t)}\big\}.$$

Hence if $\beta := \arg(A)$, show that

$$\phi(x, z, t) = \frac{\mathrm{e}^{kz}}{k}\,\mathrm{Re}\big\{(-\mathrm{i}\omega)|A|\,\mathrm{e}^{\mathrm{i}(kx-\omega t+\beta)}\big\}.$$

(b) Show that the velocity field components u and w, respectively in the x- and z-directions, are given by

$$u = \mathrm{e}^{kz}\,\omega\,|A|\,\cos(kx - \omega t + \beta) \quad \text{and} \quad w = \mathrm{e}^{kz}\,\omega\,|A|\,\sin(kx - \omega t + \beta).$$

Hence show that $u^2 + w^2$ depends on z, k and $|A|$ only.

(c) Explain why the kinetic energy per unit surface area is

$$\tfrac{1}{2}\rho \int_{-\infty}^{0} (u^2 + w^2)\,\mathrm{d}z$$

and show, for the planar waves in (a) and (b) above that it equals $\frac{1}{4}\rho g|A|^2$.

(d) Explain why a suitable expression for the potential energy per unit surface area is given by

$$\rho g \int_{0}^{\zeta} z\,\mathrm{d}z = \tfrac{1}{2}\rho g \zeta^2.$$

Hence show that if ζ has the plane wave ansatz given above then the time average, over one cycle, of the potential energy is

$$\tfrac{1}{2}\rho g \left(\frac{2\pi}{\omega}\right)^{-1} \int_{0}^{2\pi/\omega} (\zeta(t))^2\,\mathrm{d}t = \tfrac{1}{4}\rho g|A|^2.$$

(*Note:* We thus deduce that the total energy is equally divided between the kinetic and potential energies; this is termed energy equipartition.)

2.27 Water waves: Energy flow rate. We assume a similar context to that in Exercises 2.24 and 2.26: we restrict ourselves to plane waves in deep water with $\omega^2 = gk$ and velocity potential of the form $\phi(x, z, t) = \varphi(x, t)\mathrm{e}^{kz}$, with $k > 0$. We assume the following plane wave ansatz for the surface elevation of wavenumber $k \in \mathbb{R}$:

$$\zeta(x, t) = \mathrm{Re}\big\{A\,\mathrm{e}^{\mathrm{i}(kx-\omega t)}\big\},$$

where $\omega \in \mathbb{R}$ is the frequency and $A \in \mathbb{C}$ is a constant.

(a) Explain why the rate of working (power) per unit area across a surface $x =$ constant is (here $p =$ pressure and $u =$ velocity in the x-direction):

$$\int_{-\infty}^{0} pu \, \mathrm{d}z.$$

(b) Use Bernoulli's Theorem for irrotational fluids that asserts $\partial\phi/\partial t + \frac{1}{2}|\boldsymbol{u}|^2 + p/\rho + gz$ is constant throughout the flow (see Exercise 2.21) to show that for small-amplitude disturbances we have

$$p = -\rho\left(\frac{\partial\phi}{\partial t} + gz\right).$$

(c) Use part (b) to show that the work rate in (a) is given by ($\beta = \arg(A)$):

$$-\rho\int_{-\infty}^{0}\left(\frac{\partial\phi}{\partial t}\frac{\partial\phi}{\partial x} + gz\frac{\partial\phi}{\partial x}\right)\mathrm{d}z = \tfrac{1}{2}\rho\frac{g}{k}|A|^2\omega\cos^2(kx - \omega t + \beta) + \rho\frac{g}{k^2}|A|\omega\cos(kx - \omega t + \beta).$$

(d) Use part (c) to show that the time-averaged work rate over one cycle is

$$\left(\frac{2\pi}{\omega}\right)^{-1}\int_0^{2\pi/\omega}(\text{work rate})\,\mathrm{d}t = \tfrac{1}{4}\rho g|A|^2\frac{\omega}{k}.$$

(e) Explain why, dimensionally, the velocity of energy transfer is given by the ratio of the work rate per unit area above to the energy density per unit area from Exercise 2.26. Hence show that this ratio is $\frac{1}{2}(\omega/k)$ and deduce the energy in the planar waves in this context travels at half the speed of the wave crests. Compare with Example 2.57 and comments thereafter.

2.28 Water waves: Group velocity upper bound. Recall from the beginning of Section 2.9 the governing equations for small-amplitude irrotational surface waves on water, which prior to disturbance is of uniform depth h. We assume here the flow is uniform in the horizontal y-direction, and so the velocity potential $\phi = \phi(x, z, t)$ and the free surface height $\zeta = \zeta(x, t)$ satisfy $\Delta_{x,z}\phi = 0$ for all $(x, z) \in \mathbb{R} \times [-h, 0]$ and at $z = -h$ we require $\partial\phi/\partial z = 0$. For all $x \in \mathbb{R}$ at $z = 0$ we also require the respective kinematic and Bernoulli conditions

$$\frac{\partial\phi}{\partial z} = \frac{\partial\zeta}{\partial t} \quad \text{and} \quad \frac{\partial\phi}{\partial t} + g\zeta = 0.$$

(a) By substituting the ansatz

$$\begin{pmatrix}\phi\\ \zeta\end{pmatrix} = \begin{pmatrix}\phi_0(z)\\ \zeta_0\end{pmatrix}\mathrm{e}^{\mathrm{i}(kx - \omega t)}$$

where k and ω are real constants, show that necessarily for some complex constant A, we have $\phi_0(z) = A\cosh(k(z + h))$. Further, show the Bernoulli boundary condition at $z = 0$ implies $\zeta_0 = \mathrm{i}(\omega/g)\,A\cosh(kh)$ whilst the kinematic condition at $z = 0$ necessitates the dispersion relation

$$\omega^2 = gk\tanh(kh).$$

(b) For the dispersion relation in part (a), show that the corresponding group velocity $c_g := \mathrm{d}\omega/\mathrm{d}k$ has the form

$$c_g = \tfrac{1}{2}(gh)^{1/2}\left(\frac{\tanh(kh)}{kh}\right)^{1/2}\left(1+\frac{2kh}{\sinh(2kh)}\right),$$

and hence deduce an upper bound for c_g is $(gh)^{1/2}$.

2.29 Water waves: Superposition. Assume the same setting as Exercise 2.28 just above and in particular note the results of part (a) in that exercise. Given $\zeta(x,0) = a_0 \sin^3(kx)$ for some constants a_0 and k, and $\partial\zeta/\partial t = 0$ at $t = 0$ also, use linear superposition to show with $\omega(k) = (gk\tanh(kh))^{1/2}$:

$$\zeta(x,t) = \tfrac{1}{4}a_0\big(3\sin(kx)\cos(\omega(k)t) - \sin(3kx)\cos(\omega(3k)t)\big).$$

(*Hint:* Note that $4\sin^3(\theta) \equiv 3\sin(\theta) - \sin(3\theta)$.)

2.30 Water waves: Surface tension. Our goal in this exercise is to include the effect of surface tension in Section 2.9 on water waves. The Young–Laplace equation (Young 1805, Laplace 1806) tells us that due to the *surface tension*, the pressure above the surface in the air is greater than the pressure in the fluid by the amount

$$\gamma\,(R_x^{-1} + R_y^{-1}),$$

where γ is the surface tension while R_x and R_y are the *principal radii of curvature* at the surface. Let us assume the same setting as Exercise 2.28, except now with the inclusion of surface tension. In particular though, assume the flow is uniform in the horizontal y-direction. As in Exercise 2.28, the velocity potential $\phi = \phi(x,z,t)$ and the free surface height $\zeta = \zeta(x,t)$ still satisfy $\Delta_{x,z}\phi = 0$ for all $(x,z) \in \mathbb{R}\times[-h,0]$. At $z = -h$ we still have $\partial\phi/\partial z = 0$ and for all $x \in \mathbb{R}$ at $z = 0$ the kinematic condition $\partial\phi/\partial z = \partial\zeta/\partial t$ still holds. The Bernoulli condition now needs modification as follows. In the original domain $[-h,\zeta]$, since the flow is irrotational, the Bernoulli energy density H is constant throughout. At $z = \zeta(x,t)$ we have

$$\rho\frac{\partial}{\partial t}\phi(x,z,t) + \rho g z + p_{\text{fluid}} = H,$$

where we neglect quadratic terms as previously and p_{fluid} is the pressure in the fluid at the surface boundary.

(a) Since we assume the flow is uniform in the y-direction, we have $R_y^{-1} = 0$. The radius of curvature R_x is given by

$$R_x = \left|\frac{(1+(\partial\zeta/\partial x)^2)^{3/2}}{\partial^2\zeta/\partial x^2}\right|.$$

Using this expression for the radius of curvature, and neglecting quadratic terms, show, if p_{air} is the pressure in the air just above the surface, that

$$p_{\text{air}} = p_{\text{fluid}} + \gamma\frac{\partial^2\zeta}{\partial x^2}.$$

(b) Combining part (a) with the Bernoulli condition above at $z = \zeta(x,t)$, use Taylor expansion, as explained at the beginning of Section 2.9, to show the Bernoulli condition can be approximated by the following boundary condition at $z = 0$:

$$\rho\frac{\partial\phi}{\partial t} + \rho g\zeta - \gamma\frac{\partial^2\zeta}{\partial x^2} = 0,$$

where we absorbed the constants H and p_{air} into the velocity potential.

(c) By looking for a solution of the form

$$\begin{pmatrix}\phi\\ \zeta\end{pmatrix} = \begin{pmatrix}\phi_0(z)\\ \zeta_0\end{pmatrix} \mathrm{e}^{\mathrm{i}(kx-\omega t)}$$

where k and ω are real constants, show necessarily for some complex constant A that we have $\phi_0(z) = A\cosh(k(z+h))$, exactly as in the first part of Exercise 2.28(b) above. Show the kinematic condition at $z = 0$ implies $\zeta_0 = (\mathrm{i}/\omega)AK\sinh(kh)$. Now use the Bernoulli boundary condition at $z = 0$ to show necessarily the following dispersion relation must hold:

$$\omega^2 = \left(gk + \frac{\gamma k^3}{\rho}\right)\tanh(kh).$$

(d) For deep-water waves, for which $kh \to \infty$, show that the phase speed $c_{\text{p}} = \omega/k$ has a minimum value at $k = k_1$ and the group velocity $c_{\text{g}} = \mathrm{d}\omega/\mathrm{d}k$ has a minimum value at $k = k_2$, where $c_{\text{g}}(k_1) = c_{\text{p}}(k_1)$, $k_2/k_1 = (2/\sqrt{3}-1)^{1/2}$ and $c_{\text{g}}(k_2)/c_{\text{p}}(k_1) = (\sqrt{3}-1)3^{3/8}/(2(2-\sqrt{3})^{1/4})$.

(e) Sketch ω, c_{p} and c_{g} as functions of the wavenumber k. Give a physical interpretation for what you observe.

2.31 Water waves: Interface between two fluids. Consider small-amplitude waves at the interface between two uniform fluids of infinite extent, when the undisturbed state is one of rest; see Figure 2.30, assume $U_1 \equiv U_2 \equiv 0$ therein. The fluids are assumed to be incompressible and irrotational and so in the respective regions, the fluid velocity potentials ϕ_1 and ϕ_2 satisfy $\Delta_{x,z}\phi_1 = 0$ and $\Delta_{x,z}\phi_2 = 0$, with $\phi_1 \to 0$ as $z \to +\infty$ and $\phi_2 \to 0$ as $z \to -\infty$. Continuity of the displacement implies the kinematic condition at the interface $z = \zeta(x,t)$:

$$\frac{\partial\phi_1}{\partial z} = \frac{\partial\zeta}{\partial t} = \frac{\partial\phi_2}{\partial z}.$$

As in Section 2.9 we can impose this condition at $z = 0$ by Taylor approximation. The corresponding Bernoulli condition, which again can be imposed by Taylor approximation at $z = 0$, becomes

$$\rho_1\left(\frac{\partial\phi_1}{\partial t} + g\zeta\right) = \rho_2\left(\frac{\partial\phi_2}{\partial t} + g\zeta\right).$$

Look for a solutions of the form (for $j = 1,2$)

$$\phi_j(x,z,t) = f_j(z)\,\mathrm{e}^{\mathrm{i}(kx-\omega t)} \qquad \text{and} \qquad \zeta(x,t) = \zeta_0\,\mathrm{e}^{\mathrm{i}(kx-\omega t)}.$$

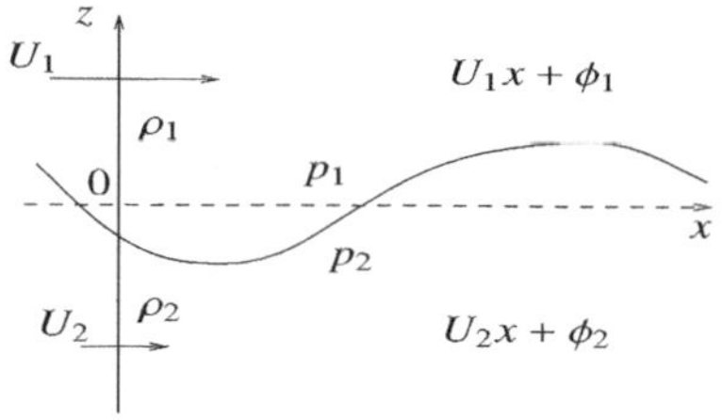

Figure 2.30 Small-amplitude waves at the interface between two uniform fluids of infinite extent, when the undisturbed state has the fluids moving with different horizontal speeds U_1 and U_2.

(a) Show that the dispersion relation is

$$\omega^2 = gk\left(\frac{\rho_2 - \rho_1}{\rho_2 + \rho_1}\right).$$

Explain why the waves are unstable if $\rho_1 > \rho_2$.

(b) Modify the Bernoulli condition at $z = 0$ to include surface tension as in Exercise 2.30 above. Show that the dispersion relation in this case is

$$\omega^2 = gk\left(\frac{\rho_2 - \rho_1}{\rho_2 + \rho_1}\right) + \frac{\gamma k^3}{\rho_2 + \rho_1}.$$

Show that if $\rho_1 > \rho_2$, instability now only occurs when $k^2 < g(\rho_1 - \rho_2)/\gamma$.

(*Note:* The instability is still present if we assume one or other fluid has finite extent. It is usually associated with Rayleigh (1900) and Taylor (1950), especially when the system is accelerated in the vertical direction, i.e. g is altered.)

2.32 Water waves: Interface between two moving fluids. Consider small-amplitude waves at the interface between two uniform fluids of infinite extent, when the undisturbed state has the fluids moving with different horizontal speeds U_1 and U_2; see Figure 2.30. As in Exercise 2.31 just above, the fluids are assumed to be incompressible and irrotational and so in the respective regions, the fluid velocity potentials ϕ_1 and ϕ_2 satisfy $\Delta_{x,z}\phi_1 = 0$ and $\Delta_{x,z}\phi_2 = 0$, with $\phi_1 \to 0$ as $z \to +\infty$ and $\phi_2 \to 0$ as $z \to -\infty$. Continuity of the displacement implies the kinematic conditions at the interface $z = \zeta(x, t)$:

$$\frac{\partial \phi_1}{\partial z} = \frac{\partial \zeta}{\partial t} + U_1 \frac{\partial \zeta}{\partial x} \qquad \text{and} \qquad \frac{\partial \phi_2}{\partial z} = \frac{\partial \zeta}{\partial t} + U_2 \frac{\partial \zeta}{\partial x}.$$

As in Exercise 2.31, we can impose these conditions at $z = 0$ by Taylor approximation. The corresponding Bernoulli condition, which again can be imposed by Taylor approximation at $z = 0$, becomes

$$\rho_1\left(\frac{\partial \phi_1}{\partial t} + U_1 \frac{\partial \phi_1}{\partial x} + g\zeta\right) = \rho_2\left(\frac{\partial \phi_2}{\partial t} + U_2 \frac{\partial \phi_2}{\partial x} + g\zeta\right).$$

As usual we have neglected quadratic terms in velocity displacements and absorbed constants into respective potentials. Look for a solution of the form (for $j = 1, 2$)

$$\phi_j(x, z, t) = f_j(z)\, \mathrm{e}^{\mathrm{i}(kx-\omega t)} \qquad \text{and} \qquad \zeta(x, t) = \zeta_0\, \mathrm{e}^{\mathrm{i}(kx-\omega t)}.$$

Show that the dispersion relation is

$$(\rho_1 + \rho_2)\left(\frac{\omega}{k}\right)^2 - 2(\rho_1 U_1 + \rho_2 U_2)\left(\frac{\omega}{k}\right) + \rho_1 U_1^2 + \rho_2 U_2^2 = (\rho_2 - \rho_1)\frac{g}{k},$$

and that waves which are short enough are unstable.

(*Note:* This short-wave instability is known as the *Kelvin–Helmholtz instability* due to Helmholtz (1868) for $\rho_1 = \rho_2$ and Kelvin (1871) for $\rho_2 > \rho_1$. Again the effect of surface tension is to tend to reduce the parameter regimes of instability. Also see Majda and Bertozzi (2002, Sec. 9.3).)

3 Two-Dimensional Irrotational Flow

3.1 Potential Solution

Herein we focus on two-dimensional irrotational flows which have a rich structure in the sense that they are intimately connected to complex variable theory and analysis. We assume the fluid flows within the region $\mathcal{D} \subseteq \mathbb{R}^2$. The flow is assumed to be incompressible so $\nabla \cdot \boldsymbol{u} = 0$, as well as irrotational in $\mathcal{D}$ so

$$\nabla \times \boldsymbol{u} = \boldsymbol{0}.$$

As we have seen, the latter condition implies that there exists a scalar potential function ϕ such that $\boldsymbol{u} = \nabla\phi$. This representation combined with incompressibility implies ϕ is *harmonic*, i.e. satisfies Laplace's equation $\Delta\phi = 0$. In the two-dimensional context here we suppose the plane polar coordinates (r, θ) parameterise $\mathcal{D} \subseteq \mathbb{R}^2$ so that Laplace's equation for $\phi = \phi(r, \theta)$ has the form

$$\frac{1}{r}\frac{\partial}{\partial r}\left(r\frac{\partial \phi}{\partial r}\right) + \frac{1}{r^2}\frac{\partial^2 \phi}{\partial \theta^2} = 0.$$

This linear partial differential equation can be solved by the method of separation of variables, revealing that the general solution is of the form

$$\phi = (A + B\log r)(C + D\theta) + \sum_{n=1}^{\infty}\left(A_n r^n + \frac{B_n}{r^n}\right)(C_n \cos(n\theta) + D_n \sin(n\theta)),$$

where the coefficients A, B, C, D, A_n, B_n, C_n and D_n for all $n \in \mathbb{N}$ are constants. We can immediately interpret some of the solution terms shown. For example, the term '$\log r$' represents a line source; this is revealed by computing the gradient of this term in polar coordinates. As we saw in Example 2.48, the term 'θ' represents a line vortex at the origin – when we imagine extending the planar flow uniformly in the orthogonal direction. The terms '$r\cos\theta$' and '$r\sin\theta$' generate uniform flow fields – we saw the former in D'Alembert's Theorem 2.49. The terms '$r^{-1}\cos\theta$' and '$r^{-1}\sin\theta$' generate line dipole (or doublet) flow fields. Let us now focus on the rich structure attributable to such flows we alluded to above.

3.2 Complex Potential Functions

If (x, y) are Cartesian coordinates parameterising $\mathcal{D}$, then since the flows herein are two-dimensional, we can represent the flow via a *stream function* $\psi = \psi(x, y)$ in addition to a potential function $\phi = \phi(x, y)$. In this circumstance we see that $\boldsymbol{u} = (u, v)^{\mathrm{T}}$ is given by

$$\begin{pmatrix} \partial\phi/\partial x \\ \partial\phi/\partial y \end{pmatrix} \equiv \begin{pmatrix} \partial\psi/\partial y \\ -\partial\psi/\partial x \end{pmatrix}.$$

These are the Cauchy–Riemann equations for the functions ϕ and ψ.

Given the complex variable $z = x + \mathrm{i}y$, we set the complex-valued function $w = w(z)$ to be

$$w(z) := \phi(x, y) + \mathrm{i}\psi(x, y).$$

Necessary and sufficient conditions for w to be (complex) differentiable at z are that ϕ and ψ satisfy the Cauchy–Riemann equations (Burkill and Burkill, 1970, p. 339)

$$\frac{\partial\phi}{\partial x} = \frac{\partial\psi}{\partial y} \qquad \text{and} \qquad \frac{\partial\phi}{\partial y} = -\frac{\partial\psi}{\partial x}.$$

Definition 3.1 (Complex derivative) The derivative of $w = w(z)$ at $z \in \mathbb{C}$ is uniquely defined by

$$w'(z) := \lim_{h \to 0} \frac{w(z + h) - w(z)}{h},$$

independent of the direction $h \in \mathbb{C}$ approaches 0 in the complex plane.

If ϕ and ψ satisfy the Cauchy–Riemann equations for all $z \in \mathcal{D}$, or equivalently $w = \phi + \mathrm{i}\psi$ is complex differentiable for all $z \in \mathcal{D}$, then we say w is *analytic* in $\mathcal{D}$. Note that if $w = w(z)$ is analytic in $\mathcal{D}$, then its derivative $w'(z)$ is continuous and indeed derivatives of w of every order exist and are analytic in $\mathcal{D}$, and in particular ϕ and ψ are harmonic in $\mathcal{D}$: $\Delta\phi = 0$ and $\Delta\psi = 0$. See e.g. Burkill and Burkill, (1970, p. 375).

Thus any two-dimensional incompressible, irrotational sufficiently smooth flow can be expressed in terms of a potential function ϕ which satisfies Laplace's equation $\Delta\phi = 0$ as well as a stream function ψ which also satisfies Laplace's equation $\Delta\psi = 0$ (since the third component of the curl of $\boldsymbol{u}$ given by $\partial v/\partial x - \partial u/\partial y$ is zero). The pair of functions satisfy the Cauchy–Riemann equations and thus there exists a *complex potential* function $w(z) := \phi(x, y) + \mathrm{i}\psi(x, y)$ which is complex differentiable. And the reverse is true (see Lemma 3.2 below): a function w analytic in $\mathcal{D}$ generates a two-dimensional incompressible, irrotational flow when we set $\phi := \mathrm{Re}\{w(z)\}$ and $\psi := \mathrm{Im}\{w(z)\}$. Whether the flow is 'physical' is another matter! Note that naturally the functions ϕ and ψ defined in this way are harmonic. Further, since the derivative w' is uniquely defined independent of the direction of approach to the origin in the difference quotient, we have

$$w' = \frac{\partial}{\partial x}(\phi + \mathrm{i}\psi) = \frac{\partial\phi}{\partial x} + \mathrm{i}\frac{\partial\psi}{\partial x} = u - \mathrm{i}v$$

or

$$w' = \frac{\partial}{\partial(\mathrm{i}y)}(\phi + \mathrm{i}\psi) = -\mathrm{i}\frac{\partial\phi}{\partial y} + \frac{\partial\psi}{\partial y} = -\mathrm{i}v + u.$$

Hence if we represent the velocity field $\boldsymbol{u} = (u, v)^{\mathrm{T}}$ by $u + \mathrm{i}v$, then we observe

$$u + \mathrm{i}v \equiv (w')^*,$$

the complex conjugate of w'. The flow speed is $|w'|$ while the velocity direction is given by '$-\arg(w')$'. Let us summarise the statements we've accrued thus far.

Lemma 3.2 (Complex potential flows) *Any two-dimensional incompressible, irrotational sufficiently smooth steady solution of the Euler equations can be expressed in terms of a complex differentiable function $w(z) := \phi(x, y) + \mathrm{i}\psi(x, y)$ with the components u and v of the velocity field given by $u + \mathrm{i}v \equiv (w')^*$. Further, any complex differentiable function $w = w(z)$ generates an analytic irrotational steady solution to the two-dimensional incompressible Euler equations.*

Proof We establish the first statement, as outlined above, by identifying ϕ and ψ respectively as the potential and stream function for the flow. For the second 'reverse' statement, given a complex differentiable function $w = w(z)$ with $z = x + \mathrm{i}y$, we identify $\phi := \mathrm{Re}\{w(z)\}$ and $\psi := \mathrm{Im}\{w(z)\}$, and by the standard complex function theory outlined above we note that $\Delta\phi = 0$ and $\Delta\psi = 0$. Hence naturally we have $J(\psi, \Delta\psi) := (\nabla^{\perp}\psi) \cdot \nabla(\Delta\psi) = 0$, where recall that the orthogonal derivative $\nabla^{\perp}$ is defined so that $\nabla^{\perp}\upsilon := (\partial\upsilon/\partial y, -\partial\upsilon/\partial x)^{\mathrm{T}}$ for any scalar field $\upsilon = \upsilon(x, y)$. Hence we can use the arguments given in the proof of Lemma 4.18 for the vorticity-stream formulation of the Euler/Navier–Stokes equations. Indeed, if we set $\boldsymbol{u} = \nabla^{\perp}\psi$ then the statement $J(\psi, \Delta\psi) = 0$ is equivalent to $\boldsymbol{u} \cdot \nabla\omega = 0$ where $\omega := \nabla^{\perp} \cdot \boldsymbol{u}$. In turn this is equivalent to the statement $\nabla^{\perp} \cdot ((\boldsymbol{u} \cdot \nabla)\,\boldsymbol{u}) = 0$. The latter condition implies that there exists a scalar function $p = p(\boldsymbol{x}, t)$ such that $(\boldsymbol{u} \cdot \nabla)\,\boldsymbol{u} = -\nabla p/\rho$, for some constant $\rho > 0$. □

Remark 3.3 In this chapter we consider steady flows only.

Remark 3.4 (Equipotential lines and streamlines are orthogonal) We observe that (recall Remark 1.22)

$$\begin{pmatrix} \partial\psi/\partial x \\ \partial\psi/\partial y \end{pmatrix} \cdot \begin{pmatrix} \partial\phi/\partial x \\ \partial\phi/\partial y \end{pmatrix} = \nabla\psi \cdot \boldsymbol{u} = \begin{pmatrix} \partial\psi/\partial x \\ \partial\psi/\partial y \end{pmatrix} \cdot \begin{pmatrix} \partial\psi/\partial y \\ -\partial\psi/\partial x \end{pmatrix} = 0.$$

In other words, *equipotential* lines and streamlines are orthogonal.

Example 3.5 (Uniform flow) Suppose $w = Az$ where $A = A_1 + \mathrm{i}A_2$ and A_1 and A_2 are real constants. Then the velocity field is given by $u + \mathrm{i}v = (w')^* = A^* = A_1 - \mathrm{i}A_2 = (A_1^2 + A_2^2)^{1/2}\mathrm{e}^{\mathrm{i}\alpha}$, where $\tan\alpha = -A_2/A_1$. The velocity field is thus constant (uniform) with the fluid flowing in the direction given by the angle α. Computing the real and imaginary parts of w we observe $\phi = A_1x - A_2y \equiv r(A_1\cos\theta - A_2\sin\theta)$ and $\psi = A_1y + A_2x$.

Example 3.6 (Wedge flow) Suppose $w = Bz^N$ where $B > 0$ and N are real constants. Writing $z = re^{i\theta}$ and using Euler's formula, we observe

$$w = \phi + i\psi = Br^N e^{iN\theta} = Br^N \cos(N\theta) + iBr^N \sin(N\theta).$$

The velocity field is given by $u+iv = (w')^* = (NBz^{N-1})^*$. Assume $\psi = 0$ on $\theta = 0$ and $\theta = \pi/N$. If $N > 1$ then $w = Bz^N$ with these boundary conditions thus corresponds to a flow in an acute or obtuse wedge and $z = 0$ is a stagnation point. If $1/2 < N < 1$ then $w = Bz^N$ corresponds to a flow around a reflexive wedge with the velocity field singular at $z = 0$; see Figure 3.1.

Example 3.7 (Line source/sink) Suppose $w = (m/2\pi)\log z$, where m is a real constant. If we write $z = re^{i\theta}$ then using the expression for the logarithm of a complex number we observe that $\phi = (m/2\pi)\log r$ and $\psi = (m/2\pi)\theta$. The corresponding velocity field is given by

$$u + iv = \frac{m}{2\pi}\frac{1}{z^*} = \frac{m}{2\pi r}(\cos\theta + i\sin\theta).$$

This corresponds to the two-dimensional velocity field $\boldsymbol{u} = (m/2\pi r)\hat{\boldsymbol{r}}$. This is the velocity field corresponding to a line source of strength m: the flux of fluid through a circle of radius r_0, where the speed '$m/2\pi r_0$' is constant and in the direction of the outward normal (if $m > 0$), is given by $2\pi r_0 \times (m/2\pi r_0) = m$. A source of negative strength is often called a sink. See Figure 3.2 (left panel – illustrated with $m > 0$).

Example 3.8 (Line dipole) Suppose μ is a real constant and the complex potential $w = w(z)$ is given by (on the right we use the representation $z = re^{i\theta}$)

$$w = \frac{\mu}{2\pi}\frac{1}{z} \qquad \Leftrightarrow \qquad \phi = \frac{\mu}{2\pi r}\cos\theta \quad \text{and} \quad \psi = -\frac{\mu}{2\pi r}\sin\theta.$$

The corresponding velocity field is given by

$$u + iv = -\frac{\mu}{2\pi}\left(\frac{1}{z^2}\right)^* = -\frac{\mu}{2\pi r^2}(\cos(2\theta) + i\sin(2\theta)).$$

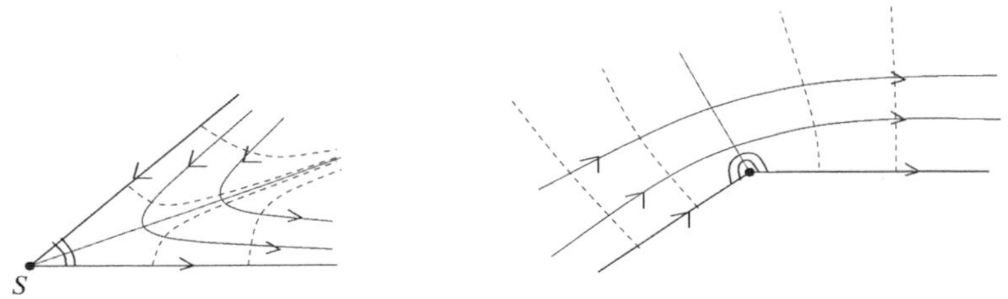

Figure 3.1 The two possible types of wedge flow from Example 3.6. The complex potential is $w = Bz^N$ and $\psi = \text{Im}\{w\}$ is zero on the straight edges shown. On the left $N > 1$, corresponding to a flow in a wedge with a stagnation point at the origin/apex. On the right $1/2 < N < 1$, corresponding to a flow around a wedge with the flow velocity infinite at the corner/origin. Dashed lines are equipotential curves.

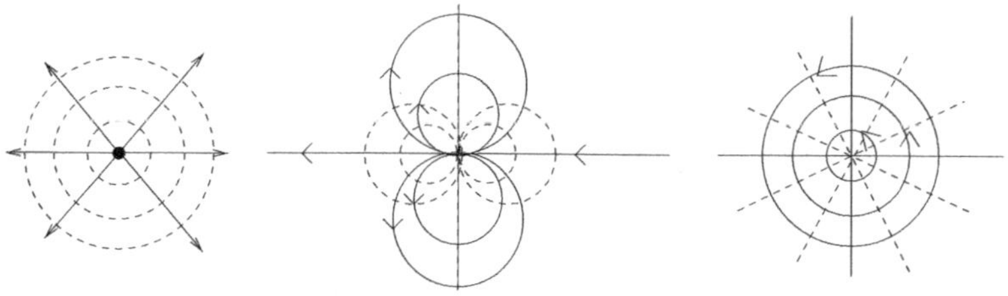

Figure 3.2 The left panel shows the flow field for the line source from Example 3.7. The middle panel shows the flow field corresponding to the line dipole or doublet from Example 3.8, with the flow on the x-axis directed towards $x = -\infty$. The right panel shows the flow field for the line vortex from Example 3.11. Dashed lines are equipotential curves.

This corresponds to the two-dimensional velocity field

$$\boldsymbol{u} = -\frac{\mu}{2\pi r^2}(\cos(\theta)\hat{\boldsymbol{r}} + \sin(\theta)\hat{\boldsymbol{\theta}}),$$

where we used that $\hat{\boldsymbol{r}} = \cos\theta\boldsymbol{i} + \sin\theta\boldsymbol{j}$ and $\hat{\boldsymbol{\theta}} = -\sin\theta\boldsymbol{i} + \cos\theta\boldsymbol{j}$. This velocity field corresponds to a line dipole or doublet directed towards the negative x-axis. Note that the streamlines $\psi = c$, with c a real constant, correspond to

$$\begin{aligned} & \frac{\mu}{2\pi r}\sin\theta = c \\ \Leftrightarrow\quad & \frac{\mu}{2\pi c}r\sin\theta = r^2 \\ \Leftrightarrow\quad & \frac{\mu}{2\pi c}y = x^2 + y^2 \\ \Leftrightarrow\quad & \frac{\mu^2}{16\pi^2c^2} = x^2 + \left(y - \frac{\mu}{4\pi c}\right)^2. \end{aligned}$$

As c varies, these curves correspond to circles with differing radii, centred at $\mu/4\pi c$ on the y-axis; see Figure 3.2 (middle panel – illustrated with $\mu > 0$).

Remark 3.9 If $\mu = \mu_0 e^{i\alpha}$ with $\mu_0 > 0$ and $\alpha \in \mathbb{R}$ then $w = \mu_0/(2\pi z e^{-i\alpha})$, corresponding to the line dipole above rotated counter-clockwise by angle α.

Remark 3.10 Suppose we had a collection of line sources represented by the complex potential function

$$w(z) = \frac{1}{2\pi}\sum_{k=1}^{N} m_k \log(z - z_k),$$

i.e. with line sources of strength m_k at positions z_k for $k = 1, \ldots, N$. Far from the sources, i.e. where $|z_k/z|$ is asymptotically small for all k, we can expand

$$w(z) \sim \frac{M}{2\pi}\log z - \frac{1}{2\pi z}\left(\sum_{k=1}^{N} m_k z_k\right) + O(|z|^{-2}),$$

where $M := \sum_{k=1}^{N} m_k$. The leading-order terms shown represent an effective line source, strength M, modified by an effective line dipole, strength $\sum_{k=1}^{N} m_k z_k$.

Example 3.11 (Line vortex) Consider the complex potential $w = w(z)$:

$$w = -\frac{\mathrm{i}\kappa}{2\pi}\log z \qquad \Leftrightarrow \qquad \phi = \frac{\kappa\theta}{2\pi} \quad \text{and} \quad \psi = -\frac{\kappa}{2\pi}\log r,$$

where κ is a real constant and we used that $z = r\mathrm{e}^{\mathrm{i}\theta}$ as well as the expression for the logarithm of $z \in \mathbb{C}$. The corresponding velocity field is given by

$$u + \mathrm{i}v = \left(-\frac{\mathrm{i}\kappa}{2\pi}\frac{1}{z}\right)^* = \frac{\kappa}{2\pi r}(-\sin\theta + \mathrm{i}\cos\theta).$$

Since $\hat{\boldsymbol{\theta}} = -\sin\theta\boldsymbol{i} + \cos\theta\boldsymbol{j}$ we observe that $\boldsymbol{u} = (\kappa/2\pi r)\hat{\boldsymbol{\theta}}$. This velocity field corresponds to a line vortex of strength κ centred at the origin; see Figure 3.2 (right panel – illustrated with $\kappa > 0$) and Example 2.48. Note that $\boldsymbol{u}$ and $\nabla \times \boldsymbol{u}$ are singular at the origin. The region $r > 0$ is not simply connected and the circulation for a closed circuit equals $2\pi\kappa N$, where N is the winding number of the circuit, i.e. the number of times the circuit winds round the origin counter-clockwise (Burkill and Burkill, 1970, p. 370).

Remark 3.12 We can obtain the complex potential function from velocity field components expressed in terms of plane polar coordinates directly as follows. If u and v represent the velocity field components in the Cartesian positive x- and y-directions, respectively, then by projecting components we know

$$u = u_r\cos\theta - u_\theta\sin\theta \qquad \text{and} \qquad v = u_r\sin\theta + u_\theta\cos\theta$$

where u_r and u_θ are the velocity field components in the $\hat{\boldsymbol{r}}$ and $\hat{\boldsymbol{\theta}}$ directions, respectively. Hence, using Euler's identity we see that $w'(z)$ is given by

$$w'(z) = u - \mathrm{i}v = \mathrm{e}^{-\mathrm{i}\theta}(u_r - \mathrm{i}u_\theta) = \mathrm{e}^{-\mathrm{i}\theta}\left(\frac{\partial\phi}{\partial r} - \mathrm{i}\frac{1}{r}\frac{\partial\phi}{\partial\theta}\right) = \mathrm{e}^{-\mathrm{i}\theta}\left(\frac{1}{r}\frac{\partial\psi}{\partial\theta} + \mathrm{i}\frac{\partial\psi}{\partial r}\right),$$

where we used the expressions for the potential function as well as the stream function in plane polar coordinates (see Section 1.8). We can thus express the complex potential function in the form $w(z) = w(r\mathrm{e}^{\mathrm{i}\theta}) = \phi(r,\theta) + \mathrm{i}\psi(r,\theta)$.

Remark 3.13 (Computing the pressure) Consider an ideal homogeneous incompressible stationary flow with no body force. If the flow is also irrotational, then the energy density $p + \frac{1}{2}\rho|\boldsymbol{u}|^2$ is constant throughout the flow; see Exercise 2.21. This means that we can compute the pressure p by computing the energy density $p_\infty + \frac{1}{2}\rho U_\infty^2$ in the far field or by computing the energy density p_S at a stagnation point. For example in the latter case, $p = p_S - \frac{1}{2}\rho|w'(z)|^2$.

3.3 Dynamic Lift

We continue to consider homogeneous two-dimensional irrotational flows. Herein we focus on the complex potential $w = w(z)$ given by

$$w = Uz + \frac{\mu}{2\pi}\frac{1}{z} - \frac{\mathrm{i}\kappa}{2\pi}\log z,$$

or equivalently

$$\begin{aligned}\phi &= Ur\cos\theta + \frac{\mu}{2\pi r}\cos\theta + \frac{\kappa\theta}{2\pi},\\ \psi &= Ur\sin\theta - \frac{\mu}{2\pi r}\sin\theta - \frac{\kappa}{2\pi}\log r,\end{aligned}$$

where U, μ and κ are real constants. The first term on the right-hand side in each of the three expressions above represents uniform flow, while the second terms represent a line dipole and the third terms a line vortex – compare with all the examples above. Using our observations in Remark 3.12, Euler's formula and $z = re^{\mathrm{i}\theta}$, we note that in plane polar coordinates we can write

$$\begin{aligned}u_r - \mathrm{i}u_\theta &= \mathrm{e}^{\mathrm{i}\theta}w'(z)\\ &= \mathrm{e}^{\mathrm{i}\theta}\left(U - \frac{\mu}{2\pi}\frac{1}{z^2} - \frac{\mathrm{i}\kappa}{2\pi}\frac{1}{z}\right)\\ &= \left(U - \frac{\mu}{2\pi r^2}\right)\cos\theta - \mathrm{i}\left(\frac{\kappa}{2\pi r} - \left(U + \frac{\mu}{2\pi r^2}\right)\sin\theta\right).\end{aligned}$$

To ensure the normal velocity field at $r = a$ is zero, i.e. $u_r(a, \theta) = 0$ for all $\theta \in [0, 2\pi)$, we observe that we must have $\mu = 2\pi Ua^2$. This corresponds to the stream function having the constant value $\psi = (-\kappa/2\pi)\log a$ on $r = a$. Substituting this choice for μ into the complex potential function above generates

$$w = Uz + \frac{Ua^2}{z} - \frac{\mathrm{i}\kappa}{2\pi}\log z,$$

which is the complex potential function for a uniform flow past a circular cylinder, of radius a, with circulation κ. The flow field in this case is given by

$$u_r - \mathrm{i}u_\theta = U\left(1 - \frac{a^2}{r^2}\right)\cos\theta - \mathrm{i}\left(\frac{\kappa}{2\pi r} - U\left(1 + \frac{a^2}{r^2}\right)\sin\theta\right).$$

Remark 3.14 Recall the setup for D'Alembert's Theorem 2.49. In that case we used spherical polar coordinates and assumed axisymmetry. The solution potential function ϕ there satisfies the uniform flow far-field boundary conditions together with the no normal-flow boundary condition on $r = a$. We see that the function form for ϕ in that case consists of a uniform flow past a solid sphere of radius a. In particular, a term analogous to the line vortex term with circulation κ present in the complex potential function above is not present in the corresponding potential function ϕ in D'Alembert's example.

Let us determine the stagnation points for this velocity flow field, i.e. where u_r and u_θ are simultaneously zero. First we observe that $u_r = 0$ iff $\cos\theta = 0$ and/or $r = a$. In either case we observe:

(a) If $r = a$, then $u_\theta = 0$ iff $\sin\theta = \kappa/(4\pi a U)$ on the cylinder surface. Two such angles exist if $0 \leqslant |\kappa/(4\pi a U)| < 1$ and they coincide if $|\kappa/(4\pi a U)| = 1$.

(b) If $\cos\theta = 0$ then either:

(i) $\theta = \pi/2$ and $u_\theta = 0$, which holds iff

$$\frac{\kappa}{2\pi r} = U\left(1 + \frac{a^2}{r^2}\right) \qquad \Leftrightarrow \qquad \frac{r}{a} = \frac{\kappa}{4\pi a U} \pm \left(\frac{\kappa^2}{16\pi^2 a^2 U^2} - 1\right)^{1/2},$$

where necessarily $\kappa > 0$. The positive square root in this quadratic equation formula represents the solution for $r/a > 1$, i.e. outside the cylinder, while the negative square root is the solution for $r/a < 1$, i.e. inside the cylinder; or

(ii) $\theta = 3\pi/2$ and $u_\theta = 0$, which holds iff

$$\frac{\kappa}{2\pi r} = -U\left(1 + \frac{a^2}{r^2}\right) \qquad \Leftrightarrow \qquad \frac{r}{a} = -\frac{\kappa}{4\pi a U} \pm \left(\frac{\kappa^2}{16\pi^2 a^2 U^2} - 1\right)^{1/2},$$

where necessarily $\kappa < 0$. Again the positive square root represents the solution for $r/a > 1$ and the negative square root is the solution for $r/a < 1$.

We are interested in the region $r > a$. See Figure 3.3 for a comparison of the different velocity flow fields around the cylinder for different values of the circulation $\kappa \geqslant 0$. Note that the circulation acts with the flow below the cylinder and against the flow above the cylinder, generating the stagnation points shown.

Let us now compute the overall force acting on the cylinder. Recall from Remark 3.13 that for irrotational flows the energy density $p + \frac{1}{2}\rho|\boldsymbol{u}|^2$ is constant throughout the flow. Hence the energy density at the cylinder surface equals that in the far field which implies (we denote the cylinder surface by C)

$$p_C + \frac{1}{2}\rho|\boldsymbol{u}_C|^2 = p_\infty + \frac{1}{2}\rho U^2,$$

where p_C and $\boldsymbol{u}_C$ are the pressure and velocity field, respectively, on the cylinder surface C and p_∞ the constant pressure in the far field. On the cylinder surface the velocity field has no normal component and is given by

$$|\boldsymbol{u}_C|^2 = u_\theta^2 = \left(\frac{\kappa}{2\pi a} - 2U\sin\theta\right)^2,$$

where we used the expression for u_θ we computed above, evaluated at $r = a$. Hence we observe

$$p_C = p_\infty + \frac{1}{2}\rho U^2 - \frac{1}{2}\rho|\boldsymbol{u}_C|^2 = c + \frac{\rho\kappa U}{\pi a}\sin\theta - 2\rho U^2\sin^2\theta,$$

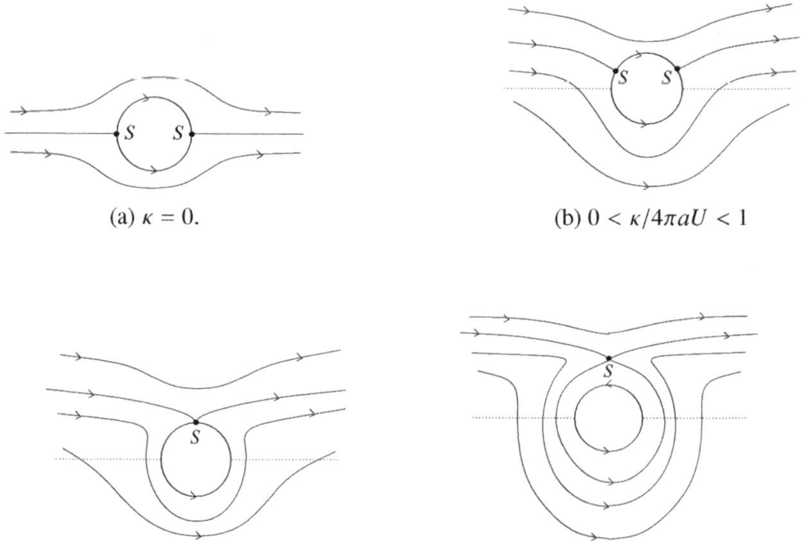

(a) $\kappa = 0$. (b) $0 < \kappa/4\pi aU < 1$

(c) $\kappa/4\pi aU = 1$ (d) $\kappa/4\pi aU > 1$

Figure 3.3 The panels show the flow field corresponding to the complex potential $w = Uz + Ua^2z^{-1} - (\mathrm{i}\kappa/2\pi)\log z$ for different values of the circulation $\kappa \geqslant 0$. We infer similar flow fields, reflected in the horizontal line through the cylinder centre, when $\kappa < 0$.

where the constant $c := p_\infty + \frac{1}{2}\rho U^2 - \rho\kappa^2/8\pi^2a^2$. Using this expression for the pressure at the cylinder surface we can compute the force components in the downstream $\boldsymbol{i}$ and the sideways $\boldsymbol{j}$ (or $\theta = \pi/2$) directions. Note that we compute the force per unit length with respect to the longitudinal direction along the axis of the cylinder. Recall from Sections 2.1 and 2.2 that the total force per unit longitudinal length, since the unit normal at the cylinder surface is $\hat{\boldsymbol{r}}$, is

$$
\begin{aligned}
\boldsymbol{F} &= -\int_0^{2\pi} p_C(\theta)\hat{\boldsymbol{r}}\; a\mathrm{d}\theta \\
&= -\int_0^{2\pi} p_C(\theta)(\boldsymbol{i}\cos\theta + \boldsymbol{j}\sin\theta)\; a\mathrm{d}\theta \\
&= -\boldsymbol{i}\int_0^{2\pi}\left(\frac{\rho\kappa U}{\pi a}\sin\theta\cos\theta - 2\rho U^2\sin^2\theta\cos\theta\right) a\mathrm{d}\theta \\
&\quad -\boldsymbol{j}\int_0^{2\pi}\left(\frac{\rho\kappa U}{\pi a}\sin^2\theta - 2\rho U^2\sin^3\theta\right) a\mathrm{d}\theta \\
&= -\boldsymbol{j}\frac{\rho\kappa U}{\pi}\int_0^{2\pi}\sin^2\theta\,\mathrm{d}\theta \\
&= -\boldsymbol{j}\frac{\rho\kappa U}{2\pi}\int_0^{2\pi}(1-\cos 2\theta)\,\mathrm{d}\theta \\
&= -\boldsymbol{j}\rho\kappa U,
\end{aligned}
$$

where we have used the standard symmetry properties and identities associated with the trigonometric functions shown. Hence there is a sideways force on the cylinder in the direction of '$-\boldsymbol{j}$' of magnitude $\rho\kappa U$. Qualitatively we can think of this as follows; see Figure 3.3. For $\kappa > 0$, the circulation acts with the flow below the cylinder and against the flow above the cylinder. And we know by Bernoulli's Theorem for irrotational flows, the energy density $p + \frac{1}{2}\rho|\boldsymbol{u}|^2$ is constant throughout the flow. Hence associated with the faster flow below the cylinder the pressure is smaller, while associated with the slower flow above the cylinder the pressure is greater. Hence, referring to Figure 3.3, there is a net downwards force on the cylinder. This is the origin of *dynamic lift* for aerofoils; see Sections 3.3 and 3.6. Even more impressively, it turns out that for homogeneous incompressible irrotational flows this force is *independent of the shape of the cylinder cross-section*! We establish this fact through the Theorem of Blasius.

Theorem 3.15 (Blasius, 1908) *The force $\boldsymbol{F} = F_x\boldsymbol{i} + F_y\boldsymbol{j}$ and counter-clockwise moment M on a cylinder with cross-section bounded by a closed curve C in a homogeneous, incompressible, irrotational, inviscid fluid flow are given by (each measured per unit longitudinal length of cylinder)*

$$F_x - \mathrm{i}F_y = \frac{\mathrm{i}\rho}{2}\oint_C (w'(z))^2\,\mathrm{d}z,$$

$$M = \mathrm{Re}\left\{-\frac{\rho}{2}\oint_C z(w'(z))^2\,\mathrm{d}z\right\}.$$

Proof The force per unit longitudinal length of the cylinder is given by

$$\boldsymbol{F} = -\oint_C p\boldsymbol{n}\,\mathrm{d}s,$$

where s parameterises the arclength around the perimeter C of the cross-section of the cylinder. Assume the origin of the Cartesian coordinates x and y to be inside C. Let ϑ be the angle a tangent line at any point on C makes with the x-axis. Then we deduce that $\boldsymbol{n} = \boldsymbol{i}\sin\vartheta - \boldsymbol{j}\cos\vartheta$. Furthermore, we observe that

$$\boldsymbol{n}\,\mathrm{d}s = \boldsymbol{i}\sin\vartheta\,\mathrm{d}s - \boldsymbol{j}\cos\vartheta\,\mathrm{d}s = \boldsymbol{i}\,\mathrm{d}y - \boldsymbol{j}\,\mathrm{d}x.$$

We now observe that using the complex coordinate $z = x + \mathrm{i}y$ we can represent the total force per longitudinal length by the integral

$$F_x - \mathrm{i}F_y = -\oint_C p\,(\mathrm{d}y + \mathrm{i}\mathrm{d}x) = -\mathrm{i}\oint_C p\,\mathrm{d}z^*,$$

where z^* is the complex conjugate of z. Using the Bernoulli Theorem for irrotational flows outlined in Remark 3.13, we know the pressure $p = c - \frac{1}{2}\rho|w'(z)|^2$ for some constant c. Hence we observe that

$$F_x - \mathrm{i}F_y = \frac{\mathrm{i}\rho}{2}\oint_C (w'(z))(w'(z))^*\,\mathrm{d}z^* - \mathrm{i}c\oint_C \mathrm{d}z^*,$$

where the last term on the right is zero via Cauchy's Theorem for a closed contour C and analytic integrand (the function which is identically one). On the bounding contour C, by direct computation we have

$$(w'(z))^* \,\mathrm{d}z^* = (u + \mathrm{i}v)(\mathrm{d}x - \mathrm{i}\mathrm{d}y) = (u - \mathrm{i}v)(\mathrm{d}x + \mathrm{i}\mathrm{d}y) + 2\mathrm{i}(v\mathrm{d}x - u\mathrm{d}y).$$

Recall that the no-normal flow boundary condition on C implies that $\boldsymbol{u} = (u, v)^{\mathrm{T}}$ is orthogonal to the tangent line at any point on C. Recall also from Section 3.2 that $\nabla\psi \cdot \boldsymbol{u} = 0$, and thus C is a streamline. If $x = x(\tau)$ and $y = y(\tau)$ parameterise the contour C for $\tau \in [0, 2\pi)$, then we observe

$$v\mathrm{d}x - u\mathrm{d}y = -(\partial_x\psi x'(\tau) + \partial_y\psi y'(\tau))\,\mathrm{d}\tau = -(\partial_x\psi\partial_y\psi - \partial_y\psi\partial_x\psi)\,\mathrm{d}\tau \equiv 0.$$

We conclude $(w'(z))^* \,\mathrm{d}z^* = w'(z)\,\mathrm{d}z$ and the first result of the theorem follows.

The total moment per unit longitudinal length of the cylinder, using our characterisation for F_x and F_y in terms of the pressure given above, is given by

$$\begin{aligned}
M &= \oint_C (x\,\mathrm{d}F_y - y\,\mathrm{d}F_x)\\
&= \oint_C (px\,\mathrm{d}x + py\,\mathrm{d}y)\\
&= C\oint_C (x\,\mathrm{d}x + y\,\mathrm{d}y) - \frac{\rho}{2}\oint_C (w'(z))(w'(z))^* \,(x\,\mathrm{d}x + y\,\mathrm{d}y)\\
&= 0 - \frac{\rho}{2}\oint_C (w'(z))(w'(z))^* \,\mathrm{Re}\{z\,\mathrm{d}z^*\}\\
&= \mathrm{Re}\left\{-\frac{\rho}{2}\oint_C z(w'(z))(w'(z))^* \,\mathrm{d}z^*\right\}\\
&= \mathrm{Re}\left\{-\frac{\rho}{2}\oint_C z(w'(z))^2\,\mathrm{d}z\right\},
\end{aligned}$$

where we used that C is a closed contour and $(w'(z))^* \,\mathrm{d}z^* = w'(z)\,\mathrm{d}z$ on C. □

With the Theorem of Blasius in hand we can now compute the total force on a cylinder whose cross-section is bounded by a general closed curve C. One way to envisage such a flow where there is a fixed closed boundary C across which there is no normal flow is to model the general shape by a suitable system of line sources and sinks; see Section 3.4 on the method of images. For now we consider a general irrotational flow prescribed by a complex potential function $w = w(z)$ outside the bounding curve C within which the origin is strictly contained. Hence, outside a circle which encloses C whose centre is the origin, we may expand $w = w(z)$ as a series as follows:

$$w(z) = Uz - \frac{\mathrm{i}\kappa}{2\pi}\log z + \frac{A_1}{z} + \frac{A_2}{z^2} + \frac{A_3}{z^3} + \cdots.$$

Here U and κ are real constants while $A_1, A_2, \ldots$ are complex constants. The first term on the right represents a uniform flow and the second term a line vortex of strength κ. When A_1 is real the third term represents a line dipole. Note in particular that there is no line source term $m/(2\pi z)$, with $m \in \mathbb{R}$, present. This is because C is closed and there is no outward net flow across it. The series above represents a Taylor–Laurent series expanded about the origin together with the log z term. Consequently we observe

$$w'(z) = U - \frac{\mathrm{i}\kappa}{2\pi z} - \frac{A_1}{z^2} - 2\frac{A_2}{z^3} + \cdots$$

$$\Rightarrow \qquad (w'(z))^2 = U^2 - \frac{\mathrm{i}\kappa U}{\pi}\frac{1}{z} - \left(2UA_1 + \frac{\kappa^2}{4\pi^2}\right)\frac{1}{z^2} + \cdots .$$

From the Theorem of Blasius the total force on the cylinder per unit longitudinal length is given by the contour integral around C of $(w'(z))^2$. From Cauchy's Theorem in complex analysis, the only non-zero contribution to the contour integral in the series expansion above is from the term z^{-1} (Burkill and Burkill, 1970, Ch. 11). Indeed we find

$$\begin{aligned}
F_x - \mathrm{i}F_y &= \frac{\mathrm{i}\rho}{2}\oint_C (w'(z))^2\,\mathrm{d}z \\
&= \frac{\mathrm{i}\rho}{2}\oint_C\left(-\frac{\mathrm{i}\kappa U}{\pi}\frac{1}{z}\right)\mathrm{d}z \\
&= \left(\frac{\mathrm{i}\rho}{2}\right)\left(-\frac{\mathrm{i}\kappa U}{\pi}\right)2\pi\mathrm{i} \\
&= \mathrm{i}\rho\kappa U.
\end{aligned}$$

Hence we deduce $F_x = 0$ and $F_y = -\rho\kappa U$, independent of the shape of the cross-section of the cylinder.

We can also compute the total moment per unit longitudinal length of the cylinder as follows, again using Cauchy's Theorem:

$$\begin{aligned}
M &= \mathrm{Re}\left\{-\frac{\rho}{2}\oint_C z(w'(z))^2\,\mathrm{d}z\right\} \\
&= \mathrm{Re}\left\{\frac{\rho}{2}\oint_C\left(2UA_1 + \frac{\kappa^2}{4\pi^2}\right)\frac{1}{z}\,\mathrm{d}z\right\} \\
&= \mathrm{Re}\left\{\left(\frac{\rho}{2}\right)\left(2UA_1 + \frac{\kappa^2}{4\pi^2}\right)2\pi\mathrm{i}\right\} \\
&= -2\pi\rho U\,\mathrm{Im}\{A_1\}.
\end{aligned}$$

We observe, unlike the total force, that the moment or torque/couple depends on the body shape through the effective net dipole term A_1/z. If there is symmetry about the direction of the background flow, which here is aligned with $\boldsymbol{i}$, then A_1 may necessarily be real. For example, consider a flat plate of length $2a$ lying perpendicular to the background flow. The complex potential for the flow around such an object is $w = U(z^2 + a^2)^{1/2}$. We observe $w'(z) = Uz(z^2 + a^2)^{-1/2}$, which we can expand for $|z| > a$ in the Taylor–Laurent series $w'(z) = U - \frac{1}{2}a^2Uz^{-2} + \cdots$. The coefficient of z^{-2} is real so the moment on the plate is zero.

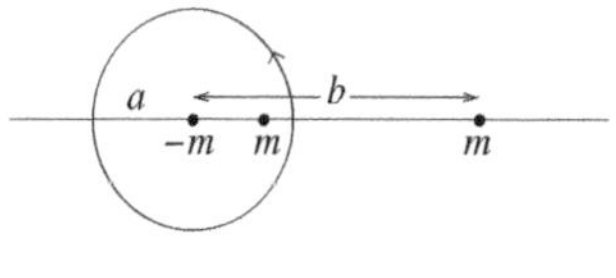

Figure 3.4 The geometric setup associated with the three line sources of $-m$, m and m respectively at $z = 0$, $z = a^2/b$ and $z = b$ with $b > a$, given in Example 3.16.

Example 3.16 Suppose there are line sources of strength $-m$, m and m, respectively at $z = 0$, $z = a^2/b$ and $z = b$ with $b > a$ real. The complex potential function for such a flow is given by

$$w(z) = \frac{m}{2\pi}\left(\log(z - b) + \log\left(z - \frac{a^2}{b}\right) - \log z\right).$$

In particular we note that

$$\psi = \mathrm{Im}\{w\} = \frac{m}{2\pi}\arg\left(\frac{(z-b)(z-a^2/b)}{z}\right).$$

If $z = a\mathrm{e}^{\mathrm{i}\theta}$ then we see that

$$\frac{(z-b)(z-a^2/b)}{z} = \left(1 - \frac{b}{a}\mathrm{e}^{-\mathrm{i}\theta}\right)\left(a\mathrm{e}^{\mathrm{i}\theta} - \frac{a^2}{b}\right) = 2a\cos\theta - \frac{a^2}{b} - b.$$

Hence $\psi = 0$ on $z = a\mathrm{e}^{\mathrm{i}\theta}$, which thus represents a streamline. We can think of $|z| = a$ as a solid boundary, and now focus on the flow for $|z| > a$. We see that

$$w'(z) = \frac{m}{2\pi}\left(\frac{1}{z-b} + \frac{1}{z-a^2/b} - \frac{1}{z}\right)$$

$$\Rightarrow \quad (w'(z))^2 = \frac{2m^2}{4\pi^2}\left(\left(\frac{1}{z-b} - \frac{1}{z}\right)\frac{1}{(z-a^2/b)} - \left(\frac{1}{z-b} - \frac{1}{z-a^2/b}\right)\frac{1}{z}\right) + \cdots,$$

where on the right we only retained the terms that will contribute to the contour integral along the contour $|z| = a$; see Figure 3.4. To be precise, we can compute the contour integral around $|z| = a$ by summing the residues at $z = 0$ and $z = a^2/b$. Indeed, this generates the total force per unit longitudinal length:

$$F_x - \mathrm{i}F_y = \frac{\mathrm{i}\rho}{2}\frac{2m^2}{4\pi^2}\left(\frac{b}{a^2-b^2} - \frac{b}{a^2} + \frac{1}{b} + \frac{b}{a^2}\right)2\pi\mathrm{i} = \frac{\rho m^2}{2\pi}\frac{a^2/b}{b^2-a^2}.$$

Hence there is a force downstream of the magnitude shown, notably irrespective of the sign of m.

3.4 Method of Images

We briefly discuss the method of images for constructing solutions to Laplace's equation in a multitude of different simple geometries and here for the case of zero

Neumann boundary conditions. The uniqueness of the solution, up to an additive constant, to Laplace's equation in such contexts plays a critical role. Here we focus on two-dimensional scenarios, though the method applies in general dimensions. Further, we implement the method via complex potential functions $w(z) := \phi(x, y) + \mathrm{i}\psi(x, y)$ with $z = x + \mathrm{i}y$. Recall that ϕ satisfies Laplace's equation for irrotational flows and $\nabla\phi \cdot \boldsymbol{n} = 0$ on any solid boundaries – corresponding to no normal flow. In addition, ψ also satisfies Laplace's equation. Further, since $\nabla\psi \cdot \boldsymbol{u} = 0$ everywhere for two-dimensional incompressible irrotational flows, we deduce that $\nabla\psi \cdot \boldsymbol{t} = 0$ on the boundary, where $\boldsymbol{t}$ is the vector tangent to the boundary. In other words ψ is constant along the boundary which is thus a streamline. We introduce the method through the following examples.

Example 3.17 (Line source and wall) Suppose there is a line source at $z = a > 0$ and a wall boundary along the imaginary axis where $x = 0$. Referring to Figure 3.5, the complex potential corresponding to this flow is given by

$$w(z) = \frac{m}{2\pi}\log(z - a) + \frac{m}{2\pi}\log(z + a) \equiv \frac{m}{2\pi}\log(z^2 - a^2).$$

The rationale that underlies this statement is as follows. We know from Example 3.7 for a single line source at the origin that the solution complex potential for all $z \in \mathbb{C}$ is $w(z) = (m/2\pi)\log z$. The solution for a corresponding line source at $z = a$ is $w(z) = (m/2\pi)\log(z - a)$. However in the physical scenario we have before us there is a solid wall along the imaginary axis at $x = 0$; see the left panel in Figure 3.5. Along this boundary we know $\nabla\phi \cdot \boldsymbol{n} = 0$ and $\nabla\psi \cdot \boldsymbol{t} = 0$. (Indeed ψ is constant along the boundary which is a streamline. Further, as mentioned above and recall from Section 1.8, stream functions are only defined up to an additive arbitrary constant so we could without loss of generality suppose $\psi = 0$ along the wall.) Let us now construct an equivalent problem – see the right panel in Figure 3.5. Keeping the line source of strength m at $z = a$, which is the original 'object', we remove the wall entirely and place a line source of strength m at $z = -a$, this is the 'image'. This construction elicits the complex potential function $w = w(z)$ above. We remark that with this w, naturally ϕ and ψ both still satisfy Laplace's equation. Furthermore we observe that on $x = 0$:

$$\nabla\phi \cdot \boldsymbol{n} = \mathrm{Re}\big\{(w'(\mathrm{i}y))^*\big\} = \frac{m}{\pi}\mathrm{Re}\bigg\{\frac{\mathrm{i}y}{y^2 + a^2}\bigg\} = 0,$$

and using that $\nabla\psi = \mathrm{i}(w'(z))^*$, then

$$\nabla\psi \cdot \boldsymbol{t} = \mathrm{Im}\big\{\mathrm{i}(w'(\mathrm{i}y))^*\big\} = \frac{m}{\pi}\mathrm{Im}\bigg\{\frac{y}{y^2 + a^2}\bigg\} = 0.$$

Hence in the original physical problem with a wall along the imaginary axis, and in the replacement problem in the whole of $\mathbb{C}$ with an image line source at $z = -a$, we have established that in the right-half plane ϕ and ψ satisfy Laplace's equation and the same boundary conditions on the imaginary axis, together with $\phi \to 0$ and $\psi \to 0$ appropriately as $|z| \to \infty$. Therefore by uniqueness, up to an additive constant,

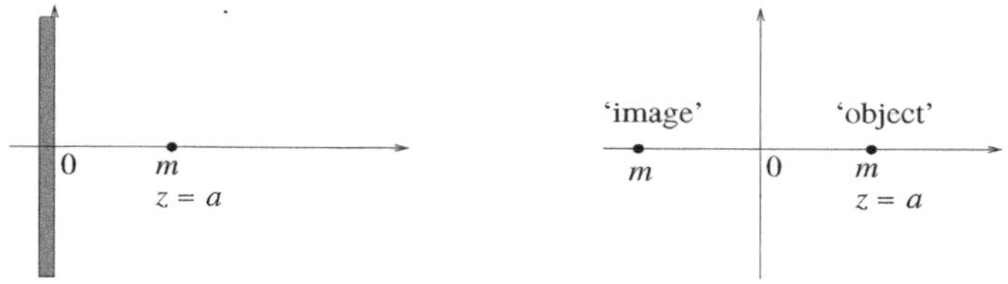

Figure 3.5 Line source with wall example using the method of images from Example 3.17. The left panel shows the original physical scenario with a line source of strength m at $z = a$ and a wall at $x = 0$. The right panel shows the 'method of images' equivalent scenario that generates the same solution for the right-hand half plane.

the solutions to the two problems must be one and the same. Whilst establishing the equivalent boundary conditions above, we used that

$$(w'(\mathrm{i}y))^* = \frac{m}{\pi}\frac{\mathrm{i}y}{y^2 + a^2},$$

representing the velocity field at the wall. Since the energy density at the wall and in the far field are equal, we observe that the pressure at the wall is

$$p = p_\infty - \tfrac{1}{2}\rho\frac{m^2}{\pi^2}\frac{y^2}{(y^2 + a^2)^2}.$$

The force per unit transverse length on the wall due to variations in pressure $\hat{p} := p - p_\infty$ across the domain is thus given by

$$\boldsymbol{F} = -\int_{-\infty}^{\infty} \hat{p}\boldsymbol{i}\,\mathrm{d}y = \boldsymbol{i}\,\tfrac{1}{2}\rho\frac{m^2}{\pi^2}\int_{-\infty}^{\infty}\frac{y^2}{(y^2 + a^2)^2}\,\mathrm{d}y.$$

In other words, the force is directed towards the line source m at $z = a$ independent of the sign of m.

Example 3.18 (Line dipole and wall) Suppose there is a line dipole at $z = a > 0$ and a wall boundary along the imaginary axis $z = \mathrm{i}y$, $y \in \mathbb{R}$. The corresponding complex potential in this case is given by

$$w(z) = \frac{\mu_0 \mathrm{e}^{\mathrm{i}\alpha}}{2\pi}\frac{1}{z - a} + \frac{\mu_0 \mathrm{e}^{\mathrm{i}(\pi - \alpha)}}{2\pi}\frac{1}{z + a},$$

where μ_0 and α are real. We argue as in the last example. We know from Remark 3.9 that the complex potential function corresponding to a line dipole of strength μ_0 at $z = a$ rotated counter-clockwise by an angle α is the first term on the right shown. To take into account that there is a physical wall along the imaginary axis on which $\nabla\phi \cdot \boldsymbol{n} = 0$ and $\nabla\psi \cdot \boldsymbol{t} = 0$, we reconstruct an equivalent problem in the whole of $\mathbb{C}$ by placing a line dipole of strength μ_0 at $z = -a$ rotated counter-clockwise by an angle $\pi - \alpha$ which is represented by the second term on the right shown. The corresponding potential and stream functions satisfy Laplace's equation in the whole two-dimensional plane, and we now just need to check that the boundary conditions hold. We observe that on $x = 0$ we have

$$\begin{aligned}\nabla\phi\cdot\boldsymbol{n} &= \mathrm{Re}\big\{(w'(\mathrm{i}y))^*\big\}\\ &= -\frac{\mu_0}{2\pi}\mathrm{Re}\bigg\{\frac{\mathrm{e}^{-\mathrm{i}\alpha}}{(a+\mathrm{i}y)^2}+\frac{\mathrm{e}^{-\mathrm{i}(\pi-\alpha)}}{(a-\mathrm{i}y)^2}\bigg\}\\ &= -\frac{\mu_0}{2\pi(a^2+y^2)^2}\big((a^2-y^2)\cos\alpha-2ay\sin\alpha\\ &\quad+(a^2-y^2)\cos(\pi-\alpha)+2ay\sin(\pi-\alpha)\big),\end{aligned}$$

and again using that $\nabla\psi = \mathrm{i}(w'(z))^*$ we have

$$\begin{aligned}\nabla\psi\cdot\boldsymbol{t} &= \mathrm{Im}\big\{\mathrm{i}(w'(\mathrm{i}y))^*\big\}\\ &= -\frac{\mu_0}{2\pi}\mathrm{Im}\bigg\{\frac{\mathrm{i}\mathrm{e}^{-\mathrm{i}\alpha}}{(a+\mathrm{i}y)^2}+\frac{\mathrm{i}\mathrm{e}^{-\mathrm{i}(\pi-\alpha)}}{(a-\mathrm{i}y)^2}\bigg\}\\ &= -\frac{\mu_0}{2\pi(a^2+y^2)^2}\big((a^2-y^2)\cos\alpha-2ay\sin\alpha\\ &\quad+(a^2-y^2)\cos(\pi-\alpha)+2ay\sin(\pi-\alpha)\big).\end{aligned}$$

We observe that both terms on the right-hand sides are the same and equal to zero as we have $\cos(\pi-\alpha)\equiv-\cos\alpha$ and $\sin(\pi-\alpha)\equiv\sin\alpha$. Hence by the uniqueness up to an additive constant for Laplace's equation for the given boundary conditions, the solution to the original problem in the right-half plane with the wall along the imaginary axis and the constructed problem with the dipoles outlined at $z=\pm a$ are one and the same. We remark that there are two stagnation points associated with this flow as follows. Observe that $(w'(z))^* = 0$ if and only if

$$\frac{z^*-a}{z^*+a} = \pm\mathrm{e}^{-\mathrm{i}\alpha} \qquad\Leftrightarrow\qquad z = a\mathrm{i}\cot(\alpha/2)\quad\text{or}\quad z=-a\mathrm{i}\tan(\alpha/2),$$

giving the location of the stagnation points on the imaginary axis.

Example 3.19 (Line vortex and wall) Suppose there is a line vortex of strength $\kappa\in\mathbb{R}$ at $z=a>0$ and a wall boundary along the imaginary axis $z=\mathrm{i}y$, $y\in\mathbb{R}$. Referring to Figure 3.6, we postulate the complex potential to be

$$w(z) = -\frac{\mathrm{i}\kappa}{2\pi}\log(z-a)+\frac{\mathrm{i}\kappa}{2\pi}\log(z+a).$$

We already know that each of the two terms present generate harmonic potential and stream functions. Our arguments in the last two examples also explain why we postulate this complex potential function as the appropriate 'method of images' combination to generate the correct boundary conditions at the wall. Direct computation confirms that indeed $\nabla\phi\cdot\boldsymbol{n} = \mathrm{Re}\{(w'(\mathrm{i}y))^*\} = 0$ and also $\nabla\psi\cdot\boldsymbol{t} = \mathrm{Im}\{\mathrm{i}(w'(\mathrm{i}y))^*\} = 0$. Let us consider the streamlines for this flow which correspond to constant ψ. Note that we have

$$\psi = \mathrm{Im}\{w(z)\} = -\frac{\kappa}{2\pi}\log\left|\frac{z-a}{z+a}\right|.$$

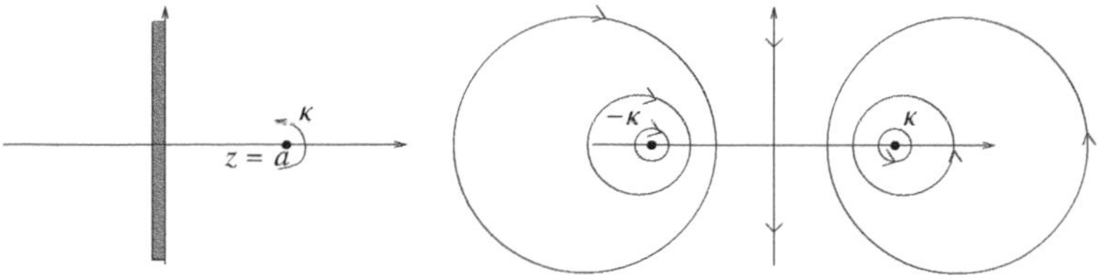

Figure 3.6 Line vortex with wall example using the method of images from Example 3.19. The left panel shows the original physical scenario with a line vortex of strength $\kappa > 0$ at $z = a$ and a wall at $x = 0$. The right panel shows the 'method of images' equivalent scenario that generates the same solution for the right-hand half plane.

Hence ψ is constant when, for some constant $B > 0$:

$$\left|\frac{z-a}{z+a}\right| = B \quad \Leftrightarrow \quad zz^* - a\frac{(1+B^2)}{(1-B^2)}(z+z^*) + a^2 = 0.$$

We look for a solution z of the form $z = A + R\mathrm{e}^{\mathrm{i}\alpha}$, where A and R are constants and $\alpha \in [0, 2\pi)$ is arbitrary. Substituting this form into the quadratic equation for z above and equating constants and also coefficients of $\cos\alpha$ reveals that

$$A = a\frac{(1+B^2)}{(1-B^2)} \quad \text{and} \quad R = \frac{2aB}{(1-B^2)} \quad \Rightarrow \quad z = a\frac{(1+B^2)}{(1-B^2)} + \frac{2aB}{(1-B^2)}\mathrm{e}^{\mathrm{i}\alpha}.$$

For $0 < B < 1$ these are circles in the right-half plane; see Figure 3.6. Note that the velocity field is given by

$$u + \mathrm{i}v = (w'(\mathrm{i}y))^* = \frac{\mathrm{i}\kappa}{2\pi}\frac{1}{z^* - a} - \frac{\mathrm{i}\kappa}{2\pi}\frac{1}{z^* + a}.$$

The effect of the line vortex at $z = -a$ on the original vortex at $z = a$, or equivalently the effect of the presence of the wall, is to induce an additional velocity $-\mathrm{i}\kappa/(4\pi a)$ on the vortex at $z = a$. This was obtained by substituting $z = a$ into the second term on the right above. Hence in the presence of the wall the vortex moves in the $-\boldsymbol{j}$ direction parallel to the wall – likewise for two counter-rotating vortices in the equivalent 'method of images' setup.

Example 3.20 (Line source in a corner) Suppose there is a line source of strength $m \in \mathbb{R}$ at $z_0 = a + ib$ with $a > 0$ and $b > 0$, and wall boundaries along the real and imaginary axes. Referring to Figure 3.7, we propose the correct complex potential to be

$$w(z) = \frac{m}{2\pi}(\log(z - z_0) + \log(z + z_0) + \log(z - z_0^*) + \log(z + z_0^*)).$$

All the terms present generate harmonic potential and stream functions. Our arguments in the examples above also explain the rationale underlying this complex potential function as the appropriate equivalent 'method of images' replacement problem. Direct computation reveals that the boundary conditions $\nabla\phi\cdot\boldsymbol{n} = 0$ and $\nabla\psi\cdot\boldsymbol{t} = 0$ are satisfied

Figure 3.7 Line source in a corner from Example 3.20. The left panel shows the original physical scenario with a line source of strength m at $z_0 = a + ib$ and walls along the real and imaginary axes. The right panel shows the 'method of images' equivalent scenario that generates the same solution for the first quadrant.

everywhere on the real and imaginary axes. Hence we have a complete prescription for the solution flow for the line source in a corner flow. Note that the image sources induce a non-zero velocity on the line source at z_0, which moves up and right (if $m > 0$) as shown in Figure 3.7.

Example 3.21 (Line vortex in a corner) Suppose there is a line vortex of strength $\kappa \in \mathbb{R}$ at $z_0 = a + ib$ with $a > 0$ and $b > 0$, and wall boundaries along the real and imaginary axes. Referring to Figure 3.8, the correct complex potential for this case using the 'method of images' should be

$$w(z) = -\frac{i\kappa}{2\pi}\big(\log(z - z_0) + \log(z + z_0) - \log(z - z_0^*) - \log(z + z_0^*)\big).$$

As we have already seen, all the terms present generate harmonic potential and stream functions, and straightforward computations establish that the boundary conditions $\nabla\phi \cdot \boldsymbol{n} = 0$ and $\nabla\psi \cdot \boldsymbol{t} = 0$ are satisfied everywhere on the real and imaginary axes. Hence the flow field is completely described by this complex potential function. Let us consider the velocity $u + iv$ induced on the line vortex at z_0 by the other three line vortices via the complex potential function

$$\overline{w}(z) := -\frac{i\kappa}{2\pi}\big(\log(z + z_0) - \log(z - z_0^*) - \log(z + z_0^*)\big).$$

Direct computation reveals the velocity field corresponding to $\overline{w}$ at $z = a + ib$:

$$u + iv = \big(\overline{w}'(a + ib)\big)^* = \frac{\kappa}{4\pi}\left(\frac{i}{a - ib} + \frac{1}{b} - \frac{i}{a}\right).$$

Separating the right-hand side into real and imaginary parts and writing $a = r\cos\theta$ and $b = r\sin\theta$, we find

$$u = \frac{\kappa}{4\pi r}\cot\theta\cos\theta \qquad \text{and} \qquad v = -\frac{\kappa}{4\pi r}\tan\theta\sin\theta$$

or equivalently

$$u_r = \frac{\kappa}{2\pi r}\cot(2\theta) \qquad \text{and} \qquad u_\theta = -\frac{\kappa}{4\pi r}.$$

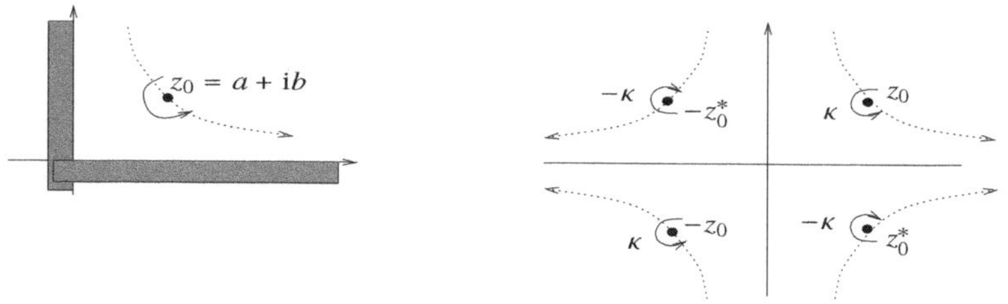

Figure 3.8 Line vortex in a corner from Example 3.21. The left panel shows the original physical scenario with a line vortex of strength $\kappa > 0$ at $z_0 = a+ib$ and walls along the real and imaginary axes. The right panel shows the 'method of images' equivalent scenario that generates the same solution for the first quadrant.

Remark 3.22 Given the instantaneous coordinate velocities at the end of Example 3.21, we can compute the streamline through $z_0 = a + ib$ as the solution of the pair of ordinary differential equations

$$\frac{\mathrm{d}r}{\mathrm{d}s} = \frac{\kappa}{2\pi r}\cot(2\theta) \qquad \text{and} \qquad r\frac{\mathrm{d}\theta}{\mathrm{d}s} = -\frac{\kappa}{4\pi r},$$

where s parameterises arclength along the streamline. By the chain rule we deduce (for some constant K):

$$\frac{1}{r}\frac{\mathrm{d}r}{\mathrm{d}\theta} = -2\cot(2\theta) \quad \Leftrightarrow \quad r\sin(2\theta) = K \quad \Leftrightarrow \quad \frac{2ab}{(a^2+b^2)^{1/2}} = K.$$

The latter two expressions thus parameterise the path of the line vortex due to the given instantaneous coordinate velocities.

Remark 3.23 Note that the 'method of images' set up in the right panel in Figure 3.8 would also be the 'method of images' equivalent of two counter-rotating line vortices at positions z_0 and $-z_0^*$ – equidistant to the imaginary axis in the top half plane – with a solid boundary along the real axis. These line vortices might represent a vortex pair shed from aircraft wingtips near a runway. As we now know, they move downwards and outwards from the aircraft path. In the equivalent scenario, the four vortices (objects and images) evolve to two +/− vortex pairs ('doublets') separating from each other with eventual doublet width given by $2ab/(a^2+b^2)^{1/2}$.

We have used the 'method of images' to find the solution flow in simple geometries. It utilises a suitable distribution of line sources/sinks/vortices and/or images outside the flow to equivalently generate the solution flow in the given region/geometry. Next we consider another method that maps the given region/geometry to another potentially more convenient/suitable region/geometry.

3.5 Conformal Mapping

We saw many examples in the last section of two-dimensional incompressible irrotational flows in simple geometries. We now consider using a method of complex transformation, conformal mapping, to map more complicated geometries to simpler ones. We consider mapping a region of the complex z-plane onto a region of the complex ζ-plane using an analytic function $\zeta = g(z)$. Suppose the original complex potential in terms of z is $w = w(z)$. Then under the map g we define the transformed complex potential $W = W(\zeta)$ by $W(\zeta) := w(z) \equiv w(g^{-1}(\zeta))$ for all ζ in the transformed region, or equivalently W is the function such that $w(z) = W(g(z))$ for all z in the original region. By the chain rule for complex functions, we have

$$\frac{\mathrm{d}w}{\mathrm{d}z} = \frac{\mathrm{d}W}{\mathrm{d}\zeta}\frac{\mathrm{d}\zeta}{\mathrm{d}z} = \frac{\mathrm{d}W}{\mathrm{d}\zeta}g'(z).$$

A point z_0 in the z-plane where $g'(z_0) \neq 0$ is called an *ordinary point* of the transformation. We observe for the complex map $\zeta = g(z)$ that since (for small increments) $\mathrm{d}\zeta = g'(z_0)\,\mathrm{d}z$ at z_0, we have

$$|\mathrm{d}\zeta| = |g'(z_0)|\,|\mathrm{d}z| \qquad \text{and} \qquad \arg(\mathrm{d}\zeta) = \arg(g'(z_0)) + \arg(\mathrm{d}z).$$

Thus under the transformation $\zeta = g(z)$, at z_0, each small line segment $\mathrm{d}z$ is scaled in magnitude by $|g'(z_0)|$ and rotated through an angle $\arg(g'(z_0))$. In particular we deduce, in the neighbourhood of z_0, that angles are preserved under the transformation g and the ratios of adjacent small line segments are preserved; see Figure 3.9. This means that 'small' geometrical figures remain 'similar' under transformation, while on larger scales this is not true. This conformal property breaks down at points z_0, where $g'(z_0) = 0$ or $g'(z_0) = \infty$. These are called *singular points* (or *critical points*) of the transformation. Near such a point we observe that

$$\zeta = g(z_0) + \tfrac{1}{2}(z - z_0)^2 g''(z_0) + \cdots,$$

and the mapping that $z \mapsto \zeta$ is no longer one-to-one. We can usually arrange any such points to be on the boundary of our flow region. Recall that with $z = x + \mathrm{i}y$ the complex potential w is defined as $w(z) = \phi(x, y) + \mathrm{i}\psi(x, y)$, in terms of the potential and stream functions ϕ and ψ, respectively. After the transformation $\zeta = g(z)$ let us denote $\zeta = \xi + \mathrm{i}\eta$. The transformed potential function Φ and stream function Ψ are thus given by

$$\Phi(\xi, \eta) + \mathrm{i}\Psi(\xi, \eta) := W(\zeta) \quad \text{where} \quad W(\zeta) = w(z) := \phi(x, y) + \mathrm{i}\psi(x, y).$$

This implies that potential lines $\phi = c$ for some constant c transform to potential lines $\Phi = c$ and similarly streamlines $\psi = c$ transform to streamlines $\Psi = c$.

How can we identify suitable maps $z \mapsto \zeta$ that map a given region in the z-plane to a simpler or more convenient region in the ζ-plane? One strong result in this direction is the *Riemann Mapping Theorem*. This states that any simply connected open subset of the complex plane can be bijectively and analytically mapped to the open unit disc (the inverse map is also analytic). The mapping concerned is thus conformal.

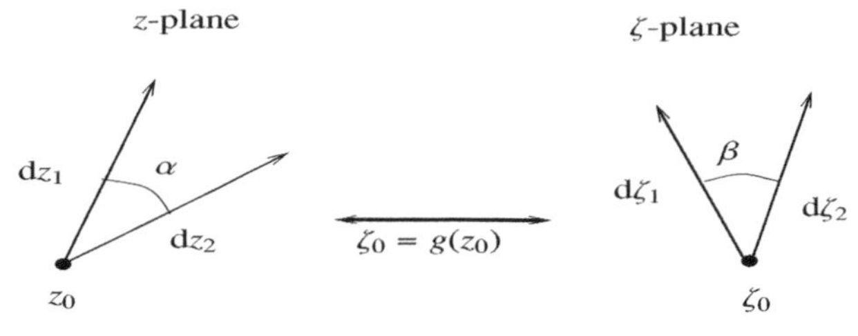

Figure 3.9 The analytic map $g\colon z \mapsto \zeta$ transforms the z-plane to the ζ-plane. The angle between adjacent small line segments is preserved so that $\alpha = \beta$ while the magnitudes of the line segments are scaled by $|g'(z_0)|$, when this is non-zero.

Unfortunately the proof is not constructive, it doesn't indicate how to construct such a map in general. However, over time and with experience, tables of mapping functions with wide applicability have become available. Let us consider some examples.

Example 3.24 (Wedge region) Consider the wedge-shaped region in the z-plane shown in the left panel in Figure 3.10. The region is contained within $0 < \theta < \pi/N$, where $N \geqslant 2$ is an integer. Consider the analytic map $z \mapsto \zeta$, where $\zeta = z^N$. It transforms the wedge-shaped region to the upper half of the ζ-plane as shown in the right panel in Figure 3.10. We note that $z = 0$ is a critical point of the transformation. If we set $z = x + \mathrm{i}y = r\mathrm{e}^{\mathrm{i}\theta}$ then

$$\zeta = R\mathrm{e}^{\mathrm{i}\Theta} \qquad \text{where} \qquad R = r^N \quad \text{and} \quad \Theta = N\theta.$$

Hence we observe $\xi = r^N \cos(N\theta)$ and $\eta = r^N \sin(N\theta)$. Consider contours of ξ, say $\xi = K$ with K a constant. We observe for such contours $\eta/K = \tan(N\theta)$ and $r^{2N} = K^2 + \eta^2$. These contours are thus parameterised in the z-plane by

$$r = \left(K^2(1 + \tan^2(N\theta))\right)^{1/2N}.$$

Consequently when $\theta = 0$ or $\theta = \pi/N$ then $r = K$. Further, as $\theta \to \pi/(2N)$ then $r \to \infty$. Contours of η, say $\eta = K$, are parameterised as follows:

$$r = \left(K^2(1 + \cot^2(N\theta))\right)^{1/2N}.$$

And when $\theta \to 0$ or $\theta \to \pi/N$ we have $r \to \infty$, while when $\theta = \pi/(2N)$ we have $r = K^{1/N}$. See Figure 3.10.

Example 3.25 (Rectangular half-strip region) Consider the rectangular half-strip region in the z-plane shown in the left panel in Figure 3.11. The region is contained within $0 \leqslant x \leqslant a$ and $y \geqslant 0$ where $a > 0$ is a constant. Consider the analytic map $z \mapsto \zeta$, where $\zeta = \cos(\pi z/a)$. This transforms the rectangular half-strip in the z-plane to the lower half of the ζ-plane shown in the right panel in Figure 3.11. Note that $z = 0$ and $z = a$ are critical points of the transformation where $g'(z) = -(\pi/a)\sin(\pi z/a)$ vanishes. With $z = x + \mathrm{i}y$ we have

$$\cos\left(\frac{\pi z}{a}\right) \equiv \cos\left(\frac{\pi x}{a}\right)\cosh\left(\frac{\pi y}{a}\right) - \mathrm{i}\sin\left(\frac{\pi x}{a}\right)\sinh\left(\frac{\pi y}{a}\right).$$

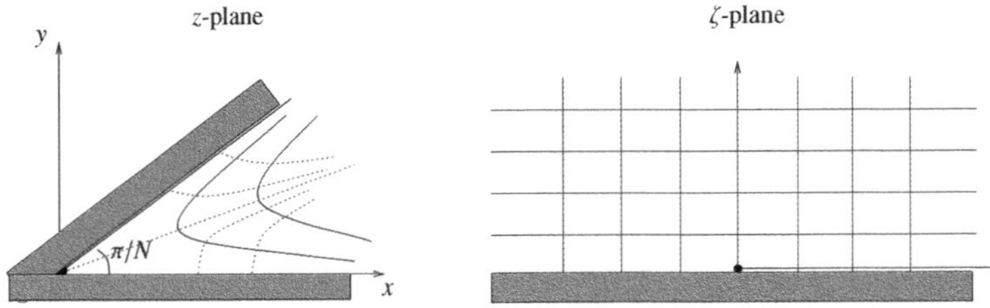

Figure 3.10 The analytic map $z \mapsto \zeta$, where $\zeta = z^N$ and $N \geqslant 2$ is an integer, transforms the wedge-shaped region in the z-plane to the upper half of the ζ-plane as shown; see Example 3.24. In both the z- and ζ-planes, dotted curves correspond to contours of ξ while solid curves correspond to contours of η.

Hence we observe that

$$\xi = \cos\left(\frac{\pi x}{a}\right)\cosh\left(\frac{\pi y}{a}\right) \qquad \text{and} \qquad \eta = -\sin\left(\frac{\pi x}{a}\right)\sinh\left(\frac{\pi y}{a}\right).$$

Contours of x, with say $x = K$ and K constant, are given by

$$\frac{\xi^2}{\cos^2(\pi K/a)} - \frac{\eta^2}{\sin^2(\pi K/a)} = 1.$$

Similarly contours of y, with say $y = K$ and K constant, are given by

$$\frac{\xi^2}{\cosh^2(\pi K/a)} + \frac{\eta^2}{\sinh^2(\pi K/a)} = 1.$$

Note that contours of x in the z-plane correspond to hyperbolae in the ζ-plane while contours of y correspond to ellipses. Further note that since $0 \leqslant x \leqslant a$, we have $\eta \leqslant 0$. The contour $x = 0$ corresponds to the section of the ξ-axis for which $\xi \geqslant 1$ and $x = a$ corresponds to the section of the ξ-axis for which $\xi \leqslant -1$. The critical points in the z-plane map to the points $\zeta = \pm 1$ in the ζ-plane and these are the foci of the ellipses in the ζ-plane representing the contours of y.

Example 3.26 (Joukowski transformation: ellipse) The Joukowski mapping $z \mapsto \zeta$, where $\zeta = z + A^2/z$, with $A > 0$ constant, plays a major role in aerofoil theory. Here we use it to transform the region exterior to the disc of radius $a \geqslant A$ in the z-plane shown in the left panel in Figure 3.12 to the region exterior to the ellipse in the ζ-plane shown in the right panel. Observe that the critical points of the Joukowski map are $z = \pm A$. With $z = r\mathrm{e}^{\mathrm{i}\theta}$ we further observe

$$\zeta = \left(r + \frac{A^2}{r}\right)\cos\theta + \mathrm{i}\left(r - \frac{A^2}{r}\right)\sin\theta.$$

Hence we deduce $\xi = (r + A^2/r)\cos\theta$ and $\eta = (r - A/r^2)\sin\theta$. Consider contours of r in the z-plane, say $r = K$ with $K \geqslant A$ constant. Then as $\theta \in [0, 2\pi)$ varies the coordinates $\xi = (K + A^2/K)\cos\theta$ and $\eta = (K - A^2/K)\sin\theta$ trace out ellipses for different values of K. When $\theta = K$ for some constant K then $\xi = (r + A^2/r)\cos K$ and $\eta = (r - A^2/r)\sin K$ trace out hyperbolae in the ζ-plane. We also observe that in

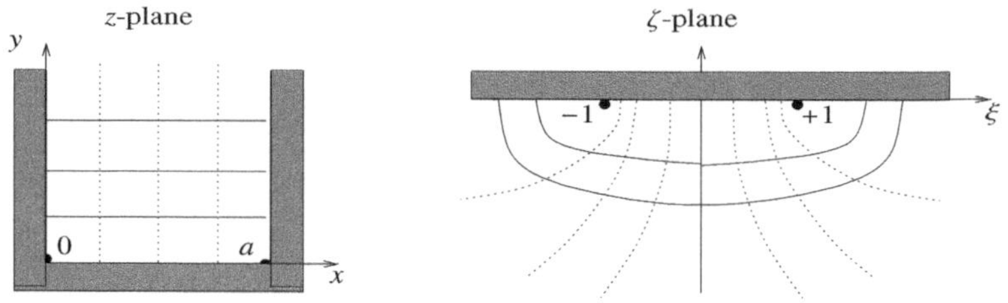

Figure 3.11 The analytic map $z \mapsto \zeta$, where $\zeta = \cos(\pi z/a)$ with $a > 0$, transforms the rectangular half-strip region in the z-plane to the lower half of the ζ-plane as shown; see Example 3.25. In both the z- and ζ-planes, dotted curves correspond to contours of x while solid curves correspond to contours of y.

the case $a = A$ the contour $r = a = A$ corresponds to $\eta = 0$ and $\xi = 2a\cos\theta$, which parameterises the line joining $\zeta = -2a$ to $\zeta = 2a$, i.e. the ellipse collapses to the line joining these two points in the ζ-plane.

Example 3.27 (Symmetric Joukowski aerofoil) We extend the last example and use the Joukowski transformation $z \mapsto \zeta$, where $\zeta = z + A^2/z$ with $A > 0$ constant, to generate a symmetric wing profile in the ζ-plane, see Figure 3.13. Note in particular the cusp in the profile at $\zeta = 2a$. To generate such a shape we start with two circles in the z-plane, one with radius $a = A$ centred at the origin while the other larger circle has radius $a + \delta$ and is centred at $z = -\delta$. The two circles are tangent to one another at $z = a$. Though any flow will take place exterior to the larger circle, it is the region between the two circles that is of interest as far as constructing the aerofoil shape is concerned; see the left panel in Figure 3.13. As we saw in Example 3.26, since the smaller circle is centred at the origin and has radius $a = A$, it is transformed to the line segment in the ζ-plane joining $\zeta = -2a$ and $\zeta = +2a$. The larger circle of radius $a + \delta$ with $\delta > 0$ and centred at $z = -\delta$ is parameterised by

$$z = -\delta + (a + \delta)\mathrm{e}^{\mathrm{i}\theta},$$

for $\theta \in [0, 2\pi)$. Under the Joukowski map $\zeta = z + a^2/z$ the trace of the larger circle in the ζ-plane, by direct substitution, is given by

$$\begin{aligned}\zeta &= -\delta + (a+\delta)\mathrm{e}^{\mathrm{i}\theta} + a^2\big(-\delta + (a+\delta)\mathrm{e}^{\mathrm{i}\theta}\big)^{-1}\\ &= \frac{2\big((a+\delta)\cos\theta - \delta\big)\big(a^2 + \delta(a+\delta)(1-\cos\theta)\big)}{(a+\delta)^2 - 2\delta(a+\delta)\cos\theta + \delta^2}\\ &\quad + \mathrm{i}\frac{2\delta(a+\delta)^2\sin\theta\,(1-\cos\theta)}{(a+\delta)^2 - 2\delta(a+\delta)\cos\theta + \delta^2}.\end{aligned}$$

We note that $\theta = 0$ corresponds to $\zeta = 2a$ while $\theta = \pi$ corresponds to $\zeta = -2(a^2 + 2\delta(a+\delta))/(a + 2\delta)$, which we can straightforwardly verify is always to the left of $\zeta = -2a$ for any $\delta > 0$. That the right-hand edge is a cusp can be interpreted from

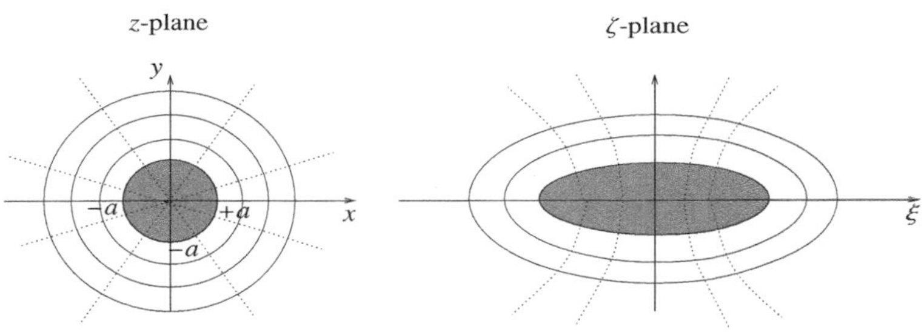

Figure 3.12 The map $z \mapsto \zeta$, where $\zeta = z + A^2/z$ with $A > 0$, transforms the region in the z-plane exterior to the disc with radius $a \geqslant A$ shown to the region in the ζ-plane exterior to the ellipse shown; see Example 3.26. In both the z- and ζ-planes, dotted curves correspond to contours of θ while solid curves correspond to contours of r. The axes of the ellipse have length $a + A^2/a$ and $a - A^2/a$.

the conformal properties of the Joukowski map away from $z = 0$, in particular the preservation of local angles.

As yet we have not applied conformal mapping to a two-dimensional incompressible irrotational flow; we have simply indicated how various regions and coordinate contours transform under typical conformal maps and the Joukowski map. One of the main applications of conformal and in particular Joukowski maps will be in aerofoil theory, coming in Section 3.6. However, let us emphasise here some useful properties/insights. Consider the conformal map $z \mapsto \zeta$ given by $\zeta = g(z)$ valid in suitable respective subregions of the z- and ζ-planes. In the interior of the subregion of the z-plane, how do line sources, line vortices and line dipoles transform? Suppose $z = z_0$ is a non-singular point of the conformal transformation. Then near $z = z_0$, indeed asymptotically as $z \to z_0$, we have

$$\zeta - \zeta_0 \sim g'(z_0)(z - z_0),$$

where $\zeta_0 = g(z_0)$. Hence we observe that

$$\log(\zeta - \zeta_0) \sim \log(z - z_0) + \log(g'(z_0)) \quad \text{and} \quad \frac{1}{\zeta - \zeta_0} \sim \frac{1}{g'(z_0)} \frac{1}{z - z_0},$$

where the term $g'(z_0)$ is a constant. Recall that the complex potential corresponding to: a line source of strength m at $z = z_0$ is $w(z) = (m/2\pi)\log(z - z_0)$; a line vortex of circulation κ at $z = z_0$ is $w(z) = -(\mathrm{i}\kappa/2\pi)\log(z - z_0)$; and a line dipole of strength μ_0, rotated counter-clockwise by angle α and centred at $z = z_0$ is $w(z) = (\mu_0 \mathrm{e}^{\mathrm{i}\alpha}/2\pi)(z - z_0)^{-1}$. Since complex potential functions are equivalent up to additive constants, we see that near $z = z_0$ and thus near $\zeta = \zeta_0$, the complex potential functions for the line source and line vortex have the same functional form under a conformal transformation, indeed respectively they transform to $W(\zeta) = (m/2\pi)\log(\zeta - \zeta_0)$ and $W(\zeta) = -(\mathrm{i}\kappa/2\pi)\log(\zeta - \zeta_0)$. In other words, nearby they respectively transform to line sources and line vortices of corresponding strengths. In addition we see that near $z = z_0$ the line dipole complex potential function becomes

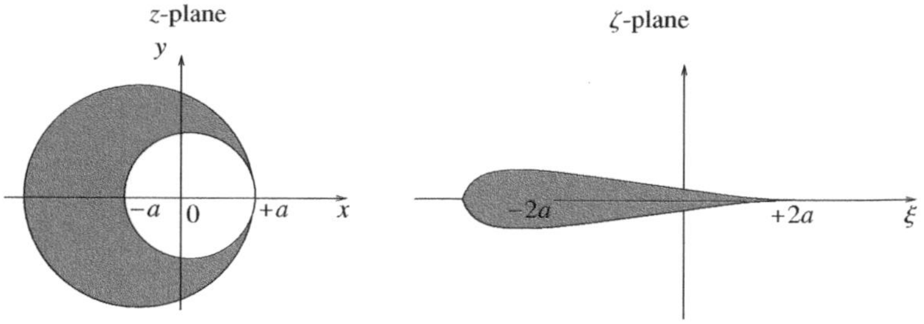

Figure 3.13 The map $z \mapsto \zeta$, where $\zeta = z + A^2/z$ with $A = a > 0$, transforms the region in the z-plane between the discs shown in the left panel to the symmetric Joukowski aerofoil shown in the ζ-plane in the right panel; see Example 3.27. In the left panel the smaller circle is centred at the origin and of radius a. The larger circle is centred at $z = -\delta$ with $\delta > 0$ and of radius $a + \delta$. The smaller circle is mapped to the interval $\zeta \in [-2a, +2a]$ in the ζ-plane shown in the right panel, while the larger circle is mapped to the boundary shown.

$W(\zeta) = (g'(z_0)\mu_0 e^{i\alpha}/2\pi)(\zeta - \zeta_0)^{-1}$, which corresponds to a rotated and stretched (by $g'(z_0)$) line dipole in the ζ-plane. More examples can be found in the exercises at the end of this chapter.

3.6 Aerofoils and Lift

Our goal herein is to understand how circulation, in homogeneous, incompressible, irrotational, inviscid flows, plays a crucial role in aerodynamic lift. We restrict ourselves to two-dimensional flows over aerofoils, i.e. cross-sectional profiles of wings. We discuss the process of designing and analysing the aerofoil design. The explanation underpinning aerodynamic lift cannot be based on inviscid theory alone and we address this issue and the role the fluid viscosity plays at the very end.

Our starting point is Section 3.3. There we considered the flow generated by the complex potential

$$w = Uz + \frac{\mu}{2\pi}\frac{1}{z} - \frac{i\kappa}{2\pi}\log z,$$

where U, μ and κ are real constants around a cylinder with circular cross-section, radius a. We computed directly via the pressure, as well as via the Theorem of Blasius, the resulting force acting per unit longitudinal length on the cylinder. Recall that for a zero normal velocity at the cylinder surface $r = a$ we require $\mu = 2\pi U a^2$. We assume this henceforth throughout this section, as well as that $U \geqslant 0$. Further recall: the first term Uz represents a uniform flow field parallel to the x-axis corresponding to the far-field flow that impinges on the cylinder; the second term $\mu/2\pi z$ represents a line dipole flow centred at the origin, again aligned with the x-axis; and the third term $-i(\kappa/2\pi)\log z$ represents a line vortex flow centred at the origin of circulation strength κ. If $\kappa > 0$ then the circulation is counter-clockwise; it is clockwise if $\kappa < 0$. The four possible characteristic flows depending on the relative strength of the circulation $\kappa > 0$ to the far-field flow speed U are summarised in Figure 3.3.

Recall also that when $\kappa < 0$ there are four further analogous characteristic flows which are upside-down versions of those shown. We computed in Section 3.3 that the total force acting on the cylinder per unit longitudinal length was $F_x = 0$, i.e. no drag, and $F_y = -\rho\kappa U$, representing a downward force if $\kappa > 0$ and a 'lift' force if $\kappa < 0$. Recall also the Theorem of Blasius tells us this force is independent of the cross-sectional shape of the cylinder. Let us begin by considering two successive modifications of such a flow through the following two examples. First we consider the same circular aerofoil geometry but now with a small angle of incidence $\alpha \geqslant 0$, i.e. the uniform far-field velocity field is tilted counter-clockwise by the angle α. Second we change the aerofoil profile to an elliptical one using the Joukowski transformation.

Example 3.28 (Circular aerofoil, incidence angle α) Recall from Example 3.5 that $w(z) = Ue^{-i\alpha}z$ represents a uniform flow of speed U in the direction characterised by the angle α; note that $u + iv = (w')^* = Ue^{-i\alpha} = U\cos\alpha - iU\sin\alpha$. Further, recall from Remark 3.9 that $w(z) = \mu_0 e^{i\alpha}/2\pi z$ represents a line dipole rotated counter-clockwise by an angle α. Lastly note the complex potential $\log z \equiv \log r + i\theta$ is invariant to complex planar rotations, so for example $\log(e^{i\alpha}z) = \log r + i(\theta + \alpha)$ generates the same flow as $\log z$ as they differ by the constant $i\alpha$. Hence the complex potential representing a flow of speed U at angle α in the far field, with a line dipole of strength $\mu_0 = 2\pi U a^2$ rotated counter-clockwise by angle α, combined with a line vortex of circulation strength κ, is given by

$$w = Ue^{-i\alpha}z + Ua^2e^{i\alpha}\frac{1}{z} - \frac{i\kappa}{2\pi}\log z.$$

Assume for the moment this is the flow field around a cylinder with circular cross-section of radius a. The four possible characteristic flow fields will look much like those in Figure 3.3 but with each of the panels rotated counter-clockwise by angle α when $\kappa \geqslant 0$ and the corresponding representative reflected flow fields rotated counter-clockwise by angle α when $\kappa \leqslant 0$. We can compute the force per longitudinal length on the cylinder using the Theorem of Blasius exactly as we did directly after the proof. The argument there applies with U replaced by $Ue^{-i\alpha}$ and so

$$F_x - iF_y = i\rho\kappa Ue^{-i\alpha} \qquad \Leftrightarrow \qquad \begin{pmatrix} F_x \\ F_y \end{pmatrix} = \rho\kappa U \begin{pmatrix} \sin\alpha \\ -\cos\alpha \end{pmatrix}.$$

We can compute the moment per unit longitudinal length on the cylinder using the Theorem of Blasius with the far-field velocity and dipole moment replaced by $Ue^{-i\alpha}$ and $Ua^2e^{i\alpha}$, respectively. Hence in this case the coefficient A_1 of z^{-1} in w is given by $A_1 = Ua^2e^{i\alpha}$. Hence following through the calculation for the moment given after the Theorem of Blasius we observe (as expected) that

$$M = \mathrm{Re}\left\{\left(\frac{\rho}{2}\right)\left(2Ue^{-i\alpha}A_1 + \frac{\kappa^2}{4\pi^2}\right)2\pi i\right\} = 2\pi\rho\,\mathrm{Im}\{U^2a^2\} = 0.$$

Example 3.29 (Elliptical aerofoil: flow) Let us now consider the transformation of the exterior flow and circular geometry of the last example under the Joukowski map $z \mapsto \zeta$, where $\zeta = z + A^2/z$. We saw in Example 3.26 how the circle of radius $a \geqslant A$ is transformed to the ellipse parameterised by $\xi = (a + A^2/a)\cos\theta$ and

$\eta = (a - A^2/a)\sin\theta$. The flow in the complex z-plane is that considered above, i.e. the flow generated by the complex potential

$$w = U\mathrm{e}^{-\mathrm{i}\alpha}z + Ua^2\mathrm{e}^{\mathrm{i}\alpha}\frac{1}{z} - \frac{\mathrm{i}\kappa}{2\pi}\log z.$$

The corresponding complex potential $W = W(\zeta)$ in terms of ζ is given by $W(\zeta) = w(z)$. To compute the corresponding flow in the ζ-plane we compute

$$W'(\zeta) = \frac{w'(z)}{\zeta'(z)},$$

i.e. we evaluate the flow in the ζ-plane by evaluating $w'(z)/\zeta'(z)$ at corresponding points in the z-plane. Note that for the Joukowksi map $\zeta'(z) = 1 - A^2/z^2$ so

$$W'(\zeta) = \frac{z^2}{(z-A)(z+A)}w'(z) = \frac{U\mathrm{e}^{-\mathrm{i}\alpha}z^2 - Ua^2\mathrm{e}^{\mathrm{i}\alpha} - \mathrm{i}\kappa z/2\pi}{(z-A)(z+A)}.$$

We observe that the velocity field $(W'(\zeta))^*$ in the ζ-plane thus has two singularities at $z = \pm A$ corresponding to $\zeta = \pm 2A$. Note that the singularity present solely in $w'(z)$ at $z = 0$ is no longer present. However, to compute the force and moment in the ζ-plane, we typically transform back the contour integrals to contour integrals in the z-plane. In particular, using that $\mathrm{d}\zeta = \zeta'(z)\,\mathrm{d}z$, we compute

$$F_\xi - \mathrm{i}F_\eta = \tfrac{1}{2}\mathrm{i}\rho\oint_{\text{ellipse}}(W'(\zeta))^2\,\mathrm{d}\zeta = \tfrac{1}{2}\mathrm{i}\rho\oint_C\frac{(w'(z))^2}{\zeta'(z)}\,\mathrm{d}z$$

and

$$M = -\tfrac{1}{2}\rho\,\mathrm{Re}\Big\{\rho\oint_{\text{ellipse}}\zeta(W'(\zeta))^2\,\mathrm{d}\zeta\Big\} = -\tfrac{1}{2}\rho\,\mathrm{Re}\Big\{\oint_C\zeta(z)\frac{(w'(z))^2}{\zeta'(z)}\,\mathrm{d}z\Big\}.$$

The integrands $(w'(z))^2/\zeta'(z)$ and $\zeta(z)(w'(z))^2/\zeta'(z)$ both have singularities at $z = \pm A$, with the latter having an additional singularity at $z = 0$. The contour C in the z-plane must be sufficiently large and contain all three singular points. The contour integrals can then be reduced by residue calculus to computing the coefficients of $1/z$, $1/(z-A)$ and $1/(z+A)$ in Taylor–Laurent series expansions of the corresponding integrands.

Remark 3.30 (Contour integration) Computing the residues associated with each of the singular points of the integrands just mentioned can be replaced by computing the coefficient of z^{-1} (only) in the Laurent expansions for the integrands centred at $z = 0$ and valid for sufficiently large z in modulus. This is because either of the contour integrals concerned, for a sufficiently large contour C containing the finite singular points, equals minus $2\pi\mathrm{i}$ times the residue for the integrand $f = f(z)$ at $z = \infty$. The residue of the integrand f at $z = \infty$ in turn equals minus the residue of $z^{-2}f(z^{-1})$ at zero, which equals the z^{-1} coefficient in the Laurent series expansion of f centred on $z = 0$. The end result of all this is that the contour integral of f around a sufficiently large contour C is $2\pi\mathrm{i}$ times the coefficient of z^{-1} in the Laurent series expansion of f centred on $z = 0$. See Ponnusamy and Silverman (2006, pp. 313–4) for more details. Also see Exercises 3.8 and 3.9, where we put this technique into action.

Remark 3.31 When the circulation is zero, i.e. $\kappa = 0$, the total force on the elliptical cylinder is zero; see Exercise 3.8 as well as Exercise 3.9.

Example 3.32 (Elliptical aerofoil: stagnation points) We have seen that the flow field characterised by $W'(\zeta)$ given in the last example above has two singular points at $z = \pm A$. In addition it has two stagnation points as well. Setting $(W'(\zeta))^*$ equal to zero, or equivalently $W'(\zeta)$ equal to zero, we see the stationary points occur where z satisfies

$$z^2 - \frac{\mathrm{i}\kappa \mathrm{e}^{\mathrm{i}\alpha}}{2\pi U} z + a^2 \mathrm{e}^{2\mathrm{i}\alpha} = 0 \quad \Leftrightarrow \quad r^2 \mathrm{e}^{2\mathrm{i}\theta} - \frac{\mathrm{i}\kappa \mathrm{e}^{\mathrm{i}\alpha}}{2\pi U} r \mathrm{e}^{\mathrm{i}\theta} + a^2 \mathrm{e}^{2\mathrm{i}\alpha} = 0,$$

where in the second equation we set $z = r\mathrm{e}^{\mathrm{i}\theta}$. Of specific interest, as we shall see, is whether there are any stagnation points on the surface of the ellipse and their locale. Since the surface of the ellipse in the ζ-plane corresponds to the surface of the circle $|z| = a$ in the z-plane, we set $r = a$ in this last equation and observe that we obtain the condition

$$\kappa = 4\pi a U \sin(\theta - \alpha),$$

characterising the stagnation points on the surface. This condition chimes with our computations at the beginning of Section 3.3 and Figure 3.3 as follows – note we should rotate the figures therein counter-clockwise by α. A solution for θ can only exist provided $|\kappa/4\pi a U| \leqslant 1$, in which case we obtain the situation in the z-plane shown in one of the cases (a), (b) or (c) in Figure 3.3 – with the reflected versions if κ is negative but $\kappa/4\pi a U > -1$. If $|\kappa/4\pi a U| > 1$ a stagnation point on the surface doesn't exist. Indeed one lies off the surface as shown in case (d) in Figure 3.3 – with the reflected version if κ is negative.

The left panel in Figure 3.14 gives a visualisation of the potential flow around the elliptical aerofoil we considered in Examples 3.29 and 3.32 above. Let us suppose that κ is negative though $\kappa/4\pi a U > -1$, so there exist two stagnation points on the ellipse surface. These two values of θ which solve the condition $\kappa/4\pi a U = \sin(\theta - \alpha)$ must lie between $-\pi + \alpha$ and α. In the left panel in Figure 3.14 we indicate the situation when α is relatively large and κ is negative though small in magnitude, so there are solutions for $\theta \in (0, \alpha)$ and θ slightly greater than $-\pi + \alpha$. With α fixed as we increase the magnitude of the negative circulation κ, we see that the first stagnation point we just highlighted moves down under the trailing edge and the second stagnation point moves further back towards the trailing edge along the underside. Note that if we would like to ensure there is a stagnation point at the trailing edge of the elliptical aerofoil where $\theta = 0$, then we must choose the circulation κ such that $\kappa = 4\pi a U \sin(-\alpha) = -4\pi a U \sin\alpha$. We remark that singular points in the flow at $z = \pm A$, or equivalently $\zeta = \pm 2A$, reside within the aerofoil cavity for $a > A$ – recall that we assumed $a \geqslant A$ in Example 3.29. Indeed, recall that the boundary surface is parameterised by $\xi = (a + A^2/a)\cos\theta$ and $\eta = (a - A^2/a)\sin\theta$, and so when $\theta = 0$ or $\theta = \pi$, respectively, $\xi = a + A^2/a > 2A$ and $\xi = -(a + A^2/a) < -2A$ since $2aA < a^2 + A^2$. We expect the flow velocity near the front and trailing edges to be high, and this is emphasised by the fact that if a is only slightly larger than A then the denominator terms $(z - A)$ and $(z + A)$ in the expression

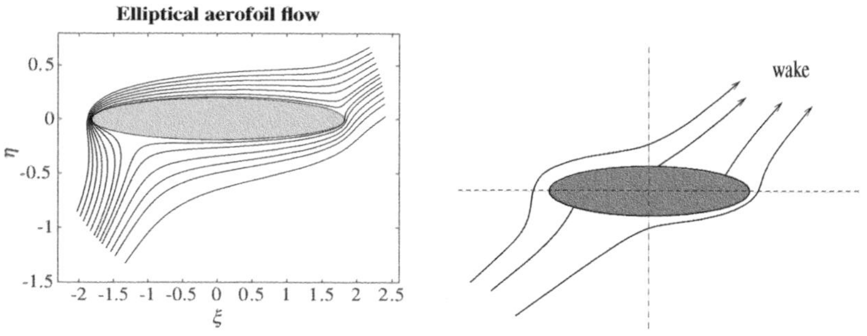

Figure 3.14 The left panel indicates the ideal potential flow around an elliptical aerofoil, as outlined in Examples 3.29 and 3.32. The parameter values are $U = 10$, $a = 1$, $A = 0.9$, $\alpha = 0.5$ and $\kappa = -36.15$. The flow velocity is very high near the front and trailing edges of the aerofoil. The right panel indicates a real flow with a separation at the trailing edge and a thin wake off the back (due to viscous effects).

for the conjugate flow velocity $W'(\zeta)$ become very small for z close to a. Indeed, in the next example we study the case $A = a$.

The right panel in Figure 3.14 shows a cartoon of a real flow around such an elliptical aerofoil. Viscous effects lead to the generation of a wake flow off the back of the trailing edge and a drag force on the aerofoil. We quote from Batchelor (1967, pp. 435–6):

'The primary requirements of an aerofoil in practice are that when in motion through fluid a side force should be exerted on it and that the drag, . . . , should be small. These requirements are both met by a flow which is irrotational everywhere except in a thin boundary layer and wake, provided a circulation around the aerofoil can be established. Avoidance of boundary-layer separation when the aerofoil is in steady motion is thus one objective, and establishment of circulation is another. We saw . . . that separation of the boundary layer from the body surface can be avoided only if the fluid just outside the boundary layer is not decelerated appreciably. The stagnation point at the rear face of a body in a two dimensional field is a source of trouble, and separation would be inevitable near the rear of a body with finite curvature. The natural suggestion is to use a slender aerofoil with a sharp cusped edge at the rear and to align the aerofoil roughly parallel to the direction of its motion.'

A first crude proposal for refining the aerofoil design to minimise drag, is to consider the limit of the elliptical aerofoil to a flat plate. Since $\xi = (a + A^2/a)\cos\theta$ and $\eta = (a - A^2/a)\sin\theta$ parameterise the elliptical aerofoil boundary, by taking the limit $A \to a$ we see that the elliptical aerofoil reduces to the flat plate parameterised by $\xi \in [-2a, 2a]$.

Example 3.33 (Flat plate aerofoil: flow) We can realise the flat-plate aerofoil as shown in the left panel of Figure 3.15 by taking the limit $A \to a$. Referring to Examples 3.29 and 3.32 and our discussion just above, we know that the Joukowski transformation becomes $\zeta = z + a^2/z$ and the circle $|z| = a$ in the z-plane collapses to the infinitesimally thin ellipse or flat plate parameterised by $\xi \in [-2a, 2a]$. Furthermore, the flow field characterised by $(W'(\zeta))^*$ is such that

$$W'(\zeta) = \frac{U e^{-i\alpha} z^2 - U a^2 e^{i\alpha} - i\kappa z/2\pi}{(z - a)(z + a)}.$$

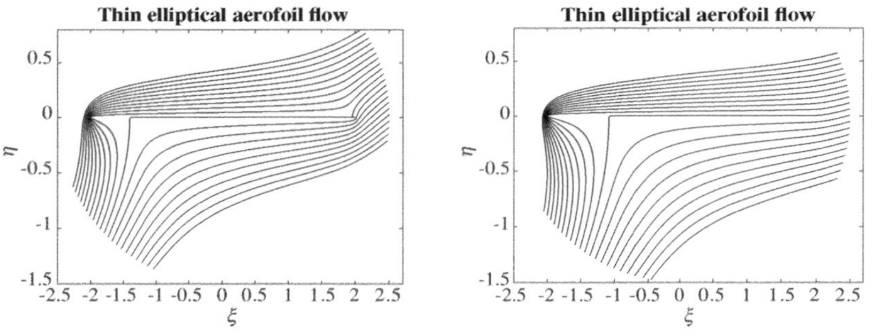

Figure 3.15 The left panel indicates the ideal potential flow around an extremely thin elliptical aerofoil, which is almost a flat plate, corresponding to the parameter values $U = 10$, $a = 1$, $A = 0.9999$, $\alpha = 0.5$ and $\kappa = -36.14$. This circulation corresponds to $\kappa/4\pi UA\sin\alpha = -0.6$. Hence the rear stagnation point is close to the trailing edge. The aerofoil is not shown in order to highlight the streamlines. The fluid velocity is very high at both the leading and trailing edges, indeed it is infinite in the limit $A \to a$. The right panel indicates the corresponding flow when the circulation is $\kappa = -60.25$, corresponding to $\kappa/4\pi UA\sin\alpha = -1$. This circulation forces the rear stagnation point to the trailing edge where the flow velocity is finite in the limit $A \to a$. See Example 3.33.

The flow of the velocity field is infinite at the leading and trailing edges of the flat plate, i.e. at $z = \pm a$ or equivalently $\zeta = \pm 2a$. The stagnation points on the surface still satisfy $\kappa = 4\pi aU\sin(\theta - \alpha)$. When $\kappa = 0$, these are located as shown in Figure 3.15 at $\theta = \alpha$ and $\theta = \pi - \alpha$. Importantly we observe that we can locate the stagnation point close to the trailing edge exactly at the trailing edge, where $\theta = 0$ if and only if the circulation takes the precise value

$$\kappa = -4\pi aU\sin\alpha.$$

A second, more refined aerofoil design proposal, more in keeping with Batchelor's suggestions, is the symmetric Joukowksi aerofoil.

Example 3.34 (Symmetric Joukowksi aerofoil: flow) Recall Example 3.27, where we outlined how the Joukowksi transformation $\zeta = z + a^2/z$ transforms the disc centred at $z = -\delta$, with $\delta > 0$, and with radius $a + \delta$ in the z-plane, to the symmetric Joukowski aerofoil with an elliptical-like front edge and a trailing cusp edge at $\zeta = +2a$ in the ζ-plane. The corresponding complex potential for the flow in the z-plane exterior to the disc boundary is given by

$$w(z) = U\mathrm{e}^{-\mathrm{i}\alpha}(z + \delta) + U(a + \delta)^2\mathrm{e}^{\mathrm{i}\alpha}\frac{1}{z + \delta} - \frac{\mathrm{i}\kappa}{2\pi}\log(z + \delta).$$

Note that the disc boundary in the z-plane is parameterised by $z = -\delta + (a + \delta)\mathrm{e}^{\mathrm{i}\theta}$ or equivalently $z + \delta = (a + \delta)\mathrm{e}^{\mathrm{i}\theta}$, for $\theta \in [0, 2\pi)$. The flow velocity in the ζ-plane is characterised by the complex conjugate of $W'(\zeta) = w'(z)/\zeta'(z)$, which explicitly takes the form

$$W'(\zeta) = z^2\frac{(U\mathrm{e}^{-\mathrm{i}\alpha} - U(a + \delta)^2\mathrm{e}^{\mathrm{i}\alpha}(z + \delta)^{-2} - (\mathrm{i}\kappa/2\pi)(z + \delta)^{-1})}{(z - a)(z + a)}.$$

We observe that the flow velocity is in general singular at $z = \pm a$. However, the singular point $z = -a$ corresponding to $\zeta = -2a$ lies within the cavity of the symmetric Joukowski aerofoil wing. The singular point $z = +a$ coincides with the trailing cusp edge of the wing at $\zeta = +2a$. Let us now locate the stagnation points on the boundary in the z-plane; hence we ignore the stagnation point at $z = 0$. From the form of $W'(\zeta)$ above there are stagnation points on the boundary, substituting $z + \delta = (a + \delta)e^{i\theta}$ into the numerator, where

$$Ue^{-i\alpha} - Ue^{i(\alpha-2\theta)} - \frac{i\kappa}{2\pi}\frac{e^{-i\theta}}{(a+\delta)} = 0.$$

Rearranging, we see that the stagnation points occur where θ satisfies

$$\kappa = 4\pi(a+\delta)U\sin(\theta - \alpha),$$

matching the condition for the elliptical aerofoil when $\delta = 0$ in Example 3.32. We thus observe that we can force the stagnation point to the trailing cusp edge of the symmetric Joukowski aerofoil by choosing (setting $\theta = 0$)

$$\kappa = -4\pi(a+\delta)U\sin\alpha.$$

Setting κ to be this value, we carefully examine $w'(a+\epsilon)/\zeta'(a+\epsilon)$ in the limit $\epsilon \to 0$ to confirm $W'(2a)$ is indeed finite, i.e. this specific choice of κ does indeed lead to a finite velocity field at the trailing cusp edge. First we observe

$$\begin{aligned} w'(a+\epsilon) &= U\left(e^{-i\alpha} - e^{i\alpha}\frac{(a+\delta)^2}{(a+\delta+\epsilon)^2} + 2i\sin\alpha\frac{(a+\delta)}{(a+\delta+\epsilon)}\right) \\ &= U\left(e^{-i\alpha} - e^{i\alpha}\left(1+\frac{\epsilon}{a+\delta}\right)^{-2} + 2i\sin\alpha\left(1+\frac{\epsilon}{(a+\delta)}\right)^{-1}\right) \\ &= \frac{2\epsilon U\cos\alpha}{a+\delta} + O(\epsilon^2). \end{aligned}$$

Second we observe, since $\zeta'(a+\epsilon) = 1 - a^2(a+\epsilon)^{-2}$, that we can write

$$\zeta'(a+\epsilon) = \epsilon\frac{(2a+\epsilon)}{(a+\epsilon)^2} = \epsilon\,(2a+\epsilon)\,a^{-2}(1+\epsilon/a)^{-2} = \frac{2}{a}\epsilon\,(1+O(\epsilon)).$$

Combining the last two expressions we see that

$$W'(a+\epsilon) = \frac{w'(a+\epsilon)}{\zeta'(a+\epsilon)} = \frac{aU\cos\alpha}{a+\delta} + O(\epsilon).$$

Hence the flow velocity at the trailing edge is indeed finite.

A third refinement in aerofoil design is the cambered Joukowksi aerofoil. Figures 3.14, 3.15 and 3.16 indicate the exact ideal flows around the corresponding aerofoils; the angle of incidence α of the impinging flow in each of the cases is large, just for emphasis. For a viscous flow, when α exceeds a few degrees, the flow separates behind the aerofoil as indicated in the right panel in Figure 3.14. The larger the angle of incidence, the wider the wake, the smaller the lift and the greater the drag on the aerofoil. The aerofoil then *stalls*. To delay the onset of stall for larger angles of incidence, which helps to increase lift, it has proven effective to give an aerofoil

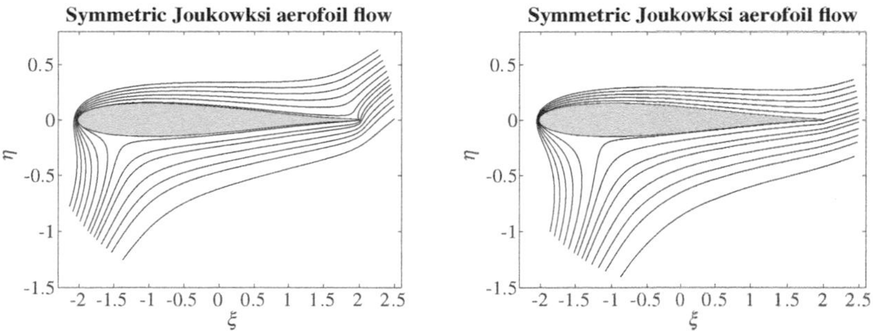

Figure 3.16 The left panel indicates the ideal potential flow around a symmetric Joukowksi aerofoil, corresponding to the parameter values $U = 10$, $a = A = 1$, $\delta = 0.06$, $\alpha = 0.5$ and $\kappa/4\pi U(a+\delta)\sin\alpha = -0.6$. The rear stagnation point is thus close to the trailing edge. The fluid velocity is infinite at the trailing edge. The right panel indicates the corresponding flow when the circulation is such that $\kappa/4\pi U(a+\delta)\sin\alpha = -1$. The rear stagnation point then coincides with the trailing edge and the flow velocity there is in fact finite for this case. See Example 3.34.

camber. This means to modify the symmetric Joukowksi aerofoil so the centre line develops an upward convex shape so the leading edge is still pointing into the stream while the bulk of the aerofoil and trailing edge are inclined to it. This is typical of a bird's wing.

Example 3.35 (Cambered Joukowksi aerofoil: flow) Recall the symmetric Joukowski aerofoil construction from Example 3.27. Here we move the centre of the larger circle in the z-plane in Figure 3.13 to the position $z = \delta e^{i\gamma}$, where $\gamma \in (\pi/2, \pi)$. The new disc intersects the real axis at $z = a$ as shown in Figure 3.17, and thus the radius R of the new circle is given by the cosine rule:

$$R^2 = a^2 + \delta^2 - 2a\delta\cos\gamma.$$

The boundary of the new circle is given by $z = \delta e^{i\gamma} + Re^{i\theta'}$, where $\theta' \in [0, 2\pi)$ is the angle the line joining the circle centre C to the point on the circle boundary makes with the x-axis. Under the Joukowski transformation $\zeta = z + a^2/z$ the boundary transforms, in the ζ-plane, to the cambered wing shown in the right panel in Figure 3.17. One direct observation revealing this is indeed the transformed aerofoil shape – the trace of the apex of the parallelogram shown in the right panel in Figure 3.17 generates the cambered aerofoil shape as follows. The dashed line from the origin O to the circle boundary gives a position $z = re^{i\theta}$ on the boundary. The dotted line from the origin represents $a^2/z = (a^2/r)\,e^{-i\theta}$. Adding the two gives the apex which must trace out the corresponding boundary on the ζ-plane.

The angle β shown in the left panel in Figure 3.17, as we see presently, contributes to the angular shift of the stagnation point on the upper trailing edge due to the camber. By considering the length of the vertical dashed line from the centre C to the x-axis shown in the left panel in Figure 3.17, we find

$$R\sin\beta = \delta\sin\gamma.$$

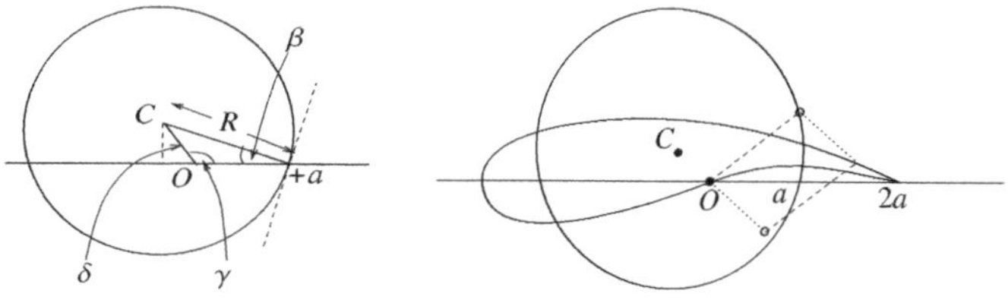

Figure 3.17 The left panel shows the disc in the z-plane centred at $z = \delta e^{i\gamma}$ with radius R. The disc boundary intersects the x-axis at $z = +a$. The right panel shows the same disc in the z-plane, however superimposed on top is the corresponding image in the ζ-plane under the Joukowski transformation $\zeta = z + a^2/z$. The dashed line from the origin O to the disc boundary gives a position $z = re^{i\theta}$ on the boundary. The dotted line from the origin represents $a^2/z = (a^2/r)\,e^{-i\theta}$. The sum of the two traces out the cambered Joukowksi aerofoil shown; see Paterson (1983). Reproduced with permission of the Licensor through PLSclear.

Further, the angle to the x-axis of the tangent to the new circle at $z = +a$ is $\theta_* := \pi/2-\beta$. The tangent line is thus parameterised by $z = a + \varrho e^{i\theta_*}$ for $\varrho \geqslant 0$. Assuming $\varrho \ll 1$, in the ζ-plane, this small section of the tangent line becomes

$$\begin{aligned}\zeta &= a + \varrho e^{i\theta_*} + a^2(a + \varrho e^{i\theta_*})^{-1}\\ &= a + \varrho e^{i\theta_*} + a\Big(1 - \frac{\varrho}{a}e^{i\theta_*} + \Big(\frac{\varrho}{a}\Big)^2 e^{2i\theta_*} + O(\varrho^3)\Big)\\ &= 2a + \frac{\varrho^2}{a}e^{2i\theta_*} + O(\varrho^3).\end{aligned}$$

Hence the argument of the corresponding small line segment in the ζ-plane through $\zeta = +2a$ is $2\theta_* = \pi - 2\beta$. This gives the incline of the trailing edge to the ξ-axis due to the camber. Hence a smooth flow over the aerofoil comes off the trailing edge at an angle -2β to the x-axis. In other words, the flow is diverted an extra 2β which should allow for larger angles of incidence, before stalling occurs. Lift is also increased, see Exercise 3.9.

The complex potential for the flow in the z-plane exterior to the large disc with centre C and radius R is

$$w(z) = Ue^{-i\alpha}(z - \delta e^{i\gamma}) + \frac{Ue^{i\alpha}R^2}{z - \delta e^{i\gamma}} - \frac{i\kappa}{2\pi}\log(z - \delta e^{i\gamma}).$$

As indicated above, the boundary of the aerofoil in the z-plane is parameterised by $z = \delta e^{i\gamma} + Re^{i\theta'}$, where $\theta' \in [0, 2\pi)$ is the angle subtended at the disc centre. Taking the complex derivative of w, setting that equal to zero to determine the stagnation points and then substituting the boundary parameterisation for z into that equation to determine the location of the stagnation points on the boundary in the z-plane, we directly obtain the following condition:

$$\kappa = -4\pi RU \sin(\alpha - \theta'),$$

for the location of the boundary stagnation points. Note, for small θ, we can see from Figure 3.17 that $\theta' < \theta$. The trailing edge is located at $z = +a$, or $\zeta = +2a$, corresponding to $\theta = 0$ or equivalently $\theta' = -\beta$. Hence the circulation that maintains the stagnation point at the trailing edge and a finite flow velocity is

$$\kappa = -4\pi RU \sin(\alpha + \beta).$$

Hence for a cambered aerofoil a larger circulation is required for the flow to come smoothly off the trailing edge. See Figure 3.18 for a visualisation of the cambered aerofoil flow for a smaller circulation (left panel) as well as the flow for the precise circulation above (right panel).

Remark 3.36 (Kutta condition and Joukowski hypothesis) The condition on the circulation in Examples 3.33, 3.34 and 3.35 that locates the stagnation point at the trailing edge, maintaining a finite flow velocity, is known as the *Kutta condition*. That in practice this precise value of the circulation is generated is the *Joukowski* or *Kutta–Joukowski hypothesis*, which we address just below.

A fourth refinement of the aerofoil design addresses the realisation of the cusp of the trailing edge. Indeed, such a cusp edge would be impossible to build in practice and a small finite wedge at the trailing edge is more realistic. The Joukowksi transformation can be tweaked to account for such a finite wedge; for more details see e.g. Paterson (1983, pp. 476–7).

With the inviscid aerofoil theory above in hand, we must now ask about the practical mechanism of aerofoil flight/lift. We give citations to the literature on 'wings' in Section 3.7. Also, viscosity plays a crucial role and we consider viscous flows next in Part II. Presently though, let us address the *Kutta–Joukowski hypothesis*. Again we quote from Batchelor (1967, p. 437):

The circulation should have that value for which, for the given orientation of the body, the rear stagnation point is located at the sharp trailing edge ... and the velocity is finite and non-zero there. It is a remarkable fact that in practice a circulation is generated round an aerofoil, owing to the convection to a non-zero amount of vorticity from the rear edge of the aerofoil at an initial stage of the motion, and then when the aerofoil is in steady motion the circulation is established with just this special value... This fortunate circumstance, that the effect of the viscosity acting on the boundary layer initially is to cause the establishment of precisely the value of the circulation that enables effects of viscosity to be ignored ... in the subsequent

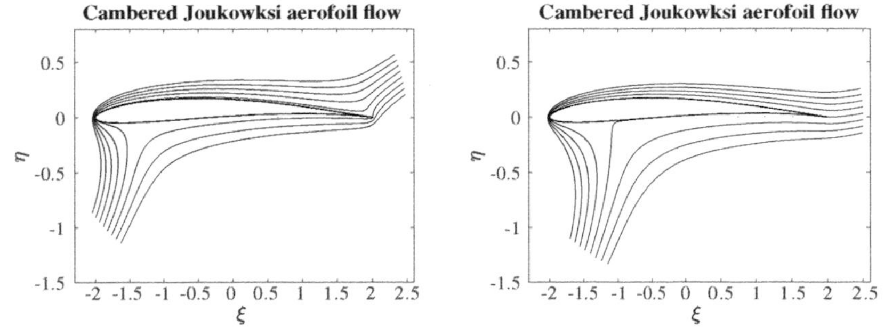

Figure 3.18 The left panel indicates the ideal potential flow around a cambered Joukowksi aerofoil, corresponding to the parameter values $U = 10$, $a = A = 1$, $\delta = 0.06$, $\alpha = 0.5$, $\gamma = \pi/2 + 0.7$ and $\kappa/(4\pi U R \sin(\alpha + \beta)) = -0.6$. The rear stagnation point is close to the trailing edge. At the trailing edge the fluid velocity is infinite. The right panel indicates the corresponding flow when the circulation is such that $\kappa/(4\pi U R \sin(\alpha + \beta)) = -1$, which forces the rear stagnation point to the trailing edge. The flow velocity at the trailing edge is then finite. See Example 3.35.

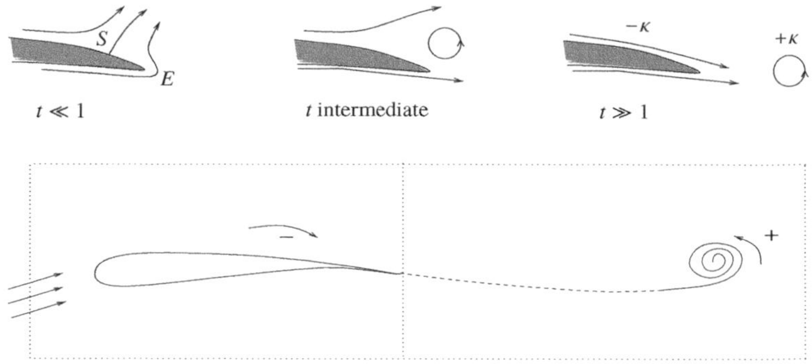

Figure 3.19 The three top panels indicate the flow sequence around the aerofoil trailing edge starting from rest through to steady motion. As the aerofoil accelerates, the flow resembles the inviscid potential flow with negligible circulation discussed in the text (left panel). A secondary flow from the stagnation point S to the trailing edge E is then initiated in the boundary layer, generating vorticity and positive circulation (middle panel) which is then shed behind the aerofoil (right panel). The top three panels are a partial reproduction of Figure XVII.47 in Paterson (1983). The bottom panel is a reproduction of Figure 6.7.4 in Batchelor (1967). Reproduced with permission of the Licensor through PLSclear. The bottom panel indicates how Kelvin's circulation theorem necessitates the generation of negative circulation around the aerofoil, and thus lift, as the vorticity of positive circulation is shed behind.

motion, is usually given the name *Joukowski's hypothesis*. It was used as an empirical rule in the early development of aerofoil theory.

Referring to Figure 3.19, let us examine the flow sequence for an aerofoil starting from rest through to the development of steady motion. At the very beginning, as the aerofoil starts to accelerate, the flow around the aerofoil resembles the inviscid potential flow with negligible circulation, revealing a stagnation point S on the upper edge near the trailing edge E; as shown in the top left panel in Figure 3.19. The flow velocity around the trailing edge E is high and thus the associated pressure field low, while near the stagnation point S the associated pressure is high. This sets up a secondary flow from S to E in the boundary layer – though exactly at the boundary of course the flow velocity matches that of the wing. The secondary flow generates vorticity with positive circulation near the trailing edge, as shown in the top middle panel in Figure 3.19. This intense vorticity is then shed from the trailing edge downstream, as shown in the top right panel. Negative circulation is consequently left around the aerofoil due to Kelvin's Circulation Theorem 2.29 for ideal flows: the circulation for any closed contour C_t is constant in time. Initially there is no circulation. Hence the circulation for the large closed contour shown in the bottom panel in Figure 3.19, i.e. the dashed boundary surrounding the entire bottom panel, is zero. Subsequently, in steady motion, positive vorticity of strength say $|\kappa|$ is shed downstream. This is the circulation corresponding to the closed contour surrounding the shed vorticity behind the aerofoil. Hence negative vorticity of equal magnitude must surround the aerofoil, corresponding to the closed contour surrounding the aerofoil only. For the cambered aerofoil at least, in steady motion, the circulation around the aerofoil is $\kappa = -4\pi RU\sin(\alpha + \beta)$; this then leads to dynamic lift as referred to in Section 3.3.

3.7 Notes

Some further details and examples are as follows.

(1) *Aerofoils: trailing wedge.* The cambered Joukowski aerofoil we considered above has a cusp at the trailing edge which cannot be attained in practice. A modification of the Joukowski transformation, the *Kármán–Trefftz transformation*, generates a wedge at the trailing edge. See e.g. Panton (2013, pp. 476–7) and Matthews (2012).

(2) *Flow separation and triple-deck theory.* We saw the phenomenon of boundary-layer separation occur as an aerofoil accelerates in the top left panel in Figure 3.19. A comprehensive treatment of boundary-layer separation requires a triple-deck structure. See Acheson, (1990, pp. 288–9), Smith (1977) and Section 6.3. There are many striking images of this phenomenon in Van Dyke (1982), notably plate 80.

(3) Wing design. We have not touched upon full three-dimensional wing design, or aircraft, helicopter or ship propeller design. For an introduction, see e.g. Milne-Thomson (1952), Batchelor (1967, Sec. 7.8), Lighthill (1986, Ch. 11), Kundu and Cohen (2008) or Anderson (2017).

(4) Supersonic flight. In Section 3.6 on aerofoil theory we treat the flow of the air as incompressible. This is a very good approximation provided the characteristic aerofoil speed is small compared with the speed of sound. However, the compressibility of air cannot be ignored once the characteristic speed approaches the speed of sound and/or transcends beyond to supersonic speeds. For an introduction to these issues, see e.g. Acheson, (1990), Kundu and Cohen (2008) or Anderson (2017).

(5) Soaring of birds (Leonardo da Vinci and Lord Rayleigh). Dynamical soaring is a flight manoeuvre performed by birds such as albatrosses. The technique exploits the gradient of the wind velocity, i.e. the wind shear, to enable sustained flight. See the recent article by Richardson (2019). Though Lord Rayleigh is 'generally accepted as being the first person to explain realistically how a bird could gain energy from wind shear for sustained soaring', Richardson explores how Leonardo da Vinci described land birds performing such manoeuvres nearly 400 years earlier. For more details, see Richardson (2019) and Rayleigh (1883).

(6) Zhukovski or Joukowski. The latter is a western transliteration of the former; see Lighthill (1986, p. 172).

3.8 Project

A suggested extended project from this chapter is as follows.

Wing design. Complete Exercises 3.8 and 3.9 and then investigate the phenomena associated with three-dimensional flows past wings including wing circulation, tip vortices and so forth. Start with the literature indicated in Section 3.7(3) and move on to different wing designs. A historical literature review starting before the Wright brothers through to the modern design of aircraft using computational fluid dynamics (CFD) simulations would also be of interest. To start, see Acheson, (1990, Ch. 4), who provides a nice summary of early aerofoil developments, as well as Milne-Thomson (1952) and Panton (2013, Sec. 18.15), and then see e.g. Johnson *et al.* (2005).

Exercises

3.1 Complex potential functions: Examples. Show that $w = z^3$ could be a complex potential, but $w = |z|$ cannot.

3.2 Complex potential functions: Non-trivial. For the complex potential given by $w = \cosh^{-1}(z/c)$ with $c > 0$ constant, sketch the streamlines and determine where the fluid velocity is zero, as well as where it is infinite.

3.3 Complex potential functions: Infinite array of vortices. Consider an infinite array of line vortices, each of strength $\kappa > 0$, which are fixed along the real axis at $z = na$, where $a > 0$ is a separation constant and $n \in \mathbb{Z}$. Using the identity

$$\frac{\sin z}{z} \equiv \prod_{n=1}^{\infty}\left(1 - \frac{z^2}{n^2\pi^2}\right),$$

show that the complex potential corresponding to this array of vortices is

$$w = -\frac{\mathrm{i}\kappa}{2\pi}\log\left(\sin\left(\frac{\pi z}{a}\right)\right).$$

Find the stream function and velocity components. Show that $(u, v) \to (\mp\kappa/2a, 0)$ as $y \to \pm\infty$.

3.4 Method of images/conformal mapping: Line source. An incompressible fluid occupies the region between walls at $x = 0, \pi$ and $y = 0$; see Example 3.25. A steady irrotational flow is generated by a line source of strength m at $z_0 = \pi/2 + a\mathrm{i}$ where $a > 0$ is a constant.

(a) Use the analytic map $z \mapsto \zeta$, where $\zeta = \cos z$, to calculate the complex potential for the flow as follows.

   (i) Show that the walled region in the z-plane transforms to the lower half complex ζ-plane, and the line source at z_0 transforms to $\zeta_0 = -\mathrm{i}\sinh(a)$.

   (ii) Use the method of images in the ζ-plane to show that the corresponding complex potential is given by $W(\zeta) = (m/2\pi)\log(\zeta^2 + \sinh^2 a)$. Then transform back to find $w = w(z)$.

(b) Find the velocity and show that as $y \to \infty$ it is $m\mathrm{i}/\pi$, for all $x \in [0, \pi]$, in accordance with fluid flux requirements.

3.5 Method of images/conformal mapping: Line vortex. A fluid occupies the region of the z-plane lying outside of the unit circle but between the lines $\theta = 0$ and $\theta = \pi/3$ within the upper right quadrant.

(a) Use a conformal transformation $z \mapsto \zeta$ to transform the region described above in the z-plane to the upper half of the ζ-plane. (*Hint:* Combine two conformal transformations, first the one from Example 3.24 for the wedge region and second the Joukowski transformation from Exercise 3.6 just below.)

(b) Now suppose a steady incompressible irrotational two-dimensional flow is generated in the wedge region outside the unit circle in the z-plane by a line vortex of strength κ at $z_0 = \sqrt{2}\mathrm{e}^{\mathrm{i}\pi/4}$, with the fluid at rest in the far field. Using the conformal transformation from part (a), find the location ζ_0 of the line vortex in the upper half ζ-plane. Use the method of images to determine the complex potential $W = W(\zeta)$ prescribing the flow generated by the line vortex at ζ_0

everywhere in the upper half ζ-plane. Hence show that the complex potential everywhere in the original region in the z-plane corresponding to the line vortex at z_0 is given by

$$w(z) = \frac{\mathrm{i}\kappa}{2\pi}\log\left(\frac{z^3 + z^{-3} + 9/4 + 7\mathrm{i}/4}{z^3 + z^{-3} + 9/4 - 7\mathrm{i}/4}\right).$$

Show that the instantaneous velocity of the vortex is given by

$$-\kappa\frac{(71 + 958\mathrm{i})}{1820\pi}.$$

3.6 Conformal mapping: Upper half plane. Consider the Joukowski transformation $z \mapsto \zeta$, where $\zeta = z + A^2/z$, with $A > 0$ constant. The goal in this problem is to use it to transform the region shown in the left panel in Figure 3.20 to the upper half of the ζ-plane shown in the right panel.

(a) Show that the critical points of the Joukowski map are $z = \pm A$.

(b) Setting $z = r\mathrm{e}^{\mathrm{i}\theta}$, show that

$$\zeta = \left(r + \frac{A^2}{r}\right)\cos\theta + \mathrm{i}\left(r - \frac{A^2}{r}\right)\sin\theta.$$

Hence deduce $\xi = (r + A^2/r)\cos\theta$ and $\eta = (r - A/r^2)\sin\theta$.

(c) Consider contours of r in the z-plane, say $r = K$ with $K \geqslant A$ constant. Show that as $\theta \in [0, 2\pi)$ varies, the coordinates ξ and η, which are the real and imaginary parts of ζ, respectively, have the form $\xi = a\cos\theta$ and $\eta = b\sin\theta$, where $a := (K + A^2/K)$ and $b := (K - A^2/K)$. Explain how these forms trace out semi-ellipses for different values of K.

(d) Now consider contours of θ, say $\theta = K$ for some constant K. Show that $\xi = (r + A^2/r)\cos K$ and $\eta = (r - A^2/r)\sin K$ and explain how these trace out semi-hyperbolae in the ζ-plane.

(e) Show that the contour $r = A$ corresponds to $\eta = 0$ and $\xi = 2A\cos\theta$ and explain how they parameterise the line joining $\zeta = -2A$ to $\zeta = 2A$. (Naturally, for $r \geqslant A$ we have $\eta \geqslant 0$.)

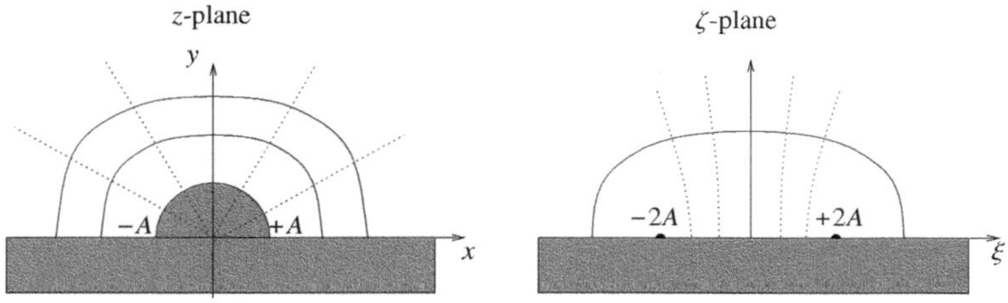

Figure 3.20 The analytic map $z \mapsto \zeta$, where $\zeta = z + A^2/z$ with $A > 0$, transforms the region shown in the z-plane to the upper half-plane region shown in the ζ-plane; see Exercise 3.6. In both the z- and ζ-planes, dotted curves correspond to contours of θ while solid curves correspond to contours of r.

3.7 Conformal mapping: Joukowski with line source. Suppose a homogeneous incompressible fluid occupies the region in the upper half complex z-plane outside of the unit circle.

(a) Use the Joukowski transformation $z \mapsto \zeta$, where $\zeta = z + 1/z$, as in Exercise 3.6, to show the region in the z-plane described above is bijectively mapped to the upper half complex ζ-plane.

(b) Suppose an irrotational flow is generated in the region described above in the z-plane by a line source of strength m situated at the point $z_0 = a\mathrm{e}^{\mathrm{i}\pi/3}$, where the constant $a > 1$. Find the location ζ_0 of the line source in the ζ-plane and then use the method of images to determine the complex potential $W = W(\zeta)$ due to the line source at ζ_0.

(c) Use the result of (b) to show that in the z-plane $z = \pm 1$ are both stagnation points. Further show that a third distinct stagnation point exists on the semi-circular boundary if and only if $a < a_0 := 2 + \sqrt{3}$.

(d) If $a = a_0$, use Bernoulli's Theorem to show that the pressure at the point $z = \mathrm{i}$ of the semi-circular boundary is less than that at infinity by an amount $8\rho m^2/169\pi^2$, where ρ is the uniform density of the fluid. (*Hint:* You may consider performing this calculation in the ζ-plane.)

3.8 Aerofoils and lift: Elliptical aerofoil. The goal herein is to compute the force and moment on an elliptical cylinder placed in a uniform ideal flow with angle of incidence α. Recall Section 3.6 and in particular Example 3.29 and the setup therein. Assume an elliptical cylinder with semi-major and minor axes a and b, respectively aligned with the real and imaginary z-axes, is placed into a stream $U\mathrm{e}^{\mathrm{i}\alpha}$, where $U > 0$ and $0 \leqslant \alpha < \pi/2$ are constants. (*Hint:* Use Example 3.29 throughout.)

(a) Using the Joukowski transformation $\zeta \mapsto z$, where $z = \zeta + A^2/\zeta$ with the constant A suitably chosen, find the solution flow in terms of the complex potential function $w = w(z)$, as follows.

  (i) Show that the boundary of the ellipse in the z-plane corresponds to the boundary of a disc of radius R in the ζ-plane if and only if $a = R + A^2/R$ and $b = R - A^2/R$. Hence deduce $R = \frac{1}{2}(a + b)$ and $A^2 = \frac{1}{4}(a^2 - b^2)$.

  (ii) Explain why we should assume the flow exterior to the disc of radius R in the ζ-plane is given by

$$W(\zeta) = U\mathrm{e}^{-\mathrm{i}\alpha}\zeta + UR^2\mathrm{e}^{\mathrm{i}\alpha}\frac{1}{\zeta} - \frac{\mathrm{i}\kappa}{2\pi}\log\zeta.$$

(b) Show that if we set $\kappa = -2\pi U(a + b)\sin\alpha$, then the trailing end of the ellipse $z = a$ is a stagnation point. (*Hint:* See Example 3.32.)

(c) Assume $\rho > 0$ is the uniform density. To use the Blasius Theorem 3.15 to compute the total force and moment on the elliptical cylinder, we must choose a suitably large contour in the ζ-plane that contains the singularities of $(W'(\zeta))^2/z'(\zeta)$ and $z(\zeta)(W'(\zeta))^2/z'(\zeta)$, respectively. Using Remark 3.30, expand these two integrands in Laurent series centred at $z = 0$ and valid sufficiently far from $z = 0$, and show that the total force and moment are respectively given by

$$F_x - \mathrm{i}F_y = \mathrm{i}\rho\kappa U\mathrm{e}^{-\mathrm{i}\alpha} \quad \text{and} \quad M = -\pi\rho U^2(a^2 - b^2)\sin\alpha\cos\alpha.$$

(*Note:* Assuming $b < a$, and since the incident flow is given by $u + \mathrm{i}v = U\mathrm{e}^{\mathrm{i}\alpha} \equiv U\cos\alpha + \mathrm{i}U\sin\alpha$, the moment is negative, turning the ellipse broadside on increasing the relative value of α. The moment is zero at $\alpha = \pi/2$ and positive just after.)

3.9 Aerofoils and lift: Camber, force and moment. This exercise is the extension of Exercise 3.8 to the cambered Joukowksi aerofoil; see Example 3.35. The goal is to compute the force and moment when such an aerofoil is placed in a uniform ideal flow with angle of incidence α. Indeed, the incidence stream is $U\mathrm{e}^{\mathrm{i}\alpha}$, where $U > 0$ and $0 \leqslant \alpha < \pi/2$ are constants. We use the setup and notation from Exercise 3.8 and Figure 3.17. To compute the force and moment on the cambered aerofoil in the ζ-plane using the Blasius Theorem 3.15, we naturally pull back the corresponding contour integrals to the z-plane via the Joukowksi transformation $z \mapsto \zeta$, where $\zeta = z + a^2/z$. In the z-plane the cambered Joukowski aerofoil corresponds to a disc centred at $z = \delta\mathrm{e}^{\mathrm{i}\gamma}$, where $\gamma \in (\pi/2, \pi)$, with radius R and so forth, as in Example 3.35. The complex potential for the flow in the z-plane exterior to the disc is

$$w(z) = U\mathrm{e}^{-\mathrm{i}\alpha}(z - \delta\mathrm{e}^{\mathrm{i}\gamma}) + \frac{U\mathrm{e}^{\mathrm{i}\alpha}R^2}{z - \delta\mathrm{e}^{\mathrm{i}\gamma}} - \frac{\mathrm{i}\kappa}{2\pi}\log(z - \delta\mathrm{e}^{\mathrm{i}\gamma}).$$

Recall from Example 3.35 that if we set $\kappa = -4\pi RU\sin(\alpha + \beta)$, where β is given by the relation $R\sin\beta = \delta\sin\gamma$, then the rear stagnation point is located at the trailing edge of the cambered Joukowski aerofoil, corresponding to $z = +a$ in the z-plane. Assume $\rho > 0$ is the uniform density of the fluid flow. We use Remark 3.30 to compute the total force and moment via the Blasius Theorem 3.15 as follows.

(a) Show that $w'(z)$ has a Laurent series expansion of the form

$$w'(z) = U\mathrm{e}^{-\mathrm{i}\alpha} - \frac{\mathrm{i}\kappa}{2\pi}\frac{1}{z} - \left(U\mathrm{e}^{\mathrm{i}\alpha}R^2 + \frac{\mathrm{i}\kappa}{2\pi}\delta\mathrm{e}^{\mathrm{i}\gamma}\right)\frac{1}{z^2} + \cdots,$$

where the terms of the form z^{-n} for $n \geqslant 3$ will not play a role in the coming contour integration.

(b) Use the Binomial series expansion to show that

$$\frac{1}{\zeta'(z)} = \left(1 - \frac{a^2}{z^2}\right)^{-1} = 1 + \frac{a^2}{z^2} + \left(\frac{a^2}{z^2}\right)^2 + \cdots.$$

(c) Using part (a), derive a Laurent series expansion for $(w'(z))^2$ and combine that with part (b) to produce a Laurent series expansion for $(w'(z))^2/\zeta'(z)$. Use the coefficient of the z^{-1} term in this last expansion to show, via the Blasius Theorem, that the total force on the cambered Joukowski aerofoil is given by (the same as for the elliptic aerofoil)

$$F_\xi - \mathrm{i}F_\eta = \mathrm{i}\rho\kappa U\mathrm{e}^{-\mathrm{i}\alpha}.$$

(d) Use parts (b) and (c) to show, retaining only the z^{-1} term, that

$$\frac{\zeta(z)}{\zeta'(z)}(w'(z))^2 = \left(2a^2U^2\mathrm{e}^{-2\mathrm{i}\alpha} - \frac{\kappa^2}{4\pi^2} - 2U^2R^2 - 2U\frac{\mathrm{i}\kappa}{2\pi}\delta\mathrm{e}^{\mathrm{i}(\gamma-\alpha)}\right)\frac{1}{z} + \cdots.$$

Hence show that the total moment on the cambered Joukowski aerofoil is

$$M = -2\pi\rho U^2 (a^2 \sin(2\alpha) - 2R\delta \sin(\alpha + \beta) \cos(\gamma - \alpha)).$$

(Compare this with Exercise 3.8(c); you will need results from part (a)(i) of that exercise to do so.)

Part II

Viscous Flow

4 Navier–Stokes Flow

4.1 Fluid Stresses

Recall our discussion on internal fluid forces in Section 2.1. Here and in the next two sections we consider the explicit form of the shear stresses and in particular the deviatoric stress matrix. This is necessary if we want to consider/model any real fluid, i.e. non-ideal fluid. We explain shear stresses as follows – see Chorin and Marsden (1990, p. 31). Imagine two neighbouring parcels of fluid P and P' as shown in Figure 4.1, with a mutual contact surface $\mathrm{d}S$. Suppose both parcels of fluid are moving parallel to $\mathrm{d}S$ and to each other, but the speed of P, say $\boldsymbol{u}$, is much faster than that of P', say $\boldsymbol{u}'$. In the kinetic theory of matter molecules jiggle about and take random walks; they diffuse into their surrounding locale and impart their kinetic energy to molecules they pass by. Hence the faster molecules in P diffuse across $\mathrm{d}S$ and impart momentum to the molecules in P'. Similarly, slower molecules from P' diffuse across $\mathrm{d}S$ to slow the fluid in P down. In regions of the flow where the velocity field changes rapidly over small length scales, this effect is important. This is the phenomenon responsible for the resistance to shearing motion real fluids have, and gives rise to shear stresses.

We now proceed more formally. An authoritative account on fluid stresses can be found in Batchelor (1967, pp. 8–10), which we essentially follow here. The force per unit area exerted across a surface (imaginary in the fluid) is called the *stress*. Let $\mathrm{d}S$ be a small imaginary surface in the fluid centred on the point $\boldsymbol{x}$ with normal $\boldsymbol{n}$ – see Figure 2.1. The force $\mathrm{d}\boldsymbol{F}$ on side (2) by side (1) of $\mathrm{d}S$ in the fluid/material is given by

$$\mathrm{d}\boldsymbol{F} = \boldsymbol{\Sigma}(\boldsymbol{n}, \boldsymbol{x}, t)\,\mathrm{d}S.$$

Here $\boldsymbol{\Sigma} = \boldsymbol{\Sigma}(\boldsymbol{n}, \boldsymbol{x}, t)$ is the local stress at the point $\boldsymbol{x}$. It is a function of position $\boldsymbol{x}$, the normal direction $\boldsymbol{n}$ to the surface $\mathrm{d}S$ and time t. In particular, for example, with reference to Figure 2.1, if $\boldsymbol{\Sigma}$ points in the same direction as $\boldsymbol{n}$, then it represents a tension.

Let us first expose how $\boldsymbol{\Sigma}$ depends on $\boldsymbol{n}$. Again, with reference to Figure 2.1, the force on side (1) by side (2) is $-\boldsymbol{\Sigma}(\boldsymbol{n}, \boldsymbol{x}, t)\,\mathrm{d}S$, which evidently from our definition above equals $\boldsymbol{\Sigma}(-\boldsymbol{n}, \boldsymbol{x}, t)\,\mathrm{d}S$. Hence $\boldsymbol{\Sigma}$ is an odd function of $\boldsymbol{n}$. Now consider the tetrahedron-shaped volume element of volume $\mathrm{d}V$ in Figure 4.2. Suppose the three orthogonal faces $\mathrm{d}S_1$, $\mathrm{d}S_2$ and $\mathrm{d}S_3$ shown have outward normals $-\boldsymbol{i}_1$, $-\boldsymbol{i}_2$ and $-\boldsymbol{i}_3$, respectively. Assume these orthogonal faces meet at position $\boldsymbol{x}$. Suppose the fourth inclined face has outward normal $\boldsymbol{n}$ and area $\mathrm{d}S$. The combined sum of all the surface forces acting

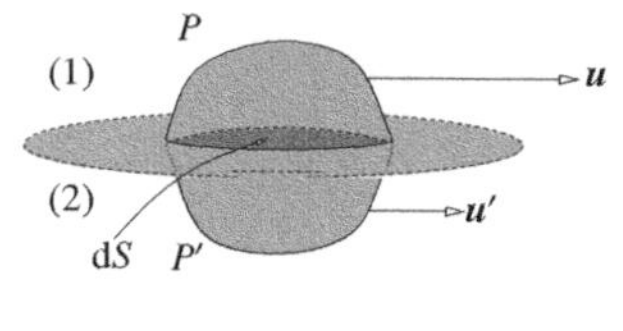

Figure 4.1 Two neighbouring parcels of fluid P and P'. Suppose $\mathrm{d}S$ is the surface of mutual contact between them. Their respective velocities are $\boldsymbol{u}$ and $\boldsymbol{u}'$ and in the same direction and parallel to $\mathrm{d}S$, but with $|\boldsymbol{u}| \gg |\boldsymbol{u}'|$. The faster molecules in P diffuse across the surface $\mathrm{d}S$ and impart momentum to P'.

on the tetrahedron is thus

$$\boldsymbol{\Sigma}(\boldsymbol{n}, \boldsymbol{x}, t)\,\mathrm{d}S + \boldsymbol{\Sigma}(-\boldsymbol{i}_1, \boldsymbol{x}, t)\,\mathrm{d}S_1 + \boldsymbol{\Sigma}(-\boldsymbol{i}_2, \boldsymbol{x}, t)\,\mathrm{d}S_2 + \boldsymbol{\Sigma}(-\boldsymbol{i}_3, \boldsymbol{x}, t)\,\mathrm{d}S_3,$$

assuming Σ is approximately constant across the tetrahedron. The higher-order corrections to this latter approximation vanish in the small volume limit of $\mathrm{d}V$. Now note that $\mathrm{d}S_1$ is the projection of $\mathrm{d}S$ onto the (x_2, x_3)-plane and thus $\mathrm{d}S_1 = (\boldsymbol{n} \cdot \boldsymbol{i}_1)\,\mathrm{d}S$. Analogous projections onto the other two orthogonal planes reveal $\mathrm{d}S_2 = (\boldsymbol{n} \cdot \boldsymbol{i}_2)\,\mathrm{d}S$ and $\mathrm{d}S_3 = (\boldsymbol{n} \cdot \boldsymbol{i}_3)\,\mathrm{d}S$. Suppose n_j are the components of $\boldsymbol{n}$ along the vectors $\boldsymbol{i}_j$ for $j = 1, 2, 3$. Then, since $\boldsymbol{\Sigma}$ is an odd function of its first argument, the combined sum of all the forces acting on the tetrahedron is

$$\Big(\boldsymbol{\Sigma}(\boldsymbol{n}, \boldsymbol{x}, t) - \big(\boldsymbol{\Sigma}(\boldsymbol{i}_1, \boldsymbol{x}, t)\,n_1 + \boldsymbol{\Sigma}(\boldsymbol{i}_2, \boldsymbol{x}, t)\,n_2 + \boldsymbol{\Sigma}(\boldsymbol{i}_3, \boldsymbol{x}, t)\,n_3\big)\Big)\,\mathrm{d}S.$$

For the tetrahedron in Figure 4.2, consider Newton's Second Law:

$$\text{mass} \times \text{acceleration} = \text{total body force} + \text{combined surface forces}.$$

The second term on the right is given just above and is proportional to $\mathrm{d}S$. The first term on the right is given by $\boldsymbol{f}\,\rho\,\mathrm{d}V$, where recall that $\boldsymbol{f}$ is the body force per unit mass and $\mathrm{d}V$ is the volume of the tetrahedron. For the term on the left, the mass is proportional to $\rho\,\mathrm{d}V$ – we assume the density and acceleration fields to be finite everywhere. Now consider the limit as the tetrahedron shrinks without changing shape to zero volume. Since the term on the left and the first term on the right decrease to zero at a rate proportional to $\mathrm{d}V$ while the second term on the right decreases to zero at a rate proportional to $\mathrm{d}S$, the coefficient of $\mathrm{d}S$ must vanish identically and we have

$$\boldsymbol{\Sigma}(\boldsymbol{n}, \boldsymbol{x}, t) = \boldsymbol{\Sigma}(\boldsymbol{i}_1, \boldsymbol{x}, t)\,n_1 + \boldsymbol{\Sigma}(\boldsymbol{i}_2, \boldsymbol{x}, t)\,n_2 + \boldsymbol{\Sigma}(\boldsymbol{i}_3, \boldsymbol{x}, t)\,n_3,$$

or in terms of matrices

$$\boldsymbol{\Sigma}(\boldsymbol{n}, \boldsymbol{x}, t) = \Big(\boldsymbol{\Sigma}(\boldsymbol{i}_1, \boldsymbol{x}, t) \quad \boldsymbol{\Sigma}(\boldsymbol{i}_2, \boldsymbol{x}, t) \quad \boldsymbol{\Sigma}(\boldsymbol{i}_3, \boldsymbol{x}, t)\Big)\begin{pmatrix} n_1 \\ n_2 \\ n_3 \end{pmatrix}.$$

The second matrix factor on the right is the vector $\boldsymbol{n}$ represented in terms of its components/projections n_1, n_2 and n_3 in the respective orthogonal $\boldsymbol{i}_1$, $\boldsymbol{i}_2$ and $\boldsymbol{i}_3$ directions. The first matrix factor on the right is the matrix with columns one, two

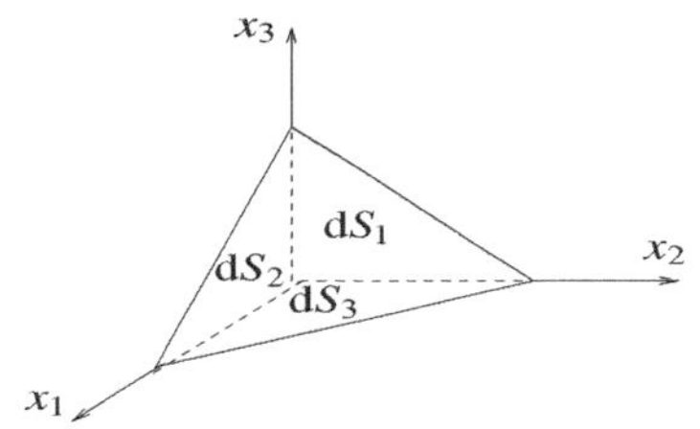

Figure 4.2 To reveal how the stress $\boldsymbol{\Sigma} = \boldsymbol{\Sigma}(\boldsymbol{n}, \boldsymbol{x}, t)$ depends on $\boldsymbol{n}$ – it depends linearly on $\boldsymbol{n}$ – we apply Newton's Second Law to the small tetrahedron-shaped volume in the fluid shown. We suppose the outward normal to the inclined face is $\boldsymbol{n}$, while the three orthogonal faces $\mathrm{d}S_1$, $\mathrm{d}S_2$ and $\mathrm{d}S_3$ have respective outward normals $-\boldsymbol{i}_1$, $-\boldsymbol{i}_2$ and $-\boldsymbol{i}_3$. Reproduced with permission of the Licensor through PLSclear.

and three given by the respective vectors $\boldsymbol{\Sigma}(\boldsymbol{i}_1, \boldsymbol{x}, t)$, $\boldsymbol{\Sigma}(\boldsymbol{i}_2, \boldsymbol{x}, t)$ and $\boldsymbol{\Sigma}(\boldsymbol{i}_3, \boldsymbol{x}, t)$. Note that these vectors are independent of $\boldsymbol{n}$ and thus so is the matrix whose columns consist of them. Given the arbitrary choice of $\boldsymbol{n}$ and the coordinate directions $\boldsymbol{i}_1$, $\boldsymbol{i}_2$ and $\boldsymbol{i}_3$ in the argument we have just provided, we deduce that $\boldsymbol{\Sigma} = \boldsymbol{\Sigma}(\boldsymbol{n}, \boldsymbol{x}, t)$ is given by

$$\boldsymbol{\Sigma}(\boldsymbol{n}, \boldsymbol{x}, t) = \sigma(\boldsymbol{x}, t)\, \boldsymbol{n},$$

for some matrix $\sigma = \sigma(\boldsymbol{x}, t)$ which is independent of $\boldsymbol{n}$. Note that $\sigma = [\sigma_{ij}]$ is a 3×3 matrix known as the *stress tensor* or *stress matrix*. The component σ_{ij} represents the 'i-component of the force per unit area exerted across a plane surface element normal to the j-direction', quoting from Batchelor (1967, p. 10). The three diagonal elements σ_{ii} for $i = 1, 2, 3$ are called the *normal stresses*. This is because, for a surface element with normal $\boldsymbol{n}$, the term σ_{ii} generates the normal contribution to the surface force component $\Sigma_i\, \mathrm{d}S$ acting across the surface element. The six off-diagonal elements σ_{ij} with $i \neq j$ and $i, j = 1, 2, 3$ are called the *tangential* or *shear stresses*. This is because they are set up by shearing motion in the fluid. Consider the rectangular cuboid fluid element in Figure 4.3. The third dimension in the figure is suppressed. Suppose the rectangular cuboid has lengths $\mathrm{d}x_1$ and $\mathrm{d}x_2$ as shown and length $\mathrm{d}x_3$ in the third direction. The surface on the right has area $\mathrm{d}x_2\, \mathrm{d}x_3$ and outward normal $\boldsymbol{n} = (1, 0, 0)^{\mathrm{T}}$. Hence the three stress components across that surface face are $\Sigma_i = \sigma_{1i}$ for $i = 1, 2, 3$. The force across the surface in the normal direction is $\sigma_{11}\, \mathrm{d}x_2\, \mathrm{d}x_3$, while the remaining force terms are orthogonal to $\boldsymbol{n} = (1, 0, 0)^{\mathrm{T}}$. One is equal to $\sigma_{21}\, \mathrm{d}x_2\, \mathrm{d}x_3$ with the direction shown in the figure, while the other is equal to $\sigma_{21}\, \mathrm{d}x_2\, \mathrm{d}x_3$ and orthogonal to the page. Similarly, for the top surface which has area $\mathrm{d}x_1\, \mathrm{d}x_3$ and outward normal $\boldsymbol{n} = (0, 1, 0)^{\mathrm{T}}$, the three stress components across that surface face are $\Sigma_i = \sigma_{2i}$ for $i = 1, 2, 3$; and so forth. Note that the left face has outward normal $\boldsymbol{n} = (-1, 0, 0)^{\mathrm{T}}$, hence the directions for $\sigma_{11}\, \mathrm{d}x_2\, \mathrm{d}x_3$ and $\sigma_{12}\, \mathrm{d}x_2\, \mathrm{d}x_3$ shown.

We now demonstrate that the stress matrix σ is in fact necessarily symmetric. A complete general account can be found in Batchelor (1967, p. 11). Here we follow Kundu and Cohen (2008, p. 91) and Ockendon and Ockendon (1995, pp. 5–6). Consider the rectangular cuboid element of fluid shown in Figure 4.3. Assume the centre of gravity to be the centre of the rectangular cuboid element which we suppose has

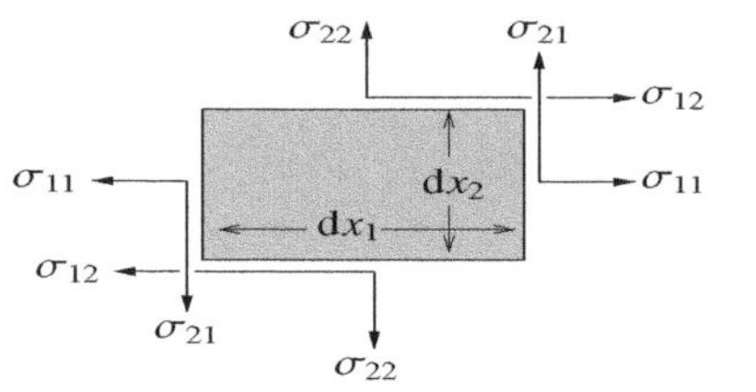

Figure 4.3 We show the forces due to the stress matrix σ acting on a rectangular cuboid element of fluid of lengths dx_1, dx_2 and dx_3. The third dimension is suppressed for simplicity. To demonstrate that the stress matrix σ is symmetric, we apply Newton's Law of Conservation of Angular Momentum to the cuboid element.

lengths dx_1, dx_2 and dx_3 in the respective directions. To leading order, we take the stress matrix σ to be uniform across the rectangular cuboid element. In the figure the third dimension, i.e. the x_3-direction, is suppressed, and for the moment we focus on the stress terms shown explicitly in the figure. The normal forces $\sigma_{11}\, dx_2\, dx_3$ and $\sigma_{22}\, dx_1\, dx_3$ due to the normal stresses σ_{11} and σ_{22} do not contribute to the couple/torque on the element (about the x_3-direction). As shown in Figure 4.3 the remaining stress terms σ_{12} and σ_{21} do, and their contribution to the couple/torque on the element is

$$\boldsymbol{G} = 2\,(\sigma_{21}\, dx_2\, dx_3)\,(\tfrac{1}{2}dx_1) - 2\,(\sigma_{12}\, dx_1\, dx_3)\,(\tfrac{1}{2}dx_2).$$

Applying Newton's Law of Conservation of Angular Momentum to the element, we know $\boldsymbol{G}$ must equal the product of the moment of inertia for the element and the angular acceleration which is dimensionless. The moment of inertia scales like the product of the mass of the element, which is proportional to $\rho\, dx_1\, dx_2\, dx_3$, and a measure of the distribution of that mass which is proportional to the square of this distance to the axis of rotation which is $((dx_1)^2 + (dx_2)^2)$. In other words, we see that $\boldsymbol{G}$ is proportional to $dx_1\, dx_2\, dx_3$, while the product of the moment of inertia and angular acceleration (assuming the latter is finite) is proportional to $dx_1\, dx_2\, dx_3\, ((dx_1)^2 + (dx_2)^2)$. Hence in the limit of vanishing rectangular cuboid element size, we deduce $\boldsymbol{G} \to \boldsymbol{0}$ and thus $\sigma_{21} = \sigma_{12}$. Since this argument can be applied in any plane, we deduce for all $i, j = 1, 2, 3$ that

$$\sigma_{ji} = \sigma_{ij},$$

so σ is symmetric. Thus σ has only six independent components.

Remark 4.1 If the fluid contains particles that might be polarised by an electric field generating a body coupling, then antisymmetric stresses would need to be considered.

Hereafter it is convenient to decompose the stress matrix σ as follows:

$$\sigma = -pI + \hat{\sigma},$$

where I is the 3×3 identity matrix. The scalar quantity $p = p(\boldsymbol{x}, t)$, which represents the fluid *pressure*, is defined to be

$$p := -\tfrac{1}{3}(\sigma_{11} + \sigma_{22} + \sigma_{33}).$$

In the decomposition, the term $-pI$ represents the stress for an inviscid fluid since if this were the only term present (recall Section 2.1),

$$\sigma = -pI \quad \Rightarrow \quad \Sigma(\boldsymbol{n}) = -p\,\boldsymbol{n} \quad \Rightarrow \quad \mathrm{d}\boldsymbol{F} = -p\,\boldsymbol{n}\,\mathrm{d}S.$$

The remaining part of the stress matrix $\hat{\sigma} = \hat{\sigma}(\boldsymbol{x}, t)$ is known as the *deviatoric stress tensor* or *deviatoric stress matrix*. It represents the additional stresses present when we consider a real fluid with viscosity. See Batchelor (1967, pp. 141–2) for more details. We examine the precise structure of $\hat{\sigma}$ in Section 4.3 after deriving Cauchy's equations of fluid motion next.

4.2 Cauchy Equations of Fluid Motion

Consider an arbitrary imaginary subregion $\mathcal{V}$ of $\mathcal{D}$ identified at time $t = 0$, as in Figure 1.3. As in our derivation of the Euler equations, let $\mathcal{V}_t$ denote the volume of the fluid occupied by the particles at $t > 0$ that originally made up $\mathcal{V}$. Recalling from Section 4.1 just above the decomposition $\mathrm{d}\boldsymbol{F} = (-pI + \hat{\sigma})\,\boldsymbol{n}\,\mathrm{d}S$, the total force exerted on the fluid inside $\mathcal{V}_t$ through the stresses exerted across its boundary $\partial\mathcal{V}_t$ is given by (see Remark 4.3 below)

$$\int_{\partial\mathcal{V}_t} (-pI + \hat{\sigma})\,\boldsymbol{n}\,\mathrm{d}S \equiv \int_{\mathcal{V}_t} (-\nabla p + \nabla \cdot \hat{\sigma})\,\mathrm{d}V.$$

If $\boldsymbol{f}$ is a body force (external force) per unit mass, which can depend on position and time, then for any parcel of fluid $\mathcal{V}_t$, the total force acting on it is

$$\int_{\mathcal{V}_t} \big(-\nabla p + \nabla \cdot \hat{\sigma} + \rho\,\boldsymbol{f}\big)\,\mathrm{d}V.$$

Hence using Newton's Second Law we have

$$\frac{\mathrm{d}}{\mathrm{d}t}\int_{\mathcal{V}_t} \rho\,\boldsymbol{u}\,\mathrm{d}V = \int_{\mathcal{V}_t} \big(-\nabla p + \nabla \cdot \hat{\sigma} + \rho\,\boldsymbol{f}\big)\,\mathrm{d}V.$$

Using the Transport Theorem with $F \equiv \boldsymbol{u}$ and that $\mathcal{V}$ and thus $\mathcal{V}_t$ are arbitrary, we see for each $\boldsymbol{x} \in \mathcal{D}$ and $t \geqslant 0$ that we can deduce the following.

Theorem 4.2 (Cauchy's equation of motion) *For given scalar mass density $\rho = \rho(\boldsymbol{x}, t)$ and pressure fields $p = p(\boldsymbol{x}, t)$, a given deviatroic stress matrix $\hat{\sigma} = \hat{\sigma}(\boldsymbol{x}, t)$ and external body force $\boldsymbol{f} = \boldsymbol{f}(\boldsymbol{x}, t)$ per unit mass, the evolution of the fluid velocity field $\boldsymbol{u} = \boldsymbol{u}(\boldsymbol{x}, t)$ is given by Cauchy's equation of motion:*

$$\rho\left(\frac{\partial \boldsymbol{u}}{\partial t} + (\boldsymbol{u} \cdot \nabla)\,\boldsymbol{u}\right) = -\nabla p + \nabla \cdot \hat{\sigma} + \rho\,\boldsymbol{f}.$$

In Section 4.4 we use Cauchy's equation of motion to derive the Navier–Stokes equations. However before we can do so, we need to know much more about the form of the deviatoric stress $\hat{\sigma}$.

Remark 4.3 Note that some care is required when using the Divergence Theorem to derive the term $\nabla \cdot \hat{\sigma}$ above. For a vector $\boldsymbol{h}$, the Divergence Theorem converts the integral of $\boldsymbol{h} \cdot \boldsymbol{n}\, \mathrm{d}S$ over $\partial \mathcal{V}_t$ to the integral of $\nabla \cdot \boldsymbol{h}$ over $\mathcal{V}_t$. The contraction in the term $\hat{\sigma}\boldsymbol{n}$ equivalent to the dot product is due to matrix multiplication across the row elements of $\hat{\sigma}$ and thus each row of $\hat{\sigma}$ is one of the eventual individual components of $\nabla \cdot \hat{\sigma}$, i.e. $[\nabla \cdot \hat{\sigma}]_i = \partial \hat{\sigma}_{ij} / \partial x_j$.

4.3 Deviatoric Stress and Deformation Matrices

Motivated by the discussion at the beginning of Section 4.1 demonstrating how shear forces naturally arise from velocity gradients in a fluid, we assume the deviatoric stress matrix $\hat{\sigma}$ is a function of the velocity gradient matrix $\nabla \boldsymbol{u}$. In this section we assume the reader has some elementary familiarity with eigenvalues, eigenvectors and eigenspaces of matrices.

Remark 4.4 Herein we systematically derive the form of the deviatoric stress matrix in terms of the velocity gradients from the three properties outlined below. The derivation we provide relies solely on matrix functions and does not require knowledge of tensors of rank 4. The argument given is based on that in Gurtin (1981, Sec. 37). Readers who wish to bypass the details of this derivation can safely fast-forward to Remark 4.11.

We make three assumptions about the deviatoric stress matrix $\hat{\sigma}$ and its dependence on the velocity gradients $\nabla \boldsymbol{u}$. These are that it is:

1. *Linear.* Each component of $\hat{\sigma}$ depends linearly on all or some of the components of the velocity gradient matrix $\nabla \boldsymbol{u}$. Indeed, we assume here this linear relationship is uniform throughout the fluid and time-independent, i.e. for each component of $\hat{\sigma}$, each linear coefficient factor of each of the components of $\nabla \boldsymbol{u}$ is independent of position $\boldsymbol{x}$ in the fluid and time t.
2. *Isotropic.* By definition, if U is any orthogonal matrix, then $\hat{\sigma}$ satisfies
$$\hat{\sigma}(U(\nabla \boldsymbol{u})U^{-1}) \equiv U\hat{\sigma}(\nabla \boldsymbol{u})U^{-1}.$$
Equivalently we might say $\hat{\sigma}$ is invariant under rigid-body rotations.
3. *Symmetric.* This means $\hat{\sigma}$ satisfies $\hat{\sigma}^{\mathrm{T}} = \hat{\sigma}$. This can be deduced as a result of balance of angular momentum, as we demonstrated in Section 4.1.

Since each component of the deviatoric stress matrix $\hat{\sigma}$ is a linear function of each of the components of the velocity gradient matrix $\nabla \boldsymbol{u}$, there is a total of 81 constants of proportionality, and all of them are independent of position $\boldsymbol{x}$ and time t. We use assumptions 2 and 3 above to systematically reduce this to two constants. We split the argument into seven steps.

Step 1: Symmetry. Recall from Section 1.9 the decomposition of the velocity gradient matrix, $\nabla \boldsymbol{u} = D + R$, where D is the symmetric part of $\nabla \boldsymbol{u}$ known as the rate of strain matrix and R is the antisymmetric part of $\nabla \boldsymbol{u}$ that generates rotation. We observe that when the fluid performs solid-body rotation, there should be no diffusion of momentum (the whole mass of fluid is behaving like a solid body) and the deviatoric stress $\hat{\sigma}$ should be zero. In other words, if $\boldsymbol{u}(\boldsymbol{x}, t) = R(t)\boldsymbol{x} \equiv \frac{1}{2}\omega(t) \times \boldsymbol{x}$ where $R(t)$ and thus $\omega(t)$ are functions of t only, recall the flows in Section 2.4, then $\hat{\sigma} = O$, the zero matrix. We thus deduce $\hat{\sigma}$ only depends on the symmetric part of $\nabla \boldsymbol{u}$, namely the rate of strain matrix D. Since $\hat{\sigma}$ is also symmetric, we can thus restrict our attention to functions from symmetric 3×3 matrices to symmetric 3×3 matrices.

Step 2: The Spectral Theorem. This plays an essential role in our argument. We quote the form given in Gurtin (1981, p. 11); also see e.g. Meyer (2000, p. 517). In preparation, let us fix some terminology and notation. The eigenspace of any matrix D corresponding to a given eigenvalue d_1 is the set of all vectors $\boldsymbol{v}$ satisfying $D\boldsymbol{v} = d_1 \boldsymbol{v}$. If $\boldsymbol{v} \in \mathbb{R}^3$ is a given vector, $\{\boldsymbol{v}\}^{\perp}$ denotes the two-dimensional subspace, i.e. plane through the origin, orthogonal to $\boldsymbol{v}$.

Theorem 4.5 (Spectral Theorem; Gurtin, 1981) *Let D be a symmetric 3×3 matrix. Then there is an orthonormal basis for $\mathbb{R}^3$ consisting entirely of eigenvectors of D. Further, for any such basis $\boldsymbol{e}_1$, $\boldsymbol{e}_2$ and $\boldsymbol{e}_3$, the corresponding eigenvalues d_1, d_2 and d_3 form the entire spectrum of D and*

$$D = d_1\, \boldsymbol{e}_1 \boldsymbol{e}_1^{\mathrm{T}} + d_2\, \boldsymbol{e}_2 \boldsymbol{e}_2^{\mathrm{T}} + d_3\, \boldsymbol{e}_3 \boldsymbol{e}_3^{\mathrm{T}}.$$

Conversely, if D can be expressed in this form with $\boldsymbol{e}_1$, $\boldsymbol{e}_2$ and $\boldsymbol{e}_3$ orthonormal, then d_1, d_2 and d_3 are eigenvalues of D with $\boldsymbol{e}_1$, $\boldsymbol{e}_2$ and $\boldsymbol{e}_3$ corresponding eigenvectors. Further:

(a) D has exactly three distinct eigenvalues if and only if the eigenspaces of D are three mutually perpendicular copies of $\mathbb{R}^1$.

(b) D has exactly two distinct eigenvalues if and only if D admits the representation

$$D = d_1\, \boldsymbol{e}\boldsymbol{e}^{\mathrm{T}} + d_2\, (I - \boldsymbol{e}\boldsymbol{e}^{\mathrm{T}}),$$

with $|\boldsymbol{e}| = 1$ and $d_1 \neq d_2$. In this case d_1 and d_2 are two distinct eigenvalues and the corresponding eigenspaces are span$\{\boldsymbol{e}\}$ *and $\{\boldsymbol{e}\}^{\perp}$, respectively. Conversely, if* span$\{\boldsymbol{e}\}$ *and $\{\boldsymbol{e}\}^{\perp}$ are the eigenspaces for D, then D can be expressed in the form involving $\boldsymbol{e}_1 \boldsymbol{e}_1^{\mathrm{T}}$ and $I - \boldsymbol{e}\boldsymbol{e}^{\mathrm{T}}$ just above.*

(c) D has exactly one eigenvalue if and only if for some scalar d_1,

$$D = d_1\, I.$$

In this case d_1 is the eigenvalue and $\mathbb{R}^3$ is the corresponding eigenspace. Conversely, if $\mathbb{R}^3$ is an eigenspace for D, then D has the form just shown.

Step 3: Wang's Lemma. This is a direct consequence of the Spectral Theorem. Its proof constitutes Exercise 4.6; also see Gurtin (1981, p. 232).

Lemma 4.6 (Wang's Lemma) *Suppose D is a symmetric 3×3 matrix. Then:*

(a) Consider the spectral decomposition

$$D = d_1\, \boldsymbol{e}_1\boldsymbol{e}_1^{\mathrm{T}} + d_2\, \boldsymbol{e}_2\boldsymbol{e}_2^{\mathrm{T}} + d_3\, \boldsymbol{e}_3\boldsymbol{e}_3^{\mathrm{T}},$$

and assume the eigenvalues d_1, d_2 and d_3 are distinct. Then the set $\{I, D, D^2\}$ is linearly independent and

$$\mathrm{span}\{I, D, D^2\} = \mathrm{span}\{\boldsymbol{e}_1\boldsymbol{e}_1^{\mathrm{T}}, \boldsymbol{e}_2\boldsymbol{e}_2^{\mathrm{T}}, \boldsymbol{e}_3\boldsymbol{e}_3^{\mathrm{T}}\}.$$

(b) Assume that D has exactly two distinct eigenvalues, so that

$$D = d_1\, \boldsymbol{e}\boldsymbol{e}^{\mathrm{T}} + d_2\,(I - \boldsymbol{e}\boldsymbol{e}^{\mathrm{T}}),$$

with $|\boldsymbol{e}| = 1$. Then the set $\{I, D\}$ is linearly independent and

$$\mathrm{span}\{I, D\} = \mathrm{span}\{\boldsymbol{e}\boldsymbol{e}^{\mathrm{T}}, I - \boldsymbol{e}\boldsymbol{e}^{\mathrm{T}}\}.$$

Step 4: The Transfer Theorem. We now lean heavily on the isotropy assumption 2; see Gurtin (1981, Sec. 37). We have the following transfer theorem.

Theorem 4.7 (Transfer theorem) *Let $\hat{\sigma}$ be a function from the set of symmetric 3×3 matrices to themselves. If $\hat{\sigma}$ is isotropic, then every eigenvector of the symmetric matrix D is an eigenvector of $\hat{\sigma}(D)$.*

Proof Let $\boldsymbol{e}$ be an eigenvector of the symmetric matrix D. Now suppose U is the orthogonal matrix denoting reflection in the plane perpendicular to $\boldsymbol{e}$, so that $U\boldsymbol{e} = -\boldsymbol{e}$. This also means that any vector perpendicular to $\boldsymbol{e}$ is invariant under U, so $D\boldsymbol{f} = \boldsymbol{f}$ if $\boldsymbol{f} \cdot \boldsymbol{e} = 0$. By the Spectral Theorem U leaves the eigenspaces of D invariant, and thus by the Commutation Theorem, see Gurtin (1981, p. 12), the matrices D and U commute so $UD = DU$, or equivalently $UDU^{-1} = D$. Thus, since $\hat{\sigma} = \hat{\sigma}(D)$ is isotropic, we have $U\hat{\sigma}U^{-1} = \hat{\sigma}(UDU^{-1}) = \hat{\sigma}(D)$ and thus $U\hat{\sigma} = \hat{\sigma}U$. Any such commuting matrices share eigenvectors. Indeed, we see that $U\hat{\sigma}\boldsymbol{e} = \hat{\sigma}U\boldsymbol{e} = -\hat{\sigma}\boldsymbol{e}$. Thus $\hat{\sigma}\boldsymbol{e}$ is also an eigenvector of the reflection transformation U corresponding to the same eigenvalue -1. Thus $\hat{\sigma}\boldsymbol{e}$ is proportional to $\boldsymbol{e}$ and so $\boldsymbol{e}$ is an eigenvector of $\hat{\sigma}$. Since $\boldsymbol{e}$ was any eigenvector of D, the statement of the theorem follows. □

Step 5: Representation Theorem for isotropic functions. For any 3×3 matrix A with eigenvalues $\lambda_1, \lambda_2, \lambda_3$, the following three scalar functions, known as the *principal invariants*, are isotropic:

$$I_1(A) := \mathrm{tr}\, A, \quad I_2(A) := \tfrac{1}{2}((\mathrm{tr}\, A)^2 - \mathrm{tr}\, A^2) \quad \text{and} \quad I_3(A) := \det A.$$

This can be checked by direct computation utilising the properties of the trace and determinant such as $\mathrm{tr}\,(BC) = \mathrm{tr}\,(CB)$ and $\det(BC) = \det(B)\det(C)$ for any commensurate matricies B and C. Indeed, these three functions are the elementary symmetric functions of the eigenvalues of A:

$$I_1(A) = \lambda_1 + \lambda_2 + \lambda_3, \quad I_2(A) = \lambda_1\lambda_2 + \lambda_2\lambda_3 + \lambda_1\lambda_3 \quad \text{and} \quad I_3(A) = \lambda_1\lambda_2\lambda_3.$$

We present the following representation result for scalar isotropic functions before one for symmetric matrix-valued isotropic functions.

Lemma 4.8 (Representation for scalar isotropic functions) *Scalar functions $\alpha = \alpha(D)$ of symmetric 3×3 matrices D are isotropic if and only if they are functions of the isotropic invariants of D only.*

Proof The 'if' part of the statement follows trivially as the isotropic invariants are isotropic. The 'only if' statement is established if, assuming the scalar function α is isotropic, we are able to show that

$$I_i(D) = I_i(D') \text{ for } i = 1, 2, 3 \qquad \Rightarrow \qquad \alpha(D) = \alpha(D').$$

Since the map between the eigenvalues of D and its isotropic invariants is bijective, if $I_i(D) = I_i(D')$ for $i = 1, 2, 3$, then D and D' have the same eigenvalues. Since the isospectral action UDU^{-1} of orthogonal matrices U on symmetric matrices D is transitive, there exists an orthogonal matrix U such that $D' = UDU^{-1}$. Since α is isotropic, $\alpha(UDU^{-1}) = \alpha(D)$, i.e. $\alpha(D') = \alpha(D)$. □

We can now prove the following more general representation theorem.

Theorem 4.9 (Representation theorem) *Let $\hat{\sigma}$ be a function from the set of symmetric 3×3 matrices to themselves. Then $\hat{\sigma}$ is isotropic if and only if*

$$\hat{\sigma}(D) = \alpha_0 I + \alpha_1 D + \alpha_2 D^2,$$

for every symmetric matrix D, where α_0, α_1 and α_2 are scalar functions that depend only on the isotropic invariants $I_1(D), I_2(D)$ and $I_3(D)$.

Proof Suppose $\hat{\sigma}$ is a symmetric matrix-valued function. The 'if' statement of the theorem follows by direct computation and Lemma 4.8 for scalar isotropic functions. Indeed, suppose we are given the representation for $\hat{\sigma} = \hat{\sigma}(D)$ shown with the coefficients $\alpha_i = \alpha_i(I_j(D))$ only for each $i = 1, 2, 3$ and all $j = 1, 2, 3$. Then $\alpha_i = \alpha_i(I_j(UDU^{-1}))$ using the representation Lemma 4.8 and thus

$$\begin{aligned} U\hat{\sigma}(D)U^{-1} &= \alpha_0(I_j(D))I + \alpha_1(I_j(D))UDU^{-1} + \alpha_2(I_j(D))UD^2U^{-1} \\ &= \alpha_0(I_j(UDU^{-1}))I + \alpha_1(I_j(UDU^{-1}))UDU^{-1} \\ &\quad + \alpha_2(I_j(UDU^{-1}))UDUU^{-1}DU^{-1} \\ &= \hat{\sigma}(UDU^{-1}). \end{aligned}$$

The 'only if' statement is proved as follows. Assume D has three distinct eigenvalues; see Exercise 4.5 for the other two possibilities. From the Spectral Theorem suppose the spectral decomposition of D is

$$D = d_1\, \boldsymbol{e}_1\boldsymbol{e}_1^{\mathrm{T}} + d_2\, \boldsymbol{e}_2\boldsymbol{e}_2^{\mathrm{T}} + d_3\, \boldsymbol{e}_3\boldsymbol{e}_3^{\mathrm{T}}.$$

By the Transfer Theorem and the Spectral Theorem, we can represent $\hat{\sigma}$ by

$$\hat{\sigma}(D) = \hat{\sigma}_1\, \boldsymbol{e}_1\boldsymbol{e}_1^{\mathrm{T}} + \hat{\sigma}_2\, \boldsymbol{e}_2\boldsymbol{e}_2^{\mathrm{T}} + \hat{\sigma}_3\, \boldsymbol{e}_3\boldsymbol{e}_3^{\mathrm{T}},$$

where $\hat{\sigma}_1$, $\hat{\sigma}_2$ and $\hat{\sigma}_3$ are the eigenvalues of $\hat{\sigma}$ and the corresponding eigenvectors $\boldsymbol{e}_1$, $\boldsymbol{e}_2$ and $\boldsymbol{e}_3$ are orthonormal. By Wang's Lemma, we know that the set $\{I, D, D^2\}$ is linearly independent and we have

$$\text{span}\{I, D, D^2\} = \text{span}\{\boldsymbol{e}_1\boldsymbol{e}_1^{\mathrm{T}}, \boldsymbol{e}_2\boldsymbol{e}_2^{\mathrm{T}}, \boldsymbol{e}_3\boldsymbol{e}_3^{\mathrm{T}}\}.$$

Hence there exist scalars α_0, α_1 and α_2 depending on D such that

$$\hat{\sigma}(D) = \alpha_0 I + \alpha_1 D + \alpha_2 D^2.$$

We now have to show that α_0, α_1 and α_2 are isotropic. This follows by direct computation, combining this last representation with the property that $\hat{\sigma}$ is isotropic. Indeed, these imply that $U(\alpha_0(D)I+\alpha_1(D)D+\alpha_2(D)D^2)U^{-1}$ equals $\alpha_0(UDU^{-1})I+\alpha_1(UDU^{-1})UDU^{-1} + \alpha_2(UDU^{-1})UD^2U^{-1}$, or equivalently,

$$\alpha_0(D)I + \alpha_1(D)D + \alpha_2(D)D^2 = \alpha_0(UDU^{-1})I + \alpha_1(UDU^{-1})D + \alpha_2(UDU^{-1})D^2.$$

Since $\{I, D, D^2\}$ is a linearly independent set, equating coefficients of I, D and D^2 reveals $\alpha_i(D) = \alpha_i(UDU^{-1})$ for $i = 1, 2, 3$. □

Remark 4.10 Note that neither the Transfer Theorem nor the Representation Theorem above require the functions $\hat{\sigma}$ from symmetric 3×3 matrices to themselves to be linear.

Step 6: Linearity. Now suppose $\hat{\sigma}$ is a linear function of D. Thus for any symmetric 3×3 matrix D it must have the form

$$\hat{\sigma}(D) = \lambda\, I + 2\mu\, D,$$

where the scalars λ and μ depend on the isotropic invariants of D. By the Spectral Theorem we have

$$D = d_1\, \boldsymbol{e}_1\boldsymbol{e}_1^{\mathrm{T}} + d_2\, \boldsymbol{e}_2\boldsymbol{e}_2^{\mathrm{T}} + d_3\, \boldsymbol{e}_3\boldsymbol{e}_3^{\mathrm{T}},$$

where d_1, d_2 and d_3 are the eigenvalues of D. Since the corresponding eigenvectors $\boldsymbol{e}_1$, $\boldsymbol{e}_2$ and $\boldsymbol{e}_3$ are orthonormal, each matrix $\boldsymbol{e}_i\boldsymbol{e}_i^{\mathrm{T}}$ for $i = 1, 2, 3$ is symmetric with an eigenvalue 1 and double eigenvalue 0. The eigenspace of $\boldsymbol{e}_i\boldsymbol{e}_i^{\mathrm{T}}$ corresponding to the eigenvalue 1 is $\text{span}\{\boldsymbol{e}_i\}$, and the eigenspace corresponding to the double eigenvalue 0 is $\{\boldsymbol{e}_i\}^{\perp} = \text{span}\{\boldsymbol{e}_j\}_{j \neq i}$. Hence, in particular, since $\hat{\sigma}$ is linear we know for each $i = 1, 2, 3$:

$$\hat{\sigma}(\boldsymbol{e}_i\boldsymbol{e}_i^{\mathrm{T}}) = \lambda\, I + 2\mu\, \boldsymbol{e}_i\boldsymbol{e}_i^{\mathrm{T}}.$$

The scalar functions λ and μ depend on the isotropic invariants $I_j(\boldsymbol{e}_i\boldsymbol{e}_i^{\mathrm{T}})$, $j = 1, 2, 3$, of $\boldsymbol{e}_i\boldsymbol{e}_i^{\mathrm{T}}$. However for all $i = 1, 2, 3$, since the eigenvalues of $\boldsymbol{e}_i\boldsymbol{e}_i^{\mathrm{T}}$ in each case are 1 and double eigenvalue 0, we have $I_1(\boldsymbol{e}_i\boldsymbol{e}_i^{\mathrm{T}}) = 1$, $I_2(\boldsymbol{e}_i\boldsymbol{e}_i^{\mathrm{T}}) = 0$ and $I_3(\boldsymbol{e}_i\boldsymbol{e}_i^{\mathrm{T}}) = 0$. Hence λ and μ are constants independent of $I_j(\boldsymbol{e}_i\boldsymbol{e}_i^{\mathrm{T}})$, $i, j = 1, 2, 3$. Since $\hat{\sigma}$ is linear, using the spectral decomposition for D above, we have

$$\hat{\sigma}(D) = d_1\, \hat{\sigma}(\boldsymbol{e}_1\boldsymbol{e}_1^{\mathrm{T}}) + d_2\, \hat{\sigma}(\boldsymbol{e}_2\boldsymbol{e}_2^{\mathrm{T}}) + d_3\, \hat{\sigma}(\boldsymbol{e}_3\boldsymbol{e}_3^{\mathrm{T}}) = \lambda(d_1 + d_2 + d_3)\, I + 2\mu\, D,$$

with λ and μ constant scalars. Further, since we assume that the linear relationship between $\hat{\sigma}$ and D is independent of position $\boldsymbol{x}$ and time t, then λ and μ are also. Recall $d_1 + d_2 + d_3 = \nabla \cdot \boldsymbol{u}$. Thus we have

$$\hat{\sigma} = \lambda(\nabla \cdot \boldsymbol{u})I + 2\mu D.$$

If we set $\zeta = \lambda + \frac{2}{3}\mu$, this last relation becomes

$$\hat{\sigma} = 2\mu(D - \tfrac{1}{3}(\nabla \cdot \boldsymbol{u})I) + \zeta(\nabla \cdot \boldsymbol{u})I,$$

where μ and ζ are the first and second *coefficients of viscosity*, respectively.

Step 7: Incompressibility. If we assume that we have a homogeneous incompressible flow so that $\nabla \cdot \boldsymbol{u} = 0$ throughout the flow, then the linear relation between $\hat{\sigma}$ and D is homogeneous, and we have the key property of what is known as a *Newtonian fluid*: the stress is proportional to the rate of strain and

$$\hat{\sigma} = 2\mu D.$$

Remark 4.11 The constitutive relation just above between the deviatoric stress and rate of strain can be derived in the following more direct fashion, starting of course from the same three assumptions of linearity, isotropy and symmetry. That the deviatoric stress $\hat{\sigma}$ depends linearly on the velocity gradients $\nabla\boldsymbol{u}$, as already intimated above, implies for each $i, j = 1, 2, 3$,

$$\hat{\sigma}_{ij} = \sum_{k,\ell=1}^{3} A_{ijk\ell}(\nabla\boldsymbol{u})_{k\ell},$$

where $A_{ijk\ell}$ is a rank 4 tensor with $3^4 = 81$ components. Recall that we also assume this linear relation, and thus $A_{ijk\ell}$, is independent of position $\boldsymbol{x}$ and time t. An isotropic rank 4 tensor must have the form $A_{ijk\ell} = \mu\delta_{ik}\delta_{j\ell} + \mu'\delta_{i\ell}\delta_{jk} + \lambda\delta_{ij}\delta_{k\ell}$, where μ, μ' and λ are constant scalars. Symmetry in $\hat{\sigma}$ implies $A_{ijk\ell} = A_{jik\ell}$, which in turn implies $\mu' = \mu$. The general linear relation above between $\hat{\sigma}$ and $\nabla\boldsymbol{u}$ thus collapses to

$$\hat{\sigma}_{ij} = \lambda(\nabla \cdot \boldsymbol{u})\delta_{ij} + 2\mu D_{ij},$$

which matches the general constitutive relation given above. It is precisely the knowledge of the form of an isotropic tensor of rank 4, utilising the symmetry of $\hat{\sigma}$ as well, that we implicity derive as part of our general arguments above. Further, since by definition $\hat{\sigma}_{11} + \hat{\sigma}_{22} + \hat{\sigma}_{33} = 0$ as we separated the trace of σ to be the pressure field, we deduce $A_{iik\ell} = 0$, with an implicit sum over the repeated i index. This implies $\lambda = -2\mu/3$ and thus that $\hat{\sigma} = 2\mu(D - \frac{1}{3}(\nabla \cdot \boldsymbol{u})I)$.

4.4 Navier–Stokes Equations

Given the constitutive relation $\hat{\sigma} = \lambda(\nabla \cdot \boldsymbol{u})I + 2\mu D$, we can now deduce the form of the term $\nabla \cdot \hat{\sigma}$ in Cauchy's equation of motion in Theorem 4.2. For convenience here set $(x_1, x_2, x_3)^{\mathrm{T}} \equiv (x, y, z)^{\mathrm{T}}$ and $(u_1, u_2, u_3)^{\mathrm{T}} \equiv (u, v, w)^{\mathrm{T}}$. Then by direct computation we observe (recall Remark 4.3)

$$[\nabla \cdot \hat{\sigma}]_i = \sum_{j=1}^{3} \frac{\partial \hat{\sigma}_{ij}}{\partial x_j}$$

$$
\begin{aligned}
&= \lambda[\nabla(\nabla \cdot \boldsymbol{u})]_i + 2\mu \sum_{j=1}^{3} \frac{\partial D_{ij}}{\partial x_j} \\
&= \lambda[\nabla(\nabla \cdot \boldsymbol{u})]_i + \mu \sum_{j=1}^{3} \frac{\partial}{\partial x_j}\left(\frac{\partial u_i}{\partial x_j} + \frac{\partial u_j}{\partial x_i}\right) \\
&= \lambda[\nabla(\nabla \cdot \boldsymbol{u})]_i + \mu \sum_{j=1}^{3} \left(\frac{\partial^2 u_i}{\partial x_j^2} + \frac{\partial^2 u_j}{\partial x_i \partial x_j}\right) \\
&= (\lambda + \mu)[\nabla(\nabla \cdot \boldsymbol{u})]_i + \mu \nabla^2 u_i.
\end{aligned}
$$

Substituting this form for $\nabla \cdot \hat{\sigma}$ into Cauchy's equation of motion, we find

$$
\rho\left(\frac{\partial \boldsymbol{u}}{\partial t} + (\boldsymbol{u} \cdot \nabla)\,\boldsymbol{u}\right) = -\nabla p + (\lambda + \mu)\nabla(\nabla \cdot \boldsymbol{u}) + \mu \Delta \boldsymbol{u} + \rho \boldsymbol{f},
$$

where $\Delta = \nabla^2$ is the Laplacian operator. These are the *Navier–Stokes equations*. If we assume we are in three-dimensional space so $d = 3$, then together with the continuity equation we have four equations, but five unknowns – namely $\boldsymbol{u}$, p and ρ. Thus for a *compressible* fluid flow, we cannot specify the fluid motion completely without specifying one more condition/relation. (We could use the principle of conservation of energy to establish an additional relation known as the *equation of state*; in simple scenarios this takes the form of a relationship between the pressure p and density ρ of the fluid.)

For a *homogeneous incompressible* flow for which the mass density ρ is constant, we get a complete set of equations as follows.

Theorem 4.12 (Navier–Stokes equations for incompressible flow) *The flow of a homogeneous incompresible Newtonian fluid is governed by the following system of equations for the velocity field $\boldsymbol{u} = \boldsymbol{u}(\boldsymbol{x}, t)$ and pressure $p = p(\boldsymbol{x}, t)$:*

$$
\begin{aligned}
\frac{\partial \boldsymbol{u}}{\partial t} + (\boldsymbol{u} \cdot \nabla)\,\boldsymbol{u} &= \nu\,\Delta \boldsymbol{u} - \frac{1}{\rho}\nabla p + \boldsymbol{f}, \\
\nabla \cdot \boldsymbol{u} &= 0.
\end{aligned}
$$

Here $\nu = \mu/\rho$ is the coefficient of kinematic viscosity.

Remark 4.13 We observe:

(i) The incompressible Navier–Stokes equations consist of a closed system of equations of four equations in four unknowns, $\boldsymbol{u}$ and p.

(ii) Often the factor $1/\rho$ is scaled into the pressure and thus omitted: since ρ is constant $(\nabla p)/\rho \equiv \nabla(p/\rho)$, and we relabel p/ρ to be p.

(iii) The pressure field p can be recovered from the velocity field $\boldsymbol{u}$ and body force $\boldsymbol{f}$ provided we can solve the Poisson equation for the given boundary conditions (more on these just below). Recall Remark 2.2(ii). The relation between the pressure field and the velocity field at any time $t \geqslant 0$ provided in that remark applies to the homogeneous incompressible Navier–Stokes equations as well, since the divergence of the term '$\nu\,\Delta \boldsymbol{u}$' is zero.

We need to specify *initial* and *boundary conditions*. Given a function, say $\boldsymbol{u}_0 \colon \mathcal{D} \to \mathbb{R}^3$, as in the case of the incompressible Euler equations, we assume initially at time $t = 0$ that $\boldsymbol{u}(\boldsymbol{x}, 0) = \boldsymbol{u}_0(\boldsymbol{x})$ for all $\boldsymbol{x} \in \mathcal{D}$. With regards to boundary conditions, however, for viscous flow we specify an additional boundary condition to that we specified for the Euler equations. This is due to the inclusion of the extra term $\nu\Delta\boldsymbol{u}$ which increases the number of spatial derivatives in the governing evolution equations from one to two. We specify that

$$\boldsymbol{u} = \boldsymbol{0}$$

everywhere on the rigid boundary, i.e. in addition to the condition that there must be no net normal flow at the boundary, we also specify there is no tangential flow there. The fluid velocity is simply zero at a rigid boundary; this is also called the *no-slip* boundary condition. Experimentally this is observed as well, to a very high degree of precision; see Chorin and Marsden (1990, p. 34). (Dye can be introduced into a flow near a boundary to observe and measure very accurately how the flow behaves.) Further, recall that in a viscous fluid flow we are incorporating the effect of molecular diffusion between neighbouring fluid parcels – see Figure 4.1. The rigid non-moving boundary should impart a zero tangential flow condition to the fluid particles right up against it. The no-slip boundary condition crucially represents the mechanism for vorticity production in nature that can be observed everywhere. For example, look at the flow of a river close to the river bank. One observes counter-clockwise eddies generated from the left bank looking downstream that propagate into the main flow. Clockwise eddies are generated at the right bank.

Remark 4.14 At a material boundary (or free surface) between two immiscible fluids, we would specify that there is no jump in the velocity across the surface boundary. This is true if there is no surface tension or at least if it is negligible – for example at the seawater–air boundary of the ocean. However at the surface of melting wax at the top of a candle, there is surface tension, and there is a jump in the stress $\sigma\,\boldsymbol{n}$ at the boundary surface. Surface tension is also responsible for the phenomenon of being able to float a needle on the surface of a bowl of water, as well as many other interesting effects such as the shape of water drops.

As for the Euler equations for a homogeneous incompressible fluid in Corollaries 2.7 and 2.8, we can establish an equivalent vorticity formulation of the incompressible Navier–Stokes equations. This requires a straightforward modification of the proof of Corollary 2.8.

Corollary 4.15 (Evolution of vorticity with viscosity) *If the velocity field $\boldsymbol{u} = \boldsymbol{u}(\boldsymbol{x}, t)$ and pressure field $p = p(\boldsymbol{x}, t)$ are solutions to the homogeneous incompressible Navier–Stokes equations with a driving body force $\boldsymbol{f}$, then the vorticity field $\boldsymbol{\omega} = \nabla \times \boldsymbol{u}$ satisfies the closed system of equations*

$$\begin{aligned}\frac{\partial \boldsymbol{\omega}}{\partial t} + (\boldsymbol{u} \cdot \nabla)\,\boldsymbol{\omega} &= \nu\Delta\boldsymbol{\omega} + D\boldsymbol{\omega} + \nabla \times \boldsymbol{f},\\ \Delta\boldsymbol{u} &= -\nabla \times \boldsymbol{\omega}.\end{aligned}$$

Conversely, when the flow domain is $\mathcal{D} = \mathbb{R}^3$ and the initial vorticity is given by $\omega_0 = \nabla \times \boldsymbol{u}_0$, assume the fields ω and $\boldsymbol{u}$ satisfy the system of equations just above. Then the field $\boldsymbol{u} = \boldsymbol{u}(\boldsymbol{x}, t)$ satisfies $\nabla \cdot \boldsymbol{u} = 0$, and there exists a scalar field $p = p(\boldsymbol{x}, t)$ such that $\boldsymbol{u}$ and p satisfy the homogeneous incompressible Navier–Stokes equations.

Proof The first statement, i.e. the 'forward' or 'necessary' direction, is a straightforward modification of the proof of Corollary 2.7. The second 'converse' or 'sufficiency' statement follows after some straightforward modifications of the arguments presented in the proof of Corollary 2.8. That $\nabla \cdot \boldsymbol{u} = 0$ follows as previously. The modification that requires particular scrutiny though is the deduction that $\nabla \cdot \omega = 0$ throughout the flow. The evolution equation for $\upsilon := \nabla \cdot \omega$ here takes the form

$$\frac{\partial \upsilon}{\partial t} + (\boldsymbol{u} \cdot \nabla)\, \upsilon = \nu \Delta \upsilon.$$

By direct computations, exactly analogous to those at the very beginning of Section 7.2, and assuming smooth fields which decay suitably in the far field, we observe the L^2-norm of υ satisfies

$$\frac{\mathrm{d}}{\mathrm{d}t}\tfrac{1}{2}\|\upsilon\|_{L^2}^2 + \nu\|\nabla\upsilon\|_{L^2}^2 = -\int_{\mathbb{R}^3} \upsilon\,(\boldsymbol{u} \cdot \nabla)\,\upsilon \,\mathrm{d}\boldsymbol{x},$$

and further, using the identity $\nabla \cdot (\phi \boldsymbol{u}) = \phi(\nabla \cdot \boldsymbol{u}) + \boldsymbol{u} \cdot \nabla\phi$ with $\phi = \frac{1}{2}\upsilon^2$, that

$$\int_{\mathbb{R}^3} \upsilon(\boldsymbol{u} \cdot \nabla)\,\upsilon\,\mathrm{d}\boldsymbol{x} = \int_{\mathbb{R}^3} (\boldsymbol{u} \cdot \nabla)(\tfrac{1}{2}\upsilon^2)\,\mathrm{d}\boldsymbol{x} = \int_{\mathbb{R}^3} \nabla \cdot (\tfrac{1}{2}\upsilon^2\boldsymbol{u})\,\mathrm{d}\boldsymbol{x} = 0.$$

Hence we deduce the L^2-norm of υ is necessarily non-increasing in time. Thus if initially $\upsilon_0 := \nabla \cdot \omega_0$ is zero everywhere in $\mathbb{R}^3$, then $\upsilon = \nabla \cdot \omega$ is zero everywhere in $\mathbb{R}^3$ thereafter. □

In the next few sections we categorise some classes of exact solution flows to both the incompressible Euler equations and Navier–Stokes equations. We have already seen certain classes of solutions that are simultaneously solutions to both the Euler equations of ideal incompressible fluid motion and the incompressible Navier–Stokes equations, namely the linear flows of Section 2.4. In the exercises at the end of this chapter we explore how to find exact solutions to the incompressible Navier–Stokes equations representing flows in pipes with circular and elliptical cross-sections, known as Poiseuille flow. Presently though, in the next section we construct shear-layer flows before going on to derive the vorticity-stream formulation of the Navier–Stokes equations in Section 4.6. Then in Section 4.7 we derive Beltrami flows (inviscid and viscous) and in particular Arnold–Beltrami–Childress flows. For more details, see in particular Majda and Bertozzi (2002).

4.5 Shear-Layer Flows

We now derive a simple class of flows that retain the three underlying mechanisms of Navier–Stokes flows: convection, vortex stretching and diffusion.

Example 4.16 (Shear-Layer Flows) Recall that the vorticity ω evolves according to the partial differential system (assume no body force is present)

$$\frac{\partial \omega}{\partial t} + (\boldsymbol{u} \cdot \nabla)\, \omega = \nu\, \Delta\omega + D\omega,$$

with $\Delta \boldsymbol{u} = -\nabla \times \boldsymbol{\omega}$. The material derivative term $\partial\omega/\partial t + (\boldsymbol{u} \cdot \nabla)\, \omega$ convects vorticity along particle paths, while the term $\nu\, \Delta\omega$ is responsible for the diffusion of vorticity and the term $D\omega$ represents vortex stretching – the vorticity ω increases/decreases when it aligns along eigenvectors of D corresponding to positive/negative eigenvalues of D. We seek an exact solution to the incompressible Navier–Stokes equations of the following form (the first two velocity components represent a strain flow):

$$\boldsymbol{u}(\boldsymbol{x}, t) = \begin{pmatrix} \gamma x \\ -\gamma y \\ w(x,t) \end{pmatrix},$$

where γ is a constant, with $p(\boldsymbol{x}, t) = -\frac{1}{2}\rho\gamma^2(x^2 + y^2)$. This represents a solution to the Navier–Stokes equations if we can determine the solution $w = w(x,t)$ to the linear diffusion equation

$$\frac{\partial w}{\partial t} + \gamma x \frac{\partial w}{\partial x} = \nu \frac{\partial^2 w}{\partial x^2},$$

with $w(x, 0) = w_0(x)$. Computing the vorticity directly we get

$$\boldsymbol{\omega} = \begin{pmatrix} 0 \\ -\partial w/\partial x \\ 0 \end{pmatrix}.$$

If we differentiate the equation above for the velocity field component w with respect to x, then if $\omega := -\partial w/\partial x$, we get

$$\frac{\partial \omega}{\partial t} + \gamma x \frac{\partial \omega}{\partial x} = \nu \frac{\partial^2 \omega}{\partial x^2} - \gamma\omega,$$

with $\omega(x, 0) = \omega_0(x)$ where $\omega_0 := -\partial w_0/\partial x$. For this simpler flow we can see simpler signatures of the three effects we want to isolate: there is the convecting velocity γx; vortex stretching from the term $-\gamma\omega$ and diffusion in the term $\nu\partial^2\omega/\partial x^2$. In the general case the velocity field $\boldsymbol{u}$ can be recovered from the vorticity field $\boldsymbol{\omega}$ by solving the Poisson equation $\Delta \boldsymbol{u} = -\nabla \times \boldsymbol{\omega}$. Here this process is much simpler, we have $\partial w/\partial x = -\omega$ and the velocity field w can be recovered from the vorticity field ω by (we assume here $w(x,t) \to 0$ as $x \to -\infty$ for all $t \geqslant 0$)

$$w(x,t) = -\int_{-\infty}^{x} \omega(x', t)\, \mathrm{d}x'.$$

Let us consider a special case, the viscous shear-layer solution where $\gamma = 0$. The partial differential equation above for ω then reduces to the heat equation

$$\frac{\partial \omega}{\partial t} = \nu \frac{\partial^2 \omega}{\partial x^2}.$$

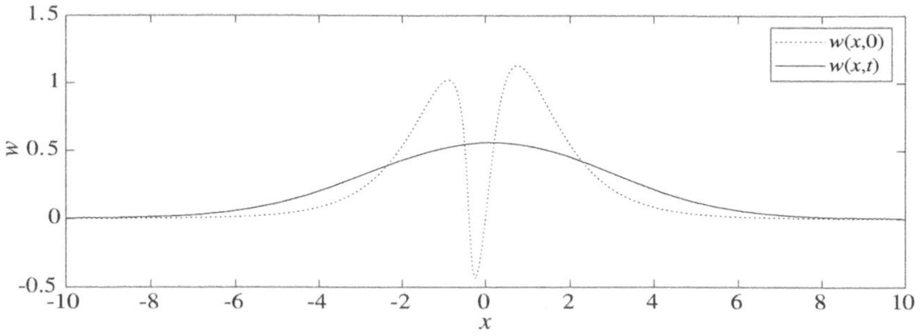

Figure 4.4 Viscous shear flow example. The effect of diffusion on the velocity field $w = w(x,t)$ is to smooth out variations in the field as time progresses.

The solution can be expressed in the form

$$\omega(x,t) = \int_{\mathbb{R}} G(x - x', \nu t)\, \omega_0(x')\, \mathrm{d}x',$$

where G is the Gaussian heat kernel $G(x,t) := (4\pi t)^{-1/2} \exp(-x^2/4t)$. That this is indeed the solution can be confirmed by direct substitution; also see Section 4.9(5). Note for any two functions f and g, for which either of the following integrals make sense, we have

$$\int_{\mathbb{R}} f(x - x')\, g(x')\, \mathrm{d}x' = \int_{\mathbb{R}} f(x')\, g(x - x')\, \mathrm{d}x'.$$

Using this result we see that the velocity field w is given by

$$w(x,t) = \int_{\mathbb{R}} G(x - x', \nu t)\, w_0(x')\, \mathrm{d}x'.$$

Note that both the vorticity field ω and the velocity field w diffuse as time evolves; see Figure 4.4. It is also possible to write down the explicit solution for the general case $\gamma \neq 0$ in terms of the Gaussian heat kernel; see Exercise 4.14.

4.6 Vorticity-Stream Formulation

We recall from Remark 2.9 in the vorticity formulation of the two-dimensional incompressible Euler equations, the 'vortex stretching' term $(\boldsymbol{\omega} \cdot \nabla)\, \boldsymbol{u}$ identically vanishes. This is of course also the case for the vorticity formulation of the two-dimensional incompressible Navier–Stokes equations. Recall the vorticity formulation of the three-dimensional Navier–Stokes equations in Corollary 4.15. We follow the arguments in Remark 2.9. We encode a two-dimensional flow as a subflow of a three-dimensional flow for which the velocity field is independent of z and $\boldsymbol{u} = (u, v, 0)^{\mathrm{T}}$, and thus $\boldsymbol{\omega} = (0, 0, \omega)^{\mathrm{T}}$ where $\omega = \partial v/\partial x - \partial u/\partial y$. Then the vorticity formulation of the two-dimensional incompressible Navier–Stokes equations, now with $\boldsymbol{u} := (u, v)^{\mathrm{T}}$, takes the form

$$\frac{\partial \omega}{\partial t} + \boldsymbol{u} \cdot \nabla \omega = \nu \Delta \omega,$$
$$\Delta \boldsymbol{u} = -\nabla^{\perp} \omega.$$

Since the two-dimensional velocity field is incompressible, we can represent it via a stream function $\psi = \psi(x, y, t)$ by $\boldsymbol{u} = \nabla^{\perp}\psi$. Recall from Definition 1.23 in Section 1.8 that $\nabla^{\perp}\psi = (\partial\psi/\partial y, -\partial\psi/\partial x)^{\mathrm{T}}$. We observe by direct substitution that $\omega = \partial v/\partial x - \partial u/\partial y \equiv -\Delta\psi$. Note that for any scalar field $v = v(x, y, t)$, we can write $\nabla^{\perp} \cdot (\nabla^{\perp} v) \equiv \Delta v$.

Notation 4.17 (Jacobian) For any two functions $\phi = \phi(x, y)$ and $\psi = \psi(x, y)$, we define the Jacobian quantity $J(\phi, \psi) := \nabla^{\perp}\phi \cdot \nabla\psi$, or equivalently,

$$J(\phi, \psi) := \frac{\partial \phi}{\partial y}\frac{\partial \psi}{\partial x} - \frac{\partial \phi}{\partial x}\frac{\partial \psi}{\partial y}.$$

The two-dimensional incompressible Navier–Stokes equations have an equivalent vorticity-stream formulation.

Lemma 4.18 (Vorticity-stream formulation) *The two-dimensional incompressible Navier–Stokes equations in $\mathcal{D} = \mathbb{R}^2$ have the equivalent formulation*

$$\frac{\partial \omega}{\partial t} + J(\psi, \omega) = \nu \Delta \omega,$$
$$\Delta \psi = -\omega.$$

Remark 4.19 That the vorticity formulation for the incompressible Navier–Stokes equations given above implies the vorticity-stream formulation in the lemma is immediate from the identification $\boldsymbol{u} = \nabla^{\perp}\psi$. The converse can also be constructed in much the same way as we did for the three-dimensional case in the proof of Corollary 4.15. However, the argument in the two-dimensional case is simpler, and we provide it here. We also refer the reader to the arguments in the proof of Lemma 3.2.

Proof In light of the remark just above, we only prove the sufficiency statement. Assume the scalar fields $\omega = \omega(x, y, t)$ and $\psi = \psi(x, y, t)$ satisfying the system of equations stated in the lemma. Then we can express the vorticity-stream formulation above in the form

$$\nabla^{\perp} \cdot \left(\frac{\partial}{\partial t}(\nabla^{\perp}\psi) + ((\nabla^{\perp}\psi) \cdot \nabla)\, \nabla^{\perp}\psi - \nu\Delta(\nabla^{\perp}\psi) \right) = 0.$$

Since $\nabla^{\perp} \cdot (\nabla v) \equiv 0$ for any scalar function $v = v(x, y, t)$, as in Corollary 2.8, we deduce that the argument of the two-dimensional curl operator $\nabla^{\perp}\cdot$ must equal the gradient of a scalar function which we set to be p/ρ for some constant $\rho > 0$. To establish the relation just above, we required the following observation, that $\nabla^{\perp} \cdot ((\nabla^{\perp}\psi) \cdot \nabla)\, \nabla^{\perp}\psi)$ equals (for clarity we also set $v_i := \nabla_i^{\perp}\psi$ for $i = 1, 2$)

$$\sum_{i,j=1}^{2} \nabla_i^{\perp}\Big((\nabla_j^{\perp}\psi)\, \nabla_j\, \nabla_i^{\perp}\psi \Big) = \sum_{i,j=1}^{2} \Big((\nabla_i^{\perp}\nabla_j^{\perp}\psi)(\nabla_j\nabla_i^{\perp}\psi) + \nabla_j^{\perp}\psi\,(\nabla_i^{\perp}\nabla_j\nabla_i^{\perp}\psi) \Big)$$

$$= \sum_{i,j=1}^{2} (\nabla_j^{\perp} v_i)(\nabla_j v_i) + \sum_{j=1}^{2} \nabla_j^{\perp}\psi\, (\nabla_j\, (\Delta\psi))$$
$$= ((\nabla^{\perp}\psi)\cdot\nabla)\,\Delta\psi,$$

where we reversed the order of the ∇ and $\nabla^{\perp}$ operators and used the identity $\nabla^{\perp} v\cdot\nabla v \equiv 0$ for any scalar-valued function $v = v(x, y, t)$. □

In the two-dimensional context in which the vorticity-stream formulation applies, we seek exact solutions, first, to the Euler equations for a homogeneous incompressible fluid (with $\nu = 0$) and second, we use these inviscid solutions to construct corresponding exact solutions to the incompressible Navier–Stokes equations. Our principal result in this direction is as follows.

Corollary 4.20 (Inviscid stationary solutions) *A given stream function $\psi = \psi(x, y)$ prescribes a stationary solution flow to the two-dimensional Euler equations for a homogeneous incompressible fluid in $\mathcal{D} \subseteq \mathbb{R}^2$ if and only if there exists a smooth function $F : \mathbb{R} \to \mathbb{R}$ such that*

$$\Delta\psi = F(\psi).$$

Proof For the stream function $\psi = \psi(x, y)$, the vorticity-stream formulation with $\nu = 0$ is equivalent to the condition in $\mathbb{R}^2$ with $\omega = -\Delta\psi$:

$$J(\psi, \omega) = \nabla^{\perp}\psi \cdot \nabla\omega = 0.$$

Recall that $\nabla\omega = \nabla\omega(x, y)$ points in the direction orthogonal to the contour line of ω through (x, y) for all $(x, y) \in \mathbb{R}^2$. However $\nabla^{\perp}\psi$ points along the contour line of ψ through (x, y) as it is orthogonal to $\nabla\psi$. The relation above asserts that at every $(x, y) \in \mathbb{R}^2$, the vector $\nabla^{\perp}\psi$ is orthogonal to $\nabla\omega$ and therefore parallel to the (tangent to the) contour line of ω through (x, y). Hence contour lines of ψ and ω are parallel. Hence they must be functionally dependent, so that there exists a function $F: \mathbb{R} \to \mathbb{R}$ such that $\omega = -F(\psi)$. Conversely, if $\omega = -F(\psi)$, then we observe that $\nabla^{\perp}\psi \cdot \nabla\omega = -F'(\psi)\nabla^{\perp}\psi \cdot \nabla\psi \equiv 0$. □

Example 4.21 (Stuart cat's-eye flow) An immediate example of Corollary 4.20 is the following one-parameter family of flows, see Stuart (1971). Assume $F(\psi) := -\mathrm{e}^{2\psi}$, then for any $c \geqslant 1$ an exact solution to $\Delta\psi = -\mathrm{e}^{2\psi}$ is

$$\psi(x, y) = -\log\Big(c \cosh y + \sqrt{(c^2 - 1)}\, \cos x\Big).$$

This represents a stationary two-dimensional incompressible Euler flow in $\mathbb{R}^2$ that is periodic in x with period 2π; see Figure 4.5 and Exercise 4.16.

Remark 4.22 (Kelvin's cat's-eyes) Kelvin's cat's-eyes solution represents the flow due to a set of point vortices placed uniformly along the x-axis. It also represents the limit as $c \to \infty$ in the Stuart solution above. For more details as well as analogous solutions, see e.g. Kelvin (1880), Stuart (1971), Holm *et al.* (1986) and Crowdy (2004), as well as Example 4.31.

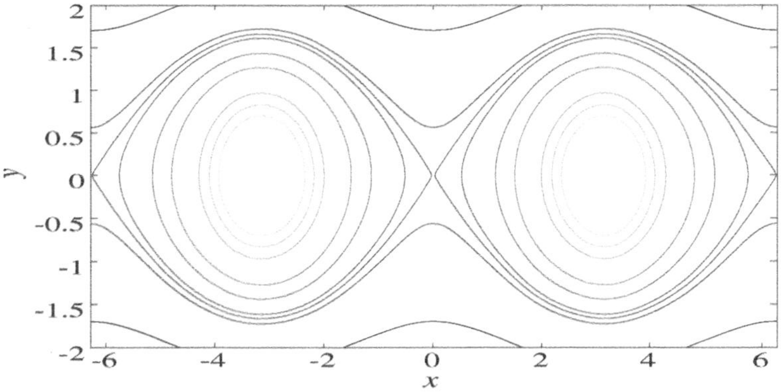

Figure 4.5 Contours of the stream function $\psi = \psi(x, y)$ corresponding to the stationary two-dimensional incompressible Euler flow in $\mathbb{R}^2$ known as Stuart cat's-eye flow. The solution is periodic in x with period 2π. In this example $c = 2$.

Example 4.23 (Inviscid eddies) Suppose we are given the radially symmetric stationary vorticity field $\omega_0 = \omega_0(r)$, where $r := (x^2 + y^2)^{1/2}$ is the radial distance in polar coordinates. Then from the vorticity-stream formulation of the Euler equations for a homogeneous incompressible fluid, i.e. setting $\nu = 0$ in Lemma 4.18, we know $J(\psi_0, \omega_0) = 0$ where $\Delta\psi_0 = -\omega_0$. From the Jacobian condition we deduce the contour lines of ψ_0 are parallel to those of ω_0 and so $\psi_0 = \psi_0(r)$ only. The relation $\Delta\psi_0 = -\omega_0$ in the radially symmetric case is

$$\frac{1}{r}\frac{\mathrm{d}}{\mathrm{d}r}\left(r\frac{\mathrm{d}\psi_0}{\mathrm{d}r}\right) = -\omega_0 \quad \Leftrightarrow \quad \psi_0'(r) = -\frac{1}{r}\int_0^r s\omega_0(s)\,\mathrm{d}s,$$

for any $\psi_0 = \psi_0(r)$ with bounded derivative at $r = 0$. The corresponding velocity field $\boldsymbol{u}_0 = \nabla^\perp\psi_0$ is given in $\mathbb{R}^2$ by

$$\boldsymbol{u}_0 = \begin{pmatrix} (\partial/\partial y)\,\psi_0(r) \\ -(\partial/\partial x)\,\psi_0(r) \end{pmatrix} = \begin{pmatrix} y/r \\ -x/r \end{pmatrix}\psi_0'(r) \equiv -\hat{\boldsymbol{\theta}}\,\psi_0'(r).$$

Hence the streamlines of the flow are circular, resembling eddy motion.

Example 4.24 (Time-dependent viscous eddies) Suppose a time-dependent solution to the vorticity-stream formulation is $\omega = \omega(r, t)$. Assume we have a solution $\psi = \psi(r, t)$ to $\Delta\psi = -\omega$ that is also radially symmetric. Indeed, as we saw in Example 4.23, we can find such a solution in the form

$$\frac{\partial\psi}{\partial r}(r,t) = -\frac{1}{r}\int_0^r s\omega(s,t)\,\mathrm{d}s \Leftrightarrow \psi(r,t) = \psi(0,t) - \int_0^r\int_0^s \frac{s'}{s}\omega(s',t)\,\mathrm{d}s'\,\mathrm{d}s,$$

for any $\psi = \psi(r, t)$ with bounded derivative at $r = 0$ for all $t \geqslant 0$. We conclude

$$\nabla\psi = \frac{\partial\psi}{\partial r}\hat{\boldsymbol{r}} \quad \text{and} \quad \nabla\omega = \frac{\partial\omega}{\partial r}\hat{\boldsymbol{r}} \quad \Rightarrow \quad \nabla^\perp\psi\cdot\nabla\omega = 0,$$

for all $r \geqslant 0$. Hence for such solutions $J(\psi, \omega) \equiv 0$. The time-evolution equation for ω is thus the heat equation

$$\frac{\partial \omega}{\partial t} = \nu \Delta \omega,$$

which we can solve explicitly to find ω at any time $t > 0$ in terms of the initial data $\omega_0 = \omega_0(r)$. Further, the velocity field $\boldsymbol{u} = \nabla^{\perp}\psi$ is given in $\mathbb{R}^2 \times [0, \infty)$ by

$$\boldsymbol{u} = \begin{pmatrix} y/r \\ -x/r \end{pmatrix} \frac{1}{r} \int_0^r s\omega(s,t)\,\mathrm{d}s \equiv \hat{\boldsymbol{\theta}} \left(\frac{1}{r} \int_0^r s\omega(s,t)\,\mathrm{d}s \right).$$

Remark 4.25 See Exercise 4.17 for further examples of stationary incompressible inviscid/viscous flows, periodic in x and y, which satisfy $J(\psi, \omega) \equiv 0$.

4.7 Beltrami Flows

We consider herein classes of three-dimensional incompressible flows for which the vorticity field is parallel to the velocity field. Under this condition the flow prescription simplifies considerably and we can find classes of exact solutions.

Definition 4.26 (Beltrami flow) A three-dimensional fluid flow is a *Beltrami flow* if the vorticity satisfies the Beltrami condition

$$\boldsymbol{\omega}(\boldsymbol{x},t) = \lambda(\boldsymbol{x},t)\,\boldsymbol{u}(\boldsymbol{x},t) \quad \Leftrightarrow \quad \boldsymbol{\omega}(\boldsymbol{x},t) \times \boldsymbol{u}(\boldsymbol{x},t) = \boldsymbol{0},$$

for all $\boldsymbol{x} \in \mathcal{D} \subseteq \mathbb{R}^3$ and $t \geqslant 0$, for some non-zero real scalar $\lambda = \lambda(\boldsymbol{x},t)$.

Recall the following identity (from the Appendix, Section A.1 #7):

$$((\boldsymbol{u} \cdot \nabla)\,\boldsymbol{u}) \equiv \tfrac{1}{2}\nabla(|\boldsymbol{u}|^2) + \boldsymbol{\omega} \times \boldsymbol{u}.$$

Hence, flows satisfying the Beltrami condition satisfy the simplified homogeneous incompressible Navier–Stokes equations:

$$\frac{\partial \boldsymbol{u}}{\partial t} + \nabla\left(\frac{p}{\rho} + \tfrac{1}{2}|\boldsymbol{u}|^2\right) = \nu\Delta\boldsymbol{u} + \boldsymbol{f},$$
$$\nabla \cdot \boldsymbol{u} = 0.$$

Since any homogeneous incompressible Navier–Stokes flow has an associated vorticity formulation, see Corollary 4.15, we have the following result.

Proposition 4.27 (Beltrami flows are equivalent to heat flows) *A homogeneous incompressible Navier–Stokes flow which satisfies the Beltrami condition satisfies the following vorticity formulation which is the heat equation:*

$$\frac{\partial \boldsymbol{\omega}}{\partial t} = \nu\Delta\boldsymbol{\omega} + \nabla \times \boldsymbol{f}.$$

Remark 4.28 We make the following observations:

(i) The Beltrami condition necessarily implies $\boldsymbol{u}\cdot\nabla\lambda \equiv 0$ since we have $0 = \nabla\cdot\boldsymbol{\omega} = \nabla\cdot(\lambda\boldsymbol{u}) = \boldsymbol{u}\cdot\nabla\lambda + \lambda\nabla\cdot\boldsymbol{u}$ and $\nabla\cdot\boldsymbol{u} = 0$.

(ii) The Beltrami condition precludes two-dimensional flows for which the vorticity vector is necessarily orthogonal to the plane of the fluid flow.

(iii) *Generalised Beltrami flows* satisfy the weaker condition $\nabla\times(\boldsymbol{\omega}\times\boldsymbol{u}) = 0$ everywhere. The vorticity field still equivalently satisfies the heat equation in Proposition 4.27. They also incorporate two-dimensional flows.

(iv) So-called *strong Beltrami flows* are characterised by λ being a constant.

(v) If the flow is inviscid, stationary and the body force conservative, then the vorticity heat equation in Proposition 4.27 is trivially satisfied.

One construction of Beltrami flows is as follows; recall Corollary 4.20.

Proposition 4.29 *Suppose $\psi = \psi(x, y)$ is a stream function for a stationary two-dimensional homogeneous incompressible Euler flow so that $\Delta\psi = F(\psi)$ for some function $F\colon \mathbb{R}\to\mathbb{R}$. Then the velocity field $\boldsymbol{u} = \boldsymbol{u}(x, y)$ given by*

$$\boldsymbol{u} := \begin{pmatrix} \partial\psi/\partial y \\ -\partial\psi/\partial x \\ W(\psi) \end{pmatrix}$$

is a three-dimensional inviscid Beltrami flow if and only if the function in the third velocity component $W\colon \mathbb{R}\to\mathbb{R}$ satisfies the ordinary differential equation

$$W'(\psi)W(\psi) = -F(\psi).$$

In which case the Beltrami coefficient $\lambda = \lambda(x, y)$ is given by $\lambda = W'(\psi)$.

Proof By direct computation, the vorticity field $\boldsymbol{\omega} = \boldsymbol{\omega}(x, y)$ associated with the velocity $\boldsymbol{u} = \boldsymbol{u}(x, y)$ shown is $\boldsymbol{\omega} := (W'(\psi)\,\partial\psi/\partial y, -W'(\psi)\,\partial\psi/\partial x, -\Delta\psi)^{\mathrm{T}}$. The Beltrami condition is $\boldsymbol{\omega} = \lambda\boldsymbol{u}$, which holds if and only if $\lambda = W'(\psi)$ and $-\Delta\psi = \lambda W(\psi)$. The latter condition is equivalent to $-F(\psi) = W'(\psi)W(\psi)$. □

Example 4.30 (Periodic Beltrami flow) Suppose we are given the following periodic stationary stream function (see Exercise 4.17):

$$\psi(\boldsymbol{x}) := \sum_{|\boldsymbol{k}|^2=\ell} \big(a_{\boldsymbol{k}}\cos(2\pi\boldsymbol{x}\cdot\boldsymbol{k}) + b_{\boldsymbol{k}}\sin(2\pi\boldsymbol{x}\cdot\boldsymbol{k})\big),$$

for a given $\ell\in\mathbb{N}$ and where $\boldsymbol{x} = (x, y)^{\mathrm{T}}$. Note that the wavenumbers $\boldsymbol{k}$ are Z^2-valued, and the $a_{\boldsymbol{k}}$ and $b_{\boldsymbol{k}}$ are constants. Further, the flow domain is $\mathcal{D} = \mathbb{T}^2$, where $\mathbb{T}^2$ is the 2-torus of period one, i.e. the function ψ given is periodic with period one in each direction. We observe by direct calculation that $\Delta\psi = -4\pi^2\ell\psi$. Hence ψ satisfies the condition $\Delta\psi = F(\psi)$, where F is the linear function $F\colon\psi\mapsto -4\pi^2\ell\psi$. Using Proposition 4.29, we can generate an inviscid Beltrami flow by solving the ordinary differential equation $W'(\psi)W(\psi) = -F(\psi)$ which here takes the form

$$\big(\tfrac{1}{2}W^2(\psi)\big)' = 4\pi^2\ell\psi \qquad\Leftrightarrow\qquad W(\psi) = 2\pi\ell^{1/2}\psi,$$

where we have taken $W(0) = 0$ for convenience. Note that $\lambda = 2\pi\ell^{1/2}$, which is constant. Hence $\boldsymbol{u} = (\partial\psi/\partial y, -\partial\psi/\partial x, 2\pi\ell^{1/2}\psi)^{\mathrm{T}}$ represents an inviscid periodic Beltrami flow.

Example 4.31 (Three-dimensional cat's-eye flow) Recall the two-dimensional inviscid Stuart cat's-eye flow from Example 4.21: the given stream function $\psi = \psi(x, y)$, 2π-periodic in x, satisfies $\Delta\psi = -\mathrm{e}^{2\psi}$ so $F(\psi) = -\mathrm{e}^{2\psi}$. The ordinary differential equation $W'(\psi)W(\psi) = -F(\psi)$ takes the form

$$\left(\tfrac{1}{2}W^2(\psi)\right)' = \mathrm{e}^{2\psi} \qquad \Leftrightarrow \qquad W(\psi) = \mathrm{e}^{\psi},$$

where we take $W(0) = 1$, again for convenience. Hence $\lambda(x, y) = \mathrm{e}^{\psi(x,y)}$ and $\boldsymbol{u} = (\partial\psi/\partial y, -\partial\psi/\partial x, \mathrm{e}^{\psi})^{\mathrm{T}}$ represents a three-dimensional cat's-eye flow.

We observe that the Beltrami condition $\boldsymbol{\omega} = \lambda\boldsymbol{u}$ is linear. Thus Beltrami flows with the same Beltrami coefficient λ constitute a linear subspace of exact solutions to the homogeneous incompressible three-dimensional Euler equations (assume no body force). For suppose the velocity field $\widehat{\boldsymbol{u}} = \widehat{\boldsymbol{u}}(\boldsymbol{x}, t)$ and corresponding vorticity field $\widehat{\boldsymbol{\omega}} = \widehat{\boldsymbol{\omega}}(\boldsymbol{x}, t)$ are given by

$$\widehat{\boldsymbol{u}} := \sum_{k=1}^{K} c_k \boldsymbol{u}_k(\boldsymbol{x}, t) \qquad \text{and} \qquad \widehat{\boldsymbol{\omega}} := \sum_{k=1}^{K} c_k \boldsymbol{\omega}_k(\boldsymbol{x}, t),$$

for some $K \in \mathbb{N}$, where the c_k are all constant coefficients. Naturally for each $k = 1, \ldots, K$, we have $\boldsymbol{\omega}_k = \nabla \times \boldsymbol{u}_k$. Here we assume for each $k = 1, \ldots, K$, the velocity fields $\boldsymbol{u}_k = \boldsymbol{u}_k(\boldsymbol{x}, t)$ are each Beltrami flows with the same Beltrami coefficient $\lambda = \lambda(\boldsymbol{x}, t)$ so that $\boldsymbol{\omega}_k = \lambda\boldsymbol{u}_k$. Thus naturally we have $\widehat{\boldsymbol{\omega}} = \lambda\widehat{\boldsymbol{u}}$ and further $\widehat{\boldsymbol{\omega}}$ also satisfies the heat equation $\partial\widehat{\boldsymbol{\omega}}/\partial t = \nu\Delta\widehat{\boldsymbol{\omega}}$. Consequently, $\widehat{\boldsymbol{\omega}}$ or equivalently $\widehat{\boldsymbol{u}}$ represent a Beltrami flow as well.

Example 4.32 (Three-dimensional inviscid Beltrami flows) Suppose the flow domain $\mathcal{D}$ is $\mathbb{T}^3$, the periodic 3-torus, and $\psi_x = \psi_x(y, z)$, $\psi_y = \psi_y(x, z)$ and $\psi_z = \psi_z(x, y)$ are eigenfunction solutions to the Laplacian operator corresponding to the constant eigenvalue $\lambda < 0$ on each of the $\mathbb{T}^2$ subspaces of $\mathbb{T}^3$, respectively parameterised by (y, z), (x, z) and (x, y). In other words, we have

$$\Delta_{y,z}\psi_x = \lambda\psi_x, \qquad \Delta_{x,z}\psi_y = \lambda\psi_y \qquad \text{and} \qquad \Delta_{x,y}\psi_z = \lambda\psi_z.$$

Then since linear combinations of Beltrami flows corresponding to the same Beltrami coefficient, here λ, are also Beltrami flows, we know that

$$\boldsymbol{u}_0 = \begin{pmatrix} \partial\psi_z/\partial y \\ -\partial\psi_z/\partial x \\ \sqrt{-\lambda}\psi_z \end{pmatrix} + \begin{pmatrix} \sqrt{-\lambda}\psi_x \\ \partial\psi_x/\partial z \\ -\partial\psi_x/\partial y \end{pmatrix} + \begin{pmatrix} -\partial\psi_y/\partial z \\ \sqrt{-\lambda}\psi_y \\ \partial\psi_y/\partial x \end{pmatrix}$$

is a three-dimensional inviscid Beltrami flow in $\mathcal{D} = \mathbb{T}^3$. Note, for example for the first component, we have $F(\psi) = \lambda\psi$ and the solution to the ordinary differential equation $W'(\psi)W(\psi) = -F(\psi)$ with $W(0) = 0$ is $W(\psi) = \sqrt{-\lambda}\psi$.

Example 4.33 (Arnold–Beltrami–Childress inviscid flows) Suppose that in Example 4.32 we choose, for some real constants A, B and C:

$$\psi_x = A\sin z, \qquad \psi_y = B\sin x \quad \text{and} \quad \psi_z = C\sin y.$$

Thus here $\lambda = -1$. Computing the linear combination given in Example 4.32, we observe the following velocity field is a three-dimensional Beltrami flow:

$$\boldsymbol{u}_0 = \begin{pmatrix} A\sin z + C\cos y \\ B\sin x + A\cos z \\ C\sin y + B\cos x \end{pmatrix}.$$

Remark 4.34 (Three-dimensional viscous Beltrami flows) Given a stationary three-dimensional strong Beltrami flow $\boldsymbol{u}_0$ with Beltrami coefficient λ, there is a solution to the three-dimensional incompressible Navier–Stokes equations of the form $\boldsymbol{u}(\boldsymbol{x},t) = \mathrm{e}^{-\lambda^2 \nu t}\boldsymbol{u}_0(\boldsymbol{x})$. To see this observe that $\boldsymbol{u}$ satisfies the Beltrami condition as $\boldsymbol{u}_0$ does, and the corresponding vorticity field satisfies the heat equation in Proposition 4.27.

4.8 Dynamical Similarity and Reynolds Number

Our goal herein is to non-dimensionalise the homogeneous incompressible Navier–Stokes equations without a body force and demonstrate their scaling properties. Recall that the velocity $\boldsymbol{u} = \boldsymbol{u}(\boldsymbol{x},t)$ and pressure $p = p(\boldsymbol{x},t)$ satisfy

$$\frac{\partial \boldsymbol{u}}{\partial t} + (\boldsymbol{u}\cdot\nabla)\,\boldsymbol{u} = \nu\Delta\boldsymbol{u} - \frac{1}{\rho}\nabla p,$$
$$\nabla\cdot\boldsymbol{u} = 0.$$

Note that two physical properties inherent to the fluid modelled are immediately apparent, the mass density ρ, which is constant throughout the flow, and the kinematic viscosity ν. Suppose the specification of the initial and boundary conditions for a homogeneous incompressible Navier–Stokes flow involves a characteristic length scale L and velocity magnitude U, i.e. two scale parameters L and U are determined by the initial and/or boundary conditions relevant to the problem. For example, we might imagine a flow past an obstacle such as a sphere whose diameter is characterised by L and the undisturbed far-field flow is uniform with speed U. These two scales naturally determine a timescale $T = L/U$. Using these scales we can introduce the dimensionless variables

$$\boldsymbol{x}' = \frac{\boldsymbol{x}}{L}, \quad t' = \frac{t}{T}, \quad \boldsymbol{u}'(\boldsymbol{x}',t') = \frac{1}{U}\boldsymbol{u}(\boldsymbol{x},t) \quad \text{and} \quad p'(\boldsymbol{x}',t') = \frac{1}{P}p(\boldsymbol{x},t).$$

Note that we have chosen to rescale the pressure field by the scalar parameter P, which as yet is undetermined. Directly substituting for $\boldsymbol{u} = U\boldsymbol{u}'$ and $p = Pp'$ and using the chain rule to replace t by t' and $\boldsymbol{x}$ by $\boldsymbol{x}'$ in the incompressible Navier–Stokes equations, we obtain

$$\frac{U}{T}\frac{\partial \boldsymbol{u}'}{\partial t'} + \frac{U^2}{L}(\boldsymbol{u}'\cdot\nabla_{\boldsymbol{x}'})\,\boldsymbol{u}' = \frac{\nu U}{L^2}\Delta_{\boldsymbol{x}'}\boldsymbol{u}' - \frac{P}{\rho L}\nabla_{\boldsymbol{x}'}p,$$

$$\nabla_{x'} \cdot u' = 0,$$

Using $T = L/U$ and dividing the momentum equation through by U^2/L, we get

$$\frac{\partial u'}{\partial t'} + (u' \cdot \nabla_{x'})\, u' = \frac{\nu}{UL} \Delta_{x'} u' - \frac{P}{\rho U^2} \nabla_{x'} p.$$

We set $P = \rho U^2$. This is the natural 'dynamic pressure' scaling in the far field. This choice here also means that the pressure term on the right has a non-dimensional scaling matching that of the inertia terms on the left in the equation above. Dropping the primes, we finally have

$$\frac{\partial u}{\partial t} + (u \cdot \nabla)\, u = \frac{1}{\mathrm{Re}} \Delta u - \nabla p,$$
$$\nabla \cdot u = 0,$$

which is the representation for the Navier–Stokes equations in dimensionless variables. The dimensionless number

$$\mathrm{Re} := \frac{UL}{\nu}$$

is the *Reynolds number*. See Table 4.1 for some typical Reynolds numbers. Its practical significance is as follows. Suppose we want to design a plane (or perhaps just a wing). It might have a characteristic scale L_1 and typically cruise at speeds U_1 with surrounding air having viscosity ν_1. Rather than build the plane to test its airflow properties, it might be cheaper to build a scale model with exactly the same shape/geometry but smaller, with characteristic scale L_2. Then we could test the airflow properties in a wind tunnel, say by using a driving impinging wind of characteristic speed U_2 and air of viscosity ν_2, so

$$\frac{U_1 L_1}{\nu_1} = \frac{U_2 L_2}{\nu_2}.$$

The Reynolds numbers in the two scenarios are the same and the dimensionless Navier–Stokes equations for the two flows identical. Hence, starting with the same initial conditions, the shape of the flows in the two scenarios will be the same. Such flows, with the same geometry, initial conditions and Reynolds number, are said to be *similar*.

Example 4.35 We could replace the wind tunnel by a water tunnel. The viscosity of air is $\nu_1 = 0.15\,\mathrm{cm}^2\,\mathrm{s}^{-1}$ and of water $\nu_2 = 0.0114\,\mathrm{cm}^2\,\mathrm{s}^{-1}$, so $\nu_1/\nu_2 \approx 13$. Hence for the same geometry and characteristic scale $L_1 = L_2$, if we want the Reynolds numbers to match, we see that we require

$$\frac{U_1 L_1}{\nu_1} = \frac{U_2 L_2}{\nu_2} \quad \Leftrightarrow \quad \frac{U_1 L_1}{U_2 L_2} = \frac{\nu_1}{\nu_2} = 13 \quad \Rightarrow \quad U_1 = 13\, U_2.$$

Example 4.36 (Aircraft Reynolds number) Assume the aircraft has length $L \approx 10\,\mathrm{m}$, and flies at speed $U \approx 700\,\mathrm{km\,h}^{-1} \approx 194\,\mathrm{m\,s}^{-1}$. Recall from Example 4.35 that the viscosity of air is $\nu_1 = 0.15\,\mathrm{cm}^2\,\mathrm{s}^{-1} = 1.5 \times 10^{-5}\,\mathrm{m}^2\,\mathrm{s}^{-1}$. Hence the Reynolds number for this flow is $\mathrm{Re} \approx 1.3 \times 10^8$. Note that wind tunnel tests for new aircraft are

Table 4.1 Some typical Reynolds numbers

Flow context	cruise ship	aircraft	blue whale	cricket ball
Reynolds number	10^9	10^8	10^8	10^5

Flow context	canine artery	nematode	capillaries	bacterium
Reynolds number	10^3	0.6	10^{-3}	10^{-4}

being increasingly supplemented by, if not replaced by, CFD simulations; see Johnson *et al.* (2005).

Remark 4.37 Recall that all the 'inertia terms' on the left of the momentum equation have scaling U^2/L, while the viscous acceleration has scaling $\nu U/L^2$ with $\nu = \mu/\rho$. We generated the Reynolds number as the reciprocal of the scaling of the viscous acceleration term once we had divided through by U^2/L. In other words, we can think of the Reynolds number as the ratio of the inertia force scaling $\rho U^2/L$ to the viscous force scaling $\mu U/L^2$.

Remark 4.38 (Pressure and coordinate direction scalings) We chose the 'dynamic pressure' scaling $P = \rho U^2$ in the non-dimesionalisation we outlined above. However, for low Reynolds-number flows such as those in Chapter 5, a different scaling may be required. For example, for the Stokes flows in Section 5.1, viscous diffusion $\nu\Delta\boldsymbol{u}$ is the dominant effect and we need to choose the pressure scaling P so as to balance the pressure term with the viscous diffusion term. If the flow is pressure-gradient-driven such as in the thin/shallow-layer or lubrication contexts we see in Section 5.2, then, in addition to considering different scalings in different coordinate directions, we must choose the pressure scaling P to balance the diffusion across the thin layer. See Chapter 5 and e.g. Kundu and Cohen (2008, Ch. 8).

4.9 Notes

Some further details are as follows.

(1) *Global regularity of the incompressible Navier–Stokes equations.* The well-posedness of the incompressible Navier–Stokes equations globally in time is an extremely important problem. In Chapter 7 we provide an introduction to the main results and issues.

(2) *Non-Newtonian fluids.* In Section 4.3 we derived the constitutive relation between the deviatoric stress and rate of strain based on the assumptions of linearity (with the linear relation independent of position and time), isotropy and symmetry, i.e. for a Newtonian fluid. There are many fluids that do not fall into this category, and for example exhibit some elastic or anisotropic behaviour. Two everyday examples are saliva and corn-starch mixed with water. Saliva is a dense soup of long-chain protein molecules which give it some solid-like behaviour. For example, it is possible to string saliva out much like a chain, as can often be observed on a trip to the dentist... The right mixture of corn-starch with water can keep children amused for hours. Pick up the mixture and it oozes like treacle through your fingers – though it is not as sticky. However, you can also roll it into a ball between the palms of your hands. There are many videos online of large

baths or pools of such mixtures, which resemble sloshing pools of a typical liquid, but which people can simply walk across. However, if they stop in the middle they sink. This mixture exhibits solid-like properties when high shear stresses are locally applied, but otherwise appears to be a liquid like any other. Further examples include thixotropic fluids or gels, such as some paints or even tomato ketchup or yoghurt. Looking ahead to Chapters 5 and 6, non-Newtonian fluids play a role in the science of drag reduction. For example, marine animals have developed mechanisms to reduce the skin friction drag of the turbulent boundary layer, including mucus secretions for fish, denticle scales for sharks and a compliant skin surface for dolphins. These have applications to drag reduction for ships, submarines, the insides of large pipes and so forth. The study of rheology and non-Newtonian fluids and their multitudinous applications can be found in many specialised books; see e.g. Shenoy (2020).

(3) *Biofluid mechanics*. There are multitudinous applications of fluid mechanics to biological systems, including blood flow involving both the shape of arteries and veins, the transport of blood cells, flow in the lungs, and surfactants and so forth. See e.g. Rubenstein *et al.* (2016).

(4) *Froude number (reprise)*. In Section 4.8 we demonstrated how to non-dimensionalise the incompressible Navier–Stokes equations. Suppose we included the force of gravity at the outset so a term $-g\boldsymbol{k}$ is included on the right-hand side of the original dimensional equations, where g is the acceleration due to gravity. If we proceed through the dimensional analysis as shown therein, and divide through by U^2/L, we obtain the same non-dimensional Navier–Stokes equations involving the Reynolds number as shown, except with an additional external forcing term $-(gL/U^2)\boldsymbol{k}$ present on the right. We observe that this additional term equals $-(\mathrm{F})^{-2}\boldsymbol{k}$, where $\mathrm{F} := U/\sqrt{gL}$ is the Froude number in this context; see Example 2.26. In other words, we can interpret the square of the Froude number as the ratio of the inertia force $\rho U^2/L$ to the gravity force ρg.

(5) *Heat equation*. There are multiple examples in this and subsequent chapters when in special circumstances the incompressible Navier–Stokes flow reduces to one described by the heat flow equation. This occurs for example in Examples 4.16 and 4.24, Section 4.7, Exercise 4.11, Exercise 4.14 and Exercise 6.11. For completeness we derive the explicit solution flow involving the Gaussian heat kernel here. Our derivation is based on the succinct account given in Evans (1998, Ch. 4). Suppose the scalar field $\upsilon = \upsilon(\boldsymbol{x}, t)$ evolves according to the heat equation

$$\frac{\partial \upsilon}{\partial t} = \nu \Delta \upsilon$$

on $\mathbb{R}^d$, with $\upsilon(\boldsymbol{x}, 0) = \upsilon_0(\boldsymbol{x})$, which decays sufficiently rapidly in the far-field, and $\nu > 0$ is a given parameter. We define the Fourier transform $\hat{\upsilon} = \hat{\upsilon}(\boldsymbol{k}, t)$ of υ and its inverse respectively as follows:

$$\hat{\upsilon}(\boldsymbol{k}, t) := \frac{1}{(2\pi)^d} \int_{\mathbb{R}^d} \upsilon(\boldsymbol{x}, t)\, \mathrm{e}^{-\mathrm{i}\boldsymbol{k}\cdot\boldsymbol{x}}\, \mathrm{d}\boldsymbol{x} \qquad \text{and} \qquad \upsilon(\boldsymbol{x}, t) := \int_{\mathbb{R}^d} \hat{\upsilon}(\boldsymbol{k}, t)\, \mathrm{e}^{\mathrm{i}\boldsymbol{k}\cdot\boldsymbol{x}}\, \mathrm{d}\boldsymbol{k}.$$

We note that for any two real parameters a and b, the transformation $z = b^{1/2}k - (a/2b^{1/2})\,\mathrm{i}$ implies

$$\int_{\mathbb{R}} \mathrm{e}^{\mathrm{i}ak - bk^2}\, \mathrm{d}k = \frac{\mathrm{e}^{-a^2/4b}}{b^{1/2}} \int_{\Gamma} \mathrm{e}^{-z^2}\, \mathrm{d}z,$$

where $\Gamma = \mathbb{R} - (a/2b^{1/2})\,\mathrm{i}$, i.e. the contour parallel to the real axis in the complex z-plane, at the height $z = -(a/2b^{1/2})\,\mathrm{i}$. Since e^{-z^2} is analytic in the z-plane, we can deform the contour Γ to the real axis and use that $\int_{\mathbb{R}} \mathrm{e}^{-z^2}\, \mathrm{d}z = \pi^{1/2}$. Hence we deduce $\int_{\mathbb{R}} \mathrm{e}^{\mathrm{i}a\xi - b\xi^2}\, \mathrm{d}\xi = (\pi/b)^{1/2}\mathrm{e}^{-a^2/4b}$ and thus

$$\int_{\mathbb{R}^d} \mathrm{e}^{\mathrm{i}\boldsymbol{x}\cdot\boldsymbol{k} - \nu t\,|\boldsymbol{k}|^2}\, \mathrm{d}\boldsymbol{k} = \prod_{j=1}^{d} \int_{\mathbb{R}} \mathrm{e}^{-\mathrm{i}x_j k_j - \nu t\, k_j k_j}\, \mathrm{d}k_j = \left(\frac{\pi}{\nu t}\right)^{d/2} \mathrm{e}^{-|\boldsymbol{x}|^2/4\nu t}.$$

Now taking the Fourier transform of the heat equation above for υ, we find

$$\frac{\partial \hat{\upsilon}}{\partial t} = -\nu|\boldsymbol{k}|^2\hat{\upsilon} \qquad \Leftrightarrow \qquad \hat{\upsilon}(\boldsymbol{k}, t) = \mathrm{e}^{-\nu t\,|\boldsymbol{k}|^2}\,\hat{\upsilon}_0(\boldsymbol{k}),$$

where $\hat{\upsilon}_0$ is the Fourier transform of υ_0. By standard Fourier theory, see e.g. Evans (1998, Ch. 4), if $\hat{G}(\boldsymbol{k}, t) := \mathrm{e}^{-\nu t\,|\boldsymbol{k}|^2}$ so the solution $\hat{\upsilon}$ above is the real product of $\hat{G}$ and $\hat{\upsilon}_0$, then $\upsilon = \upsilon(\boldsymbol{x}, t)$ is the scaled convolution of G and υ_0, i.e. we have

$$\upsilon(\boldsymbol{x}, t) = \frac{1}{(2\pi)^d} \int_{\mathbb{R}^d} G(\boldsymbol{x} - \boldsymbol{y}, t)\, \upsilon_0(\boldsymbol{y})\, \mathrm{d}\boldsymbol{y}.$$

The inverse Fourier transform of $\hat{G}$, i.e. the function $G(\boldsymbol{x}, t)$, is given by the integral of $\mathrm{e}^{\mathrm{i}\boldsymbol{x}\cdot\boldsymbol{k}-\nu t\,|\boldsymbol{k}|^2}$ over $\boldsymbol{k} \in \mathbb{R}^d$ shown above. Substituting the closed-form solution shown there for this integral into the convolution integral just above generates the desired solution:

$$v(\boldsymbol{x}, t) = \frac{1}{(4\pi\nu t)^{d/2}} \int_{\mathbb{R}^d} \mathrm{e}^{-|\boldsymbol{x}-\boldsymbol{y}|^2/4\nu t}\, v_0(\boldsymbol{y})\, \mathrm{d}\boldsymbol{y}.$$

(6) *Brownian motion.* The Newtonian viscous dissipation in the incompressible Navier–Stokes equations, represented by the term '$\nu\Delta\boldsymbol{u}$', can be represented probabilistically as follows. We can augment the equations for the Lagrangian particle paths/trajectories by including an additive Wiener/Brownian motion increment $\mathrm{d}W$, scaled by $\sqrt{2\nu}$, on the right-hand side of the trajectory equations. In other words, the particle paths in Definition 1.1 are replaced by $\mathrm{d}\boldsymbol{x}/\mathrm{d}t = \boldsymbol{u}(\boldsymbol{x}, t) + \sqrt{2\nu}\,\mathrm{d}W$. The solution $\boldsymbol{u} = \boldsymbol{u}(\boldsymbol{x}, t)$ to the incompressible Navier–Stokes equation can then be shown to satisfy an implicit Feynman–Kac-type formula. See Busnello *et al.* (2005) and Constantin and Iyer (2008) for more details. Indeed, the physics may be more subtle than this, see Franosh *et al.* (2011) for details on hydrodynamic memory effects.

(7) *Stochastically driven Navier–Stokes equations/turbulence.* Navier–Stokes equations driven by space-time white noise and their connection to theories of turbulence, building on the seminal work of Kolmogorov in 1941, is a very large area of research. For a recent review, see Bianchi and Flandoli (2020) for stochastically driven Navier–Stokes equations and/or Nickelsen (2017) for turbulence models.

(8) *Environmental fluid mechanics.* Naturally, major applications of the Navier–Stokes equations are concerned with the multitudinous aspects of environmental fluid mechanics. Briefly, these include: the study of pollutant dispersal in rivers, oceans and the atmosphere; suspended sediment dynamics in river and estuary systems; coastal erosion and/or flooding; wind and ocean/tidal/river energy harnessing, see Section 6.4(7); rogue waves and last, but not least, climate change!

Exercises

4.1 Fluid stresses: Viscous force. Consider a steady two-dimensional incompressible Navier–Stokes flow prescribed by the stream function

$$\psi = a_0 x^3 + a_1 x^2 y + a_2 x y^2 + a_3 y^3,$$

where the a_i, $i = 0, 1, 2, 3$ are all real constants. Naturally the velocity field is given by $u = \partial\psi/\partial y$ and $v = -\partial\psi/\partial x$.

(a) Recall for such a flow that the deviatoric stress is given by $\hat{\sigma} = 2\mu D$, where D is the rate of strain matrix which is the symmetric part of $\nabla\boldsymbol{u}$. Show for the flow prescribed by ψ above that we have ($\hat{\sigma}_{21} = \hat{\sigma}_{12}$)

$$\begin{aligned}\hat{\sigma}_{11} &= 4\mu\,(a_1 x + a_2 y),\\ \hat{\sigma}_{12} &= \mu\,(2a_2 x + 6a_3 y - 6a_0 x - 2a_1 y),\\ \hat{\sigma}_{22} &= -4\mu\,(a_1 x + a_2 y).\end{aligned}$$

(b) Show that the deviatoric stress force on a volume $\mathcal{V}$ given by

$$\int_{\partial\mathcal{V}} \hat{\sigma}\,\boldsymbol{n}\,\mathrm{d}S = \int_{\mathcal{V}} \nabla\cdot\hat{\sigma}\,\mathrm{d}V$$

is equal to $2\mu\,|\mathcal{V}|\,(a_1 + 3a_3, -3a_0 - a_2)^{\mathrm{T}}$.

(c) Show that the deviatoric stress force in part (b) vanishes if the flow is irrotational.

4.2 Fluid stresses: Rate of strain in cylindrical polar coordinates. The goal of this question is to express the rate of strain matrix, i.e. the symmetric part of $\nabla\boldsymbol{u}$, in cylindrical polar coordinates.

(a) Show that if we rotate the Cartesian coordinates $\boldsymbol{x} = (x, y, z)^{\mathrm{T}}$ through an angle θ about the z-axis to obtain the coordinates $\boldsymbol{x}' = (x', y', z)^{\mathrm{T}}$, then $(\boldsymbol{x}')^{\mathrm{T}} = \boldsymbol{x}^{\mathrm{T}} L$ where L is the matrix

$$L = \begin{pmatrix} \cos\theta & -\sin\theta & 0 \\ \sin\theta & \cos\theta & 0 \\ 0 & 0 & 1 \end{pmatrix}.$$

(b) Suppose $\hat{\boldsymbol{r}}$, $\hat{\boldsymbol{\theta}}$ and $\hat{\boldsymbol{z}}$ are unit vectors in the corresponding cylindrical coordinate directions and the velocity field $\boldsymbol{u} = u_r\,\hat{\boldsymbol{r}} + u_\theta\,\hat{\boldsymbol{\theta}} + u_z\,\hat{\boldsymbol{z}}$. If u, v and w are the velocity field components in the usual respective Cartesian directions x, y and z, show

$$u_r = u\cos\theta + v\sin\theta \qquad \text{and} \qquad u_\theta = -u\sin\theta + v\cos\theta,$$

as well as the inverse mapping

$$u = u_r\cos\theta - u_\theta\sin\theta \qquad \text{and} \qquad v = u_r\sin\theta + u_\theta\cos\theta.$$

(c) Use the chain rule to show that $D_{11} := \partial u/\partial x$ is given by

$$D_{11} = \frac{\partial u}{\partial r}\cos\theta - \frac{1}{r}\frac{\partial u}{\partial \theta}\sin\theta.$$

Further, using the chain rule, find analogous formulae for D_{12} and D_{22}.

(d) Use that $D^{\mathrm{cyl}}_{rr} = [L^{\mathrm{T}} D L]_{11}$, where D^{cyl} represents the rate of strain matrix expressed in cylindrical polar coordinates, and the results of part (c) to show $D^{\mathrm{cyl}}_{rr} = \partial u_r/\partial r$.

(e) Similarly, using the chain rule to derive formulae for D_{13} and D_{23} as in part (c) above, use that $D^{\mathrm{cyl}}_{\theta z} = [L^{\mathrm{T}} D L]_{23}$ to show

$$D^{\mathrm{cyl}}_{\theta z} = \tfrac{1}{2}\left(\frac{\partial u_\theta}{\partial z} + \frac{1}{r}\frac{\partial u_z}{\partial \theta}\right).$$

(f) By similar techniques determine $D^{\mathrm{cyl}}_{r\theta}$, $D^{\mathrm{cyl}}_{\theta\theta}$, D^{cyl}_{rz} and D^{cyl}_{zz}.
(*Note:* See the Appendix Section A.2 for the explicit expressions.)

4.3 Fluid stresses: Deviatoric stress, alternative formulation. The goal of this question is to investigate an alternative expression for the deviatoric stress matrix $\hat{\sigma}$.

(a) Recall that $\nabla\boldsymbol{u}$ can be additively decomposed, $\nabla\boldsymbol{u} = D + R$, into its symmetric D and antisymmetric parts R, see Section 1.9. Use this fact to show that, if $\boldsymbol{\omega}$ is the vorticity field, for any vector $\boldsymbol{h} \in \mathbb{R}^3$ we have

$$D\boldsymbol{h} \equiv (\boldsymbol{h}\cdot\nabla)\,\boldsymbol{u} + \tfrac{1}{2}\boldsymbol{h}\times\boldsymbol{\omega}.$$

(*Note:* You can use that $(\nabla\boldsymbol{u})\,\boldsymbol{h} \equiv (\boldsymbol{h}\cdot\nabla)\,\boldsymbol{u}$.)

(b) Assume no-slip boundary conditions $\boldsymbol{u} = \boldsymbol{0}$ on a rigid boundary. At a given point on the boundary, let $\boldsymbol{n}$ denote the unit outward normal and $\boldsymbol{\tau}_1$ and $\boldsymbol{\tau}_2$ denote two mutually orthogonal unit vectors tangential to the boundary at that point.

  (i) Explain why the directional derivatives $\partial\boldsymbol{u}/\partial\tau_1 = 0$ and $\partial\boldsymbol{u}/\partial\tau_2 = 0$ in the respective directions $\boldsymbol{\tau}_1$ and $\boldsymbol{\tau}_2$.

  (ii) Using the incompressibilty condition expressed in the form

$$\frac{\partial u}{\partial \tau_1} + \frac{\partial v}{\partial \tau_2} + \frac{\partial w}{\partial n} = 0,$$

where u, v and w are the velocity field components in the respective directions $\boldsymbol{\tau}_1, \boldsymbol{\tau}_2$ and $\boldsymbol{n}$, deduce $\partial w/\partial n = 0$ at the boundary.

(iii) Since for any vector $\boldsymbol{h}$ the directional derivative $\partial w/\partial h \equiv \boldsymbol{h} \cdot \nabla w$, and for any two constant/fixed vectors $\boldsymbol{h}_1$ and $\boldsymbol{h}_2$ we have

$$(\boldsymbol{h}_1 \cdot \nabla)(\boldsymbol{u} \cdot \boldsymbol{h}_2) \equiv ((\boldsymbol{h}_1 \cdot \nabla)\, \boldsymbol{u}) \cdot \boldsymbol{h}_2 \equiv \boldsymbol{h}_2^{\mathrm{T}}(\nabla \boldsymbol{u})\, \boldsymbol{h}_1,$$

deduce that at the boundary

$$(\boldsymbol{n} \cdot \nabla)(\boldsymbol{u} \cdot \boldsymbol{n}) = 0.$$

(iv) Use part (a) to show $\boldsymbol{n} \cdot (D\boldsymbol{n}) = 0$ at the boundary and thus deduce the deviatoric stress is wholly tangential.

4.4 Fluid stresses: Rate of dissipation of energy. The goal of this question is to compute the rate of dissipation of energy defined to be

$$E := 2\mu \int_{\mathcal{D}} \operatorname{tr} D^2 \,\mathrm{d}V,$$

for an incompressible fluid inside a vessel $\mathcal{D}$ and derive, first, an alternative expression in the case of a stationary rigid boundary $\partial\mathcal{D}$, and second, an explicit formula for the component of the energy dissipation due to the boundary $\partial\mathcal{D}$ for fluid rotating in an axisymmetric vessel. See Chapter 7 for more details.

(a) Assume the incompressible fluid is contained within a stationary vessel $\mathcal{D}$ with rigid boundary $\partial\mathcal{D}$. Show that the following two identities hold for an incompressible velocity field $\boldsymbol{u}$:

$$\operatorname{tr} D^2 \equiv \tfrac{1}{2}\big(\operatorname{tr}(\nabla \boldsymbol{u})^2 + \nabla \cdot ((\boldsymbol{u} \cdot \nabla)\, \boldsymbol{u})\big) \quad \text{and} \quad |\omega|^2 \equiv \operatorname{tr}(\nabla \boldsymbol{u})^2 + \nabla \cdot ((\boldsymbol{u} \cdot \nabla)\, \boldsymbol{u}).$$

Assuming no-slip boundary conditions at the boundary $\partial\mathcal{D}$, deduce

$$E := \mu \int_{\mathcal{D}} |\omega|^2 \,\mathrm{d}V.$$

What is the energy dissipation when the flow is irrotational?

(b) Now suppose the vessel $\mathcal{D}$ is axisymmetric and rotates about its axis of symmetry, which coincides with the z-axis, with constant angular velocity Ω. The contribution to the energy dissipation E from the boundary due to the term $\nabla \cdot ((\boldsymbol{u} \cdot \nabla)\, \boldsymbol{u})$ in part (a) above is no longer zero. We can compute this contribution explicitly as follows.

(i) Suppose $\boldsymbol{\Omega} := (0, 0, \Omega)^{\mathrm{T}}$ and $\boldsymbol{r}$ is the vector $\boldsymbol{r} := (x, y, 0)^{\mathrm{T}}$. Explain why the velocity field on $\partial\mathcal{D}$ is given by $\boldsymbol{u} = \boldsymbol{\Omega} \times \boldsymbol{r}$.

(ii) If $\boldsymbol{n} = (n_1, n_2, n_3)^{\mathrm{T}}$ denotes the unit outer normal to the surface $\partial\mathcal{D}$, show, since $((\boldsymbol{u} \cdot \nabla)\, \boldsymbol{u}) \cdot \boldsymbol{n} = \boldsymbol{n}^{\mathrm{T}}(\nabla \boldsymbol{u})\, \boldsymbol{u}$, on the surface $\partial\mathcal{D}$ we have

$$((\boldsymbol{u} \cdot \nabla)\, \boldsymbol{u}) \cdot \boldsymbol{n} = \Omega\, (n_1(\boldsymbol{r}^{\perp} \cdot \nabla)\, u + n_2(\boldsymbol{r}^{\perp} \cdot \nabla)\, v + n_3(\boldsymbol{r}^{\perp} \cdot \nabla)\, w),$$

where $\boldsymbol{r}^{\perp} = (-y, x, 0)^{\mathrm{T}}$.

(iii) Noting that the contribution to the energy dissipation from the cap ends of $\mathcal{D}$ is zero, and that w is zero on the curved part of $\partial\mathcal{D}$, substitute for u and v from $\boldsymbol{u} = \boldsymbol{\Omega} \times \boldsymbol{r}$ to show that the contribution to the energy dissipation E from the boundary is given by $-4\mu\Omega^2|\mathcal{D}|$.

4.5 Deviatoric stress: Symmetric matrix-valued isotropic functions. The goal of this exercise is to complete the proof of the Representation Theorem 4.9 for the 'only if' statement when the symmetric 3×3 matrix D does not necessarily have three distinct eigenvalues.

(a) Suppose D has two distinct eigenvalues. Hence by the Spectral Theorem we know that D has the spectral decomposition

$$D = d_1\, \boldsymbol{e}\boldsymbol{e}^{\mathrm{T}} + d_2\,(I - \boldsymbol{e}\boldsymbol{e}^{\mathrm{T}}),$$

with $|\boldsymbol{e}| = 1$, and the eigenspaces corresponding to the eigenvalues d_1 and d_2 are $\mathrm{span}\{\boldsymbol{e}\}$ and $\{\boldsymbol{e}\}^{\perp}$, respectively. Further, by Wang's Lemma the set $\{I, D\}$ is linearly independent and $\mathrm{span}\{I, D\} = \mathrm{span}\{\boldsymbol{e}\boldsymbol{e}^{\mathrm{T}}, I - \boldsymbol{e}\boldsymbol{e}^{\mathrm{T}}\}$.

(i) Using the Transfer Theorem, explain why the eigenspaces $\mathrm{span}\{\boldsymbol{e}\}$ and $\{\boldsymbol{e}\}^{\perp}$ must be contained in an eigenspace for $\hat{\sigma}(D)$.

(ii) Explain why there are only two possibilities from part (i), either: (A) $\hat{\sigma}(D)$ has the eigenspaces $\mathrm{span}\{\boldsymbol{e}\}$ and $\{\boldsymbol{e}\}^{\perp}$ as well; or (B) the eigenspace of $\hat{\sigma}(D)$ is $\mathbb{R}^3$.

(iii) Using the Spectral Theorem explain why we can deduce

$$\hat{\sigma}(D) = \hat{\sigma}_1\, \boldsymbol{e}\boldsymbol{e}^{\mathrm{T}} + \hat{\sigma}_2\,(I - \boldsymbol{e}\boldsymbol{e}^{\mathrm{T}}),$$

in either case (A) or (B) from part (ii). State whether $\hat{\sigma}_1$ is distinct from or equal to $\hat{\sigma}_2$ in each case (A) or (B).

(iv) Using the result of Wang's Lemma in our case here, explain why we can thus deduce, with $\alpha_2(D) = 0$:

$$\hat{\sigma}(D) = \alpha_0(D)\, I + \alpha_1(D)\, D + \alpha_2(D)\, D^2.$$

(v) The goal now is to show $\alpha_0(D)$ and $\alpha_1(D)$ are isotropic. You may take as given that if the principal invariants of any two symmetric matrices D and D' match, then they have the same spectrum. Using that the principle invariants of D are isotropic, explain why D and UDU^{-1} have the same spectrum, and thus each have two distinct eigenvalues.

(vi) Using the representation in part (iv) just above and that we are assuming $\hat{\sigma}(D)$ is isotropic so $U\hat{\sigma}(D)U^{-1} = \hat{\sigma}(UDU^{-1})$, or equivalently $\hat{\sigma}(D) = U^{-1}\hat{\sigma}(UDU^{-1})U$, show that

$$\begin{aligned} &\alpha_0(D)\, I + \alpha_1(D)\, D + \alpha_2(D)\, D^2 \\ &= \alpha_0(UDU^{-1})\, I + \alpha_1(UDU^{-1})\, D + \alpha_2(UDU^{-1})\, D^2. \end{aligned}$$

(vii) Use Wang's Lemma to deduce from part (vi) that $\alpha_2(UDU^{-1}) = 0$, $\alpha_1(D) = \alpha_1(UDU^{-1})$ and $\alpha_0(D) = \alpha_0(UDU^{-1})$, and hence deduce the final representation result for $\hat{\sigma}(D)$.

(b) Suppose D has only one distinct eigenvalue. Hence by the Spectral Theorem it has the representation, $D = d_1 I$, and the eigenspace corresponding to the single eigenvalue d_1 is $\mathbb{R}^3$.
  (i) Explain why the Transfer Theorem implies the eigenspace of $\hat{\sigma}(D)$ is $\mathbb{R}^3$.
  (ii) Using the Spectral Theorem and part (i), explain why $\hat{\sigma}(D)$ has the representation, the same form as previously but with $\alpha_1(D) = \alpha_2(D) = 0$:

$$\hat{\sigma}(D) = \hat{\sigma}_1(D)\, I.$$

  (iii) Using part (ii) and an argument analogous to that in part (a)(vii) just above, use that $\hat{\sigma}(D)$ is isotropic to show

$$\alpha_0(D) = \alpha_0(UDU^{-1}),$$

while all other coefficients are zero. Hence deduce the corresponding representation result for $\hat{\sigma}(D)$.

4.6 Deviatoric stress: Proof of Wang's Lemma. The goal of this question is to prove Wang's Lemma 4.6 for the case of a symmetric 3×3 matrix D with three distinct eigenvalues. Recall from the Spectral Theorem 4.5 the spectral decomposition for such a matrix D, namely

$$D = d_1\, \boldsymbol{e}_1\boldsymbol{e}_1^{\mathrm{T}} + d_2\, \boldsymbol{e}_2\boldsymbol{e}_2^{\mathrm{T}} + d_3\, \boldsymbol{e}_3\boldsymbol{e}_3^{\mathrm{T}}.$$

(a) The first part of Wang's Lemma is to show that the set $\{I, D, D^2\}$ is linearly independent. Hence assume for some constant scalars γ_0, γ_1 and γ_2 we have, with O here denoting the zero matrix,

$$\gamma_2 D^2 + \gamma_1 D + \gamma_0 I = O.$$

  (i) By applying the matrix $\gamma_2 D^2 + \gamma_1 D + \gamma_0 I$ to each of the eigenvectors $\boldsymbol{e}_i$ for $i = 1, 2, 3$ in turn, show that the scalars γ_0, γ_1 and γ_2 must necessarily satisfy the system of equations

$$\begin{pmatrix} 1 & d_0 & d_0^2 \\ 1 & d_1 & d_1^2 \\ 1 & d_2 & d_2^2 \end{pmatrix} \begin{pmatrix} \gamma_0 \\ \gamma_1 \\ \gamma_2 \end{pmatrix} = \mathbf{0}.$$

  (ii) Explain why the coefficient matrix on the left involving the eigenvalues d_1, d_2 and d_3 of D is non-singular and use this fact to deduce that the set $\{I, D, D^2\}$ is linearly independent.

(b) The second part of Wang's Lemma is to show that $\mathrm{span}\{I, D, D^2\}$ equals $\mathbb{E} := \mathrm{span}\{\boldsymbol{e}_1\boldsymbol{e}_1^{\mathrm{T}}, \boldsymbol{e}_2\boldsymbol{e}_2^{\mathrm{T}}, \boldsymbol{e}_3\boldsymbol{e}_3^{\mathrm{T}}\}$.
  (i) Explain why we know the dimension of $\mathbb{E}$ is three.
  (ii) Explain why the spectral decomposition of D^2 is

$$D^2 = d_1^2\, \boldsymbol{e}_1\boldsymbol{e}_1^{\mathrm{T}} + d_2^2\, \boldsymbol{e}_2\boldsymbol{e}_2^{\mathrm{T}} + d_3^2\, \boldsymbol{e}_3\boldsymbol{e}_3^{\mathrm{T}}.$$

  (iii) Using the spectral decompositions for D and D^2, explain why $D, D^2 \in \mathbb{E}$.
  (iv) Recall that the eigenvectors $\boldsymbol{e}_i$ for $i = 1, 2, 3$ are orthonormal. By applying the matrix $\boldsymbol{e}_1\boldsymbol{e}_1^{\mathrm{T}} + \boldsymbol{e}_2\boldsymbol{e}_2^{\mathrm{T}} + \boldsymbol{e}_3\boldsymbol{e}_3^{\mathrm{T}}$ to an arbitrary vector $\boldsymbol{v} \in \mathbb{R}^3$, show that the result necessarily equals

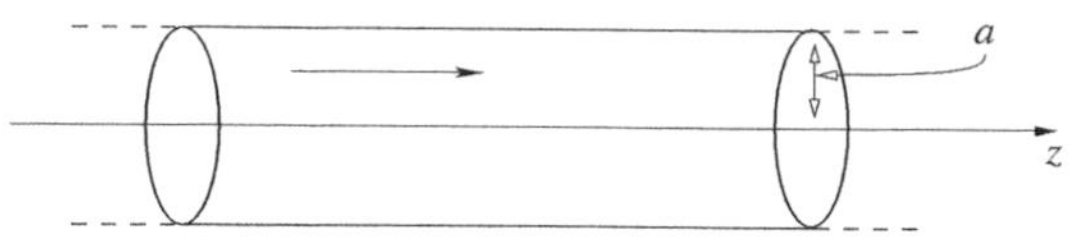

Figure 4.6 Poiseuille flow. A viscous fluid flows along an infinite horizontal pipe with circular cross-section of radius a, whose centre line is aligned along the z-axis. A constant pressure gradient is assumed, as well as axisymmetry, no radial flow and no swirl.

$$\boldsymbol{e}_1(\boldsymbol{e}_1^{\mathrm{T}}\boldsymbol{v}) + \boldsymbol{e}_2(\boldsymbol{e}_2^{\mathrm{T}}\boldsymbol{v}) + \boldsymbol{e}_3(\boldsymbol{e}_3^{\mathrm{T}}\boldsymbol{v}).$$

Using this result, explain why we can deduce $\boldsymbol{e}_1\boldsymbol{e}_1^{\mathrm{T}} + \boldsymbol{e}_2\boldsymbol{e}_2^{\mathrm{T}} + \boldsymbol{e}_3\boldsymbol{e}_3^{\mathrm{T}} = I$.

(v) Using part (iv), deduce $I \in \mathbb{E}$. Using the result of part (a) and results of parts (iii) and (iv) just above, explain why we can deduce span$\{I, D, D^2\}$ equals $\mathbb{E}$.

4.7 Navier–Stokes equations: Flow in an infinite pipe, Poiseuille flow. Consider an infinite pipe with circular cross-section of radius a, whose centre line is aligned along the z-axis (see Figure 4.6). Assume *no-slip* boundary conditions at $r = a$, for all z, i.e. on the inside surface of the cylinder. Using cylindrical polar coordinates, look for a stationary solution to the fluid flow in the pipe of the following form. Assume there is no radial flow, $u_r = 0$, and no swirl, $u_\theta = 0$. Further assume there is a constant pressure gradient down the pipe, i.e. that $p = -Cz$ for some constant C. Lastly, suppose the flow down the pipe, the velocity component u_z, has the form $u_z = u_z(r)$, i.e. it is a function of r only.

(a) Using the Navier–Stokes equations, show that

$$C = -\rho\nu\Delta u_z = -\rho\nu\left(\frac{1}{r}\frac{\partial}{\partial r}\left(r\frac{\partial u_z}{\partial r}\right)\right).$$

(b) Show that integrating the equation above yields

$$u_z = -\frac{C}{4\rho\nu}r^2 + A\log r + B,$$

where A and B are constants. We naturally require that the solution be bounded. Explain why this implies $A = 0$. Now use the no-slip boundary condition to determine B. Hence show that

$$u_z = \frac{C}{4\rho\nu}(a^2 - r^2).$$

(c) Show that the *mass-flow rate* across any cross-section of the pipe is given by (known as the *fourth power law*)

$$\int \rho u_z \,\mathrm{d}S = \pi C a^4/8\nu.$$

4.8 Navier–Stokes equations: Elliptical pipe flow. Consider an infinite horizontal pipe with constant elliptical cross-section, whose centre line is aligned along the z-axis. Assume *no-slip* boundary conditions at

$$\frac{x^2}{a^2} + \frac{y^2}{b^2} = 1,$$

where a and b are the semi-axis lengths of the elliptical cross-section. Using the incompressible Navier–Stokes equations for a homogeneous fluid in Cartesian coordinates, look for a stationary solution to the fluid flow in the pipe of the following form. Assume there is no flow transverse to the axial direction of the pipe, so that if u and v are the velocity components in the coordinate x- and y-directions, respectively, then $u = 0$ and $v = 0$. Further assume there is a constant pressure gradient down the pipe, i.e. the pressure $p = -Gz$ for some constant G, and there is no body force. Lastly, suppose that for the flow down the pipe the velocity component $w = w(x, y)$ only.

(a) Using the Navier–Stokes equations, show that w must satisfy

$$\frac{\partial^2 w}{\partial x^2} + \frac{\partial^2 w}{\partial y^2} = -\frac{G}{\mu}.$$

(b) Show that, assuming A, B and C are constants,

$$w = Ax^2 + By^2 + C$$

is a solution to the partial differential equation in part (a) provided

$$A + B = -\frac{G}{2\mu}.$$

(c) Use the no-slip boundary condition to show

$$A = -\frac{G}{2\mu} \cdot \frac{b^2}{a^2 + b^2}, \qquad B = -\frac{G}{2\mu} \cdot \frac{a^2}{a^2 + b^2} \qquad \text{and} \qquad C = \frac{G}{2\mu} \cdot \frac{a^2 b^2}{a^2 + b^2}.$$

(d) Explain why the volume flux across any cross-section of the pipe is given by $\int w \, \mathrm{d}S$, and then show that it is given by

$$\frac{\pi a^3 b^3 G}{4(a^2 + b^2)\mu}.$$

(*Hint:* To compute the integral you may find the substitutions $x = ar\cos\theta$ and $y = br\sin\theta$, together with the fact that the infinitesimal area element '$\mathrm{d}x\,\mathrm{d}y$' transforms to '$ab\,r\,\mathrm{d}r\,\mathrm{d}\theta$', useful.)

(e) Explain why for a given elliptical cross-sectional area, the optimal choice of a and b to maximise the volume flow rate is $a = b$.

4.9 Navier–Stokes equations: Air blown through a wet tube. Consider a long circular tube which has a layer of liquid of uniform thickness adhering to its inner surface. In order to remove the liquid, air is blown through the tube by applying a pressure difference between the two ends. The goal of this question is to determine the ratio of the steady volume fluxes of the air and the liquid. Suppose the viscosity of the air is μ_0 while for the liquid it is μ_1, with $\mu_1 \gg \mu_0$. Further suppose the air blows through the central circular region $0 \leqslant r < r_0$ while the liquid lies between $r_0 \leqslant r < r_1$, where r_1 is the radius of the circular tube. Suppose the pressure gradient applied across the ends of the tube is '$-G$'.

(a) Explain why the velocity fields $u_0 = u_0(r)$ and $u_1 = u_1(r)$, respectively for the air and liquid flows, satisfy the equations for $j = 0, 1$:

$$\frac{1}{r}\frac{\mathrm{d}}{\mathrm{d}r}\left(r\frac{\mathrm{d}u_j}{\mathrm{d}r}\right) = -\frac{G}{\mu_j},$$

and the boundary conditions: (i) $u_0 = u_1$ at $r = r_0$; (ii) $\partial u_0/\partial r = \partial u_1/\partial r$ at $r = r_0$ (continuous shear) and (iii) $u_1 = 0$ at $r = r_1$.

(b) By successive integration of the equation in part (a), show

$$u_j = -\frac{G}{4\mu_j}r^2 + A_j \log r + B_j,$$

for $j = 0, 1$, where A_j and B_j are arbitrary constants. Explain why $A_0 = 0$ and use the boundary conditions in part (a) to show

$$A_1 = 0, \quad B_1 = \frac{Gr_1^2}{4\mu_1} \quad \text{and} \quad B_0 = \frac{G}{4\mu_1}(r_1^2 - r_0^2) + \frac{Gr_0^2}{4\mu_0}.$$

(c) By explicitly computing the volume fluxes of air and liquid,

$$Q_0 := \int_0^{r_0} u_0(r)2\pi r \, \mathrm{d}r \quad \text{and} \quad Q_1 := \int_{r_0}^{r_1} u_1(r)2\pi r \, \mathrm{d}r,$$

respectively, show

$$\frac{Q_0}{Q_1} = \frac{\mu_1}{\mu_0}\frac{r_0^4}{(r_1^2 - r_0^2)^2} + \frac{2r_0^2}{(r_1^2 - r_0^2)}.$$

Why in practice can we ignore the second term on the right above?

4.10 Navier–Stokes equations: Two-dimensional channel flow. A channel $(x, y) \in \mathbb{R} \times [0, L]$ for some $L > 0$ is filled with a viscous incompressible fluid which is initially at rest. A uniform pressure gradient '$-G$' is applied at time $t = 0$ along the channel, i.e. in the x-direction, and maintained.

(a) Show that the velocity field component $u = u(y, t)$ in the x-direction satisfies the partial differential equation

$$\frac{\partial u}{\partial t} = \nu \frac{\partial^2 u}{\partial y^2} + \frac{G}{\rho},$$

together with the boundary conditions $u = 0$ at $y = 0$ and $y = L$, and the initial condition $u = 0$ at $t = 0$.

(b) Setting $u = Gy(L - y)/(2\mu) - v$ show that $v = v(y, t)$ satisfies the heat equation. What boundary and initial conditions does v satisfy?

(c) Using the method of separation of variables, find the solution to the initial boundary value problem for v from part (b), in the form of a Fourier series on $[0, L]$. Hence show that at leading order for $\nu t/L^2 \gg 1$,

$$u = \frac{G}{2\mu}y(L - y) - \frac{4GL^2}{\pi^3 \mu} \sin\left(\frac{\pi y}{L}\right) \exp\left(-\frac{\pi^2 \nu t}{L^2}\right).$$

4.11 Navier–Stokes equations: Oscillating plane. A viscous incompressible fluid is contained between two infinite parallel planes, a distance d apart, one of which is oscillating in its own plane with velocity $U \cos \omega t$, and the other of which is at rest.

(a) Let $y \in [0, d]$ parameterise the orthogonal distance between the planes, and assume the oscillating plane is at $y = 0$. Explain why it is sufficient to look for a solution in terms of a horizontal velocity field $u = u(y, t)$ and that we can assume u satisfies the heat equation

$$\frac{\partial u}{\partial t} = \nu \frac{\partial^2 u}{\partial y^2},$$

and the boundary conditions $u = U \cos \omega t$ at $y = 0$ and $u = 0$ at $y = d$.

(b) Looking for a solution of the form $u = \mathrm{Re}\{U f(y) \mathrm{e}^{\mathrm{i}\omega t}\}$, show that $f = f(y)$ satisfies the ordinary differential equation

$$f'' - \left(\frac{\mathrm{i}\omega}{\nu}\right) f = 0,$$

together with the boundary conditions $f = 1$ at $y = 0$ and $f = 0$ at $y = d$. Set $\sigma := \omega d^2/\nu$ and $z := y/d$. Solve for f and show that the solution to the problem in part (a) can be expressed in the form

$$u = \mathrm{Re}\left\{\frac{U \sinh\left((\mathrm{i}\sigma)^{1/2}(1-z)\right) \mathrm{e}^{\mathrm{i}\omega t}}{\sinh (\mathrm{i}\sigma)^{1/2}}\right\}.$$

(c) Show that if $\sigma \ll 1$ then $u \sim U(1-z)\cos(\omega t)$. Explain why in this context of slow oscillations of the lower plane, the long-time solution has this 'steady-state' type form.

(d) Show that for any z we have

$$\frac{\sinh\left((\mathrm{i}\sigma)^{1/2}(1-z)\right)}{\sinh (\mathrm{i}\sigma)^{1/2}} \equiv \frac{\mathrm{e}^{-(\mathrm{i}\sigma)^{1/2}z} - \mathrm{e}^{-(\mathrm{i}\sigma)^{1/2}(2-z)}}{1 - \mathrm{e}^{-2(\mathrm{i}\sigma)^{1/2}}}.$$

Now suppose $\sigma \gg 1$.

(i) Using the identity above, show that unless $\sigma^{1/2} z$ is asymptotically small then $u \to 0$ for $\sigma \gg 1$.

(ii) Assume z is such that $\sigma^{1/2} z \ll 1$ for $\sigma \gg 1$. Show in this case

$$u \sim U \mathrm{e}^{-\sigma^{1/2} z/\sqrt{2}} \cos\left(\omega t - \sigma^{1/2} z/\sqrt{2}\right).$$

(*Note:* When the frequency of oscillations is high, there exists a Stokes layer adjacent to the lower plate. Vorticity of equal and opposite sign is generated at the lower plate over a short period, and cancels out over the distance of a Stokes layer thickness $\delta = (2/\sigma)^{1/2} d = (2\nu/\omega)^{1/2} \ll d$.)

4.12 Navier–Stokes equations: Porous channel flow. An incompressible viscous fluid is contained between two plates at $y = 0$ and $y = h$. The plates are at rest and a pressure gradient acts along the channel so the horizontal fluid velocity is

$$u^* = \frac{G}{2\mu} y(h - y).$$

Now suppose the plates are modified so they are porous, and that a uniform cross-channel velocity of speed V is imposed. The goal of this question is to determine the flow solution in the porous context and investigate its nature when Vh/ν is both asymptotically small and large. Here Vh/ν is an example of a Reynolds number; see Section 4.8.

(a) Assuming the cross-channel flow is uniformly $v = -V$, use the incompressibility condition to explain why for the horizontal velocity field u we know $\partial u/\partial x = 0$.

(b) Assuming the flow is steady, show that the governing equation is

$$\frac{d^2u}{dy^2} + \frac{V}{\nu}\frac{du}{dy} = \frac{G}{\mu},$$

together with the boundary conditions $u = 0$ at $y = 0$ and $y = h$. Show that the solution is given by

$$u = \frac{Gh}{\rho V}\left(\frac{1 - e^{-Vy/\nu}}{1 - e^{-Vh/\nu}} - \frac{y}{h}\right).$$

(c) Assume $Vh/\nu \ll 1$ which implies $Vy/\nu \ll 1$. Show in this limit $u \sim u^*$ and interpret this result.

(d) Now assume $Vh/\nu \gg 1$.

(i) If y is large so Vy/ν is large, show that $u \sim u_{\text{out}}$ where

$$u_{\text{out}} = \frac{Gh}{\rho V}\left(1 - \frac{y}{h}\right),$$

which only satisfies the no-slip boundary condition at $y = h$.

(ii) If y is small, show that $u \sim u_{\text{in}}$ where

$$u_{\text{in}} = \frac{Gh}{\rho V}(1 - e^{-Vy/\nu}).$$

(*Note:* The solution u_{in} represents a boundary layer present near $y = 0$ which satisfies the no-slip boundary condition at $y = 0$. Further note that u_{out} at $y = 0$ and u_{in} at $y = \infty$ match. See Chapter 6 for more details.)

4.13 Navier–Stokes equations: Two and a half dimensional flows. Suppose that $(\boldsymbol{v}, p)$, where $\boldsymbol{v} = \boldsymbol{v}(\boldsymbol{x}, t)$ and $p = p(\boldsymbol{x}, t)$ with $\boldsymbol{x} = (x, y)^{\mathrm{T}} \in \mathbb{R}^2$, represent a solution to the two-dimensional homogeneous incompressible Navier–Stokes equations ($\rho > 0$ is the constant mass density):

$$\frac{\partial \boldsymbol{v}}{\partial t} + (\boldsymbol{v} \cdot \nabla_{\boldsymbol{x}})\boldsymbol{v} = -\frac{1}{\rho}\nabla_{\boldsymbol{x}} p + \nu \Delta_{\boldsymbol{x}} \boldsymbol{v},$$

$$\nabla_{\boldsymbol{x}} \cdot \boldsymbol{v} = 0.$$

In addition, suppose that the field $w = w(\boldsymbol{x}, t)$ satisfies the evolution equation

$$\frac{\partial w}{\partial t} + (\boldsymbol{v} \cdot \nabla_{\boldsymbol{x}})\, w = \nu \Delta_{\boldsymbol{x}} w.$$

Show that if $\boldsymbol{u}$ is the three-vector whose first two components are $\boldsymbol{v}$ and third component is w, then $(\boldsymbol{u}, p)$ satisfy the three-dimensional incompressible Navier–Stokes equations, uniform in $z \in \mathbb{R}$.

4.14 Shear-layer flows: Explicit solutions. In Section 4.5 we presented the following shear-layer flow expressed in terms of the vorticity $\omega = \omega(x, t)$:

$$\frac{\partial \omega}{\partial t} + \gamma x \frac{\partial \omega}{\partial x} = \nu \frac{\partial^2 \omega}{\partial x^2} - \gamma \omega,$$

with $\omega(x, 0) = \omega_0(x)$, where $\omega_0 \in \mathbb{R}$ is given. The goal of this question is to derive explicit solutions, first in the case when $\gamma \neq 0$ and $\nu = 0$, and second in the case when $\gamma \neq 0$ and $\nu > 0$.

(a) First consider the case $\gamma \neq 0$ and $\nu = 0$. In this case the evolution equation for $\omega = \omega(x,t)$ is

$$\frac{\partial \omega}{\partial t} + \gamma x \frac{\partial \omega}{\partial x} = -\gamma \omega.$$

We can solve this first-order partial differential equation using the method of characteristics as follows.

(i) Define the characteristic curves $x = x(a,t)$ that start at $x = a \in \mathbb{R}$ when $t = 0$ by

$$\frac{\mathrm{d}}{\mathrm{d}t} x(a,t) = \gamma x(a,t),$$

with $x(a,0) = a$. Show that the characteristic solution curves are given by

$$x(a,t) = \mathrm{e}^{\gamma t}\, a.$$

(ii) Show that along the characteristic curves $\omega = \omega(x(a,t),t)$ satisfies

$$\frac{\mathrm{d}}{\mathrm{d}t}\omega(x(a,t),t) = \gamma\, \omega(x(a,t),t).$$

(iii) Hence, given $x \in \mathbb{R}$ and $t \geqslant 0$, show that $\omega(x,t) = \mathrm{e}^{\gamma t}\omega_0(\mathrm{e}^{-\gamma t}\, x)$ solves the first-order partial differential equation for ω above.

(b) Second consider the case $\gamma \neq 0$ and $\nu > 0$. In this case the evolution equation for $\omega = \omega(x,t)$ is that given at the beginning of the question above. We can solve this linear second-order parabolic partial differential equation as follows.

(i) Show that if we set $\tilde{\omega} := \mathrm{e}^{\gamma t}\omega$, then $\tilde{\omega}$ satisfies the parabolic equation

$$\frac{\partial \tilde{\omega}}{\partial t} + \gamma x \frac{\partial \tilde{\omega}}{\partial x} = \nu \frac{\partial^2 \tilde{\omega}}{\partial x^2}.$$

(ii) Consider the following change of variables from (x,t) to (ξ,τ):

$$\xi := \mathrm{e}^{-\gamma t}\, x \quad \text{and} \quad \tau := (\nu/2\gamma)\,(1 - \mathrm{e}^{2\gamma t}) \quad \text{with} \quad \omega^*(\xi,\tau) := \tilde{\omega}(x,t).$$

Using the chain rule, show that $\omega^* = \omega^*(\xi,\tau)$ satisfies the heat equation

$$\frac{\partial \omega^*}{\partial \tau} = \frac{\partial^2 \omega^*}{\partial \xi^2}.$$

Then using the form of the solution for the heat equation given in Example 4.16 in terms of the Gaussian heat kernel $G = G(x,t)$, show that

$$\omega^*(\xi,\tau) = \int_{\mathbb{R}} G(\xi - \xi',\tau)\,\omega_0^*(\xi')\,\mathrm{d}\xi'.$$

(iii) Hence show that the solution to the linear second-order parabolic equation given at the beginning of this question is given explicitly by

$$\omega(x,t) = \mathrm{e}^{-\gamma t}\int_{\mathbb{R}} G\big(\mathrm{e}^{-\gamma t}\, x - \xi', (\nu/2\gamma)\,(1 - \mathrm{e}^{2\gamma t})\big)\,\omega_0(\xi')\,\mathrm{d}\xi'.$$

(*Note:* See Majda and Bertozzi, 2002, p. 18.)

4.15 Shear-layer flows: Axisymmetric flow. Consider an axisymmetric flow in cylindrical polar coordinates (r, θ, z) prescribed as follows. If u_r, u_θ and u_z denote the velocity components in the corresponding coordinate directions, then the flow is given by $u_r = -\alpha r/2$, $u_\theta = u_\theta(r,t)$ only and $u_z = \alpha z$, where α is a constant.

(a) Show that the given flow is incompressible.

(b) Show that the vorticity corresponding to the flow has the form $(0, 0, \omega)^{\mathrm{T}}$, where $\omega = \omega(r,t)$ is given by

$$\omega = \frac{1}{r}\frac{\partial}{\partial r}(r u_\theta).$$

(c) Recall that if the velocity field $\boldsymbol{u} = \boldsymbol{u}(\boldsymbol{x},t)$ satisfies the Navier–Stokes equations, then the corresponding vorticity field $\boldsymbol{\omega}$ satisfies

$$\frac{\partial \boldsymbol{\omega}}{\partial t} + (\boldsymbol{u} \cdot \nabla)\,\boldsymbol{\omega} = \nu \Delta \boldsymbol{\omega} + \boldsymbol{\omega} \cdot \nabla \boldsymbol{u}.$$

Using the flow ansatz and the form for the only non-zero component of vorticity ω in part (b), show that ω satisfies the following evolution equation in cylindrical polar coordinates:

$$\frac{\partial \omega}{\partial t} + u_r \frac{\partial \omega}{\partial r} = \nu \frac{1}{r}\frac{\partial}{\partial r}\left(r \frac{\partial \omega}{\partial r}\right) + \omega \frac{\partial u_z}{\partial z}.$$

(d) Assume in this part that the fluid is *inviscid*. Also assume that at time $t = 0$ the initial data for the vorticity has the form $\omega(r) = \omega_0 f(r)$ for some function $f = f(r)$ and some constant ω_0. Verify that

$$\omega(r,t) = \omega_0\, \mathrm{e}^{\alpha t} f(r \mathrm{e}^{\alpha t/2})$$

is a solution for all $t > 0$ to the evolution equation for ω in part (c).
What does this solution ansatz say about vorticity and its evolution for an inviscid flow?

(e) Assume in this part that the fluid is *viscous* but steady. Using part (c) write down the governing equation for ω and show that it is satisfied by

$$\omega(r,t) = \omega_0 \mathrm{e}^{-\alpha r^2/4\nu}.$$

Why is a steady flow possible in this case – what is the dominant physical balance in the flow?

4.16 Vorticity-stream formulation: Stuart cat's-eye flow. By direct substitution, show that the stream function $\psi = \psi(x,y)$ given in Example 4.21, for the two-dimensional incompressible ideal flow known as the Stuart cat's-eye flow, is indeed a solution to $\Delta\psi = -\mathrm{e}^{2\psi}$ for any $c \geqslant 1$.

4.17 Vorticity-stream formulation: Inviscid/viscous periodic flows. Suppose the vorticity field $\omega = \omega(\boldsymbol{x})$ is periodic in $\boldsymbol{x} = (x,y)^{\mathrm{T}}$ in both directions. Indeed, suppose the flow domain is $\mathcal{D} = \mathbb{T}^2$ where $\mathbb{T}^2$ is the 2-torus of period one in each direction.

(a) Show for each $\boldsymbol{k} \in \mathbb{Z}^2$ that the functions (with $a_{\boldsymbol{k}}$ and $b_{\boldsymbol{k}}$ constants)

$$\hat{\psi}_{\boldsymbol{k}}(\boldsymbol{x}) := a_{\boldsymbol{k}} \cos(2\pi \boldsymbol{x} \cdot \boldsymbol{k}) + b_{\boldsymbol{k}} \sin(2\pi \boldsymbol{x} \cdot \boldsymbol{k})$$

are eigenfunctions of the Laplacian operator Δ in $\mathbb{T}^2$, in particular they satisfy $\Delta\hat{\psi}_{\boldsymbol{k}} = -4\pi^2|\boldsymbol{k}|^2\hat{\psi}_{\boldsymbol{k}}$.

(b) Now consider the vorticity-stream formulation for homogeneous incompressible stationary ideal flows from Corollary 4.20. Recall that a given function $\psi = \psi(\boldsymbol{x})$ represents the stream function for such a flow if and only if there exists a function $F\colon \mathbb{R} \to \mathbb{R}$ such that $\Delta\psi = F(\psi)$. By considering linear combinations of eigenfunctions $\hat{\psi}_{\boldsymbol{k}} = \hat{\psi}_{\boldsymbol{k}}(\boldsymbol{x})$ from part (a) such that $|\boldsymbol{k}|^2 = \ell$, show that for any $\ell \in \mathbb{N}$ the stream function

$$\psi_0 := \sum_{|\boldsymbol{k}|^2=\ell} \hat{\psi}_{\boldsymbol{k}}$$

satisfies $\Delta\psi_0 = F(\psi_0)$, where F is the linear function $F\colon \psi \mapsto -4\pi^2\ell\psi$. Deduce that ψ_0 generates a periodic incompressible stationary ideal flow.

(c) Suppose $\boldsymbol{u}_0 = \boldsymbol{u}_0(\boldsymbol{x})$ is an initial velocity field associated with the stream function ψ_0 from part (b) so that $\boldsymbol{u}_0 = \nabla^{\perp}\psi_0$. Show that the scalar vorticity field $\omega = \omega(\boldsymbol{x}, t)$ corresponding to the velocity field

$$\boldsymbol{u}(\boldsymbol{x}, t) := \mathrm{e}^{-4\pi^2\ell\nu t}\boldsymbol{u}_0(\boldsymbol{x})$$

satisfies the vorticity formulation $\partial\omega/\partial t + \boldsymbol{u}\cdot\nabla\omega = \nu\Delta\omega$ for the incompressible Navier–Stokes equations and hence represents an exact incompressible viscous flow in $\mathbb{T}^2$.

(*Note:* For every Lagrangian path $\boldsymbol{x} = \boldsymbol{x}(\boldsymbol{a}, t)$ emanating from $\boldsymbol{a} \in \mathbb{T}^2$,

$$\frac{\mathrm{d}}{\mathrm{d}t}\boldsymbol{x}(\boldsymbol{a}, t) = \boldsymbol{u}(\boldsymbol{x}(\boldsymbol{a}, t), t) = \mathrm{e}^{-4\pi^2\ell\nu t}\boldsymbol{u}_0(\boldsymbol{x}(\boldsymbol{a}, t)),$$

demonstrating the effect of viscosity on the flow.)

5 Low Reynolds-Number Flow

5.1 Stokes Flow

Consider the homogeneous incompressible Navier–Stokes equations with no body force in dimensionless form:

$$\frac{\partial \boldsymbol{u}'}{\partial t'} + \underbrace{(\boldsymbol{u}' \cdot \nabla_{\boldsymbol{x}'})\, \boldsymbol{u}'}_{\text{inertia term}} = \underbrace{(\mathrm{Re})^{-1} \Delta_{\boldsymbol{x}'} \boldsymbol{u}'}_{\text{diffusion term}} - \nabla_{\boldsymbol{x}'} p',$$

with $\nabla_{\boldsymbol{x}'} \cdot \boldsymbol{u}' = 0$ and where Re is the Reynolds number. Note that we have reverted to using primes to indicate that the variables shown are non-dimensional – recall their definitions from Section 4.8. We wish to consider the small Reynolds number limit $\mathrm{Re} \to 0$. Naively this means viscous diffusion will be the dominant effect and the diffusion term in the equation above will be the dominant term, in particular dominant over the advective inertia term shown. However, we also want to maintain incompressibility, i.e. $\nabla_{\boldsymbol{x}'} \cdot \boldsymbol{u}' = 0$. Since the pressure field is the Lagrange multiplier that maintains the incompressibility constraint, we should attempt to maintain it in the limit $\mathrm{Re} \to 0$. Hence we suppose

$$p' = (\mathrm{Re})^{-1} q,$$

for a scaled pressure q. Further, the flow may evolve on a slow timescale so that

$$t' = (\mathrm{Re})\, \tau.$$

Using the chain rule and direct substitution to make these changes of variables in the non-dimensional incompressible Navier–Stokes equations, they become

$$(\mathrm{Re})^{-1} \frac{\partial \boldsymbol{u}'}{\partial \tau} + (\boldsymbol{u}' \cdot \nabla_{\boldsymbol{x}'})\, \boldsymbol{u}' = (\mathrm{Re})^{-1} \Delta_{\boldsymbol{x}'} \boldsymbol{u}' - (\mathrm{Re})^{-1} \nabla_{\boldsymbol{x}'} q.$$

Multiplying through by the Reynolds number and taking $\mathrm{Re} \to 0$ generates

$$\frac{\partial \boldsymbol{u}'}{\partial \tau} = \Delta_{\boldsymbol{x}'} \boldsymbol{u}' - \nabla_{\boldsymbol{x}'} q.$$

These are the non-dimensional Stokes equations. The original *dimensional* pressure p and time t variables are given by

$$p = \rho U^2 p' = \frac{\rho \nu U}{L} q \qquad \text{and} \qquad t = \frac{L}{U} t' = \frac{L^2}{\nu} \tau.$$

In other words, instead of rescaling the pressure $p' = p/P$ by the 'dynamic pressure' scaling $P = \rho U^2$, we have rescaled it to $q = p/P$ with $P = \rho\nu U/L$. Recall that when we non-dimensionalised the incompressible Navier–Stokes equations in Section 4.8, the scaling factor of the viscous diffusion and pressure terms was $\nu U/L^2$ and $P/\rho L$, respectively. Note that to ensure these terms are of the same order, we should choose $P/\rho L = \nu U/L^2$, i.e. $P = \rho\nu U/L$. Similarly, instead of rescaling the time $t' = t/T$ with $T = L/U$, we rescaled it to $\tau = t/T$ with $T = L^2/\nu$. Again this ensures that the inertia term $\partial\boldsymbol{u}/\partial t = (U/T)\partial\boldsymbol{u}'/\partial t'$ and the viscous diffusion term $\nu\Delta\boldsymbol{u} = (\nu U/L^2)\Delta_{\boldsymbol{x}'}\boldsymbol{u}'$ in the non-dimensionalised Navier–Stokes equations in Section 4.8, have the same order as $U/T = \nu U/L^2 \Leftrightarrow T = L^2/\nu$. Reverting the non-dimensional Stokes equation to full-dimensional variables $t = Tt'$, $\boldsymbol{x} = L\boldsymbol{x}'$, $\boldsymbol{u} = U\boldsymbol{u}'$, etc., we find

$$
\begin{aligned}
& & \frac{\partial \boldsymbol{u}'}{\partial \tau} &= \Delta_{\boldsymbol{x}'}\boldsymbol{u}' - \nabla_{\boldsymbol{x}'} q \\
&\Leftrightarrow & \frac{L^2}{\nu}\frac{1}{U}\frac{\partial \boldsymbol{u}}{\partial t} &= \frac{L^2}{U}\Delta_{\boldsymbol{x}}\boldsymbol{u} - L\frac{L}{\rho\nu U}\nabla_{\boldsymbol{x}} p \\
&\Leftrightarrow & \rho\frac{\partial \boldsymbol{u}}{\partial t} &= \mu\Delta\boldsymbol{u} - \nabla p.
\end{aligned}
$$

We have thus arrived at the following system of fluid flow equations.

Definition 5.1 (Stokes flow) A *Stokes flow* is a homogeneous incompressible fluid flow for which the velocity field $\boldsymbol{u} = \boldsymbol{u}(\boldsymbol{x}, t)$ and pressure field $p = p(\boldsymbol{x}, t)$ satisfy the system of equations given by

$$
\begin{aligned}
\rho\frac{\partial \boldsymbol{u}}{\partial t} &= \mu\Delta\boldsymbol{u} - \nabla p, \\
\nabla \cdot \boldsymbol{u} &= 0,
\end{aligned}
$$

together with no-slip boundary conditions for $\boldsymbol{u}$.

Remark 5.2 We note the following points: (i) more commonly the stationary versions of these equations are nominated as *Stokes flows*; (ii) Stokes flows are also known as *slow flows* or *creeping flows*; and (iii) taking the divergence of the Stokes momentum equation and using the incompressibility condition reveals that the pressure satisfies $\Delta p = 0$.

Remark 5.3 (Stokes equations: well-posedness) For interior or exterior domains with smooth boundaries, and smooth initial data, the Stokes flow equations above are well-posed globally in time. In other words, a global smooth solution exists, is unique and depends continously on the data. See e.g. Ladyzhenskaya (1969) and Hieber and Saal (2018). For periodic boundary conditions, this can be established via the arguments in Chapter 7, modified rather helpfully by removing the nonlinear term '$(\boldsymbol{u} \cdot \nabla)\,\boldsymbol{u}$' throughout.

Recall from Corollary 4.15 and Section 4.6 that the homogeneous incompressible Navier–Stokes equations have an equivalent formulation in terms of the vorticity field. Here we have a special case. Taking the curl of the Stokes equations establishes the equivalent formulation in terms of the vorticity field (we have also divided through by the constant density ρ):

$$\frac{\partial \omega}{\partial t} = \nu \Delta \omega.$$

We now consider a result useful for our subsequent Stokes flow analysis.

Definition 5.4 (Solenoidal fields) A field $\boldsymbol{v}: \mathcal{D} \to \mathbb{R}^3$ which is such that $\nabla \cdot \boldsymbol{v} \equiv 0$ in $\mathcal{D} \subseteq \mathbb{R}^3$ is said to be *solenoidal* in $\mathcal{D}$.

Lemma 5.5 (Vector potentials) *Any solenoidal field* $\boldsymbol{v}: \mathbb{R}^3 \to \mathbb{R}^3$ *can be expressed as the curl of a* vector potential *field, say* $\boldsymbol{\psi}: \mathbb{R}^3 \to \mathbb{R}^3$, *i.e. we know that* $\nabla \cdot \boldsymbol{v} \equiv 0$ *if and only if there exists a vector-valued function* $\boldsymbol{\psi}: \mathbb{R}^3 \to \mathbb{R}^3$ *such that* $\boldsymbol{v} = \nabla \times \boldsymbol{\psi}$ *everywhere in* $\mathbb{R}^3$.

Proof The 'if' case is a standard calculus identity; see the Appendix Section A.1 #4. For the 'only if' case, assume $\nabla \cdot \boldsymbol{v} \equiv 0$ in $\mathbb{R}^3$. For any $\boldsymbol{x} \in \mathbb{R}^3$, set $\boldsymbol{\psi} := -\Delta^{-1}(\nabla \times \boldsymbol{v})$ or more explicitly, using the integral kernel formulation for Δ^{-1}, see e.g. Evans (1998, p. 23), we set

$$\boldsymbol{\psi}(\boldsymbol{x}) := \frac{1}{4\pi} \int_{\mathbb{R}^3} \frac{(\nabla_{\boldsymbol{y}} \times \boldsymbol{v})(\boldsymbol{y})}{|\boldsymbol{x} - \boldsymbol{y}|} \, \mathrm{d}\boldsymbol{y}.$$

Note the notation in the numerator in the integrand. It is to be interpreted as the curl of $\boldsymbol{v}$ evaluated at $\boldsymbol{y}$. A change of variables shows, equivalently, that we have

$$\boldsymbol{\psi}(\boldsymbol{x}) = \frac{1}{4\pi} \int_{\mathbb{R}^3} \frac{(\nabla_{\boldsymbol{y}} \times \boldsymbol{v})(\boldsymbol{x} - \boldsymbol{y})}{|\boldsymbol{y}|} \, \mathrm{d}\boldsymbol{y}.$$

We observe by direct computation that

$$\begin{aligned}
(\nabla_{\boldsymbol{x}} \times \boldsymbol{\psi})(\boldsymbol{x}) &= \frac{1}{4\pi} \int_{\mathbb{R}^3} \frac{(\nabla_{\boldsymbol{x}} \times (\nabla_{\boldsymbol{y}} \times \boldsymbol{v}))(\boldsymbol{x} - \boldsymbol{y})}{|\boldsymbol{y}|} \, \mathrm{d}\boldsymbol{y} \\
&= -\frac{1}{4\pi} \int_{\mathbb{R}^3} \frac{(\nabla_{\boldsymbol{y}} \times (\nabla_{\boldsymbol{y}} \times \boldsymbol{v}))(\boldsymbol{x} - \boldsymbol{y})}{|\boldsymbol{y}|} \, \mathrm{d}\boldsymbol{y} \\
&= \frac{1}{4\pi} \int_{\mathbb{R}^3} \frac{(\Delta_{\boldsymbol{y}} \boldsymbol{v})(\boldsymbol{x} - \boldsymbol{y})}{|\boldsymbol{y}|} \, \mathrm{d}\boldsymbol{y} \\
&= (\Delta^{-1}(\Delta \boldsymbol{v}))(\boldsymbol{x}),
\end{aligned}$$

giving $\boldsymbol{v}$ and thus the required result. Note that $\nabla \cdot \boldsymbol{v} \equiv 0$ in $\mathbb{R}^3$ was used in the assertion $\nabla \times (\nabla \times \boldsymbol{v}) \equiv -\Delta \boldsymbol{v}$. □

Remark 5.6 (Vector potentials for smooth domains) An essential ingredient in the proof of existence of a vector potential was the solution of Poisson's equation in $\mathbb{R}^3$. For interior domains $\mathcal{D} \subseteq \mathbb{R}^3$ with smooth boundaries $\partial \mathcal{D}$, or corresponding exterior domains, the existence of a Greens' kernel and a solution to Poisson's equation is more nuanced. We treat such examples on a case-by-case basis. Indeed, our examples below all involve a coordinate symmetry and thus the existence of a scalar stream function.

Suppose we know that a *stream function* exists for the flow under consideration, i.e. the flow is incompressible and an additional symmetry allows us to eliminate one spatial coordinate and one velocity component. For example, suppose we have a *stationary* two-dimensional flow $\boldsymbol{u} = (u, v)^{\mathrm{T}}$ in Cartesian coordinates $\boldsymbol{x} = (x, y)^{\mathrm{T}}$. Then there exists a stream function $\psi = \psi(x, y)$ given by

$$u = \frac{\partial \psi}{\partial y} \qquad \text{and} \qquad v = -\frac{\partial \psi}{\partial x}.$$

Hence the vorticity of such a flow is $\omega = (0, 0, -\Delta\psi)^{\mathrm{T}}$ in Cartesian coordinates. The equations for a stationary Stokes flow, $\Delta\omega = 0$, are thus equivalent to

$$\Delta(\Delta\psi) = 0.$$

In other words, the stream function satisfies the *biharmonic equation*.

In *cylindrical polar coordinates*, assume we have a stationary axisymmetric Stokes flow with no swirl. In other words, the velocity field is explicitly independent of time t and the azimuthal angle θ, and we have $u_\theta \equiv 0$. Hence we know $u_r = u_r(r, z)$ and $u_z = u_z(r, z)$ only. If the flow under consideration is in $\mathbb{R}^3$, then from Lemma 5.5 above we know there exists a $\boldsymbol{\psi}\colon \mathbb{R}^3 \to \mathbb{R}^3$ of the form $\boldsymbol{\psi} = \psi_r\,\hat{\boldsymbol{r}} + \psi_\theta\,\hat{\boldsymbol{\theta}} + \psi_z\,\hat{\boldsymbol{z}}$ such that $\boldsymbol{u} = \nabla \times \boldsymbol{\psi}$, where (see the Appendix, Section A.2)

$$\nabla \times \boldsymbol{\psi} = \left(\frac{1}{r}\frac{\partial \psi_z}{\partial \theta} - \frac{\partial \psi_\theta}{\partial z}\right)\hat{\boldsymbol{r}} + \left(\frac{\partial \psi_r}{\partial z} - \frac{\partial \psi_z}{\partial r}\right)\hat{\boldsymbol{\theta}} + \left(\frac{1}{r}\frac{\partial}{\partial r}(r\psi_\theta) - \frac{1}{r}\frac{\partial \psi_r}{\partial \theta}\right)\hat{\boldsymbol{z}}.$$

Since $u_r = u_r(r, z)$ only, the $\hat{\boldsymbol{r}}$ component above must be a function of (r, z) only and thus $\partial\psi_z/\partial\theta$ must be θ-independent. Hence ψ_z must have the form $\psi_z = \psi_1(r, z)\,\theta + \psi_2(r, z)$, which can only be periodic in θ if $\psi_1 \equiv 0$. Hence $\psi_z = \psi_z(r, z)$ only. A similar argument with the $\hat{\boldsymbol{z}}$ component above implies $\psi_r = \psi_r(r, z)$ only. Further, due to no swirl, the component of $\hat{\boldsymbol{\theta}}$ is zero. Consequently we see that ψ_r and ψ_z play no role in the prescription of $\boldsymbol{u}$ and without loss of generality we can set both of them to zero. Hence we deduce $\boldsymbol{\psi} = \psi_\theta\,\hat{\boldsymbol{\theta}}$, with $\psi_\theta = \psi_\theta(r, z)$, and the velocity field $\boldsymbol{u} = \nabla \times \boldsymbol{\psi}$ is given by

$$\nabla \times \boldsymbol{\psi} = \left(-\frac{\partial \psi_\theta}{\partial z}\right)\hat{\boldsymbol{r}} + \left(\frac{1}{r}\frac{\partial}{\partial r}(r\psi_\theta)\right)\hat{\boldsymbol{z}},$$

and so $u_r = -\partial\psi_\theta/\partial z$ and $u_z = r^{-1}\partial(r\psi_\theta)/\partial r$. The vorticity field $\omega = \nabla \times \boldsymbol{u}$ in cylindrical coordinates, from Section A.2 in the Appendix, has the form $\omega = \omega_\theta\,\hat{\boldsymbol{\theta}}$ and indeed

$$\omega = \left(\frac{\partial u_r}{\partial z} - \frac{\partial u_z}{\partial r}\right)\hat{\boldsymbol{\theta}}.$$

Substituting for u_r and u_z in terms of $\boldsymbol{\psi}$ from above, the vorticity ω is given by

$$\begin{aligned}
\nabla \times (\nabla \times \boldsymbol{\psi}) &= -\left(\frac{\partial^2 \psi_\theta}{\partial z^2} + \frac{\partial}{\partial r}\left(\frac{1}{r}\frac{\partial}{\partial r}(r\psi_\theta)\right)\right)\hat{\boldsymbol{\theta}} \\
&= -\frac{1}{r}\left(\frac{\partial^2}{\partial z^2}(r\psi_\theta) + r\frac{\partial}{\partial r}\left(\frac{1}{r}\frac{\partial}{\partial r}(r\psi_\theta)\right)\right)\hat{\boldsymbol{\theta}}
\end{aligned}$$

$$= -\frac{1}{r}\mathrm{D}^2_{\mathrm{cyl}}(r\psi_\theta)\,\hat{\boldsymbol{\theta}},$$

where $\mathrm{D}^2_{\mathrm{cyl}}$ is the second-order partial differential operator defined by

$$\mathrm{D}^2_{\mathrm{cyl}} := \frac{\partial^2}{\partial z^2} + r\frac{\partial}{\partial r}\left(\frac{1}{r}\frac{\partial}{\partial r}\right).$$

For a stationary Stokes flow, $\Delta\boldsymbol{\omega} = 0$. Using that for a divergence-free vector field $\boldsymbol{\omega}$ we have $\Delta\boldsymbol{\omega} = -\nabla\times\nabla\times\boldsymbol{\omega}$ and $\boldsymbol{\omega} = \omega_\theta\,\hat{\boldsymbol{\theta}}$, then repeating the arguments above, replacing ψ_θ by ω_θ, we observe that

$$\begin{aligned}
\nabla\times\nabla\times\boldsymbol{\omega} &= -\left(\frac{\partial^2\omega_\theta}{\partial z^2} + \frac{\partial}{\partial r}\left(\frac{1}{r}\frac{\partial}{\partial r}(r\omega_\theta)\right)\right)\hat{\boldsymbol{\theta}} \\
&= -\frac{1}{r}\left(\frac{\partial^2}{\partial z^2}(r\omega_\theta) + r\frac{\partial}{\partial r}\left(\frac{1}{r}\frac{\partial}{\partial r}(r\omega_\theta)\right)\right)\hat{\boldsymbol{\theta}} \\
&= -\frac{1}{r}\mathrm{D}^2_{\mathrm{cyl}}(r\omega_\theta)\,\hat{\boldsymbol{\theta}} \\
&= \frac{1}{r}\mathrm{D}^2_{\mathrm{cyl}}\mathrm{D}^2_{\mathrm{cyl}}(r\psi_\theta)\,\hat{\boldsymbol{\theta}}.
\end{aligned}$$

Hence we have the following result.

Proposition 5.7 (Stokes flow: cylindrical polar coordinates) *A stationary axisymmetric Stokes flow with no swirl is prescribed by the transformed stream function* $\Psi := r\psi_\theta$ *in cylindrical polar coordinates which satisfies*

$$\mathrm{D}^2_{\mathrm{cyl}}(\mathrm{D}^2_{\mathrm{cyl}}\Psi) = 0.$$

Let us now consider the same flow in *spherical polar coordinates*, again, stationary, axisymmetric and with no swirl. Here this means the velocity field is explicitly independent of time t and the azimuthal angle φ, so $u_r = u_r(r,\theta)$ and $u_\theta = u_\theta(r,\theta)$ only. No swirl corresponds to $u_\varphi \equiv 0$. If the flow domain is $\mathbb{R}^3$, then as above, from Lemma 5.5 there exists a $\boldsymbol{\psi}\colon \mathbb{R}^3 \to \mathbb{R}^3$ of the form $\boldsymbol{\psi} = \psi_r\,\hat{\boldsymbol{r}} + \psi_\varphi\,\hat{\boldsymbol{\varphi}} + \psi_\theta\,\hat{\boldsymbol{\theta}}$ such that $\boldsymbol{u} = \nabla\times\boldsymbol{\psi}$ where (see Section A.3 in the Appendix)

$$\nabla\times\boldsymbol{\psi} = \frac{1}{\varrho}\left(\frac{\partial}{\partial\theta}(\sin\theta\,\psi_\varphi) - \frac{\partial\psi_\theta}{\partial\varphi}\right)\hat{\boldsymbol{r}} + \frac{1}{r}\left(\frac{\partial}{\partial r}(r\psi_\theta) - \frac{\partial\psi_r}{\partial\theta}\right)\hat{\boldsymbol{\varphi}} + \left(\frac{1}{\varrho}\frac{\partial\psi_r}{\partial\varphi} - \frac{1}{r}\frac{\partial}{\partial r}(r\psi_\varphi)\right)\hat{\boldsymbol{\theta}},$$

where for convenience we set $\varrho := r\sin\theta$. By arguments analogous to those for the cylindrical polar coordinates case above, since $u_r = u_r(r,\theta)$ and $u_\theta = u_\theta(r,\theta)$ only, we deduce that to adhere to periodicity in φ, necessarily we have $\psi_r = \psi_r(r,\theta)$ and $\psi_\theta = \psi_\theta(r,\theta)$ only. These facts, together with the no-swirl condition, implies that we can assume ψ_r and ψ_θ are both zero. Hence we have $\boldsymbol{\psi} = \psi_\varphi\,\hat{\boldsymbol{\varphi}}$, with $\psi_\varphi = \psi_\varphi(r,\theta)$, and the velocity field $\boldsymbol{u} = \nabla\times\boldsymbol{\psi}$ is given by

$$\nabla\times\boldsymbol{\psi} = \left(\frac{1}{\varrho}\frac{\partial}{\partial\theta}(\sin\theta\,\psi_\varphi)\right)\hat{\boldsymbol{r}} + \left(-\frac{1}{r}\frac{\partial}{\partial r}(r\psi_\varphi)\right)\hat{\boldsymbol{\theta}}.$$

Noting the form for the velocity components u_r and u_θ from this expression, from Section A.3 of the Appendix, the vorticity field $\boldsymbol{\omega} = \nabla \times \boldsymbol{u}$ is given by

$$
\begin{aligned}
\nabla \times (\nabla \times \boldsymbol{\psi}) &= -\frac{1}{r}\left(\frac{\partial^2}{\partial r^2}(r\psi_\varphi) + \frac{\partial}{\partial \theta}\left(\frac{1}{\varrho}\frac{\partial}{\partial \theta}(\sin\theta\,\psi_\varphi)\right)\right)\hat{\boldsymbol{\varphi}} \\
&= -\frac{1}{\varrho}\left(\frac{\partial^2}{\partial r^2}(\varrho\psi_\varphi) + \frac{\sin\theta}{r}\frac{\partial}{\partial \theta}\left(\frac{1}{\varrho}\frac{\partial}{\partial \theta}(\varrho\psi_\varphi)\right)\right)\hat{\boldsymbol{\varphi}} \\
&= -\frac{1}{\varrho}\mathrm{D}^2_{\mathrm{sph}}(\varrho\,\psi_\varphi)\,\hat{\boldsymbol{\varphi}},
\end{aligned}
$$

where $\mathrm{D}^2_{\mathrm{sph}}$ is the second-order partial differential operator given by

$$
\mathrm{D}^2_{\mathrm{sph}} := \frac{\partial^2}{\partial r^2} + \frac{\sin\theta}{r}\frac{\partial}{\partial \theta}\left(\frac{1}{\varrho}\frac{\partial}{\partial \theta}\right).
$$

For a stationary Stokes flow, $\Delta\boldsymbol{\omega} = 0$. Again using that $\Delta\boldsymbol{\omega} = -\nabla\times\nabla\times\boldsymbol{\omega}$ and $\boldsymbol{\omega} = \omega_\varphi\,\hat{\boldsymbol{\varphi}}$, repeating the arguments above with ψ_φ replaced by ω_φ, we find

$$
\begin{aligned}
\nabla \times (\nabla \times \boldsymbol{\omega}) &= -\frac{1}{r}\left(\frac{\partial^2}{\partial r^2}(r\omega_\varphi) + \frac{\partial}{\partial \theta}\left(\frac{1}{\varrho}\frac{\partial}{\partial \theta}(\sin\theta\,\omega_\varphi)\right)\right)\hat{\boldsymbol{\varphi}} \\
&= -\frac{1}{\varrho}\mathrm{D}^2_{\mathrm{sph}}(\varrho\,\omega_\varphi)\,\hat{\boldsymbol{\varphi}} \\
&= \frac{1}{\varrho}\mathrm{D}^2_{\mathrm{sph}}\mathrm{D}^2_{\mathrm{sph}}(\varrho\,\psi_\varphi)\,\hat{\boldsymbol{\varphi}}.
\end{aligned}
$$

Hence we have established the following equivalent result.

Proposition 5.8 (Stokes flow: spherical polar coordinates) *A stationary axisymmetric Stokes flow with no swirl is prescribed by the transformed stream function $\Psi := r\sin\theta\,\psi_\varphi$ in spherical polar coordinates which satisfies*

$$
\mathrm{D}^2_{\mathrm{sph}}(\mathrm{D}^2_{\mathrm{sph}}\Psi) = 0.
$$

Remark 5.9 (Stokes stream function) Since respectively, in cylindrical and spherical polar coordinates, r and $\varrho := r\sin\theta$ represent the orthogonal distance from the z-axis, and equivalent $\theta = 0$ axis, the stream function Ψ is the same in both sets of coordinates. It is known as the *Stokes stream function.*

As an application of stationary axisymmetric Stokes flows with no swirl, we compute the Stokes flow viscous drag on a sphere. Thus, consider a uniform incompressible Stokes flow of velocity magnitude U, into which we place a spherical obstacle, of radius a. The physical setup is similar to that shown in Figure 2.18. Our goal is to compute the total axial force on the sphere due to the surrounding background flow. We use spherical polar coordinates (r, θ, φ) to represent the flow with the south–north pole axis passing through the centre of the sphere and aligned with the uniform flow U at infinity. We assume the flow around the sphere is a stationary axisymmetric Stokes flow with no swirl, so the flow is independent of the azimuthal angle φ and $u_\varphi = 0$. Hence the velocity field $\boldsymbol{u} = \boldsymbol{u}(r,\theta)$ and pressure field $p = p(r,\theta)$ satisfy

$$\mu\Delta\boldsymbol{u} = \nabla p,$$
$$\nabla\cdot\boldsymbol{u} = 0,$$

where μ is the constant viscosity. As we saw above, such a Stokes flow is prescribed by the Stokes stream function $\Psi = \Psi(r,\theta)$, which satisfies

$$\mathrm{D}^2_{\mathrm{sph}}(\mathrm{D}^2_{\mathrm{sph}}\,\Psi) = 0,$$

where $\mathrm{D}^2_{\mathrm{sph}}$ is the second-order partial differential operator given above. Note that the velocity field components are given by

$$u_r = \frac{1}{r\sin\theta}\frac{\partial}{\partial\theta}(\sin\theta\,\psi_\varphi) \qquad \text{and} \qquad u_\theta = -\frac{1}{r}\frac{\partial}{\partial r}(r\psi_\varphi),$$

and thus, in terms of the Stokes stream function $\Psi := r\sin\theta\,\psi_\varphi$, by

$$u_r = \frac{1}{r^2\sin\theta}\frac{\partial\Psi}{\partial\theta} \qquad \text{and} \qquad u_\theta = -\frac{1}{r\sin\theta}\frac{\partial\Psi}{\partial r}.$$

We have the following result.

Theorem 5.10 (Stokes drag on a sphere) *The axial drag force on a sphere of radius a placed into a uniform Stokes flow field of speed U is given by*

$$6\pi\mu Ua.$$

Proof We prove the result in the following five steps.

Step 1: Determine the boundary conditions. First let us consider the far-field boundary conditions which are $\boldsymbol{u} \sim U\hat{\boldsymbol{z}}$ as $r \to \infty$. Here $\hat{\boldsymbol{z}}$ is the unit vector in the axial direction aligned with the flow. Projecting $\hat{\boldsymbol{z}}$ along $\hat{\boldsymbol{r}}$ and $\hat{\boldsymbol{\theta}}$, we see that $\hat{\boldsymbol{z}} = \cos\theta\,\hat{\boldsymbol{r}} - \sin\theta\,\hat{\boldsymbol{\theta}}$ and so the far-field velocity field $U\hat{\boldsymbol{z}}$ decomposes into $U\hat{\boldsymbol{z}} = U\cos\theta\,\hat{\boldsymbol{r}} - U\sin\theta\,\hat{\boldsymbol{\theta}}$. Hence in the far-field limit we have

$$\frac{1}{r^2\sin\theta}\frac{\partial\Psi}{\partial\theta} \sim U\cos\theta \qquad \text{and} \qquad \frac{1}{r\sin\theta}\frac{\partial\Psi}{\partial r} \sim U\sin\theta.$$

Solving this pair of first-order partial differential equations in the far field reveals the far-field boundary condition in terms of Ψ, namely, as $r \to \infty$,

$$\Psi \sim \tfrac{1}{2}Ur^2\sin^2\theta.$$

Second, consider the boundary conditions on the surface of the sphere. The no-slip condition on $r = a$ implies

$$\frac{1}{a^2\sin\theta}\frac{\partial\Psi}{\partial\theta} = 0 \qquad \text{and} \qquad \frac{1}{a\sin\theta}\frac{\partial\Psi}{\partial r} = 0.$$

Hence Ψ is independent of r and θ along the boundary $r = a$. Since any stream function is only defined up to an additive constant, here we choose the additive constant so as to ensure that the value of Ψ, which is constant on $r = a$, is such that $\Psi = 0$ there.

In addition, we retain the second boundary condition above, that $\partial\Psi/\partial r = 0$ on $r = a$. Thus, to summarise, the boundary conditions are

$$\Psi \to \tfrac{1}{2}Ur^2\sin^2\theta \text{ as } r \to \infty \qquad \text{and} \qquad \Psi = \frac{\partial\Psi}{\partial r} = 0 \text{ on } r = a.$$

Step 2: Solve the biharmonic equation. Motivated by the boundary conditions, we look for a solution to the biharmonic equation $\mathrm{D}^2_{\mathrm{sph}}(\mathrm{D}^2_{\mathrm{sph}}\Psi) = 0$ of the form $\Psi = Uf(r)\sin^2\theta$. First, computing $\mathrm{D}^2_{\mathrm{sph}}\Psi$ gives

$$\begin{aligned}\mathrm{D}^2_{\mathrm{sph}}\Psi &= \mathrm{D}^2_{\mathrm{sph}}(Uf(r)\sin^2\theta)\\ &= U\Big(f''(r) - \frac{2}{r^2}f(r)\Big)\sin^2\theta.\end{aligned}$$

For all $r \geqslant a$ we set

$$F(r) := f''(r) - \frac{2}{r^2}f(r).$$

Now we compute $\mathrm{D}^2_{\mathrm{sph}}(\mathrm{D}^2_{\mathrm{sph}}\Psi)$, which gives

$$\begin{aligned}\mathrm{D}^2_{\mathrm{sph}}(\mathrm{D}^2_{\mathrm{sph}}\Psi) &= \mathrm{D}^2_{\mathrm{sph}}(UF(r)\sin^2\theta)\\ &= U\Big(F''(r) - \frac{2}{r^2}F(r)\Big)\sin^2\theta,\end{aligned}$$

simply mimicking the preceding result. Hence $\mathrm{D}^2_{\mathrm{sph}}(\mathrm{D}^2_{\mathrm{sph}}\Psi) = 0$ if and only if

$$F''(r) - \frac{2}{r^2}F(r) = 0.$$

This is a linear second-order ordinary differential equation whose two independent solutions are r^2 and $1/r$. Thus $F(r)$ is a linear combination of these two solutions. However, we require $f(r)$ which satisfies

$$f''(r) - \frac{2}{r^2}f(r) = F(r).$$

This is a non-homogeneous linear second-order ordinary differential equation, again whose two independent homogeneous solutions are r^2 and $1/r$. The particular integral for the non-homogeneous component $F(r)$, which is a linear combination of r^2 and $1/r$, has the form $Ar^4 + Cr$ for some constants A and C. This is because the second derivative of r^4, and the ratio of r^4 over r^2, both generate a constant multiple of r^2 which is one component of the non-homogeneous term. And similarly the second derivative of r is zero whilst its ratio over r^2 generates a constant multiple of $1/r$, which is the other component of the non-homogeneous term. Hence $f = f(r)$ necessarily has the form

$$f(r) = Ar^4 + Br^2 + Cr + \frac{\hat{D}}{r},$$

where B and $\hat{D}$ are two further constants.

Step 3: Use the boundary conditions. First, as $r \to \infty$ we require

$$U\left(Ar^4 + Br^2 + Cr + \frac{\hat{D}}{r}\right)\sin^2\theta \sim \tfrac{1}{2}Ur^2\sin^2\theta.$$

We deduce $A = 0$ and $B = \frac{1}{2}$. Second, on $r = a$ for all $\theta \in [0, 2\pi)$, we require

$$U\left(\frac{a^2}{2} + Ca + \frac{\hat{D}}{a}\right)\sin^2\theta = 0 \quad \text{and} \quad U\left(a + C - \frac{\hat{D}}{a^2}\right)\sin^2\theta = 0.$$

Thus we have simultaneous equations $a^2/2 + Ca + \hat{D}/a = 0$ and $a + C - \hat{D}/a^2 = 0$ for C and $\hat{D}$. Their solution is $C = -3a/4$ and $\hat{D} = a^3/4$. Hence we have

$$\Psi = \tfrac{1}{4}Ua^2\left(\frac{2r^2}{a^2} - \frac{3r}{a} + \frac{a}{r}\right)\sin^2\theta.$$

Step 4: Compute the pressure. We can compute the pressure field from the stationary Stokes equation $\mu\Delta\boldsymbol{u} = \nabla p$, once we have computed the velocity field $\boldsymbol{u}$, i.e. the components u_r and u_θ, from the stream function Ψ. First, using the relations for u_r and u_θ in terms of Ψ preceding the theorem, we find

$$u_r = \tfrac{1}{2}Ua^2\left(\frac{2}{a^2} - \frac{3}{ar} + \frac{a}{r^3}\right)\cos\theta$$

and

$$u_\theta = -\tfrac{1}{4}Ua^2\left(\frac{4}{a^2} - \frac{3}{ar} - \frac{a}{r^3}\right)\sin\theta.$$

Second, we compute $\nabla\times\boldsymbol{u}$ and then $-\nabla\times(\nabla\times\boldsymbol{u})$, which equals $\Delta\boldsymbol{u}$ for incompressible fields. Since $\boldsymbol{u} = u_r\hat{\boldsymbol{r}} + u_\theta\hat{\boldsymbol{\theta}}$ with $u_r = u_r(r,\theta)$ and $u_\theta = u_\theta(r,\theta)$, then from the form for the curl operator in spherical polar coordinates given in Section A.3 of the Appendix, we have $\nabla\times\boldsymbol{u} = \omega_\varphi\,\hat{\boldsymbol{\varphi}}$ where

$$\begin{aligned}\omega_\varphi &= \frac{1}{r}\frac{\partial}{\partial r}(ru_\theta) - \frac{1}{r}\frac{\partial}{\partial\theta}(u_r)\\ &= -\frac{3Ua}{2r^2}\sin\theta,\end{aligned}$$

after some cancellation. Then we have

$$\begin{aligned}-\nabla\times(\nabla\times\boldsymbol{u}) &= -\nabla\times(\omega_\varphi\hat{\boldsymbol{\varphi}})\\ &= -\left(\frac{1}{r\sin\theta}\frac{\partial}{\partial\theta}(\sin\theta\,\omega_\varphi)\right)\hat{\boldsymbol{r}} + \left(\frac{1}{r}\frac{\partial}{\partial r}(r\omega_\varphi)\right)\hat{\boldsymbol{\theta}}\\ &= \frac{3Ua}{2r^3}((2\cos\theta)\,\hat{\boldsymbol{r}} + (\sin\theta)\,\hat{\boldsymbol{\theta}}).\end{aligned}$$

Hence, to find the pressure p we solve the pair of partial differential equations

$$\frac{\partial p}{\partial r} = \frac{3Ua\mu}{r^3}\cos\theta \qquad \text{and} \qquad \frac{1}{r}\frac{\partial p}{\partial\theta} = \frac{3Ua\mu}{2r^3}\sin\theta.$$

If p_∞ is the constant far-field pressure, the pressure $p = p(r, \theta)$ is given by

$$p = p_\infty - \frac{3Ua\mu}{2r^2}\cos\theta.$$

Step 5: Compute the axial force. For a small patch of area $\mathrm{d}S$ on the surface of the sphere, the force is given by $\mathrm{d}\boldsymbol{F} = \sigma(\boldsymbol{x})\hat{\boldsymbol{n}}\,\mathrm{d}S$. Since the stress matrix $\sigma = -p\mathrm{I} + \hat{\sigma}$ where p is the pressure and $\hat{\sigma}$ is the deviatoric stress matrix, the total force on the surface of the sphere $r = a$ is given by

$$\boldsymbol{F} = \iint (-p\mathrm{I} + \hat{\sigma})\hat{\boldsymbol{n}}\,\mathrm{d}S.$$

Now use that the deviatoric stress is a linear function of the deformation matrix D, indeed $\hat{\sigma} = 2\mu D$. Further, the normal $\hat{\boldsymbol{n}}$ to the surface $r = a$ is $\hat{\boldsymbol{r}}$, the unit vector in the r direction. For any vector $\boldsymbol{v} = (\boldsymbol{v}\cdot\hat{\boldsymbol{r}})\,\hat{\boldsymbol{r}} + (\boldsymbol{v}\cdot\hat{\boldsymbol{\theta}})\,\hat{\boldsymbol{\theta}} + (\boldsymbol{v}\cdot\hat{\boldsymbol{\varphi}})\,\hat{\boldsymbol{\varphi}}$ we have

$$\begin{aligned} D\boldsymbol{v} = &\Big(D_{rr}(\boldsymbol{v}\cdot\hat{\boldsymbol{r}}) + D_{r\theta}(\boldsymbol{v}\cdot\hat{\boldsymbol{\theta}}) + D_{r\varphi}(\boldsymbol{v}\cdot\hat{\boldsymbol{\varphi}})\Big)\,\hat{\boldsymbol{r}} \\ &+ \Big(D_{\theta r}(\boldsymbol{v}\cdot\hat{\boldsymbol{r}}) + D_{\theta\theta}(\boldsymbol{v}\cdot\hat{\boldsymbol{\theta}}) + D_{\theta\varphi}(\boldsymbol{v}\cdot\hat{\boldsymbol{\varphi}})\Big)\,\hat{\boldsymbol{\theta}} \\ &+ \Big(D_{\varphi r}(\boldsymbol{v}\cdot\hat{\boldsymbol{r}}) + D_{\varphi\theta}(\boldsymbol{v}\cdot\hat{\boldsymbol{\theta}}) + D_{\varphi\varphi}(\boldsymbol{v}\cdot\hat{\boldsymbol{\varphi}})\Big)\,\hat{\boldsymbol{\varphi}}. \end{aligned}$$

Hence on the surface $r = a$ with $\boldsymbol{v} = \hat{\boldsymbol{r}}$, and using that all partial derivatives with respect to φ are zero and $u_\varphi \equiv 0$, we have

$$\hat{\sigma}\hat{\boldsymbol{n}} = 2\mu D\hat{\boldsymbol{r}} = 2\mu(D_{rr}\hat{\boldsymbol{r}} + D_{r\theta}\hat{\boldsymbol{\theta}}).$$

Due to axisymmetry, the area integral over the sphere surface $r = a$ can be decomposed into a single integral over concentric rings of radius $a\sin\theta$ and width $a\,\mathrm{d}\theta$, with the area of each ring given by $(2\pi a\sin\theta)(a\,\mathrm{d}\theta)$. Thus we find

$$\boldsymbol{F} = 2\pi a^2\int_0^\pi \Big(-p\hat{\boldsymbol{r}} + 2\mu D_{rr}\hat{\boldsymbol{r}} + 2\mu D_{r\theta}\hat{\boldsymbol{\theta}}\Big)\sin\theta\,\mathrm{d}\theta.$$

The *axial* component of the force F_{ax} along the direction of the far-field flow, i.e. in the direction $\hat{\boldsymbol{z}}$, can be computed as follows. The projection of $\hat{\boldsymbol{r}}$ along $\hat{\boldsymbol{z}}$ is $\cos\theta\,\hat{\boldsymbol{z}}$ while the projection of $\hat{\boldsymbol{\theta}}$ along $\hat{\boldsymbol{z}}$ is $-\sin\theta\,\hat{\boldsymbol{z}}$. Hence the axial component F_{ax} of $\boldsymbol{F}$ along $\hat{\boldsymbol{z}}$ is

$$F_{\mathrm{ax}} = 2\pi a^2\int_0^\pi \Big((-p + 2\mu D_{rr})\cos\theta - 2\mu D_{r\theta}\sin\theta\Big)\sin\theta\,\mathrm{d}\theta.$$

From the formulae in Section A.3 of the Appendix, we find

$$D_{rr} = \frac{\partial u_r}{\partial r} \qquad\text{and}\qquad D_{r\theta} = \tfrac{1}{2}\bigg(\frac{\partial u_\theta}{\partial r} - \frac{u_\theta}{r} + \frac{1}{r}\frac{\partial u_r}{\partial\theta}\bigg).$$

The no-slip boundary condition on the sphere surface implies $u_r = u_\theta = 0$ on $r = a$. And from our expressions for u_r and u_θ from Step 4, we see on $r = a$:

$$\frac{\partial u_r}{\partial r} = \frac{\partial u_r}{\partial\theta} = 0 \qquad\text{and}\qquad \frac{\partial u_\theta}{\partial r} = -\frac{3U}{2a}\sin\theta.$$

Substituting these into D_{rr} and $D_{r\theta}$ just above, and those into the expression for F_{ax}, as well as using the expression for the pressure p from Step 4, we find

$$\begin{aligned} F_{\text{ax}} &= 2\pi a^2 \int_0^{\pi} \left(\left(-p_\infty + \frac{3U\mu}{2a} \cos\theta \right) \cos\theta - 2\mu \left(-\frac{3U}{4a} \sin\theta \right) \sin\theta \right) \sin\theta \, \mathrm{d}\theta \\ &= 2\pi a^2 \int_0^{\pi} \left(-p_\infty \cos\theta + \left(\frac{3U\mu}{2a} \right) (\cos^2\theta \sin\theta + \sin^3\theta) \right) \mathrm{d}\theta \\ &= 3\pi U\mu a \int_0^{\pi} \sin\theta \, \mathrm{d}\theta \\ &= 6\pi U\mu a, \end{aligned}$$

giving the required result. □

Remark 5.11 (Oseen's equations) We derived the axial drag force on a sphere placed in a uniform flow field under the assumption that the flow is a Stokes flow. This assumes the advective inertia term is negligible compared to the viscous diffusion term. However, this assumption breaks down at large distances from the body. Oseen (1910) provided an *ad hoc* correction to Stokes flow that leads to a correction to the axial drag force we computed above in Theorem 5.10. Independently, Kaplun (1957) and Proudman and Pearson (1957) subsequently provided a rationale for Oseen's flow and drag force correction via matched asymptotic expansions. We explore these developments in Exercises 5.5 to 5.8, as well as in a project in Section 5.3.

Example 5.12 (Viscous corner flow) Consider a steady incompressible *viscous* corner flow as shown in Figure 5.1. The fluid is trapped between two plates, where one is horizontal, while the other plate lies above the horizontal plate at an acute angle α. The flat edge of the upper plate almost touches the horizontal plate; there is a small gap between the two. The horizontal plate moves with a speed U to the left, and perpendicular to the imaginary line of intersection between the two plates; the upper plate remains fixed. We assume the trapped flow between the two plates to be a Stokes flow, i.e. we have

$$\begin{aligned} \mu \Delta \boldsymbol{u} &= \nabla p, \\ \nabla \cdot \boldsymbol{u} &= 0, \end{aligned}$$

where p is the pressure field, $\boldsymbol{u}$ the velocity field and μ is the viscosity. Further, we assume the flow is uniform in the direction given by the imaginary line of intersection between the plates – we denote this line of intersection the z-axis. We also assume no flow component in the z-direction. If we utilise cylindrical polar coordinates, these assumptions correspond to $u_r = u_r(r,\theta)$ and $u_\theta = u_\theta(r,\theta)$, only, with $u_z \equiv 0$. Recall from the material preceding Proposition 5.7, since the flow is incompressible, in general there exists a function $\boldsymbol{\psi} = \boldsymbol{\psi}(r,\theta,z)$ with $\boldsymbol{\psi} = \psi_r \hat{\boldsymbol{r}} + \psi_\theta \hat{\boldsymbol{\theta}} + \psi_z \hat{\boldsymbol{z}}$ such that $\boldsymbol{u} = \nabla \times \boldsymbol{\psi}$, where

$$\nabla \times \boldsymbol{\psi} = \left(\frac{1}{r} \frac{\partial \psi_z}{\partial \theta} - \frac{\partial \psi_\theta}{\partial z} \right) \hat{\boldsymbol{r}} + \left(\frac{\partial \psi_r}{\partial z} - \frac{\partial \psi_z}{\partial r} \right) \hat{\boldsymbol{\theta}} + \left(\frac{1}{r} \frac{\partial}{\partial r}(r\psi_\theta) - \frac{1}{r} \frac{\partial \psi_r}{\partial \theta} \right) \hat{\boldsymbol{z}}.$$

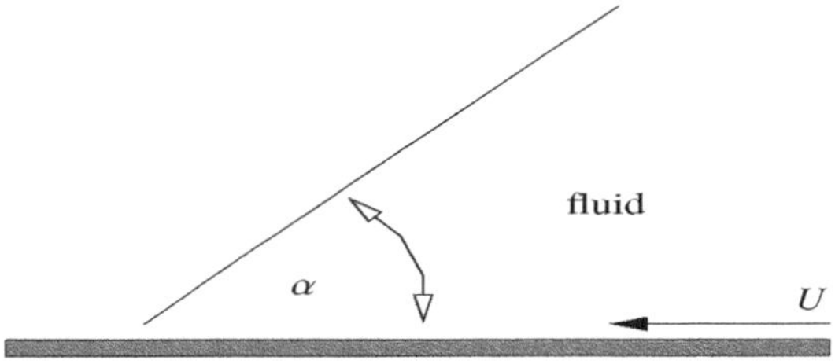

Figure 5.1 Corner flow. An incompressible viscous fluid is trapped between two plates, one horizontal, while the other lies above at an acute angle α. The flat edge of the upper plate almost touches the horizontal plate; there is a small gap between the two. The horizontal plate moves with a speed U to the left, perpendicular to the imaginary line of intersection between the two plates; the upper plate remains fixed.

Since here $u_r = u_r(r,\theta)$ and $u_\theta = u_\theta(r,\theta)$, only, the terms involving ψ_θ and ψ_r, respectively in the $\hat{\boldsymbol{r}}$ and $\hat{\boldsymbol{\theta}}$ components of $\boldsymbol{\psi}$, are zero. And the remaining terms in those respective components, namely $r^{-1}(\partial\psi_z/\partial\theta)$ which equals u_r and $-\partial\psi_z/\partial r$ which equals u_θ, are necessarily functions of r and θ only. Integrating these two first-order partial differential equations for ψ_z reveals necessarily that $\psi_z = \psi_z(r,\theta)$ only as well. Further, since in addition $u_z \equiv 0$, we see that the functions ψ_r and ψ_θ play no role in the prescription of $\boldsymbol{u}$ and thus without loss of generality we can set both of them equal to zero. Consequently, we deduce that $\boldsymbol{\psi} = \psi_z\,\hat{\boldsymbol{z}}$ with $\psi_z = \psi_z(r,\theta)$, so there exists a stream function $\psi := \psi_z$ such that

$$\boldsymbol{u} = \nabla \times (\psi\,\hat{\boldsymbol{z}}),$$

where $\psi = \psi(r,\theta)$ only. Taking the curl of the viscous diffusion term in the Stokes flow equation, we see

$$\begin{aligned}
\nabla \times (\mu\Delta\boldsymbol{u}) &= \mu\Delta(\nabla \times \boldsymbol{u}) \\
&= \mu\Delta\left(\nabla \times \nabla \times (\psi\,\hat{\boldsymbol{z}})\right) \\
&= \mu\Delta\,(-\Delta)(\psi\,\hat{\boldsymbol{z}}) \\
&= -\mu(\Delta(\Delta\psi))\,\hat{\boldsymbol{z}},
\end{aligned}$$

where we used that $\nabla \cdot (\psi\,\hat{\boldsymbol{z}}) \equiv 0$ since $\psi = \psi(r,\theta)$ only, with the divergence for cylindrical polar coordinates given in Section A.2 of the Appendix. Since $\nabla\times(\nabla p) \equiv \boldsymbol{0}$, we deduce that the stream function $\psi = \psi(r,\theta)$ for this Stokes flow satisfies the biharmonic equation

$$\Delta(\Delta\psi) - 0.$$

Next we determine the boundary conditions. Henceforth we use plane polar coordinates, dropping the z coordinate with respect to which the flow is uniform, and in the direction of which there is no flow. The relations between the velocity components u_r and u_θ and the stream function ψ, as we have seen, are

$$u_r = \frac{1}{r}\frac{\partial \psi}{\partial \theta} \qquad \text{and} \qquad u_\theta = -\frac{\partial \psi}{\partial r}.$$

First, using the no-slip boundary conditions on the plate along $\theta = 0$ which is moving towards the origin at speed U, we see on $\theta = 0$ that we have

$$\frac{\partial \psi}{\partial r} = 0 \qquad \text{and} \qquad \frac{1}{r}\frac{\partial \psi}{\partial \theta} = -U.$$

Second, for the plate at angle $\theta = \alpha$, the no-slip boundary conditions imply that, on $\theta = \alpha$, we have

$$\frac{\partial \psi}{\partial r} - 0 \qquad \text{and} \qquad \frac{\partial \psi}{\partial \theta} = 0.$$

Given these boundary conditions, in particular that on $\theta = 0$ we must have $\partial\psi/\partial\theta = -Ur$, we look for a solution to the biharmonic equation $\Delta(\Delta\psi) = 0$ of the form $\psi = Urf(\theta)$. Using the form for the Laplacian in cylindrical polar coordinates from Section A.2 of the Appendix, by direct computation we have

$$\begin{aligned}\Delta(Urf(\theta)) &= U\left(\frac{\partial^2}{\partial r^2} + \frac{1}{r}\frac{\partial}{\partial r} + \frac{1}{r^2}\frac{\partial^2}{\partial \theta^2}\right)(rf(\theta))\\ &= \frac{U}{r}(f(\theta) + f''(\theta)).\end{aligned}$$

Then we directly compute

$$\begin{aligned}\Delta\big(\Delta(Urf(\theta))\big) &= \Delta\left(\frac{U}{r}(f(\theta) + f''(\theta))\right)\\ &= U\left(\frac{\partial^2}{\partial r^2} + \frac{1}{r}\frac{\partial}{\partial r} + \frac{1}{r^2}\frac{\partial^2}{\partial \theta^2}\right)\left(\frac{1}{r}(f(\theta) + f''(\theta))\right)\\ &= \frac{U}{r^3}(f(\theta) + 2f''(\theta) + f''''(\theta)).\end{aligned}$$

Hence $\Delta(\Delta\psi) = 0$ with $\psi = Urf(\theta)$ if and only if $f = f(\theta)$ satisfies

$$f'''' + 2f'' + f = 0.$$

This is a linear homogeneous constant-coefficient fourth-order ordinary differential equation. Solutions of the form $f(\theta) = \exp(\lambda\theta)$ exist, provided the parameter λ satisfies the auxiliary degree-four polynomial equation $(\lambda^2 + 1)^2 = 0$. This equation has four solutions $\lambda = \pm\mathrm{i}$, each repeated. Hence, for some constants A, B, C and D, the general solution has the form

$$f(\theta) = A\sin\theta + B\cos\theta + C\theta\sin\theta + D\theta\cos\theta.$$

Now consider the boundary conditions. Note that the form of the solution implies $u_r = Uf'(\theta)$ and $u_\theta = -Uf(\theta)$. First consider the boundary conditions on $\theta = 0$:

$$u_r = -U \quad \Leftrightarrow \quad f'(0) = -1 \qquad \text{and} \qquad u_\theta = 0 \quad \Leftrightarrow \quad f(0) = 0.$$

Second consider the boundary conditions on $\theta = \alpha$:

$$u_r = 0 \quad \Leftrightarrow \quad f'(\alpha) = 0 \qquad \text{and} \qquad u_\theta = 0 \quad \Leftrightarrow \quad f(\alpha) = 0.$$

In other words, $f = f(\theta)$ must satisfy

$$f(0) = f(\alpha) = f'(\alpha) = 0 \qquad \text{and} \qquad f'(0) = -1.$$

We use these four boundary conditions to determine the constants A, B, C and D. The condition $f(0) = 0$ implies $B = 0$ and the condition $f'(0) = -1$ implies $A + D = -1$, allowing us to express A in terms of D. Hence we can express the remaining two boundary conditions as a pair of simultaneous linear equations:

$$(\alpha \sin \alpha)\, C + (\alpha \cos \alpha - \sin \alpha)\, D = \sin \alpha,$$
$$(\sin \alpha + \alpha \cos \alpha)\, C - (\alpha \sin \alpha)\, D = \cos \alpha,$$

for C and D. The solution satisfying all four boundary conditions is

$$f(\theta) = \frac{\theta \sin \alpha \sin(\alpha - \theta) - \alpha(\alpha - \theta) \sin \theta}{\alpha^2 - \sin^2 \alpha}.$$

Remark 5.13 (Region of validity of Stokes flow assumption) We should ask ourselves, what is the distance from the origin within which the solution we sought above is consistent with our assumption of Stokes flow? Consider the incompressible Navier–Stokes equations in the corner flow context above, i.e. where in cylindrical polar coordinates the flow is independent of z and $u_z \equiv 0$. From Section A.2 of the Appendix, the advective inertia and viscous diffusion terms, respectively $(\boldsymbol{u} \cdot \nabla)\, \boldsymbol{u}$ and $\nu \Delta \boldsymbol{u}$, have the forms, for the components u_r and u_θ:

$$\left(u_r \frac{\partial u_r}{\partial r} + \frac{u_\theta}{r} \frac{\partial u_r}{\partial \theta} - \frac{u_\theta^2}{r} \right) \quad \text{and} \quad \nu \left(\frac{1}{r} \frac{\partial}{\partial r} \left(r \frac{\partial u_r}{\partial r} \right) + \frac{1}{r^2} \frac{\partial^2 u_r}{\partial \theta^2} - \frac{u_r}{r^2} - \frac{2}{r^2} \frac{\partial u_\theta}{\partial \theta} \right),$$

$$\left(u_r \frac{\partial u_\theta}{\partial r} + \frac{u_\theta}{r} \frac{\partial u_\theta}{\partial \theta} - \frac{u_r u_\theta}{r} \right) \quad \text{and} \quad \nu \left(\frac{1}{r} \frac{\partial}{\partial r} \left(r \frac{\partial u_\theta}{\partial r} \right) + \frac{1}{r^2} \frac{\partial^2 u_\theta}{\partial \theta^2} - \frac{u_\theta}{r^2} + \frac{2}{r^2} \frac{\partial u_r}{\partial \theta} \right).$$

We introduce the non-dimensional variables $r' := r/R$, $u_r'(r', \theta) := u_r(r, \theta)/U$ and $u_\theta'(r', \theta) := u_\theta(r, \theta)/U$. Note that the angle variable θ is already dimensionless. In these nominations, R is a scale characterising radial distance from the origin, where the origin is at the intersection of the two planes. The natural choice for the velocity scale U is the horizontal speed of the lower plate. Transforming the terms above into dimensionless variables by direct substitution and the chain rule reveals the exact same expressions as above, except that all the variables shown are replaced by their primed counterparts, and the two advective inertia terms on the left now include the scaling factor U^2/R while the viscous diffusion terms on the right now include the scaling factor $\nu U/R^2$. The Stokes flow assumption is that the advective inertia term is asymptotically small compared to the viscous inertia term. This is equivalent to the assumption that the Reynolds number for the flow is asymptotically small. Indeed, we observe

$$\frac{U^2/R}{\nu U/R^2} \ll 1 \qquad \Leftrightarrow \qquad \frac{UR}{\nu} \ll 1.$$

We conclude that our Stokes flow solution is valid provided $R \ll \nu/U$.

Remark 5.14 For the original investigation on this problem, see Taylor (1962).

5.2 Lubrication Theory

In *lubrication theory* we consider the following scenario. Suppose a homogeneous incompressible viscous fluid occupies a *shallow layer* whose depth is characterised by the length scale H and whose horizontal extent is characterised by the length scale L; see Figure 5.2 for the setup. Let x and y denote horizontal Cartesian coordinates and z the vertical coordinate. Suppose (u, v, w) are the fluid velocity components in the three coordinate directions x, y and z, respectively. Further, let p denote the pressure and ρ denote the constant density of the fluid. We also assume that the horizontal velocity fields (u, v) are characterised by the horizontal velocity scale U while the vertical velocity field w is characterised by the vertical velocity field scale W. As we have already hinted, we assume the ratio of the vertical length scale H to the horizontal length scale L, denoted ϵ, is asymptotically small, i.e. we assume

$$\epsilon := \frac{H}{L} \ll 1.$$

Our goal is to systematically derive a reduced set of simpler equations from the homogeneous incompressible Navier–Stokes equations, that represent a very accurate approximation to the flow in the context presented. We begin by introducing the following dimensionless variables:

$$x' := \frac{x}{L}, \qquad y' := \frac{y}{L}, \qquad z' := \frac{z}{H} \qquad \text{and} \qquad t' := \frac{t}{T},$$

and with $\boldsymbol{x}' := (x', y', z')^{\mathrm{T}}$ we set

$$u'(\boldsymbol{x}', t') := \frac{1}{U}u(\boldsymbol{x}, t), \;\; v'(\boldsymbol{x}', t') := \frac{1}{U}v(\boldsymbol{x}, t) \text{ and } w'(\boldsymbol{x}', t') := \frac{1}{W}w(\boldsymbol{x}, t).$$

In addition we set $p'(\boldsymbol{x}', t') = p(\boldsymbol{x}, t)/P$, where the parameter P is a characteristic pressure scale which we determine presently. First, consider the incompressibility condition. If we directly substitute $u = Lu'$, $v = Lv'$ and $w = Hw'$ and use the chain rule to replace partial derivatives with respect to x, y and z by those with respect to x', y' and z', we find, since $\epsilon = H/L$:

$$\frac{U}{L}\left(\frac{\partial u'}{\partial x'} + \frac{\partial v'}{\partial y'}\right) + \frac{W}{H}\left(\frac{\partial w'}{\partial z'}\right) = 0 \quad \Leftrightarrow \quad \epsilon\frac{U}{W}\left(\frac{\partial u'}{\partial x'} + \frac{\partial v'}{\partial y'}\right) + \frac{\partial w'}{\partial z'} = 0.$$

In lubrication phenomena, incompressibility of the flow is an essential characteristic, and for this to be maintained we choose $W := \epsilon U$ so that in particular, $\epsilon(U/W) = 1$. With this choice for W, we note that $\epsilon = H/L = W/U$, or equivalently $L/U = H/W$, and there is a natural choice for the timescale T:

$$T := \frac{L}{U} = \frac{H}{W}.$$

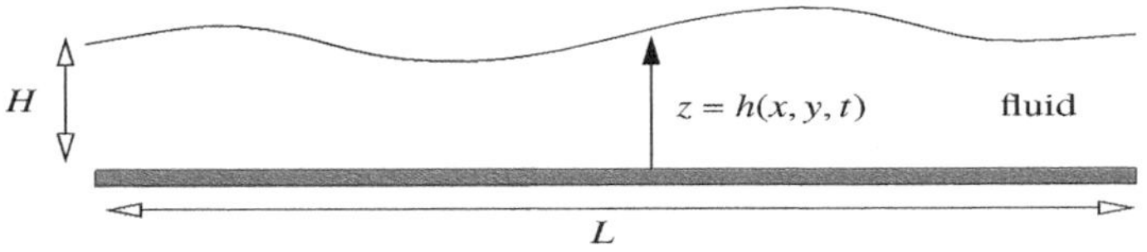

Figure 5.2 Lubrication theory. A homogeneous incompressible viscous fluid occupies a shallow layer whose typical depth is H and horizontal extent is L. In the asymptotic limit $H \ll L$, the Navier–Stokes equations reduce to the lubrication model (or shallow-layer) equations.

Second, now consider the Navier–Stokes momentum equations for the velocity components u and v transformed into non-dimensional variables. Using the fact that we know $WU/H = \epsilon U^2/H = U^2/L$, we find

$$\frac{U^2}{L}\left(\frac{\partial u'}{\partial t'} + u'\frac{\partial u'}{\partial x'} + v'\frac{\partial u'}{\partial y'} + w'\frac{\partial u'}{\partial z'}\right) = \nu\frac{U}{L^2}(\Delta_{x',y'}u') + \nu\frac{U}{H^2}\left(\frac{\partial^2 u'}{\partial z'^2}\right) - \frac{P}{\rho L}\left(\frac{\partial p'}{\partial x'}\right),$$

$$\frac{U^2}{L}\left(\frac{\partial v'}{\partial t'} + u'\frac{\partial v'}{\partial x'} + v'\frac{\partial v'}{\partial y'} + w'\frac{\partial v'}{\partial z'}\right) = \nu\frac{U}{L^2}(\Delta^2_{x',y'}v') + \nu\frac{U}{H^2}\left(\frac{\partial^2 v'}{\partial z'^2}\right) - \frac{P}{\rho L}\left(\frac{\partial p'}{\partial y'}\right),$$

where $\Delta_{x',y'}$ is the Laplacian operator with respect to the variables x' and y'. If we divide both equations through by $\nu U/H^2$, we obtain the following equations:

$$(\mathrm{Re})\,\epsilon^2\left(\frac{\partial u'}{\partial t'} + u'\frac{\partial u'}{\partial x'} + v'\frac{\partial u'}{\partial y'} + w'\frac{\partial u'}{\partial z'}\right) = \epsilon^2\,(\Delta_{x',y'}u') + \frac{\partial^2 u'}{\partial z'^2} - \frac{PH^2}{\mu UL}\left(\frac{\partial p'}{\partial x'}\right),$$

$$(\mathrm{Re})\,\epsilon^2\left(\frac{\partial v'}{\partial t'} + u'\frac{\partial v'}{\partial x'} + v'\frac{\partial v'}{\partial y'} + w'\frac{\partial v'}{\partial z'}\right) = \epsilon^2\,(\Delta^2_{x',y'}v') + \frac{\partial^2 v'}{\partial z'^2} - \frac{PH^2}{\mu UL}\left(\frac{\partial p'}{\partial y'}\right),$$

where 'Re' is the Reynolds number $\mathrm{Re} := UL/\nu$. Recall from Section 5.1 that the pressure field mediates the incompressibility condition. We should thus attempt to maintain it in the small ϵ limit coming presently. We thus choose

$$P := \frac{\mu UL}{H^2}.$$

Hence in the asymptotic limit $\epsilon \to 0$, in which we also demand $(\mathrm{Re})\,\epsilon^2 \to 0$, the equations for the dimensionless variables u' and v' become

$$\frac{\partial p'}{\partial x'} = \frac{\partial^2 u'}{\partial z'^2},$$

$$\frac{\partial p'}{\partial y'} = \frac{\partial^2 v'}{\partial z'^2}.$$

Third, now consider the Navier–Stokes momentum equation for the velocity component w, transformed into non-dimensional variables. Using the fact that $W^2/H = UW/L$, since $W/H = U/L$, we find

$$\frac{UW}{L}\left(\frac{\partial w'}{\partial t'} + u'\frac{\partial w'}{\partial x'} + v'\frac{\partial w'}{\partial y'} + w'\frac{\partial w'}{\partial z'}\right) = \nu\frac{W}{L^2}(\Delta^2_{x',y'}w') + \nu\frac{W}{H^2}\left(\frac{\partial^2 w'}{\partial z'^2}\right) - \frac{P}{\rho H}\left(\frac{\partial p'}{\partial z'}\right).$$

Dividing this equation by $\nu W/H^2$, using $P = \mu UL/H^2$, we find that w' satisfies

$$(\mathrm{Re})\,\epsilon^2\left(\frac{\partial w'}{\partial t'} + u'\frac{\partial w'}{\partial x'} + v'\frac{\partial w'}{\partial y'} + w'\frac{\partial w'}{\partial z'}\right) = \epsilon^2\,(\Delta^2_{x',y'}w') + \frac{\partial^2 w'}{\partial z'^2} - \epsilon^{-2}\left(\frac{\partial p'}{\partial z'}\right).$$

If we multiply this equation by ϵ^2 and take the limit $\epsilon \to 0$, while demanding $(\mathrm{Re})\,\epsilon^2 \to 0$, then the third momentum equation reduces to $\partial p'/\partial z' = 0$. Hence, back in dimensional variables, and together with the incompressibility condition, we have arrived at the *lubrication model equations* which are also known as the *thin-layer* or *shallow-layer* equations.

Definition 5.15 (Lubrication model equations) Suppose the z-direction is the thin restricted direction. The *lubrication model equations* for the velocity field components $u = u(x, y, z, t)$, $v = v(x, y, z, t)$ in the non-restricted directions, the velocity field component $w = w(x, y, z, t)$ in the restricted direction and the pressure field $p = p(x, y, z, t)$ are

$$\frac{\partial p}{\partial x} = \mu\frac{\partial^2 u}{\partial z^2}, \quad \frac{\partial p}{\partial y} = \mu\frac{\partial^2 v}{\partial z^2}, \quad \frac{\partial p}{\partial z} = 0 \quad \text{and} \quad \frac{\partial u}{\partial x} + \frac{\partial v}{\partial y} + \frac{\partial w}{\partial z} = 0,$$

together with no-slip boundary conditions for u, v and w at any rigid boundary, or matching conditions at any free boundary.

Remark 5.16 (Modified Reynolds number) To derive the lubrication model equations, we took the limit $\epsilon \to 0$ while demanding $(\mathrm{Re})\,\epsilon^2 \to 0$. Note that it is not necessary that the Reynolds number 'Re' is asymptotically small, but that the *modified Reynolds number* $(\mathrm{Re})\,\epsilon^2 = UL\epsilon^2/\nu$ be asymptotically small. This is also sometimes referred to as the *reduced Reynolds number*.

Remark 5.17 The third lubrication model equation implies the pressure $p = p(x, y, t)$ only.

Example 5.18 (Reynolds lubrication equation) As shown in Figure 5.2, suppose the fluid occupies a region between a lower rigid plate at $z = 0$ and a top surface at $z = h(x, y, t)$. We assume here that the top surface is free, so for example it represents a fluid–air boundary. However, we can straightforwardly specialise our subsequent analysis to a rigid-lid upper boundary; see Example 5.19. We assume no-slip boundary conditions at $z = 0$ while we suppose the velocity components at the surface height $z = h(x, y, t)$ are $u(x, y, h, t) = U$ and $v(x, y, h, t) = V$, with U and V given fixed velocities. For this scenario, starting with the shallow-layer (lubrication model) equations, we can derive a closed-form equation for the evolution of the surface height $z = h(x, y, t)$ as follows. Since $p = p(x, y, t)$ only, using the third shallow-layer equation, we deduce from the first two shallow-layer equations that u and v are quadratic functions of z. Indeed, consider the first shallow-layer equation for $u = u(x, y, z, t)$. Integrating with respect to the vertical coordinate z twice successively reveals that $u = u(x, y, z, t)$ must have the form

$$u = \frac{1}{2\mu}\left(\frac{\partial p}{\partial x}\right)z^2 + Cz + D,$$

where $C = C(x, y, t)$ and $D = D(x, y, t)$ are arbitrary functions of x, y and t in general. The no-slip boundary condition at $u(x, y, 0, t) = 0$ implies $D \equiv 0$. The boundary condition $u(x, y, h, t) = U$ at the free surface implies

$$C = \frac{U}{h} - \frac{1}{2\mu}\left(\frac{\partial p}{\partial x}\right) h.$$

Thus, after integrating the shallow-layer equation for $v = v(x, y, z, t)$ in a similar manner, we find

$$u = \frac{1}{2\mu}\left(\frac{\partial p}{\partial x}\right) z(z - h) + \frac{U}{h} z,$$
$$v = \frac{1}{2\mu}\left(\frac{\partial p}{\partial y}\right) z(z - h) + \frac{V}{h} z.$$

Further, since the fluid is incompressible, integrating with respect to the vertical coordinate between $z = 0$ and $z = h(x, y, t)$, we find

$$w(x, y, h, t) = -\int_0^h \left(\frac{\partial u}{\partial x} + \frac{\partial v}{\partial y}\right) \mathrm{d}z,$$

where we have used the no-slip boundary conditions at $z = 0$, which implies $w(x, y, 0, t) = 0$. Let us label all the particles of the fluid at time $t = 0$ by the three-dimensional coordinates $\boldsymbol{a} = (a, b, c)^{\mathrm{T}}$, where a and b are horizontal coordinates and c is the vertical coordinate. At any time $t \geqslant 0$ later, the corresponding particle has coordinates $\boldsymbol{x} = \boldsymbol{x}(\boldsymbol{a}, t)$ and naturally $\boldsymbol{x}(\boldsymbol{a}, 0) = \boldsymbol{a}$. Recall that $\boldsymbol{x} = \boldsymbol{x}(\boldsymbol{a}, t)$ is the fluid flow; see Section 1.6. Recall, for example, that the vertical velocity of the particle $\boldsymbol{x} = \boldsymbol{x}(\boldsymbol{a}, t)$ is given by

$$\frac{\mathrm{d}}{\mathrm{d}t} z(\boldsymbol{a}, t) = w(x(\boldsymbol{a}, t), y(\boldsymbol{a}, t), z(\boldsymbol{a}, t), t).$$

We can identify those particles starting at the free surface as those particles $\boldsymbol{a}$ which satisfy $c = h(a, b, 0)$. The same particles at any time $t \geqslant 0$ satisfy $z(\boldsymbol{a}, t) = h(x(\boldsymbol{a}, t), y(\boldsymbol{a}, t), t)$ and thus at the free surface $z = h$:

$$\frac{\mathrm{d}z}{\mathrm{d}t} = \frac{\partial h}{\partial t} + \frac{\mathrm{d}x}{\mathrm{d}t}\frac{\partial h}{\partial x} + \frac{\mathrm{d}y}{\mathrm{d}t}\frac{\partial h}{\partial y} = \frac{\partial h}{\partial t} + U\frac{\partial h}{\partial x} + V\frac{\partial h}{\partial y},$$

using the free-surface boundary conditions. In other words, at $z = h$ we know

$$w = \frac{\partial h}{\partial t} + U\frac{\partial h}{\partial x} + V\frac{\partial h}{\partial y}.$$

Further, we also have the following calculus identity, since $h = h(x, y, t)$:

$$\frac{\partial}{\partial x}\int_0^h u \,\mathrm{d}z = \int_0^h \frac{\partial u}{\partial x} \mathrm{d}z + \left(\frac{\partial}{\partial h}\int_0^h u \,\mathrm{d}z\right)\left(\frac{\partial h}{\partial x}\right) = \int_0^h \frac{\partial u}{\partial x} \mathrm{d}z + U\frac{\partial h}{\partial x},$$

where we used that $u(x, y, h, t) = U$. Similarly we have

$$\frac{\partial}{\partial y}\int_0^h v\,\mathrm{d}z = \int_0^h \frac{\partial v}{\partial y}\,\mathrm{d}z + V\frac{\partial h}{\partial y}.$$

Rearranging these two results, we can substitute for the terms on the right, and using the expression for w at the free surface we can substitute for the term on the left, in the vertically integrated incompressibility constraint above. We find

$$\frac{\partial h}{\partial t} = -\frac{\partial}{\partial x}\int_0^h u\,\mathrm{d}z - \frac{\partial}{\partial y}\int_0^h v\,\mathrm{d}z,$$

where the terms $U\partial h/\partial x$ and $V\partial h/\partial y$ cancel. If we substitute our expressions for u and v above into the right-hand side and integrate, we establish a closed-form evolution equation for h known as the *Reynolds lubrication equation*:

$$\frac{\partial h}{\partial t} = \frac{\partial}{\partial x}\left(\frac{h^3}{12\mu}\left(\frac{\partial p}{\partial x}\right) - \tfrac{1}{2}Uh\right) + \frac{\partial}{\partial y}\left(\frac{h^3}{12\mu}\left(\frac{\partial p}{\partial y}\right) - \tfrac{1}{2}Vh\right),$$

or equivalently

$$12\mu\frac{\partial h}{\partial t} + 6\mu\left(\frac{\partial}{\partial x}(Uh) + \frac{\partial}{\partial y}(Vh)\right) = \frac{\partial}{\partial x}\left(h^3\left(\frac{\partial p}{\partial x}\right)\right) + \frac{\partial}{\partial y}\left(h^3\left(\frac{\partial p}{\partial y}\right)\right).$$

Example 5.19 (Squeeze film) Suppose a homogeneous incompressible viscous liquid occupies a region between two parallel plates that are very close together – a squeeze film. The upper plate is a disc with radius L. Assume that the volume occupied by the liquid is a flat cylindrical shape with circular cross-section; see Figure 5.3 for the setup. Suppose (r, θ, z) are cylindrical polar coordinates with the plane $z = 0$ corresponding to the lower fixed plate and the origin lying directly below the centre of the upper disc plate. Let (u_r, u_θ, u_z) be the fluid velocity components in the three coordinate directions r, θ and z, respectively. We assume throughout that the flow is axisymmetric and thus independent of θ. We also assume there is no swirl, so the velocity component $u_\theta \equiv 0$. Further, while the lower plate remains fixed at $z = 0$, the height of the parallel upper disc plate is given by $z = h(t)$ and changes with time. No-slip boundary conditions apply on both plates, i.e. $u_r = 0$ and $u_z = 0$ at $z = 0$, while $u_r = 0$ and $u_z = h'(t)$ at $z = h(t)$. Our goal is to compute the total force on the upper disc plate. We assume that the characteristic scale H of the height $z = h(t)$ of the upper disc plate is asymptotically small compared to the lateral characterising scale L of the fluid volume, i.e. $H \ll L$, and the lubrication model equations apply. Since we assume axisymmetry and no swirl, these are

$$\frac{\partial p}{\partial r} = \mu\frac{\partial^2 u_r}{\partial z^2}, \quad \frac{\partial p}{\partial z} = 0 \quad \text{and} \quad \frac{\partial u_r}{\partial r} + \frac{\partial u_z}{\partial z} = 0,$$

where the last equation is the incompressibility condition, and $u_r = u_r(r, z, t)$ and $u_z = u_z(r, z, t)$. The second lubrication model equation above implies $p = p(r, t)$ only. If we integrate the first lubrication model equation twice successively with respect to z, we find

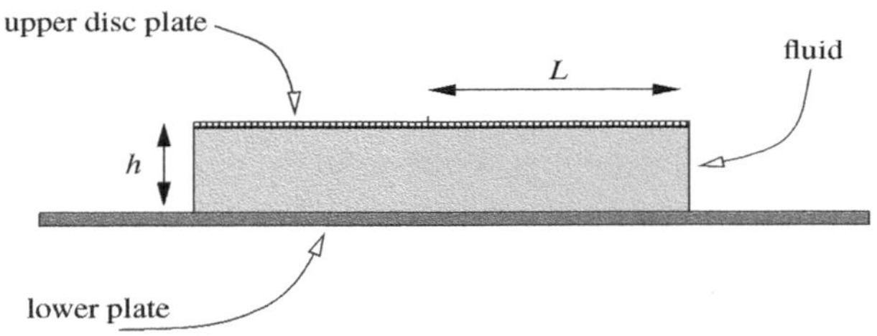

Figure 5.3 Lubrication theory. A homogeneous incompressible viscous liquid occupies a region between two plates that are very close together – a squeeze film. The upper plate is a disc with radius L at height $z = h(t)$ above the lower fixed plate. We assume that the volume occupied by the liquid is a flat cylindrical shape.

$$u_r = \frac{1}{2\mu}\left(\frac{\partial p}{\partial r}\right) z^2 + Cz + D,$$

where $C = C(r, t)$ and $D = D(r, t)$ are arbitrary functions. The no-slip boundary condition $u_r(r, 0, t) = 0$ on the lower plate implies $D \equiv 0$. The no-slip boundary condition $u_r(r, h, t) = 0$ at $z = h(t)$ implies $C = -(1/2\mu)(\partial p/\partial r)\, h$, and thus

$$u_r = \frac{1}{2\mu}\left(\frac{\partial p}{\partial r}\right) z(z - h).$$

Fix a radial distance $r > 0$ and consider the cylindrical volume of fluid $\mathcal{V}_t := [0, r] \times [0, 2\pi] \times [0, h(t)]$. Given that $h = h(t)$ and the fluid velocity field $u_r = u_r(r, z, t)$ just computed are time-dependent, the volume of this cylinder, with r fixed, is time-dependent. The total volume of $\mathcal{V}_t$ is $\pi r^2 h(t)$ and the rate of change of this total volume is $\pi r^2 h'(t)$. Since the liquid is incompressible, as $h = h(t)$ varies, the total volume inside $\mathcal{V}_t$ changes as a result of a flux of fluid through the curved surface $\mathcal{S}_t := \{r\} \times [0, 2\pi] \times [0, h(t)]$. For a small patch of area '$r\,\mathrm{d}\theta\,\mathrm{d}z$' at position (r, z) on $\mathcal{S}_t$, the volume flux of fluid through this patch is $u_r(r, z, t)\, r\,\mathrm{d}\theta\,\mathrm{d}z$, i.e. this is the volume of fluid passing through this patch per unit time. Note that u_r is the normal velocity through the patch '$r\,\mathrm{d}\theta\,\mathrm{d}z$'. Hence the total volume flux through the surface $\mathcal{S}_t$ is given by

$$\begin{aligned}
\int_0^h \int_0^{2\pi} u_r\, r\,\mathrm{d}\theta\,\mathrm{d}z &= 2\pi r \int_0^h u_r\,\mathrm{d}z \\
&= \frac{\pi r}{\mu}\left(\frac{\partial p}{\partial r}\right) \int_0^h z(z - h)\,\mathrm{d}z \\
&= -\frac{\pi r}{\mu}\left(\frac{\partial p}{\partial r}\right) \frac{h^3}{6},
\end{aligned}$$

where we substituted the expression for $u_r = u_r(r, z, t)$ computed above. Since $-\pi r^2 h'(t)$ is the rate of decrease of the total volume of $\mathcal{V}_t$, and by incompressibility this must match the outward volume flux just computed, we have

$$-\pi r^2 h'(t) = -\frac{\pi r}{\mu}\left(\frac{\partial p}{\partial r}\right)\frac{h^3}{6} \qquad \Leftrightarrow \qquad \frac{\partial p}{\partial r} = \frac{6\mu r}{h^3} h'.$$

Assuming the pressure at the boundary $r = L$ of the cylindrical volume of fluid is atmospheric pressure P_0, the pressure $p = p(r,t)$ can be computed by integrating the expression above for $\partial p/\partial r$ between r and L, so that

$$P_0 - p(r,t) = \frac{6\mu}{h^3} h' \int_r^L \tilde{r}\, \mathrm{d}\tilde{r} = \frac{3\mu}{h^3} h'(L^2 - r^2).$$

Hence the pressure $p = p(r,t)$ at radial distance r, uniform in z, is given by

$$p(r,t) = P_0 - \frac{3\mu}{h^3} h'(L^2 - r^2).$$

We observe that if $h'(t) < 0$ so the upper disc plate is squeezing the liquid between the plates, then the pressure in the liquid is greater than atmospheric pressure. In particular, there is a very strong additional pressure proportional to h^{-3}, creating a lubrication effect. On the other hand, if $h'(t) > 0$ so the upper disc is lifting away from the fixed lower plate, then the pressure in the liquid is less than atmospheric pressure. And indeed that additional negative pressure is also proportional to h^{-3}, and creates a strong suction effect. To compute the total force on the upper disc plate, we start by computing the force on a small patch of area '$r\,\mathrm{d}\theta\,\mathrm{d}r$' at position (r,θ) on the plate. Recall that the force $\mathrm{d}\boldsymbol{F}$ across this patch of area due to the pressure p is given by '$-p\hat{z}\, r\,\mathrm{d}\theta\,\mathrm{d}r$', where $\hat{z}$ is the unit vector in the vertical direction which represents the upward/outward normal direction associated with the patch. Hence the total force $\boldsymbol{F}$ on the disc is given by

$$\begin{aligned}
\boldsymbol{F} &= -\int_0^L \int_0^{2\pi} p(r,t)\hat{z}\, r\,\mathrm{d}\theta\,\mathrm{d}r \\
&= -\left(2\pi \int_0^L p(r,t)\, r\,\mathrm{d}r\right)\hat{z} \\
&= -\left(\pi L^2 P_0 - \frac{6\pi\mu}{h^3} h' \int_0^L (L^2 - r^2)\, r\,\mathrm{d}r\right)\hat{z} \\
&= -\left(\pi L^2 P_0 - \frac{3\pi\mu L^4}{2h^3} h'\right)\hat{z}.
\end{aligned}$$

Finally, we can approximate the time for an external constant vertical force F_* to pull the parallel plates apart if the initial separation is h_0. Assuming the mass of the upper disc plate is negligible, we have

$$\begin{aligned}
F_* = \frac{3\pi\mu L^4}{2h^3} h' - \pi L^2 P_0 \quad &\Leftrightarrow \quad t\,(F_* + \pi L^2 P_0) = \frac{3\pi\mu L^4}{4}(h_0^{-2} - h^{-2}) \\
&\Leftrightarrow \quad t = \frac{3\pi\mu L^4}{4(F_* + \pi L^2 P_0)}(h_0^{-2} - h^{-2}),
\end{aligned}$$

where in the first step we integrated with respect to time. Hence $h \to \infty$ if and only if $t \to 3\pi\mu L^4/4h_0^2(F_* + \pi L^2 P_0)$, which is the time it takes to pull the parallel plates apart.

Remark 5.20 There are a plethora of applications of the lubrication effects we outlined above. The squeeze film example above helps explain the lubrication effect of oil in car engines and gear boxes, or the slipperiness of smooth wet surfaces. A dangerous example of the latter is black ice. A thin and smooth layer of ice on a road or pavement becomes dangerous whether initially wet or not – the car tyre pressure will locally increase the temperature of the ice immediately beneath it and create a thin water film between the car tyre and smooth ice surface. And then there is the opposite suction effect, when you try to pull two smooth surfaces apart which have a thin liquid film between them. For example, two flat swimming aid floats in a swimming pool stuck together. For another example consider a toddler's sticky mat. One side is very smooth and sits on a smooth table surface while the other side has a suction cup to stick to the underside of the bowl. The mat steadfastly remains stuck to the table – though in this case the layer between it and the table is air!

5.3 Projects

Some suggested extended projects from this chapter are as follows.

(1) Stokes drag correction. Recall from Theorem 5.10 that the Stokes drag on the sphere of radius a placed in a stream of uniform velocity magnitude U is $6\pi\mu Ua$. However, this is derived under the assumption that the flow is a Stokes flow. Complete Exercises 5.7 and 5.8 where better inner and outer solution approximations, to first order in the Reynolds number $\text{Re} := Ua/\nu$, are respectively derived. Show that the first-order correction to the axial drag force on the sphere is given by

$$F_{\text{ax}} = 6\pi\mu Ua\Big(1 + \tfrac{3}{8}(\text{Re})\Big),$$

as follows. The inner solution from Exercise 5.7 contains the unknown constant C and is valid provided the radial distance $R \ll 1/(\text{Re})$. The Oseen outer solution from Exercise 5.8 is valid for $r \gg 1$, and for the expansion of the exponential term therein, valid for $r(\text{Re}) \ll 1$. This asymptotic condition is consistent with the statement regarding the wake therein. Since the regions $R \ll 1/(\text{Re})$ and $1 \ll r \ll 1/(\text{Re})$ overlap, we can match the inner Ψ_{in} and outer Ψ_{Oseen} solutions from the respective exercises. In particular, show by matching the terms of order '$r^2(\text{Re})$' that $C = 3/8$. Then, using the form for Ψ_{in}, compute the corresponding velocity components, pressure field and non-trivial shear stress terms. Finally, integrate over the surface of the sphere to deduce the result quoted above. See Kundu and Cohen (2008, p. 331) for more details.

(2) Lubrication problems. Lubrication theory forms a part of tribology which, quoting from Bayada and Vázquez (2007), is 'concerned with the study of close interacting surfaces, including the study of friction, lubrication, wear and erosion'. Sticking to the sub-discipline of lubrication theory though, conduct a survey of the range of lubrication problems in science and technology. You might like to proceed as follows. First, complete Exercises 5.9–5.12. Second, use the comprehensive 'survey on the mathematical aspects of lubrication problems' by Bayada and Vázquez (2007), and the references therein, and also Batchelor (1967, Sec. 4.8) and Ockendon and Ockendon (1995, Ch. 4) as starting points. Third, consider including quantitative details on some or all of the following topics: (a) slider bearings, journal bearings and models that account for the practical problem of cavitation formation; (b) thin films with free surfaces that also account for surface tension; (c) further aspects of Hele–Shaw flow incorporating fluid injection or a fluid sink, etc.; (d) aspects related to land, cliff or beach erosion, and so forth. . .

Exercises

5.1 Stokes flow: Between hinged plates. We determine the viscous flow between two hinged plates. The setup is as follows. Assume two semi-infinite plates are hinged along the z-axis (see Figure 5.4). One of the plates is stationary and coincides with the span of the z-axis and positive x-axis. The second plate coincides, via its hinge, with the z-axis and is rotating with a steady angular velocity Ω. A viscous incompressible fluid lies between the two plates. The goal of this question is to determine the instantaneous solution flow between the plates via a Stokes flow approximation when the angle between the two plates is, say, β. We assume the flow is uniform in the z-direction with no z-component flow velocity and thus two-dimensional. We can thus express the flow via a stream function ψ.

(a) Assume the flow between the plates is a Stokes flow; see Definition 5.1. Recall the derivation of the biharmonic equation for the stream function ψ at the beginning of Section 5.1, or alternatively, in the first part of Example 5.12. Note that the plate at angle $\theta = \beta$ moves steadily at a constant angular velocity, however we are interested in the instantaneous Stokes flow solution. Explain why the viscous boundary conditions imply that the instantaneous flow is stationary at the plate at $\theta = \beta$, and thus it is reasonable to look for a stationary Stokes flow solution to this problem, and thus a solution to the biharmonic equation for the stream function ψ. In plane polar coordinates this has the form

$$\Delta(\Delta\psi) = 0,$$

where

$$\Delta = \frac{1}{r}\frac{\partial}{\partial r}\left(r\frac{\partial}{\partial r}\right) + \frac{1}{r^2}\frac{\partial^2}{\partial \theta^2}.$$

(b) Explain why the boundary conditions for this problem are

$$u_r = u_\theta = 0 \text{ on } \theta = 0 \quad \text{and} \quad u_r = 0,\ u_\theta = \Omega r \text{ on } \theta = \beta,$$

where

$$u_r = \frac{1}{r}\frac{\partial\psi}{\partial\theta} \qquad \text{and} \qquad u_\theta = -\frac{\partial\psi}{\partial r}.$$

(c) Using part (b), explain why it is thus reasonable to look for a solution $\psi = \psi(r, \theta)$ to the biharmonic equation in part (a) of the form

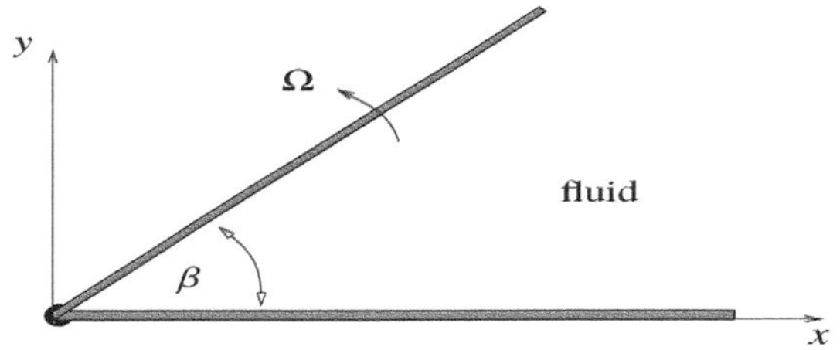

Figure 5.4 Flow between hinged plates. An incompressible viscous fluid is trapped between two infinite plates which are hinged along the z-axis. The lower horizontal plate is stationary while the upper plate is rotating with a steady angular velocity Ω. The goal in Exercise 5.1 is to determine the instantaneous Stokes flow solution when the angle between the plates is β.

$$\psi = \Omega r^2 f(\theta).$$

Show in this case $\Delta\psi = \Omega(f''(\theta) + 4f(\theta))$. Setting $F(\theta) := f''(\theta) + 4f(\theta)$, show that $\Delta(\Omega F) = 0$ if and only if $F''(\theta) = 0$. Hence deduce that the solution form for ψ shown is a solution to the biharmonic equation provided we can find a suitable form for $f = f(\theta)$ of the form

$$f(\theta) = \tfrac{1}{4}A + \tfrac{1}{4}B\theta + C\cos 2\theta + D\sin 2\theta,$$

where A, B, C and D are arbitrary constants, such that ψ satisfies the boundary conditions in part (b).

(d) Show, given the solution form in part (c), that the first three boundary conditions in part (b) are equivalent to the following conditions for f:

$$f(0) = 0, \quad f'(0) = 0 \quad \text{and} \quad f'(\beta) = 0.$$

Show that the first boundary condition implies $C = -A/4$, the second implies $D = -B/8$, while the third generates the condition

$$A = -\tfrac{1}{2}B\tan\beta.$$

Further, with these relations in hand, show that

$$f(\beta) = \tfrac{1}{4}B(\beta - \tan\beta).$$

(e) Show that the fourth boundary condition from part (b), i.e. that $u_\theta = \Omega r$ on $\theta = \beta$, implies the boundary condition $f(\beta) = -1/2$. Using the relation for $f(\beta)$ at the end of part (d), find B and consequently explicit expressions for A, C and D. Hence show

$$\psi = \tfrac{1}{2}\Omega r^2 \frac{(\sin\theta\cos(\beta - \theta) - \theta\cos\beta)}{(\beta - \tan\beta)\cos\beta}.$$

(f) By following Remark 5.13 on the region of validity of the Stokes assumption and noting the characteristic velocity magnitude in this example is $U = \Omega R$, or otherwise, show in this example that the Stokes flow approximation is valid provided $R \ll (\nu/\Omega)^{1/2}$.

(g) Note, as $\tan\beta \to \beta$, which occurs when $\beta \approx 257^{\circ}$, the solution in part (e) breaks down. Show in this case that there exists a solution $\psi = \Omega r^2 f(\theta)$ with $f(0) = f'(0) = f'(\beta) = f(\beta) = 0$. How do you interpret this result?
(*Hint:* See Exercise 5.3 on Moffatt vortices below. In addition, also see Hancock *et al.*, 1981, Sec. 5 where the corresponding instantaneous solution is sought to the case when the hinged plates are both moving apart at the relative angular velocity Ω and they are at $\theta = \pm\frac{1}{2}\beta$. By using the form for $f = f(\theta)$ from part (c) above and adapting the boundary conditions to this slightly different setup, the solution given therein can be straightforwardly determined.)

5.2 Stokes flow: Rotating sphere, couple exerted. Consider a rotating sphere placed into a fluid at rest at infinity. The goal of this question is to estimate the couple exerted on the fluid by the sphere. We approximate the flow around the rotating sphere by an incompressible Stokes flow and compute the solution flow and then couple exerted under this assumption. Suppose the sphere has radius a and its rotation is prescribed by the angular velocity vector $\boldsymbol{\Omega}$. We use spherical polar coordinates (r, θ, φ) and assume the axis to which the co-latitude angle θ is measured is aligned with, and in the same direction as, the angular velocity vector $\boldsymbol{\Omega}$; see Figure 5.5. We assume a stationary flow and the velocity components in the respective coordinate directions are $u_r = 0$ and $u_\theta = 0$ and $u_\varphi = u_\varphi(r, \theta)$ only. The steady Stokes flow equations from Definition 5.1 for the velocity field $\boldsymbol{u}$ and pressure p are

$$\mu\Delta\boldsymbol{u} = \nabla p \qquad \text{and} \qquad \nabla\cdot\boldsymbol{u} = 0,$$

where μ is the first coefficient of viscosity which is constant. We naturally assume no-slip boundary conditions for $\boldsymbol{u}$. Note from Section A.3 in the Appendix that the Laplacian operator in spherical polar coordinates is given by

$$\Delta = \frac{1}{r^2}\frac{\partial}{\partial r}\left(r^2\frac{\partial}{\partial r}\right) + \frac{1}{r^2\sin\theta}\frac{\partial}{\partial\theta}\left(\sin\theta\frac{\partial}{\partial\theta}\right) + \frac{1}{r^2\sin^2\theta}\frac{\partial^2}{\partial\varphi^2}.$$

(a) Using the fact that we assume the velocity field $\boldsymbol{u}$ has the form $\boldsymbol{u} = u_\varphi(r, \theta)\,\hat{\boldsymbol{\varphi}}$, and from Exercise 2.14 we know that the coordinate vector $\hat{\boldsymbol{\varphi}}$ can be expressed in the form $\hat{\boldsymbol{\varphi}} \equiv -\sin\varphi\,\boldsymbol{i} + \cos\varphi\,\boldsymbol{j}$, show

$$\Delta(u_\varphi\,\hat{\boldsymbol{\varphi}}) = \left(\Delta_{r,\theta}u_\varphi - \frac{1}{r^2\sin^2\theta}u_\varphi\right)\hat{\boldsymbol{\varphi}},$$

where $\Delta_{r,\theta}$ is the operator

$$\Delta_{r,\theta} = \frac{1}{r^2}\frac{\partial}{\partial r}\left(r^2\frac{\partial}{\partial r}\right) + \frac{1}{r^2\sin\theta}\frac{\partial}{\partial\theta}\left(\sin\theta\frac{\partial}{\partial\theta}\right).$$

(b) Using the fact that in spherical polar coordinates

$$\nabla p = \hat{\boldsymbol{r}}\frac{\partial p}{\partial r} + \hat{\boldsymbol{\theta}}\frac{1}{r}\frac{\partial p}{\partial\theta} + \hat{\boldsymbol{\varphi}}\frac{1}{r\sin\theta}\frac{\partial p}{\partial\varphi}$$

and part (a), show that the pressure field must be such that $p = p(\varphi)$ only and the steady Stokes equations collapse to

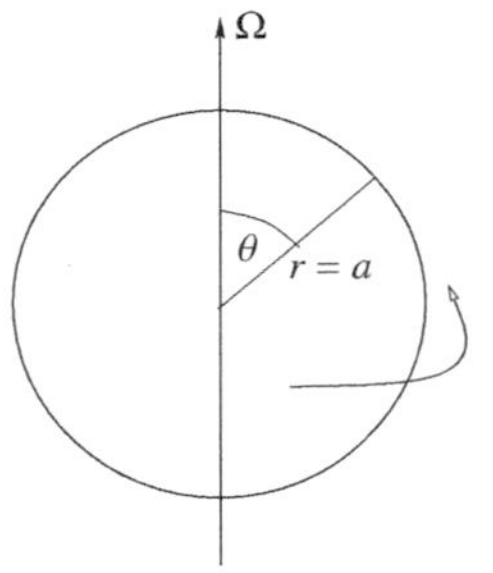

Figure 5.5 Stokes flow, couple exerted by rotating sphere. A sphere of radius a rotates as shown with angular velocity $\mathbf{\Omega}$. In spherical polar coordinates the co-latitude angle θ is measured as shown, with respect to the axis through the centre of the sphere aligned with, and in the same direction as, the angular velocity vector $\mathbf{\Omega}$. The couple exerted on the fluid by the sphere is estimated to be $8\pi\mu a^3\mathbf{\Omega}$.

$$\Delta_{r,\theta}\, u_\varphi - \frac{1}{r^2\sin^2\theta}u_\varphi = \frac{1}{r\sin\theta}\frac{\partial p}{\partial\varphi}.$$

Explain why since $p = p(\varphi)$ is a function that cannot be multivalued, then necessarily the steady Stokes equations become

$$\Delta_{r,\theta}\, u_\varphi = \frac{1}{r^2\sin^2\theta}u_\varphi.$$

(c) Explain why, the boundary conditions for this problem are

$$u_\varphi \to 0 \quad \text{as} \quad r \to \infty \qquad \text{and} \qquad u_\varphi = \Omega a\sin\theta \quad \text{on} \quad r = a,$$

where $\Omega = |\mathbf{\Omega}|$ is the magnitude of the angular velocity. Explain why a suitable guess for the solution u_φ to the final equation in part (b) is

$$u_\varphi = \Omega a f(r)\sin\theta,$$

where $f = f(r)$ is some function of r which must decay as $r \to \infty$. By substituting this guess for the solution into the final equation in part (b), show that $f = f(r)$ necessarily satisfies the ordinary differential equation

$$r^2 f'' + 2rf' - 2f = 0.$$

(d) By looking for a solution to the Cauchy–Euler ordinary differential equation for $f = f(r)$ in part (c) of the form $f = r^\lambda$, where λ is a parameter to be determined, or otherwise, show that f must have the form

$$f(r) = \frac{A}{r^2} + Br,$$

where A and B are arbitrary constants. Use the boundary conditions from part (c) to show that $B = 0$ and $A = a^2$, and thus

$$u_\varphi = \Omega\frac{a^3}{r^2}\sin\theta.$$

(e) Explain why the total couple $\boldsymbol{G}$ exerted by the fluid on the sphere is given by

$$\boldsymbol{G} = \iint \boldsymbol{x} \times (\sigma \hat{\boldsymbol{n}})\,\mathrm{d}S,$$

where the integral is over the surface of the sphere $r = a$, and the stress matrix $\sigma = -pI + \hat{\sigma}$, where p is the pressure field and $\hat{\sigma}$ is the deviatoric stress matrix. Note that at the sphere surface $\hat{\boldsymbol{n}} = \hat{\boldsymbol{r}}$. Recall from Section 4.3 that for an incompressible Newtonian flow we have $\hat{\sigma} = 2\mu D$, where D is the rate of strain or deformation matrix. The components of the deformation matrix D in spherical polar coordinates can be found at the end of Section A.3 in the Appendix. Compute the total couple $\boldsymbol{G}$ as follows.

(i) Show that the only non-zero components of the deformation matrix D are $D_{r\varphi}$ and $D_{\theta\varphi}$ and compute the explicit form for $D_{r\varphi} = D_{\varphi r}$. Recall the expression for $D\boldsymbol{v}$ for any vector $\boldsymbol{v}$ from Step 5 in the proof of Theorem 5.10. Using the expression for $D\boldsymbol{v}$, show in this case that

$$\hat{\sigma}\hat{\boldsymbol{n}} = 2\mu D_{r\varphi}\,\hat{\boldsymbol{\varphi}} = -\left(3\mu\Omega\frac{a^3}{r^3}\sin\theta\right)\hat{\boldsymbol{\varphi}}.$$

(ii) Using part (i) just above and that on the surface of the sphere $\boldsymbol{x} = a\,\hat{\boldsymbol{r}}$ and $\hat{\boldsymbol{r}} \times \hat{\boldsymbol{\varphi}} = -\hat{\boldsymbol{\theta}}$, show

$$\boldsymbol{G} = 3\mu a\Omega \iint \sin\theta\,\hat{\boldsymbol{\theta}}\,\mathrm{d}S.$$

(iii) Recall from Exercise 2.14 that $\hat{\boldsymbol{\theta}} \equiv \cos\theta\cos\varphi\,\boldsymbol{i} + \cos\theta\sin\varphi\,\boldsymbol{j} - \sin\theta\,\boldsymbol{k}$. Using this form for $\hat{\boldsymbol{\theta}}$ in the expression for $\boldsymbol{G}$ in part (ii) just above, show that the only non-zero component of $\boldsymbol{G}$ is the $\boldsymbol{k}$ component, i.e. that of $\hat{\boldsymbol{\Omega}}$, and that in fact

$$\boldsymbol{G} = -6\pi\mu a^3\boldsymbol{\Omega}\int_0^{\pi}\sin^3\theta\,\mathrm{d}\theta.$$

Hence by computing the integral, show $\boldsymbol{G} = -8\pi\mu a^3\boldsymbol{\Omega}$.

(*Note*: The couple $\boldsymbol{G}$ computed is the couple exerted on the sphere by the fluid – the fluid resists the rotation in the positive $\hat{\boldsymbol{\varphi}}$ direction. Hence if the sphere rotates at a constant angular velocity $\boldsymbol{\Omega}$, the couple exerted on the fluid by the sphere is $8\pi\mu a^3\boldsymbol{\Omega}$.)

(f) Explain why this estimate holds asymptotically for $a^2\Omega/\nu \ll 1$.
(*Hint:* This is the realisation in this context of the conditions under which a Stokes flow applies, for example also see Exercise 5.1.)

5.3 Stokes flow: Moffatt vortices. Recall that for a stationary incompressible two-dimensional Stokes flow, the governing equation in terms of a stream function ψ is the biharmonic equation

$$\Delta(\Delta\psi) = 0.$$

Assume plane polar coordinates (r, θ) hereafter.

(a) Show that solutions to the biharmonic equation of the form $\psi = r^{n+1}f(\theta)$ can be found for $n \neq 0, \pm 1$ if (A, B, C and D are arbitrary constants)

$$f(\theta) = A\cos((n+1)\,\theta) + B\sin((n+1)\,\theta) + C\cos((n-1)\,\theta) + D\sin((n-1)\,\theta).$$

(*Hint:* Show $\Delta\psi = r^{n-1}((n+1)^2 f + f'')$ and set $F := (n+1)^2 f + f''$. Deduce $\Delta(\Delta\psi) = r^{n-3}((n-1)^2 F + F'')$, set this equal to zero and successively solve the two second-order linear equations for F and then f.)

(b) Now suppose the Stokes flow occurs in the corner between two rigid boundaries at $\theta = \pm\alpha$. We look for solutions for ψ of the form above which are symmetric, and then antisymmetric in θ.

(i) Explain why suitable boundary conditions for f are $f(\alpha) = 0$ and $f'(\alpha) = 0$.

(ii) In the case of solutions ψ symmetric in θ, explain why necessarily $B = D = 0$ and n satisfies $\sin(2n\alpha) = -n\sin(2\alpha)$.

(iii) In the case of solutions ψ antisymmetric in θ, explain why necessarily $A = C = 0$ and n satisfies $\sin(2n\alpha) = n\sin(2\alpha)$.

(c) There are an infinite number of roots to the equations $\sin(2n\alpha) = \pm n\sin(2\alpha)$ from parts (b)(ii) and (b)(iii) just above. They are complex if 2α is less than about 146°. For example, set $n = \lambda/\alpha$ and suppose $\alpha \ll 1$ so the equation becomes $\sin(2\lambda) \approx \pm 2\lambda$, which only has complex solutions. Show that when $n = p + \mathrm{i}q$ is complex and we look for real solutions $\psi = r^{p+1}(r^{\mathrm{i}q} f(\theta) + r^{-\mathrm{i}q} f^*(\theta))$, then

$$\psi = 2r^{p+1}\Big(\cos(q\log r)\,\mathrm{Re}\{f(\theta)\} - \sin(q\log r)\,\mathrm{Im}\{f(\theta)\}\Big).$$

(d) Figure 5.6 demonstrates symmetric (top panel) and antisymmetric (bottom panel) examples of the stream function ψ from part (c); they are contour plots. The flow lines indicate vortices. These are known as Moffatt vortices; see Moffatt (1964). In the figure, α corresponds to 20°. In the top panel $n = 6.0578 + 3.0954\mathrm{i}$, which is one of the four smallest (one exists in each quadrant) non-trivial solutions to $\sin(2n\alpha) = -n\sin(2\alpha)$. In the bottom panel $n = 10.7541 + 3.8431\mathrm{i}$, which is one of the four smallest non-trivial solutions to $\sin(2n\alpha) = n\sin(2\alpha)$. Neighbouring vortices rotate in opposite directions to ensure continuity. Consider the form for ψ in part (c), and in particular, for example, the symmetric case.

(i) Recall in plane polar coordinates $u_\theta = -\partial\psi/\partial r$. Using the general form for ψ from part (c) and the symmetric form for f from part (a), with $B = D = 0$, explain why we can express (for two real constants β and γ)

$$u_\theta(r, 0) = \gamma r^p \sin(q\log r + \beta).$$

(ii) Explain why we know that there exist an infinite number of vortices, which decrease in size and approach the origin, with centres at $r = r_m$, where for any $m = 0, 1, 2, \ldots$

$$q\log r_m + \beta = -m\pi \quad \Leftrightarrow \quad r_m = \mathrm{e}^{-\beta/q}\mathrm{e}^{-m\pi/q}.$$

(iii) Show that

$$\frac{r_m}{r_{m+1}} = \frac{r_m - r_{m+1}}{r_{m+1} - r_{m+2}} = \mathrm{e}^{\pi/q}.$$

Deduce that the sizes of successive vortices fall off in a geometric progression. What is the common ratio?

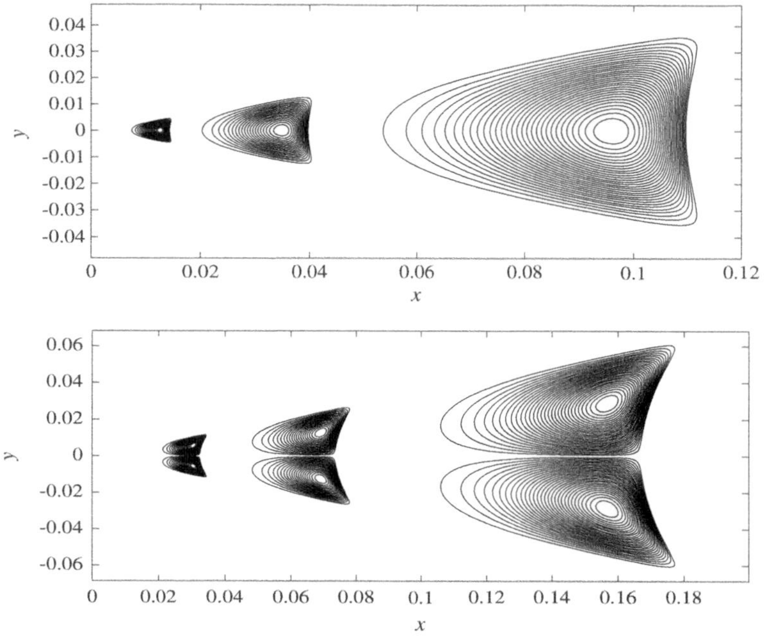

Figure 5.6 Moffatt vortices with α corresponding to 20°. Both panels show contour lines for the biharmonic stream function $\psi = r^{n+1} f(\theta)$. The top panel demonstrates the symmetric case with $n = 6.0578 + 3.0954\mathrm{i}$ one of the non-trivial solutions to $\sin(2n\alpha) = -n\sin(2\alpha)$. The bottom panel demonstrates the antisymmetric case with $n = 10.7541 + 3.8431\mathrm{i}$ one of the non-trivial solutions to $\sin(2n\alpha) = n\sin(2\alpha)$. Neighbouring vortices rotate in opposite directions to ensure continuity.

(iv) Explain why $u_\theta(r, 0)$ has local absolute maxima at

$$r_{m+1/2} := \mathrm{e}^{-\beta/q}\mathrm{e}^{-(m+1/2)\pi/q},$$

and the velocity at these maxima is $u_\theta(r_{m+1/2}, 0) = \gamma r^p_{m+1/2}$. Why does this velocity represent a measure of the 'intensity' of successive vortices? Show that the intensities of the vortices fall off in a geometric progression with common ratio

$$\left(\frac{r_{m+1/2}}{r_{m+3/2}}\right)^p = \mathrm{e}^{p\pi/q}.$$

5.4 Stokes flow: Flow between co-axial cones. Consider a steady incompressible slow flow in spherical polar coordinates (r, θ, φ) of the form $\boldsymbol{u} = u_\varphi\,\hat{\boldsymbol{\varphi}}$, with $u_\varphi = r\sin\theta\, f(\theta)$ and the pressure field p constant.

(a) Show that if this velocity field $\boldsymbol{u}$ is such a slow flow, then necessarily $\Delta(u_\varphi\,\hat{\boldsymbol{\varphi}}) = \mathbf{0}$ or equivalently

$$\mathrm{D}^2_{\mathrm{sph}}(r\sin\theta\, u_\varphi) = 0,$$

where

$$\mathrm{D}^2_{\mathrm{sph}} := \frac{\partial^2}{\partial r^2} + \frac{\sin\theta}{r^2}\frac{\partial}{\partial\theta}\left(\frac{1}{\sin\theta}\frac{\partial}{\partial\theta}\right).$$

(b) Show that the explicit form $u_\varphi = r \sin\theta\, f(\theta)$ is a solution to the equation in part (a) provided f satisfies the ordinary differential equation

$$f'' + 3\cot\theta\, f' = 0.$$

(c) Solve the linear differential equation for f in part (b) as follows. Using the integrating factor method, or other means, first show

$$f' = \frac{A}{\sin^3\theta},$$

where A is an arbitrary constant. Then show

$$f = \tfrac{1}{2}A\left(\log\left(\tan\tfrac{1}{2}\theta\right) - \frac{\cos\theta}{\sin^2\theta}\right) + B$$

satisfies the equation for f' above, where B is another arbitrary constant.

(d) Now suppose this slow flow is restricted to the domain between two co-axial inverted cones with the same vertex but different semi-vertex angles α and β. In other words, one narrower cone lies inside another cone, upside down, with their vertices coinciding. The slow flow takes place between the two cones. Assume that both cones are rotating about their common axis of symmetry with respective angular velocities Ω_α and Ω_β. Explain why the appropriate boundary conditions are given by $u_\varphi = \Omega_\alpha r \sin\alpha$ on $\theta = \alpha$ and $u_\varphi = \Omega_\beta r \sin\beta$ on $\theta = \beta$. Use these boundary conditions to determine the values for the constants A and B in the form of f in part (c) in this case.

5.5 Stokes flow: Stokes Paradox. Consider a two-dimensional steady flow around a cylinder with circular cross-section of radius a, placed in a uniform stream of velocity magnitude U. Use plane polar coordinates (r, θ), centred at the middle of the circular cross-section, to parameterise this two-dimensional exterior problem. Assume the uniform stream is aligned with the $\theta = 0$ direction. Since the flow is incompressible and two-dimensional, there exists a stream function $\psi = \psi(r, \theta)$ and the governing equations can be expressed in vorticity form as

$$(\boldsymbol{u} \cdot \nabla)\,\omega = \nu\Delta\omega,$$

where $\omega = -\Delta\psi$; recall this formulation from Section 4.6. The Reynolds number here is $\mathrm{Re} := aU/\nu$.

(a) Introducing the non-dimensional coordinates

$$r' = \frac{r}{a} \quad \text{and} \quad \psi' = \frac{\psi}{aU},$$

show, using the forms for $\boldsymbol{u} \cdot \nabla$ and Δ given in Section A.2 of the Appendix, that the vorticity-stream formulation of the governing equations above is equivalent to the following equation for $\psi' = \psi'(r', \theta)$:

$$\frac{(\mathrm{Re})}{r'}\left(\frac{\partial\psi'}{\partial\theta}\frac{\partial}{\partial r'} - \frac{\partial\psi'}{\partial r'}\frac{\partial}{\partial\theta}\right)\Delta'\psi' = \Delta'^2\psi',$$

where

$$\Delta' = \frac{1}{r'}\frac{\partial}{\partial r'}\left(r'\frac{\partial}{\partial r'}\right) + \frac{1}{r'^2}\frac{\partial^2}{\partial \theta^2}.$$

Explain why suitable boundary conditions are: at $r' = 1$ we have $\psi' = 0$ and $\partial\psi'/\partial r' = 0$ and as $r' \to \infty$ we have $\psi' \sim r' \sin\theta$.

(b) Assume ψ' has the form $\psi' \sim \psi_0' + (\mathrm{Re})\,\psi_1'$. Substituting this into the governing equations in part (a), show that at leading order ψ_0', which corresponds to Stokes flow, satisfies the biharmonic equation (the quantity ψ_1' plays a role in the next Exercise 5.6)

$$\Delta'^2\psi_0' = 0.$$

(c) Look for a solution to the biharmonic equation for ψ_0' from part (b) of the form $\psi_0' = f(r') \sin\theta$ and show that $f = f(r')$ necessarily satisfies the fourth-order linear ordinary differential equation

$$\left(\frac{\mathrm{d}^2}{\mathrm{d}r'^2} + \frac{1}{r'}\frac{\mathrm{d}}{\mathrm{d}r'} - \frac{1}{r'^2}\right)^2 f = 0,$$

together with the boundary conditions $f = 0$ and $\mathrm{d}f/\mathrm{d}r' = 0$ at $r' = 1$, and $f \to r'$ as $r' \to \infty$. This equation is of Cauchy–Euler form. Hence, seeking solutions of the form $f = A(r')^n$, where A is a constant, show that necessarily the only possible values for n are 1, 1, −1 and 3. The first root is repeated. Hence deduce that f must have the form

$$f = Ar'^3 + Br' + Cr' \log r' + \frac{D}{r'},$$

where A, B, C and D are arbitrary constants. Further show that the two boundary conditions at $r' = 1$ are satisfied if and only if

$$A + B + D = 0 \qquad \text{and} \qquad 3A + B + C - D = 0.$$

Given these conditions, and considering that matching the far-field condition $f \to r'$ as $r' \to \infty$ necessarily requires $A = 0$ and $C = 0$, deduce that necessarily $f = 0$ and the only solution to the biharmonic equation in part (b) is $\psi_0' = 0$. This is the Stokes Paradox. (*Note:* Quoting from Ockendon and Ockendon, 1995, pp. 55–66 'it can be proved that there is no [non-trivial] solution to the problem posed' above 'and our failure to find a solution is not the result of assuming a particular form for ψ. Further, it can be shown that there is no solution to the problem of uniform slow flow past a cylinder of arbitrary cross-section'.)

5.6 Stokes flow: Matched asymptotics for a cylinder. We saw in the last Exercise 5.5 that the two-dimensional Stokes flow associated with a cylinder of circular cross-section placed in a uniform stream is necessarily trivial, i.e. $\psi_0' = 0$. In other words, in the limit $\mathrm{Re} \to 0$ in which inertia terms play no part, in the context stated, this is the leading order steady flow satisfying all the boundary conditions. Stokes flow is a good approximation close to the cylinder where no-slip boundary conditions are consistent with low Reynolds-number flow. The issue is that it is no longer a good approximation 'far' from the cylinder. Hence we can employ the method of matched asymptotic expansions to match a Stokes flow close to the cylinder, the *inner* solution, with a solution to the full steady incompressible two-dimensional Navier–Stokes equations 'far' from the cylinder, the *outer* solution. This is the goal of this question. Note, consider the form of the governing equations in Exercise 5.5, part (a). If we had non-dimensionalised r as $r' = r/R$ for some characteristic length scale instead, then we would have the same equations as in Exercise 5.5, part (a), but with the modified Reynolds number $RU/\nu = (R/a)(\text{Re})$ as a factor of the inertia terms instead. Hence we regard the inner region as that corresponding to $(R/a)(\text{Re}) \ll 1$, i.e. the region for which radial distances R satisfy $R/a \ll 1/(\text{Re})$.

(a) Suppose the inner solution ψ'_{in} satisfies the steady Stokes equation $\Delta'^2\psi'_{\text{in}} = 0$ and the boundary conditions $\psi'_{\text{in}} = 0$ and $\partial\psi'_{\text{in}}/\partial r' = 0$ on $r' = 1$, and does not grow as fast as r'^3 as $r' \to \infty$, i.e. we retain both the linear r' and the $r' \log r'$ terms. Show that the solution in this context, see Exercise 5.5, part (c), is

$$\psi'_{\text{in}} = C\left(r' \log r' - \tfrac{1}{2}r' + \frac{1}{2r'}\right) \sin\theta,$$

where we expect C to depend on Re.

(b) By writing $r' = \hat{r}/(\text{Re})$, show when r' is of order $1/(\text{Re})$, we have

$$\psi'_{\text{in}} \sim \frac{C}{(\text{Re})} \log\left(\frac{1}{(\text{Re})}\right) \hat{r} \sin\theta.$$

(c) Consider the outer solution ψ'_{out} satisfying the equations in Exercise 5.5, part (a). Write $\psi'_{\text{out}}(r', \theta) = \hat{\psi}(\hat{r}, \theta)/(\text{Re})$ and show that $\hat{\psi}$ satisfies

$$\frac{1}{\hat{r}}\left(\frac{\partial\hat{\psi}}{\partial\theta}\frac{\partial}{\partial\hat{r}} - \frac{\partial\hat{\psi}}{\partial\hat{r}}\frac{\partial}{\partial\theta}\right)\hat{\Delta}\hat{\psi} = \hat{\Delta}^2\hat{\psi},$$

where the operator $\hat{\Delta}$ is of the same form as Δ' but with r' replaced by $\hat{r}$. Explain why $\hat{\psi} \sim C\log(1/(\text{Re}))\,\hat{y}$ as $\hat{r} \to 0$ and $\hat{\psi} \to \hat{y}$ as $\hat{r} \to \infty$, where $\hat{y} = \hat{r}\sin\theta$. Deduce that $C = 1/\log(1/(\text{Re}))$.

(d) Consider the expansion $\hat{\psi} = \hat{y} + \epsilon\hat{\psi}_1 + \cdots$, where $\epsilon := 1/\log(1/(\text{Re}))$, to show $\hat{\psi}_1$ satisfies the equation

$$\frac{\partial}{\partial\hat{x}}(\hat{\Delta}\hat{\psi}_1) = \hat{\Delta}^2\hat{\psi}_1.$$

(e) Show that the equation in part (d) satisfied by $\hat{\psi}_1$ in the outer region is equivalent to Oseen's equation in Exercise 5.8.

(*Note:* See Proudman and Pearson, 1957, who used matched asymptotic expansions to justify Oseen's equation.)

5.7 Stokes flow: Inner correction for a sphere. In this and Exercise 5.8 we see the next-order corrections to axisymmetric Stokes flow around a sphere, in particular the next-order correction in terms of the Reynolds number. A project in Section 5.3 examines the next-order correction to the Stokes drag derived in Theorem 5.10. Consider a sphere of radius a placed in a uniform stream of velocity magnitude U. Assume the three-dimensional steady flow around the sphere is axisymmetric and there is no swirl. Use spherical polar coordinates (r, θ, φ) centred in the middle of the sphere. Assume the uniform flow is aligned with the $\theta = 0$ direction, as in Theorem 5.10. Consider the vorticity formulation of the incompressible Navier–Stokes equations given in Corollary 4.15. For steady flow the vorticity $\boldsymbol{\omega} = \nabla \times \boldsymbol{u}$ satisfies

$$(\boldsymbol{u} \cdot \nabla)\,\boldsymbol{\omega} = (\mathrm{Re})^{-1}\Delta\boldsymbol{\omega} + (\boldsymbol{\omega} \cdot \nabla)\,\boldsymbol{u},$$

in non-dimensional coordinates, where $\Delta\boldsymbol{u} = -\nabla \times \boldsymbol{\omega}$. Recall from the proof of Corollary 2.7 that $(\boldsymbol{\omega} \cdot \nabla)\,\boldsymbol{u} \equiv D\boldsymbol{\omega}$, where D is the symmetric part of $\nabla\boldsymbol{u}$. Further recall from Exercise 2.14 in the case of axisymmetry and no swirl that the nonlinear terms in the above formulation have the form

$$(\boldsymbol{u} \cdot \nabla)\,\boldsymbol{\omega} = \left(u_r \frac{\partial \omega_\varphi}{\partial r} + \frac{u_\theta}{r}\frac{\partial \omega_\varphi}{\partial \theta}\right)\hat{\boldsymbol{\varphi}}$$

and

$$(\boldsymbol{\omega} \cdot \nabla)\,\boldsymbol{u} = \frac{\omega_\varphi}{r\sin\theta}(u_r \sin\theta + u_\theta \cos\theta)\,\hat{\boldsymbol{\varphi}},$$

where $\boldsymbol{u} = \boldsymbol{u}(r,\theta)$ has components $\boldsymbol{u} = u_r\,\hat{\boldsymbol{r}} + u_\theta\,\hat{\boldsymbol{\theta}}$ and $\boldsymbol{\omega} = \boldsymbol{\omega}(r,\theta)$ has only one non-zero component $\boldsymbol{\omega} = \omega_\varphi\,\hat{\boldsymbol{\varphi}}$. Such a flow can be represented by the Stokes stream function $\Psi = \Psi(r,\theta)$; see Remark 5.9. The non-dimensional form of the Stokes stream function here is given by the dimensional Stokes stream function divided by Ua^2. Since our goal in this question is to obtain a better approximation to the Stokes flow solution derived in the proof of Theorem 5.10, we focus on the *inner* solution. By arguments exactly analogous to those in Exercise 5.6, the inner region is identified by radial distances R such that $R \ll 1/(\mathrm{Re})$ – note that in non-dimensional coordinates the sphere has unit radius. The boundary conditions for the inner region, as for the Stokes flow case though here non-dimensionalised, are

$$\Psi = \frac{\partial \Psi}{\partial r} = 0 \quad \text{on} \quad r = 1.$$

(a) Set $c := \cos\theta$. Using the derivations preceding Proposition 5.8 for axisymmetric flow with no swirl in spherical polar coordinates and the chain rule, show

$$u_r = -\frac{1}{r^2}\frac{\partial \Psi}{\partial c}, \qquad u_\theta = -\frac{1}{r\sin\theta}\frac{\partial \Psi}{\partial r} \qquad \text{and} \qquad \omega_\varphi = -\frac{1}{r\sin\theta}\mathrm{D}^2\Psi,$$

where

$$\mathrm{D}^2 := \frac{\partial^2}{\partial r^2} + \frac{1}{r^2}(1-c^2)\frac{\partial^2}{\partial c^2},$$

is the form for $\mathrm{D}^2_{\mathrm{sph}}$ converted to the variable c instead of θ.

(b) Again using the derivations preceding Proposition 5.8, show

$$\Delta\boldsymbol{\omega} = -\left(\frac{1}{r\sin\theta}\mathrm{D}^2(\mathrm{D}^2\Psi)\right)\hat{\boldsymbol{\varphi}},$$

where D^2 is the operator from part (a).

(c) Substituting the forms for u_r, u_θ and ω_φ in part (a) into the expressions for $(\boldsymbol{u}\cdot\nabla)\,\omega$ and $(\omega\cdot\nabla)\,\boldsymbol{u}$ above, and using part (b), show that the vorticity formulation of the steady incompressible Navier–Stokes equations takes the form

$$\mathrm{D}^2(\mathrm{D}^2\Psi) = \frac{(\mathrm{Re})}{r^2}\left(J(\Psi,\mathrm{D}^2\Psi) + 2(\mathrm{D}^2\Psi)\left(\frac{1}{r}\frac{\partial\Psi}{\partial c} + \frac{c}{1-c^2}\frac{\partial\Psi}{\partial r}\right)\right),$$

where $J(\Psi,\mathrm{D}^2\Psi)$ is the Jacobian

$$J(\Psi,\mathrm{D}^2\Psi) := \frac{\partial\Psi}{\partial r}\frac{\partial}{\partial c}(\mathrm{D}^2\Psi) - \frac{\partial\Psi}{\partial c}\frac{\partial}{\partial r}(\mathrm{D}^2\Psi).$$

(d) Taking $(\mathrm{Re})\to 0$ in the equation in part (c), the equation reverts to the steady Stokes flow $\mathrm{D}^2(\mathrm{D}^2\Psi)=0$ with the Stokes stream function

$$\Psi_0 := \tfrac{1}{4}\left(2r^2 - 3r + \frac{1}{r}\right)(1-c^2),$$

satisfying the boundary conditions at $r=1$. To find a better approximation in the inner region we write

$$\Psi_{\text{in}} = \Psi_0 + \Psi_1\,(\mathrm{Re}) + O((\mathrm{Re})^2),$$

where $O((\mathrm{Re})^2)$ denotes terms bounded by a constant times $(\mathrm{Re})^2$ in absolute value as $(\mathrm{Re})\to 0$. Substitute Ψ_{in} into the equation in part (c), divide by (Re) and take the limit $(\mathrm{Re})\to 0$ to show that Ψ_1 satisfies

$$\mathrm{D}^2(\mathrm{D}^2\Psi_1) = \frac{1}{r^2}\left(J(\Psi_0,\mathrm{D}^2\Psi_0) + 2(\mathrm{D}^2\Psi_0)\left(\frac{1}{r}\frac{\partial\Psi_0}{\partial c} + \frac{c}{1-c^2}\frac{\partial\Psi_0}{\partial r}\right)\right).$$

By substituting for Ψ_0 further show that Ψ_1 satisfies

$$\mathrm{D}^2(\mathrm{D}^2\Psi_1) = -\frac{9}{4}\frac{1}{r^2}\left(2 - \frac{3}{r} + \frac{1}{r^3}\right)c(1-c^2).$$

(e) Solve the equation in part (d) as follows. The form for Ψ_0, or any constant multiple of it, is a homogeneous solution. Show that

$$\Psi_{\text{part}} := -\frac{3}{32}\left(2r^2 - 3r + 1 - \frac{1}{r} + \frac{1}{r^2}\right)c(1-c^2)$$

is a particular integral for the equation in part (d). Hence deduce, for any constant C, that

$$\Psi_1 := \frac{C}{4}\left(2r^2 - 3r + \frac{1}{r}\right)(1-c^2) - \frac{3}{32}\left(2r^2 - 3r + 1 - \frac{1}{r} + \frac{1}{r^2}\right)c(1-c^2)$$

solves the equation and satisfies the boundary conditions at $r=1$. The constant C is fixed by matching this inner solution to the outer solution derived in Exercise 5.8. See Kundu and Cohen (2008, pp. 329–31) for more details and also the first project in Section 5.3.

5.8 Stokes flow: Oseen correction for a sphere. We saw in Exercises 5.5 and 5.6 Stokes' Paradox in the case of two-dimensional Stokes flow around a cylinder and how a suitable modification of Stokes flow resolved the paradox. The suitable modification corresponded to an ad hoc correction to Stokes flow known as the Oseen flow. Here we consider Oseen flow around a sphere placed in a uniform stream of velocity U. Assume the three-dimensional steady flow around the sphere is axisymmetric and there is no swirl. Suppose the sphere has radius a. Use spherical polar coordinates (r, θ, φ) centred in the middle of the sphere. Assume the uniform flow is aligned with the $\theta = 0$ direction, exactly as in Theorem 5.10 and Exercise 5.7 just above. Note that the $\theta = 0$ direction coincides with the positive Cartesian z-direction. The Oseen equations for the flow are

$$U\frac{\partial \boldsymbol{u}}{\partial z} = -\frac{1}{\rho}\nabla p + \nu\Delta\boldsymbol{u},$$
$$\nabla \cdot \boldsymbol{u} = 0,$$

where $\boldsymbol{u} = \boldsymbol{u}(r, \theta)$ and $p = p(r, \theta)$ are the axisymmetric velocity and pressure fields, respectively. Such a flow can be represented by a stream function, and here we use the Stokes stream function $\Psi = \Psi(r, \theta)$; see Remark 5.9. The boundary conditions, as for the Stokes flow case, are

$$\Psi \sim \tfrac{1}{2}Ur^2\sin^2\theta \text{ as } r \to \infty \qquad \text{and} \qquad \Psi = \frac{\partial\Psi}{\partial r} = 0 \text{ on } r = a.$$

(a) Show that the Oseen equations can be justified as follows. In Cartesian coordinates, with the z-direction indicated above, substitute $u = \hat{u}$, $v = \hat{v}$ and $w = U + \hat{w}$ into the full steady incompressible Navier–Stokes equations. By neglecting terms quadratic in $\hat{u}$, $\hat{v}$ and $\hat{w}$, i.e. the $(\hat{\boldsymbol{u}} \cdot \nabla)\hat{\boldsymbol{u}}$ term, show that the equation above is satisfied by $\hat{\boldsymbol{u}} = (\hat{u}, \hat{v}, \hat{w})^{\mathrm{T}}$. Then by substituting back for $\boldsymbol{u} = (u, v, w)^{\mathrm{T}}$ in place of $\hat{\boldsymbol{u}}$, still neglecting the quadratic terms, show that $\boldsymbol{u}$ satisfies the *linear* Oseen equations above.

(b) Taking the curl, the Oseen equations become $U(\partial\boldsymbol{\omega}/\partial z) = \nu\Delta\boldsymbol{\omega}$ for the vorticity field $\boldsymbol{\omega} = \boldsymbol{\omega}(r, \theta)$. Recall from the derivations preceding Proposition 5.8 for axisymmetric flow with no swirl in spherical polar coordinates that we know

$$\boldsymbol{\omega} = -\left(\frac{1}{r\sin\theta}(\mathrm{D}^2_{\mathrm{sph}}\Psi)\right)\hat{\boldsymbol{\varphi}} \quad \text{and} \quad \Delta\boldsymbol{\omega} = -\left(\frac{1}{r\sin\theta}\mathrm{D}^2_{\mathrm{sph}}(\mathrm{D}^2_{\mathrm{sph}}\Psi)\right)\hat{\boldsymbol{\varphi}}.$$

Using the chain rule to re-express $\partial/\partial z$ in terms of $\partial/\partial r$ and $\partial/\partial\theta$, show that $\Psi = \Psi(r, \theta)$ satisfies the equation

$$U\left(\cos\theta\frac{\partial}{\partial r}(\mathrm{D}^2_{\mathrm{sph}}\Psi) - \frac{1}{r}\sin\theta\frac{\partial}{\partial\theta}(\mathrm{D}^2_{\mathrm{sph}}\Psi)\right) = \nu\,\mathrm{D}^2_{\mathrm{sph}}(\mathrm{D}^2_{\mathrm{sph}}\Psi).$$

(c) Define the non-dimensional variables $r' := r/a$ and $\Psi' := \Psi/Ua^2$. Further set $c := \cos\theta$. By representing the equation in part (b) in terms of these non-dimensional variables as well as replacing θ by c, show that the non-dimensional form of the equation, after dropping primes, is

$$c\frac{\partial}{\partial r}(\mathrm{D}^2\Psi) + \frac{1}{r}(1 - c^2)\frac{\partial}{\partial c}(\mathrm{D}^2\Psi) = (\mathrm{Re})^{-1}\,\mathrm{D}^2(\mathrm{D}^2\Psi),$$

where $\mathrm{Re} := Ua/\nu$ is the Reynolds number and

$$\mathrm{D}^2 := \frac{\partial^2}{\partial r^2} + \frac{1}{r^2}(1 - c^2)\frac{\partial^2}{\partial c^2}$$

represents the non-dimensional form for $\mathrm{D}^2_{\text{sph}}$ expressed in terms of c.

(d) Show that the following form is a solution to the linear Oseen equation in part (c):

$$\Psi_1 := (1+c)\Big(1-\mathrm{e}^{-\frac{1}{2}(\mathrm{Re})r(1-c)}\Big).$$

First show that

$$\mathrm{D}^2\Psi_1 = -(\mathrm{Re})(1-c^2)\big((\mathrm{Re})/2+1/r\big)\,\mathrm{e}^{-\frac{1}{2}(\mathrm{Re})r(1-c)},$$

then show that the left and right sides of the Oseen equation equal

$$(\mathrm{Re})(1-c^2)\big(3c/r^2-(\mathrm{Re})/2r+3c(\mathrm{Re})/2r+(c-1)(\mathrm{Re})^2/4\big)\,\mathrm{e}^{-\frac{1}{2}(\mathrm{Re})r(1-c)}.$$

(e) Define the following two functions:

$$\tilde{\Psi}_0 := \frac{1}{4}\Big(2r^2+\frac{1}{r}\Big)(1-c^2) \quad\text{and}\quad \Psi_{\text{Stokes}} := \frac{1}{4}\Big(2r^2-3r+\frac{1}{r}\Big)(1-c^2).$$

(i) Show that $\mathrm{D}^2\tilde{\Psi}_0 = 0$ and in fact this holds individually for each term.
(ii) Using part (i), deduce $\mathrm{D}^2(\mathrm{D}^2(1/r)) = 0$.
(iii) Deduce that $\tilde{\Psi}_0$ trivially satisfies the Oseen equation in part (c).
(iv) Using parts (i) and (ii), deduce that $\mathrm{D}^2(\mathrm{D}^2\Psi_{\text{Stokes}}) = 0$ and thus Ψ_{Stokes} corresponds to the Stokes flow solution around the sphere.

(f) Define the function Ψ_{Oseen} by

$$\Psi_{\text{Oseen}} := \tilde{\Psi}_0 - \frac{3}{2}(\mathrm{Re})^{-1}\Psi_1.$$

By using the series expansion for the exponential, show that as $(\mathrm{Re}) \to 0$,

$$\Psi_1 \sim \tfrac{1}{2}r(1-c^2)\,(\mathrm{Re}) - \tfrac{1}{8}r^2(1-c^2)(1-c)\,(\mathrm{Re})^2 + O(r^3(\mathrm{Re})^3)$$

and thus

$$\Psi_{\text{Oseen}} \sim \Psi_{\text{Stokes}} + \frac{3}{16}r^2(1-c^2)(1-c)\,(\mathrm{Re}) + O(r^3(\mathrm{Re})^2)$$

Here $O(r^m(\mathrm{Re})^n)$ for integers m and n denotes terms bounded by a constant times $r^m(\mathrm{Re})^n$ in absolute value as $(\mathrm{Re}) \to 0$. Further, show that at the sphere boundary $r=1$, in non-dimensional coordinates, we have $\Psi_{\text{Oseen}} = O((\mathrm{Re}))$ and $(\partial/\partial r)\Psi_{\text{Oseen}} = O((\mathrm{Re}))$. Also show that $\Psi_{\text{Oseen}} \sim \frac{1}{2}r^2(1-c^2)$ as $r \to \infty$ and thus Ψ_{Oseen} satisfies the far-field boundary condition. Hence Ψ_{Oseen} represents Oseen's correction to the Stokes flow Ψ_{Stokes} around the sphere. Note that in the region of the wake of the sphere where $|\theta| \ll 1$, the first correction term to the Stokes flow, i.e. the second term in the expression above, is only appreciable if $r^2\theta^2(\mathrm{Re}) = O(1)$, or in other words if $r^2 = O(1/(\theta^2(\mathrm{Re})))$. For more details see Ockendon and Ockendon (1995, p. 63) and Kundu and Cohen (2008, p. 331). Also see the first project in Section 5.3.

5.9 Lubrication theory: Shear stress. Consider a homogeneous incompressible fluid of viscosity μ which is contained between $z=0$ and $z=h(x)$ for $0 \leqslant x \leqslant L$. Here z is a vertical coordinate while x and y are horizontal coordinates; however the flow can be considered to be *independent* of y. Let u and w be the velocity components in the coordinate directions x and z, respectively. Assume a typical horizontal velocity scale is U and a typical vertical velocity scale is W.

(a) Assume throughout that $h \ll L$.

(i) Using that the fluid is incompressible, explain how we can deduce that W is asymptotically smaller than U.

(ii) Using part (i), show using lubrication theory that the incompressible three-dimensional Navier–Stokes equations reduce to the shallow-layer equations (neglecting all body forces)

$$\frac{\partial p}{\partial x} = \mu \frac{\partial^2 u}{\partial z^2} \quad \text{and} \quad \frac{\partial p}{\partial z} = 0.$$

(b) Now integrate the shallow-layer equations as follows.

(i) Using the shallow-layer equations from part (a)(ii), show that the velocity field u is given by

$$u = -\frac{1}{2\mu}\frac{\partial p}{\partial x} z(h - z).$$

(ii) Assume that the volume flux through the layer is constant so that

$$\int_0^h u \, \mathrm{d}z = Q,$$

where Q is a constant. Use this property to deduce that

$$\frac{\partial p}{\partial x} = \frac{A}{(h(x))^3},$$

where A is the constant given by $A = -12\mu Q$.

(c) Now suppose that, $h = h(x)$ is given by $h = C\mathrm{e}^{-Bx}$, where B and C are positive constants. Assuming that the fluid pressure at $x = 0$ exceeds that at $x = L$ by an amount $\hat{p}$, i.e. $\hat{p} = p(0) - p(L)$, show:

(i) By integrating the equation for the pressure in part (b)(ii) with respect to x between 0 and L, that the volume flux Q is given by

$$\frac{\hat{p}}{4\mu}\left(\frac{BC^3}{\mathrm{e}^{3BL} - 1}\right).$$

(ii) Using part (b)(i), the shear stress on the plane $z = 0$ is given by

$$\mu \frac{\partial u}{\partial z} = \tfrac{3}{2}\hat{p}\left(\frac{BC\mathrm{e}^{2Bx}}{\mathrm{e}^{3BL} - 1}\right).$$

(iii) The maximum shear stress on the plane occurs at $x = L$ and is given by

$$\tfrac{3}{2}\hat{p}\left(\frac{BC\mathrm{e}^{2BL}}{\mathrm{e}^{3BL} - 1}\right).$$

5.10 Lubrication theory: Shallow layer. A homogeneous incompressible viscous fluid occupies a shallow layer whose typical depth is H and horizontal extent is L; see Figure 5.2 for the setup. Suppose for this question that x and y are horizontal Cartesian coordinates and z is the vertical coordinate. Let u, v and w be the fluid velocity components in the three coordinate directions x, y and z, respectively. Suppose p is the pressure. Assume throughout that a typical horizontal velocity scale for u and v is

U and a typical vertical velocity scale for w is W. We denote by ρ, the constant density of the fluid.

(a) Assume throughout that $H \ll L$.

(i) Using that the fluid is incompressible, explain how we can deduce that W is asymptotically smaller than U.

(ii) The body force ρg in this example is due to gravity g, which we suppose acts in the negative z-direction (i.e. downwards). Using part (i), show that the three-dimensional Navier–Stokes equations reduce to the *shallow-layer* equations

$$\frac{\partial P}{\partial x} = \mu \frac{\partial^2 u}{\partial z^2}, \qquad \frac{\partial P}{\partial y} = \mu \frac{\partial^2 v}{\partial z^2} \qquad \text{and} \qquad \frac{\partial P}{\partial z} = 0,$$

where $P := p + \rho g z$ is the modified pressure and $\mu = \rho \nu$ is the first coefficient of viscosity.

(b) For the shallow fluid layer shown in Figure 5.2, assume no-slip boundary conditions on the rigid lower layer. Further suppose that the pressure p is constant (indeed take it to be zero) along the top surface of the layer at $z = h(x, y, t)$. Further suppose that surface stress forces are applied to the surface $z = h(x, y, t)$ so that

$$\mu \frac{\partial u}{\partial z} = \frac{\partial \Gamma}{\partial x} \qquad \text{and} \qquad \mu \frac{\partial v}{\partial z} = \frac{\partial \Gamma}{\partial y},$$

where $\Gamma = \Gamma(x, y)$ is a given function. Using the shallow-layer equations in part (a) and the boundary conditions, show that the height function $h(x, y, t)$ satisfies the partial differential equation

$$2\mu \frac{\partial h}{\partial t} + \frac{\partial}{\partial x}\left(h^2 \frac{\partial}{\partial x}\left(\Gamma - \tfrac{1}{3}\rho g h^2\right)\right) + \frac{\partial}{\partial y}\left(h^2 \frac{\partial}{\partial y}\left(\Gamma - \tfrac{1}{3}\rho g h^2\right)\right) = 0.$$

5.11 Lubrication theory: Hele–Shaw cell. A homogeneous incompressible fluid occupies the region between two horizontal rigid parallel planes, which are a distance h apart, and outside a rigid cylinder of radius a which intersects the planes normally; see Figure 5.7 for the setup. Suppose that for this question x and y are horizontal coordinates and z is the vertical coordinate. Assume throughout that a typical horizontal velocity scale for u and v is U and a typical vertical velocity scale for w is W.

(a) Explain very briefly why a is a typical horizontal scale for (x, y) and h a typical vertical scale for z.

(b) Hereafter further assume that $h \ll a$ and that

$$\rho U h^2 \ll a\mu,$$

where $\mu = \rho\nu$ is the first coefficient of viscosity. Using these assumptions, show that the Navier–Stokes equations for a homogeneous incompressible fluid reduce to the system of equations

$$\mu \frac{\partial^2 u}{\partial z^2} = \frac{\partial p}{\partial x}, \qquad \mu \frac{\partial^2 v}{\partial z^2} = \frac{\partial p}{\partial y} \qquad \text{and} \qquad \frac{\partial p}{\partial z} = 0,$$

for a steady flow, where $p = p(x, y, z)$ is the pressure.

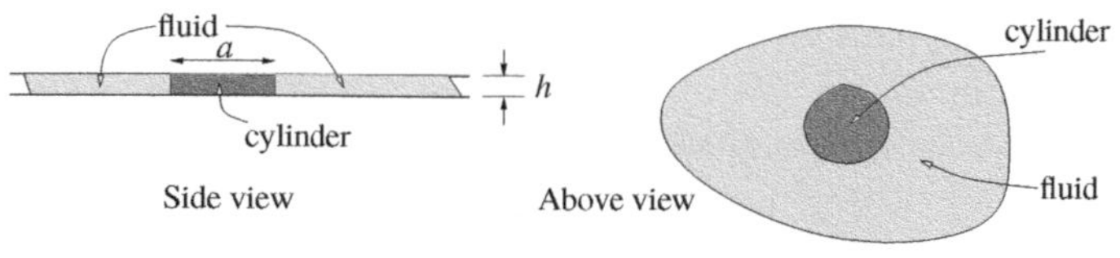

Figure 5.7 Hele–Shaw cell. A homogeneous incompressible fluid occupies the region between two parallel planes, a distance h apart, and outside the cylinder of radius a orthogonal to the planes.

(c) We define the vertically averaged velocity components $\overline{u} = \overline{u}(x, y)$ and $\overline{v} = \overline{v}(x, y)$ for the flow in part (b) by

$$\overline{u}(x, y) := \frac{1}{h}\int_0^h u(x, y, z)\,\mathrm{d}z \quad \text{and} \quad \overline{v}(x, y) := \frac{1}{h}\int_0^h v(x, y, z)\,\mathrm{d}z.$$

Use the incompressibility of $(u, v, w)^{\mathrm{T}}$ to show that the vertically averaged velocity field $(\overline{u}, \overline{v})^{\mathrm{T}}$ is *incompressible*. Use that $\partial p/\partial z = 0$ to show that $(\overline{u}, \overline{v})^{\mathrm{T}}$ is *irrotational*.

5.12 Lubrication theory: Flow between parallel cylinders. Two identical solid cylinders of radius a lie parallel to each other a distance b apart, i.e. their centres are a distance $2a + b$ apart. Oil of constant density ρ and viscosity μ is forced by a pressure difference P through the narrow gap between the parallel cylinders. Assume that $b/a \ll 1$ and $\rho b^3 P/a\mu^2 \ll 1$. Suppose the axes of symmetry of the two solid cylinders are parameterised by $\boldsymbol{x} = (0, \pm\frac{1}{2}b \pm a, z)^{\mathrm{T}}$ for $z \in \mathbb{R}$. See Figure 5.8.

(a) Assume the cylinders are fixed.

(i) Using lubrication theory, explain why the governing equation for the velocity field u in the x-direction is

$$\frac{\mathrm{d}p}{\mathrm{d}x} = \mu\frac{\partial^2 u}{\partial y^2},$$

where $p = p(x)$ is the pressure field. Explain why associated appropriate boundary conditions are $u = 0$ on $y = \pm\frac{1}{2}d$, where $d = d(x)$ is given by $d = b + 2(a - (a^2 - x^2)^{1/2})$.

(ii) Solve the boundary value problem in part (i) to show

$$u = -\frac{1}{2\mu}\left(\frac{\partial p}{\partial x}\right)\left((\tfrac{1}{2}d)^2 - y^2\right).$$

(iii) Show the total flux of oil between the cylinders $Q := \int_{-d/2}^{d/2} u\,\mathrm{d}y$ is

$$Q = -\frac{d^3}{12\mu}\left(\frac{\partial p}{\partial x}\right).$$

(iv) Hence deduce that the pressure $p = p(x)$ is given by (here A is an arbitrary constant)

$$p(x) = A - 12\mu Q\int_{-\infty}^{x}\left(b + 2(a - (a^2 - \xi^2)^{1/2})\right)^{-3}\mathrm{d}\xi.$$

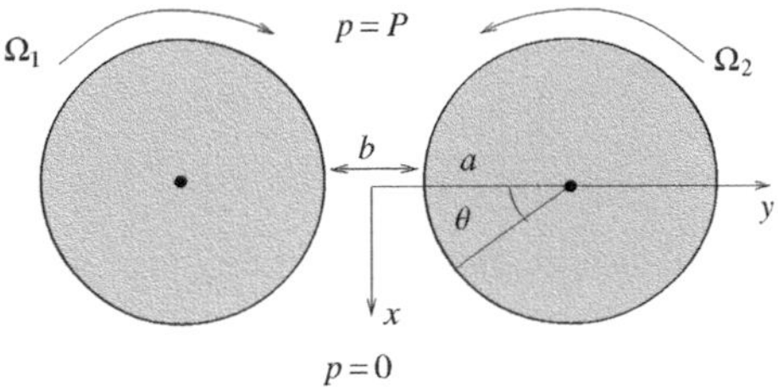

Figure 5.8 Two parallel solid cylinders lie side by side. The goal in Exercise 5.12 is to compute the volume flux for a flow between the cylinders when a pressure difference P is applied: (a) when the cylinders are fixed and $\Omega_1 = \Omega_2 = 0$; and (b) when they rotate with angular speeds Ω_1 and Ω_2 as shown.

(v) Assume that the dominant contribution to the pressure p in part (iv) will be from regions where x is small, in particular $x/a \ll 1$. Hence by approximating $(1 - \xi^2/a^2)^{1/2}$ by a Taylor series, show that approximately

$$p(x) = A - \frac{12\mu Q}{b^3} \int_{-\infty}^{x} (1 + \xi^2/(ab))^{-3} \, \mathrm{d}\xi.$$

(vi) Note that the dominant contribution to the integral in part (v) will come from the region where $|\xi| \approx (ab)^{1/2} \ll a$, justifying the Taylor approximation made there. Explain why to the same approximation we have

$$p(x) = P - \frac{12\mu Q}{b^3} \int_{-\infty}^{x} (1 + \xi^2/(ab))^{-3} \, \mathrm{d}\xi.$$

Hence by using the substitution $\xi = (ab)^{1/2} \tan\theta$ or otherwise, show that the volume flux is approximately given by

$$Q = \frac{2b^{5/2}P}{9\pi\mu a^{1/2}}.$$

(*Note:* A characteristic x-scale is $(ab)^{1/2}$ while a characteristic y-scale is b. If velocity scale U characterises the velocity magnitude between the cylinders, let $Q^* = Ub$ represent a characteristic flux scale, which must also scale like $b^{5/2}P/\mu a^{1/2}$. Then the Reynolds number here is

$$\mathrm{Re} = \frac{Ub}{\nu} = \frac{Q^*}{\nu} = \frac{\rho b^{5/2}P}{\mu^2 a^{1/2}}.$$

In lubrication theory we assume $(\mathrm{Re})\,\epsilon^2 \ll 1$, where here $\epsilon = b/(ab)^{1/2} = (b/a)^{1/2}$. Hence we see that we require $(\mathrm{Re})\,\epsilon^2 = (\rho b^3 P/\mu^2 a)(b/a)^{1/2} \ll 1$. It is sufficient to assume that the first factor is asymptotically small.)

(b) Now suppose the two cylinders rotate with angular velocities Ω_1 and Ω_2 in opposite senses.

(i) Explain why the boundary conditions are now $u = \Omega_2 a \cos\theta$ at $y = +\frac{1}{2}d$ and $u = \Omega_1 a \cos\theta$ at $y = -\frac{1}{2}d$, where θ parameterises the anticlockwise rotation angle shown in Figure 5.8.

(ii) Show that the velocity field u in this case is given by ($\overline{\Omega} := \frac{1}{2}(\Omega_1 + \Omega_2)$)

$$u = -\frac{1}{2\mu}\left(\frac{\partial p}{\partial x}\right)\left(\left(\tfrac{1}{2}d\right)^2 - y^2\right) + \frac{1}{d}(\Omega_2 - \Omega_1)\,(a\cos\theta)\,y + \overline{\Omega}\,a\cos\theta.$$

(iii) Hence show that the total flux is given by

$$Q = -\frac{d^3}{12\mu}\left(\frac{\partial p}{\partial x}\right) + a\overline{\Omega}\,\cos\theta\,d.$$

(iv) Explain why we know

$$P = 12\mu\int_{-\infty}^{\infty}\frac{Q - a\overline{\Omega}\cos\theta\,d}{d^3}\,\mathrm{d}\xi.$$

(v) Using methods analogous to those in part (a) above, explain why

$$P \approx \frac{9\mu\pi a^{1/2}Q}{2b^{5/2}} - \frac{6\mu\pi a^{3/2}\overline{\Omega}}{b^{3/2}},$$

and the change in volume flux from the case in part (a) is $\frac{2}{3}ab(\Omega_1 + \Omega_2)$.

6 Boundary-Layer Theory

The Reynolds numbers associated with flows past aircraft or ships are typically large, indeed of the order of 10^8 of 10^9; recall Remark 4.36 and Table 4.1. For individual wings or fins, the Reynolds number may be an order of magnitude or two smaller. However, such Reynolds numbers are still large and the flow around a wing for example is well approximated by Euler flow. We can imagine the flow over the top of the wing of an aircraft has a high relative velocity tangential to the surface wing directed towards the rear edge of the wing. This would appear to be consistent with no-flux boundary conditions we apply for the Euler equations – in particular there is no boundary restriction on the tangential component of the fluid velocity field. However, air or water flow is viscous. The fluid particles on the wing must satisfy viscous boundary conditions, i.e. exactly at the surface they must adhere to the wing and thus have zero velocity relative to the wing. The reconciliation of this conundrum is that there must be a *boundary layer* on the surface of the wing. By this we mean a special thin fluid layer exists between the wing and the fast-moving Euler flow past the wing. The velocity profile of the flow past the wing across the boundary layer as one measures continuously from the wing surface to the top of the boundary layer must change extremely rapidly. Indeed it must change from zero velocity relative to the wing surface to fast relative velocity (the speed of the aircraft) towards the rear of the wing; see Figure 6.1.

6.1 Prandtl Equations

We now derive an accurate model, deduced from the full incompressible Navier–Stokes equations, for the scenario just described. We assume a two-dimensional flow around an object of typical size L, for example the length from wing tip to the trailing edge. The setup is shown in Figure 6.1. We focus on the flow in a section on the wing as shown in the right panel of the figure. We assume the bulk flow is governed by the incompressible Euler equations, which is a good approximation given it is a very high Reynolds-number flow, as discussed above. We assume this bulk Euler flow is prescribed by the horizontal and vertical velocity fields $U_{\mathrm{E}} = U_{\mathrm{E}}(x, y, t)$ and $V_{\mathrm{E}} = V_{\mathrm{E}}(x, y, t)$, respectively, where $x \in \mathbb{R}$ is the horizontal parameter, positive towards the right, and $y \geqslant 0$ is the vertical parameter, positive in the vertical upwards direction. The whole flow, including the boundary layer, takes place in the upper half-plane region above the lower boundary which coincides with $y = 0$. We suppose the bulk fluid horizontal velocity field U_{E},

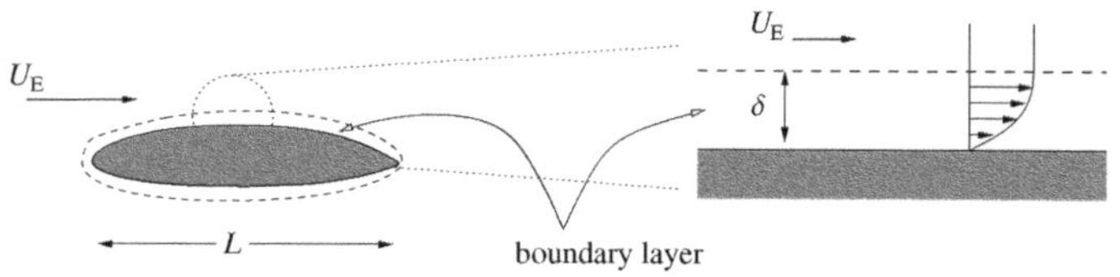

Figure 6.1 Boundary-layer theory. The flow over the wing is well approximated by an Euler flow as it is a high Reynolds-number flow. However the fluid is ultimately viscous and fluid particles at the wing surface must adhere to it. Hence there is a boundary layer between the wing and bulk flow, across which the horizontal velocity field rapidly changes from zero velocity relative to the wing to the bulk velocity U_E past the wing.

which is the relative speed of the object, has characteristic scale U. A natural time scale is thus $T = L/U$. Further, a natural underlying Reynolds number is

$$\mathrm{Re} := \frac{UL}{\nu},$$

which we assume is asymptotically large, i.e. $\mathrm{Re} \gg 1$. We return to discussing the bulk flow once we have derived the boundary-layer equations. Let us now focus on the boundary layer itself. We assume the flow is two-dimensional homogeneous incompressible Navier–Stokes flow. Suppose the velocity components $u = u(x, y, t)$ and $v = v(x, y, t)$ in the horizontal and vertical directions, respectively, have characteristic scales U and V. Here U is the same characteristic scale as that for U_E above. Let δ denote the characteristic width scale of the boundary layer. We introduce the following non-dimensional variables:

$$x' := \frac{x}{L}, \qquad y' := \frac{y}{\delta} \qquad \text{and} \qquad t' := \frac{t}{T},$$

as well as

$$u'(x', y', t') := \frac{1}{U}u(x, y, t) \quad \text{and} \quad v'(x', y', t') := \frac{1}{V}v(x, y, t).$$

We also set $p'(x', y', t') := p(x, y, t)/P$, where we are yet to determine the characteristic pressure scale P. Transforming to non-dimensional variables, the incompressibility condition becomes

$$\frac{U}{L}\left(\frac{\partial u'}{\partial x'}\right) + \frac{V}{\delta}\left(\frac{\partial v'}{\partial y'}\right) = 0 \qquad \Leftrightarrow \qquad \frac{\partial u'}{\partial x'} + \frac{LV}{\delta U}\left(\frac{\partial v'}{\partial y'}\right) = 0.$$

To maintain incompressibility we choose the vertical velocity scaling V such that $V = U\delta/L$. We now consider the Navier–Stokes momentum equation for the velocity component u, which in non-dimensional variables is given by

$$\frac{U^2}{L}\left(\frac{\partial u'}{\partial t'} + u'\frac{\partial u'}{\partial x'}\right) + \frac{VU}{\delta}\left(v'\frac{\partial u'}{\partial y'}\right) = \nu\frac{U}{L^2}\left(\frac{\partial^2 u'}{\partial x'^2}\right) + \nu\frac{U}{\delta^2}\left(\frac{\partial^2 u'}{\partial y'^2}\right) - \frac{P}{\rho L}\left(\frac{\partial p'}{\partial x'}\right).$$

Note that since $V = U\delta/L$, all the inertia terms have the same scaling. In the boundary layer we anticipate that the inertia terms and viscous vertical diffusion term are essential and should be retained. In particular, the horizontal velocity field in the boundary layer varies rapidly with respect to the vertical coordinate. Consider dividing the above equation through by $\nu U/\delta^2$. This gives

$$(\mathrm{Re})\frac{\delta^2}{L^2}\left(\frac{\partial u'}{\partial t'} + u'\frac{\partial u'}{\partial x'} + v'\frac{\partial u'}{\partial y'}\right) = \frac{\delta^2}{L^2}\left(\frac{\partial^2 u'}{\partial x'^2}\right) + \frac{\partial^2 u'}{\partial y'^2} - \frac{P\delta^2}{\mu L U}\left(\frac{\partial p'}{\partial x'}\right).$$

As mentioned above, we wish to retain the inertia terms and to do this we choose δ such that $\delta^2 = L^2/(\mathrm{Re})$. In other words, the natural characteristic scale of the boundary layer is

$$\delta = \frac{L}{(\mathrm{Re})^{1/2}}.$$

The viscous horizontal diffusion term thus has the scaling $\delta^2/L^2 = (\mathrm{Re})^{-1}$, which is asymptotically small as we assume $\mathrm{Re} \gg 1$. To maintain the pressure term we now choose P such that

$$P = \frac{\mu L U}{\delta^2} = \frac{\mu U}{L}(\mathrm{Re}),$$

since $\delta^2 = L^2/(\mathrm{Re})$. Consequently, the equation for u' becomes

$$\frac{\partial u'}{\partial t'} + u'\frac{\partial u'}{\partial x'} + v'\frac{\partial u'}{\partial y'} = (\mathrm{Re})^{-1}\left(\frac{\partial^2 u'}{\partial x'^2}\right) + \frac{\partial^2 u'}{\partial y'^2} - \frac{\partial p'}{\partial x'}.$$

In the limit $\mathrm{Re} \to \infty$ the viscous horizontal diffusion term vanishes and we deduce that at leading order $u' = u'(x', y', t')$ satisfies

$$\frac{\partial u'}{\partial t'} + u'\frac{\partial u'}{\partial x'} + v'\frac{\partial u'}{\partial y'} = \frac{\partial^2 u'}{\partial y'^2} - \frac{\partial p'}{\partial x'}.$$

Now consider the Navier–Stokes momentum equation for the velocity component v, which in non-dimensional variables is given by

$$\frac{UV}{L}\left(\frac{\partial v'}{\partial t'} + u'\frac{\partial v'}{\partial x'}\right) + \frac{V^2}{\delta}\left(v'\frac{\partial v'}{\partial y'}\right) = \nu\frac{V}{L^2}\left(\frac{\partial^2 v'}{\partial x'^2}\right) + \nu\frac{V}{\delta^2}\left(\frac{\partial^2 v'}{\partial y'^2}\right) - \frac{P}{\rho\delta}\left(\frac{\partial p'}{\partial y'}\right).$$

Note that since $V = \delta U/L$ we see all the inertia terms have the same scaling, namely UV/L. And since $P = \mu U(\mathrm{Re})/L$, we see that the pressure term has the scaling

$$\frac{P}{\rho\delta} = \frac{\mu U(\mathrm{Re})}{L\rho\delta} = \frac{U^2}{\delta},$$

using that $\mathrm{Re} := UL/\nu$ and $\nu = \mu/\rho$. Now we divide the equation for v' through by U^2/δ. Using $V = \delta U/L$ and $\delta^2/L^2 = (\mathrm{Re})^{-1}$, the scalings of the inertia, viscous horizontal and vertical diffusion terms therein respectively become

$$\left(\frac{UV}{L}\right)\left(\frac{\delta}{U^2}\right) = \frac{\delta V}{LU} = \frac{\delta^2}{L^2} = (\mathrm{Re})^{-1},$$

$$\left(\nu\frac{V}{L^2}\right)\left(\frac{\delta}{U^2}\right)=\left(\frac{\nu}{UL}\right)\left(\frac{\delta V}{LU}\right)=(\mathrm{Re})^{-2},$$
$$\left(\nu\frac{V}{\delta^2}\right)\left(\frac{\delta}{U^2}\right)=\nu\frac{V}{\delta U^2}=\frac{\nu}{UL}=(\mathrm{Re})^{-1}.$$

Hence the equation for v' becomes

$$(\mathrm{Re})^{-1}\left(\frac{\partial v'}{\partial t'}+u'\frac{\partial v'}{\partial x'}+v'\frac{\partial v'}{\partial y'}\right)=(\mathrm{Re})^{-2}\left(\frac{\partial^2 v'}{\partial x'^2}\right)+(\mathrm{Re})^{-1}\left(\frac{\partial^2 v'}{\partial y'^2}\right)-\frac{\partial p'}{\partial y'}.$$

Thus, using the scalings we established from the equation for u' for V, δ and P to respectively retain thc incompressibility condition, inertia and pressure terms therein, we arrive at the corresponding scalings above for the equation for v'. Compared to the pressure term, all the other terms are asymptotically small and vanish in the limit $\mathrm{Re}\to\infty$. Indeed, at leading order we deduce

$$\frac{\partial p'}{\partial y'}=0.$$

Hence in the limit $\mathrm{Re}\to\infty$, at leading order, in the boundary layer of characteristic thickness $\delta=L/(\mathrm{Re})^{1/2}$, the equations for u' and v', together with the incompressibility condition, constitute the following system.

Definition 6.1 (Prandtl boundary layer equations; Prandtl, 1904) In the boundary layer, the non-dimensional horizontal and vertical velocity field components $u'=u'(x',y',t')$ and $v'=v'(x',y',t')$, respectively, and the pressure field $p'=p'(x',y',t')$ satisfy the *Prandtl boundary layer equations*:

$$\frac{\partial u'}{\partial t'}+u'\frac{\partial u'}{\partial x'}+v'\frac{\partial u'}{\partial y'}=\frac{\partial^2 u'}{\partial y'^2}-\frac{\partial p'}{\partial x'},$$
$$\frac{\partial p'}{\partial y'}=0,$$
$$\frac{\partial u'}{\partial x'}+\frac{\partial v'}{\partial y'}=0,$$

together with the boundary conditions $u'(x',0,t')=0$ and $v'(x',0,t')=0$.

Remark 6.2 We also assume, for given data $u_0'=u_0'(x',y')$ and $v_0'=v_0'(x',y')$, the initial conditions $u_0'(x',y',0)=u_0'(x',y')$ and $v_0'(x',y',0)=v_0'(x',y')$. For consistency we require $v_0'(x',0)=0$, but do not assume a similar constraint on u_0' as we do not wish to preclude examples where the lower boundary is set in impulsive horizontal motion at $t=0$.

The second Prandtl boundary-layer equation implies the pressure field in the boundary layer is independent of y', so $p=p(x',t')$ only. Hence if the pressure field or more particularly $\partial p'/\partial x'$ can be determined at the top boundary of the boundary layer, then it is determined inside the boundary layer. With this knowledge, we note that the system of equations represented by the first and third Prandtl equations above is third order with respect to y' as the first equation is a second-order partial differential equation with respect to y' for u' while the third equation is first order with respect to y' for v'. We have already specified one boundary condition for u' at $y' = 0$ as well as one boundary condition for v' at $y' = 0$. We thus anticipate that we require an additional boundary condition in the y' direction for u'. This is naturally provided by matching the boundary layer flow for u' with the bulk Euler flow outside the boundary layer.

The bulk incompressible Euler flow outside the boundary layer, in dimensional variables, is given by

$$\begin{aligned}\frac{\partial U_{\mathrm{E}}}{\partial t} + U_{\mathrm{E}}\frac{\partial U_{\mathrm{E}}}{\partial x} + V_{\mathrm{E}}\frac{\partial U_{\mathrm{E}}}{\partial y} &= -\frac{1}{\rho}\frac{\partial P_{\mathrm{E}}}{\partial x},\\ \frac{\partial V_{\mathrm{E}}}{\partial t} + U_{\mathrm{E}}\frac{\partial V_{\mathrm{E}}}{\partial x} + V_{\mathrm{E}}\frac{\partial V_{\mathrm{E}}}{\partial y} &= -\frac{1}{\rho}\frac{\partial P_{\mathrm{E}}}{\partial y},\\ \frac{\partial U_{\mathrm{E}}}{\partial x} + \frac{\partial V_{\mathrm{E}}}{\partial y} &= 0,\end{aligned}$$

where $P_{\mathrm{E}} = P_{\mathrm{E}}(x, y, t)$ is the pressure field in the bulk Euler flow. We assume 'no normal flow' boundary conditions at $y = 0$, so $V_{\mathrm{E}}(x, 0, t) = 0$. For given data $U_0^{\mathrm{E}} = U_0^{\mathrm{E}}(x, y)$ and $V_0^{\mathrm{E}} = V_0^{\mathrm{E}}(x, y)$ with $V_0^{\mathrm{E}}(x, 0) = 0$, we assume $U_{\mathrm{E}}(x, y, 0) = U_0^{\mathrm{E}}(x, y)$ and $V_{\mathrm{E}}(x, y, 0) = V_0^{\mathrm{E}}(x, y)$. In principle we can solve the initial boundary value problem for the two-dimensional incompressible Euler equations above for U_{E}, V_{E} and P_{E}. Thus we can determine the fields $U_{\mathrm{E}}(x, 0, t)$, $V_{\mathrm{E}}(x, 0, t)$ and $P_{\mathrm{E}}(x, 0, t)$ at the lower boundary $y = 0$. At the lower boundary $V_{\mathrm{E}}(x, 0, t) = 0$ and these fields, $U_{\mathrm{E}}(x, 0, t)$, $V_{\mathrm{E}}(x, 0, t)$ and $P_{\mathrm{E}}(x, 0, t)$, must satisfy the compatability conditions (see e.g. Hunter, 2004)

$$\begin{aligned}\frac{\partial U_{\mathrm{E}}}{\partial t} + U_{\mathrm{E}}\frac{\partial U_{\mathrm{E}}}{\partial x} &= -\frac{1}{\rho}\frac{\partial P_{\mathrm{E}}}{\partial x},\\ 0 &= -\frac{1}{\rho}\frac{\partial P_{\mathrm{E}}}{\partial y},\\ \frac{\partial U_{\mathrm{E}}}{\partial x} + \frac{\partial V_{\mathrm{E}}}{\partial y} &= 0.\end{aligned}$$

We can now complete *matching* the Prandtl boundary layer flow with the bulk flow. In terms of the non-dimensional variables x', y' and t' given above, we set $U'_{\mathrm{E}}(x', y', t') := U_{\mathrm{E}}(x, y, t)/U$. Recall the non-dimensionalised coordinate $y' = (\mathrm{Re})^{1/2} y/L$. In the limit $\mathrm{Re} \to \infty$, the top boundary corresponds to $y' \to \infty$. Thus the third boundary condition for the Prandtl boundary layer equations we impose is, as $y' \to \infty$,

$$u' \to U'_{\mathrm{E}}(x', 0, t').$$

In principle we can thus now solve the Prandtl boundary-layer equations. Some further observations are helpful. We have already remarked that in the boundary-layer equations $p' = p'(x', t')$ only. In the limit $y' \to \infty$ from the Prandtl equation for u' we must have the compatability condition

$$\frac{\partial U'_{\mathrm{E}}}{\partial t'} + U'_{\mathrm{E}}\frac{\partial U'_{\mathrm{E}}}{\partial x'} = -\frac{\partial p'}{\partial x'},$$

where here $U_E' = U_E'(x', 0, t')$. Since $p' = p'(x', t')$ only, the information on the left explicitly determines $\partial p'/\partial x'$. From another perspective, set $P_E'(x', y', t') := P_E(x, y, t)/P$, where $P = \mu U(\text{Re})/L$. Since p' and $\partial p'/\partial x'$ are independent of y', they are determined by their limiting values as $y' \to \infty$, i.e. we must have $\partial p'/\partial x' = (\partial P_E'/\partial x')(x', 0, t')$. The compatibility condition for $U_E(x, 0, t)$ above, converted to non-dimensional coordinates, has the form

$$\frac{\partial U_E'}{\partial t'} + U_E' \frac{\partial U_E'}{\partial x'} = -\frac{\partial P_E'}{\partial x'}.$$

We reach the same conclusion as just above, that $\partial p'/\partial x'$ is completely determined by $U_E'(x, 0, t)$, or equivalently $U_E(x, 0, t)$, via the expression on the left above. We can thus express the Prandtl boundary layer equations in the form

$$\frac{\partial u'}{\partial t'} + u' \frac{\partial u'}{\partial x'} + v' \frac{\partial u'}{\partial y'} = \frac{\partial^2 u'}{\partial y'^2} + \frac{\partial U_E'}{\partial t'} + U_E' \frac{\partial U_E'}{\partial x'},$$

$$\frac{\partial u'}{\partial x'} + \frac{\partial v'}{\partial y'} = 0,$$

where here $U_E' = U_E'(x', 0, t')$. The boundary conditions are $u'(x', 0, t') = 0$, $v'(x', 0, t') = 0$ and $u' \to U_E'(x', 0, t')$ as $y' \to \infty$.

Remark 6.3 We implicitly assumed the lower surface is flat. If it has curvature then $\partial p'/\partial y'$ is not zero, corresponding to some centripetal acceleration. We discuss some important consequences of this (separation) presently.

Remark 6.4 (Vertical velocity at the top boundary) A natural unresolved question so far concerns the character of the vertical velocity v' close to the top of the boundary layer, i.e. in the limit $y' \to \infty$. For the bulk incompressible Euler flow as $y \to 0$ we know the following. Using the Taylor expansion for $V_E = V_E(x, y, t)$ about $y = 0$, combined with the boundary condition $V_E(x, 0, t) = 0$, and using the incompressibility compatibility condition above for $U_E(x, 0, t)$ and $V_E(x, 0, t)$, we see as $y \to 0$, for the bulk flow we have

$$V_E(x, y, t) \sim -\left(\frac{\partial U_E}{\partial x}\right) y,$$

where $U_E = U_E(x, 0, t)$. Hence, to first order for the bulk flow, the normal velocity decays linearly with y to zero in the limit $y \to 0$, i.e. as we approach the boundary. Now consider, from the Prandtl boundary-layer equations, how v' behaves near the same boundary, corresponding to $y' \to \infty$. Following Panton (2013, pp. 557–8), using the incompressibilty condition from the Prandtl equations, we observe since $v'(x', 0, t') = 0$ that $v' = v'(x', y', t')$ is given by

$$v' = \int_0^{y'} \frac{\partial v'}{\partial \tilde{y}'} \, d\tilde{y}' = -\int_0^{y'} \frac{\partial u'}{\partial x'} \, d\tilde{y}' = \frac{\partial}{\partial x'} \int_0^{y'} (U_E' - u') \, d\tilde{y}' - \left(\frac{\partial U_E'}{\partial x'}\right) y',$$

where here $U_E' = U_E'(x', 0, t')$. Hence in the limit $y' \to \infty$ we observe

$$v'(x', y', t') \sim -\left(\frac{\partial U_E'}{\partial x'}\right) y' + V^*(x', t'),$$

where

$$V^*(x',t') := \frac{\partial}{\partial x'}\int_0^\infty (U'_{\mathrm{E}} - u')\,\mathrm{d}\tilde{y}'.$$

Note at first order the term '$-(\partial U'_{\mathrm{E}}/\partial x')\,y'$' matches, once converted to dimensional coordinates, the behaviour of the bulk flow given just above as $y \to 0$. The term $V^*(x',t')$ represents a correction term which can be related to the so-called *displacement thickness* δ^* as follows; see Panton (2013, pp. 538–9) and Lighthill (1958). We focus on the boundary layer flow and use the non-dimensional boundary-layer coordinates (x',y',t'). Due to viscosity the bulk far-field horizontal velocity field $U'_{\mathrm{E}}(x',0,t')$ at $y'=\infty$ is smoothly retarded to zero approaching the boundary $y'=0$. Hence there is a loss of horizontal flux of fluid compared to the situation of there being no viscous effects at all and the bulk inviscid velocity field $U'_{\mathrm{E}}(x',0,t')$ extended all the way down to $y'=0$. The loss of horizontal fluid flux due to viscosity is finite and given by

$$\int_0^\infty \big(U'_{\mathrm{E}}(x',0,t') - u'(x',y',t')\big)\,\mathrm{d}\tilde{y}'.$$

Solely in terms of the bulk inviscid velocity field $U'_{\mathrm{E}}(x',0,t')$, the loss of horizontal fluid flux due to viscosity is represented by a fluid flux

$$\int_0^{\delta^*} U'_{\mathrm{E}}(x',0,t')\,\mathrm{d}\tilde{y}' = \delta^*\,U'_{\mathrm{E}}(x',0,t'),$$

for some displacement thickness δ^*; see Figure 6.2. We can also interpret the displacement thickness δ^* as, from Kundu and Cohen (2008, p. 346): 'the distance by which the wall would have to to be displaced outward in a hypothetical frictionless flow so as to maintain the same mass flux as in the actual flow'. Equating the last two fluxes above, we deduce

$$\delta^* = \int_0^\infty \left(1 - \frac{u'(x',y',t')}{U'_{\mathrm{E}}(x',0,t')}\right)\mathrm{d}\tilde{y}'.$$

Hence, going back to our characterisation of v' as $y' \to \infty$, we see that

$$v'(x',y',t') \sim -\left(\frac{\partial U'_{\mathrm{E}}}{\partial x'}\right) y' + \frac{\partial}{\partial x'}(U'_{\mathrm{E}}\,\delta^*),$$

where $U'_{\mathrm{E}} = U'_{\mathrm{E}}(x',0,t')$. The second term above involving the displacement thickness is 'often viewed as a correction effect that the boundary layer imposes on the inviscid flow'; see Panton (2013, p. 558).

Remark 6.5 (Steady Prandtl boundary-layer equations) In the case of steady Prandtl boundary-layer flow, the boundary-layer equations look much like those in Definition 6.1 though the time-derivative term $\partial u'/\partial t'$ is absent in the equation for u'. The fields u' and v' still satisfy no-slip boundary conditions at $y'=0$ and we match u' in the limit $y' \to \infty$ to the steady bulk incompressible Euler flow involving U_{E}, i.e. we assume $u' \to U'_{\mathrm{E}}$ as $y' \to \infty$, where $U'_{\mathrm{E}} = U'_{\mathrm{E}}(x',0)$. The pressure gradient term is

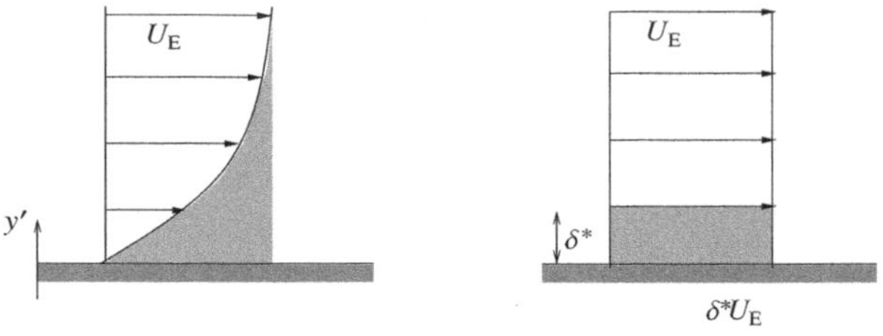

Figure 6.2 The decrease in the fluid volume flux in the boundary layer due to viscous effects, compared to there being no viscosity and thus boundary layer, is represented by the shaded region in the left panel. The corresponding equal fluid volume flux loss in a hypothetical inviscid fluid flow that stretches down to the lower boundary is represented by the shaded region on the right. The thickness of the region is the displacement thickness δ^*.

given by $\mathrm{d}p'/\mathrm{d}x' = -U'_E\,(\partial U'_E/\partial x')$. The equation for u' now has a parabolic character with the x' variable having a time-like character. Hence we need an 'initial condition' in the sense that we need to impose for some finite x'_0 that $u'(x'_0, y') = u'_0(y')$ for a given function $u'_0 = u'_0(y')$.

Remark 6.6 (Matching procedures) A brief survey of possible matching strategies can be found in Chorin and Marsden (1990, Sec. 2.2). See the end of Section 6.3 for a scenario when matching with three layers is required.

Remark 6.7 (Dimensional Prandtl equations) It is common, as demonstrated in some of the later exercises at the end of this chapter, to revert to the dimensional form for the Prandtl equations, namely:

$$\frac{\partial u}{\partial t} + u\frac{\partial u}{\partial x} + v\frac{\partial u}{\partial y} = \nu\frac{\partial^2 u}{\partial y^2} + \frac{\partial U_E}{\partial t} + U_E\frac{\partial U_E}{\partial x},$$

$$\frac{\partial u}{\partial x} + \frac{\partial v}{\partial y} = 0,$$

together with the boundary conditions $u(x, 0, t) = 0$ and $v(x, 0, t) = 0$. Then a solution $u = u(x, \eta, t)$ is sought which depends on t, x and $\eta := y/\delta$, with the matching requirement $u \to U_E$ as $\eta \to \infty$. See the next Section 6.2 for an immediate example of this. Typically the bulk velocity $U_E(x, 0, t)$ is assumed to be steady and abbreviated to $U_E = U_E(x)$.

6.2 Blasius Problem

The following problem was considered by Blasius (1908). Consider a steady boundary-layer flow on a semi-infinite flat plate as shown in Figure 6.3. The leading edge of the plate coincides with the origin while the plate itself lies along the positive x-axis. The y-axis is orthogonal to the plate, as shown in the figure. The plate and flow is assumed to be uniform in the direction which is orthogonal to both the x- and y-directions. We focus on the flow in the $x \geqslant 0$ and $y \geqslant 0$ region. We suppose a uniform horizontal

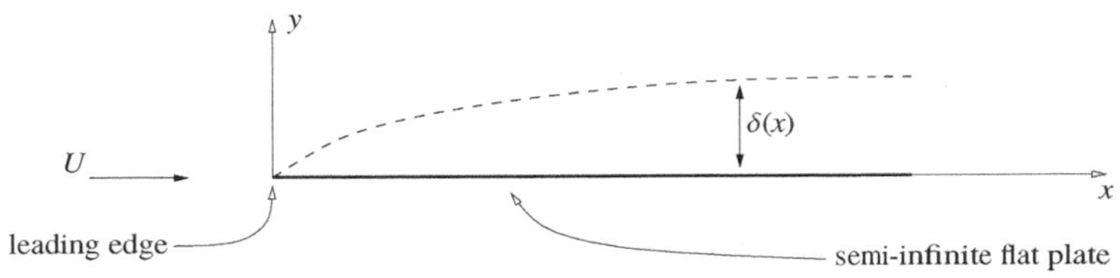

Figure 6.3 Blasius problem. This considers a steady boundary-layer flow over a semi-infinite flat plate, uniform in the direction orthogonal to the x- and y-axes shown.

flow of velocity U in the positive x-direction washes over the plate. We assume a steady boundary layer flow develops on the plate, with the extent of the boundary layer as shown in the figure. Since the plate is semi-infinite, coinciding with $x \geqslant 0$, there is no imposed horizontal scale. In this scenario we follow the arguments in the general theory above, but with L replaced by X. Here we think of X as a characteristic horizontal scale. It also independently parameterises horizontal displacement. The Reynolds number here is thus $\mathrm{Re} := UX/\nu$. To maintain incompressibility we choose $V = U\delta/X$. Further we should choose δ so

$$\delta := \frac{X}{(\mathrm{Re})^{1/2}} = \left(\frac{\nu X}{U}\right)^{1/2}.$$

Since the Euler flow outside the boundary is uniformly U, the pressure gradient $\partial p/\partial x$ is zero outside the boundary layer. By the arguments in the general theory above, it is also zero inside the boundary layer. Hence the Prandtl boundary-layer equations in dimensional form for $u = u(x, y)$ and $v = v(x, y)$ are

$$u\frac{\partial u}{\partial x} + v\frac{\partial u}{\partial y} = \nu\frac{\partial^2 u}{\partial y^2},$$

$$\frac{\partial u}{\partial x} + \frac{\partial v}{\partial y} = 0.$$

The boundary conditions on the flat plate, for all $x \geqslant 0$, are

$$u(x,0) = 0 \qquad \text{and} \qquad v(x,0) = 0.$$

In the boundary layer, the natural non-dimensional vertical coordinate is

$$\eta := \frac{y}{\delta},$$

where here we have $\delta = \delta(x)$ with $\delta(x) := (\nu x/U)^{1/2}$. As the flow is incompressible and two-dimensional there is a stream function $\psi = \psi(x, y)$ such that $u = \partial\psi/\partial y$ and $v = -\partial\psi/\partial x$. We seek a similarity solution of the form

$$\psi = U\delta f(\eta).$$

Directly computing the partial derivatives, we find

$$u = Uf' \qquad \text{and} \qquad v = U\delta'(\eta f' - f),$$

where we emphasise that the primes shown here are ordinary derivatives, either with respect to x in the case of δ, or with respect to η in the case of f. Further direct computation using the chain rule implies

$$\frac{\partial u}{\partial x} = -\frac{yUf''\delta'}{\delta^2}, \qquad \frac{\partial u}{\partial y} = \frac{Uf''}{\delta} \qquad \text{and} \qquad \frac{\partial^2 u}{\partial y^2} = \frac{Uf'''}{\delta^2}.$$

Substituting the forms for u, v, $\partial u/\partial x$, $\partial u/\partial y$ and $\partial^2 u/\partial y^2$ into the first Prandtl boundary-layer equation above, we find that $f = f(\eta)$ satisfies the third-order nonlinear ordinary differential equation

$$f''' + \tfrac{1}{2} f f'' = 0,$$

where we used that $U\delta'\delta/\nu = 1/2$ since $\delta = (\nu x/U)^{1/2}$. This is supplemented with the boundary conditions on the flat plate corresponding to $f(0) = 0$ and $f'(0) = 0$. The final boundary condition from the Prandtl boundary-layer equations is $u \sim U$ as $\eta \to \infty$, where recall $\eta := y/\delta$ is the non-dimensionalised vertical coordinate which is equivalent to y' in Section 6.1. Hence this boundary condition corresponds to $f' \to 1$ as $\eta \to \infty$. We can solve this ordinary differential boundary value problem for f numerically. See Figure 6.4, which shows the solution f and its derivatives f' and f'', satisfying all three boundary conditions.

Remark 6.8 (Region of validity) In the Blasius problem we used Prandtl's boundary-layer equations, which are valid provided $\mathrm{Re} \gg 1$, or in other words provided $X \gg \nu/U$. Hence the solution is only valid sufficiently downstream of the leading edge. See e.g. Kundu and Cohen (2008, p. 352).

Remark 6.9 (Falkner–Skan similarity solutions) When the driving bulk flow $U_{\mathrm{E}}(x, 0, t)$ is non-uniform though stationary, i.e. independent of t, do classes of similarity solutions to the Prandtl boundary-layer equations still exist? This question was investigated by Falkner and Skan (1931). See e.g. Panton (2013, Sec. 20.5) or Ockendon and Ockendon (1995, pp. 32–3) for more details, as well as Exercise 6.7 below.

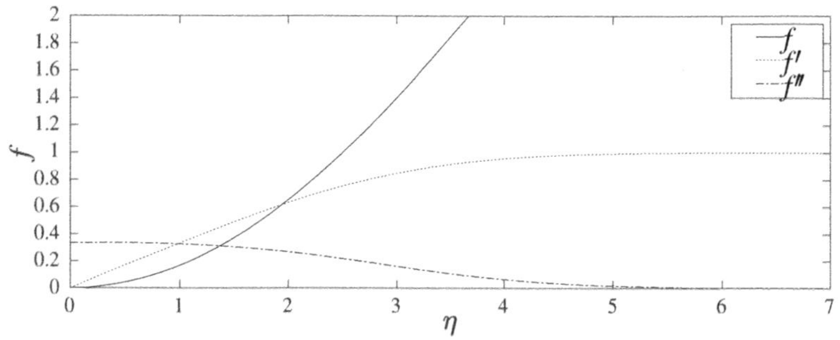

Figure 6.4 Blasius problem. We plot the solution $f = f(\eta)$, and its derivatives, to the ordinary differential boundary value problem arising in the search for a similarity solution $\psi = U\delta f$ to the Blasius problem.

6.3 Separation

From the beginning, Prandtl's boundary-layer theory has been very successful and a major tool in fluid mechanics. However, it does have its limitations. One is that it does not encompass the phenomenon of boundary-layer separation. We briefly address this issue here. For more general issues on the well-posedness of Prandtl's equations and their emulation of a full incompressible Navier–Stokes flow, see Section 6.4 at the end of this chapter.

Boundary-layer separation is an everyday phenomenon not captured by the boundary-layer theory outlined above. Consider a boundary-layer problem in which the bulk Euler flow pressure $P_{\mathrm{E}}(x,0,t)$ varies with x. We expect this to happen, for example, when the surface of the body is curved. Imagine a setup as shown in Figure 6.5, and in particular the profile of the horizontal boundary-layer velocity u' impinging on the body, monotonically increasing from zero to $U'_{\mathrm{E}}(x',0,t')$ as $y' \to \infty$. Upstream of the curved surface of the body the streamlines in the bulk Euler flow converge, resulting in an increase in $U_{\mathrm{E}}(x,0,t)$ and thus a decrease in the bulk flow pressure, so $\partial P_{\mathrm{E}}/\partial x < 0$ in the upstream region. Downstream of the curved surface of the body the streamlines in the bulk flow diverge, resulting in a decrease in $U_{\mathrm{E}}(x,0,t)$ and thus an increase in the bulk flow pressure, so $\partial P_{\mathrm{E}}/\partial x > 0$. At the body surface, which we denote by $y' = 0$, using the no-slip boundary conditions, the Prandtl boundary-layer equation for $u' = u'(x',0,t')$ becomes

$$\frac{\partial^2 u'}{\partial y'^2} = \frac{\partial p'}{\partial x'},$$

where $p' = p'(x',0,t')$ and $\partial p'/\partial x'$ is given by

$$\frac{\partial p'}{\partial x'} = \frac{\partial U'_{\mathrm{E}}}{\partial t'} + U'_{\mathrm{E}}\frac{\partial U'_{\mathrm{E}}}{\partial x'},$$

and $U'_{\mathrm{E}} = U'_{\mathrm{E}}(x',0,t')$. Note that we use x' to parameterise distance along the body surface. We observe from the equation for u' at the body surface above that the sign of $\partial p'/\partial x'$ coincides with that of $\partial^2 u'/\partial y'^2$. The profile of $u' = u'(x',y',t')$ with respect to y' always has the characteristic that it is zero at $y' = 0$ and smoothly matches $U'_{\mathrm{E}}(x',0,t') = U_{\mathrm{E}}(x,0,t)/U$ for large y'. Examining the profile of the impinging horizontal velocity field u', we see that it monotonically increases to U'_{E} as y' increases. In particular, $\partial u'/\partial y'$ is always positive but decreasing with increasing y' and so $\partial^2 u'/\partial y'^2 < 0$.

Recall that in the boundary layer $\partial p'/\partial x'$ everywhere matches $\partial P'_{\mathrm{E}}/\partial x'$ from the bulk Euler flow, where $P'_{\mathrm{E}}(x',t') = P_{\mathrm{E}}(x,t)/P$. Indeed, if $\partial P_{\mathrm{E}}/\partial x < 0$ then $\partial p'/\partial x' < 0$ and at the body surface $y' = 0$ we know $\partial^2 u'/\partial y'^2 < 0$. All the signs are reversed if $\partial P_{\mathrm{E}}/\partial x > 0$. From our arguments above for a curved body surface, the pressure gradient $\partial P_{\mathrm{E}}/\partial x$ changes sign from negative upstream to positive downstream. This means that $\partial^2 u'/\partial y'^2$ at $y' = 0$ must change sign at some x'. For large y' close to the boundary-layer edge, we always expect $\partial u'/\partial y' > 0$ but $\partial^2 u'/\partial y'^2 < 0$. See for example the sequence u' profiles in Figure 6.5. We conclude that an *inflection point* where $\partial^2 u'/\partial y'^2 = 0$ at some finite height $y' \geqslant 0$ must develop.

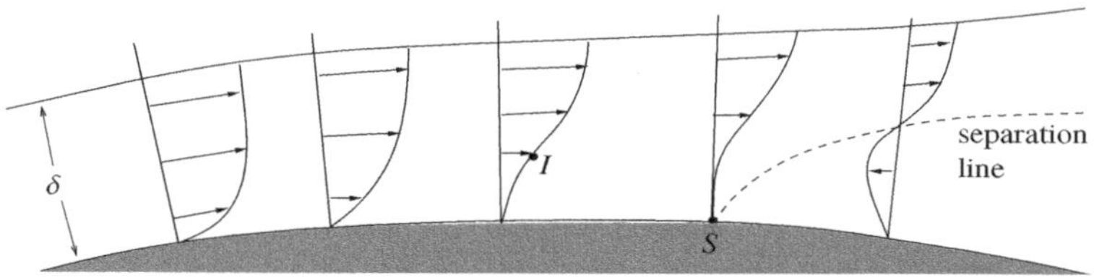

Figure 6.5 An illustration of how boundary-layer separation occurs. An inviscid bulk flow drives the boundary-layer flow at the top of the boundary layer, with no-slip boundary conditions maintained on the curved surface of the body. The development of the flow illustrated by the y'-profile of the horizontal velocity field u' with distance x' along the body surface is shown. Once an adverse, i.e. positive, pressure gradient $\partial P_E/\partial x$ develops in the bulk flow downstream of the curved surface of the body, the inflection point I develops. When the adverse pressure gradient is strong enough, a reverse flow develops. A heuristic locale for the separation point is the point S shown, where $\partial u'/\partial y' = 0$ on the body surface $y' = 0$. To the left $\partial u'/\partial y' > 0$ on the body surface, while to the right of S we observe $\partial u'/\partial y' < 0$ on the body surface.

Also note that as $\partial P_E/\partial x$ changes sign as the flow traverses the curved body, the boundary-layer thickness increases; see Figure 6.5. We can also infer this from the continuity equation, see Kundu and Cohen (2008, p. 366), which implies

$$v'(x', y', t') = -\int_0^{y'} \frac{\partial u'}{\partial x'}(x', \tilde{y}', t')\,\mathrm{d}\tilde{y}'.$$

As the bulk flow enters the section downstream, U_E starts to decrease as a function of x, i.e. $\partial U_E/\partial x$ becomes negative. Due to the viscous effects in the boundary layer, this influences the velocity field u' in the boundary layer, and $-\partial u'/\partial x'$ becomes positive. Hence from the formula above v' will become positive and thus directed away from the surface of the body. Not only does the boundary layer become thicker due to viscous diffusion, in this section of the flow there is also advection away from the surface and thus an increase in boundary-layer thickness with x'; see Kundu and Cohen (2008, pp. 364–8).

The development of the inflection point in the y' profile of u' implies a slow-down in u' in the region close to the body surface; see Figure 6.5. If the adverse pressure gradient in the bulk flow becomes sufficiently strong, i.e. the positive pressure gradient becomes large enough, then the flow u' close to the body surface will necessarily reverse direction. This is explained as follows. Note of course that at the body surface $y' = 0$, we have $u' = 0$ and $\partial p'/\partial x' = \partial^2 u'/\partial y'^2$. Recall Prandtl's equation for $u' = u'(x', y', t')$, which is

$$\frac{\partial u'}{\partial t'} + u'\frac{\partial u'}{\partial x'} + v'\frac{\partial u'}{\partial y'} = \frac{\partial^2 u'}{\partial y'^2} - \frac{\partial p'}{\partial x'}.$$

Away from the body surface where $y' > 0$, the term $\partial^2 u'/\partial y'^2$ necessarily decreases, indeed it becomes zero at the inflection point. However $\partial p'/\partial x'$ remains unchanged as it is uniform across the boundary layer. Hence if $\partial p'/\partial x'$ is sufficiently large and positive so $-\partial p'/\partial x'$ becomes large and negative, the acceleration terms on the left in

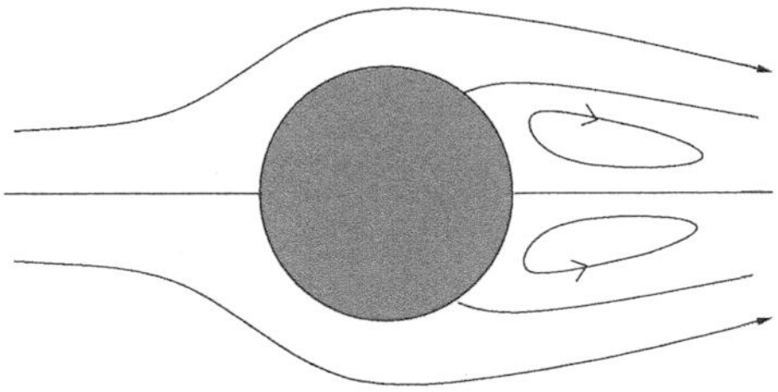

Figure 6.6 For a flow past a blunt body such as a circular cylinder, where strong adverse pressure gradients may develop in the bulk incompressible Euler flow past the body, boundary-layer separation occurs relatively quickly. In contrast, for a streamlined body with a thin trailing body surface, such strong adverse pressure gradients in the bulk flow can be avoided and no separation may occur at all.

Prandtl's equation for u' above will represent deceleration terms. If the deceleration is large enough and applied sufficiently long enough, the fluid particles will decelerate to zero horizontal velocity and then indeed reverse their direction of motion. Hence there is a region of backwards flow near the body surface – u' will be negative for some interval $[0, y'_*]$ of the y' domain. Along the body surface the backwards flow will meet the forwards flow at some point S, where the 'fluid is transported out into the mainstream'; see Kundu and Cohen (2008, p. 366). In other words, the boundary-layer flow 'separates' from the body surface at S. From Figure 6.5 we see that S coincides with the point on the body surface where

$$\left(\frac{\partial u'}{\partial y'}\right)(x', 0, t') = 0.$$

Since at the body surface $\partial v'/\partial x' \equiv 0$, and the deviatoric stress $\hat{\sigma} = 2\mu(\partial u'/\partial y')$, at the separation point the deviatoric stress is zero.

Remark 6.10 Assuming full knowledge of the boundary-layer flow (u', v'), we have given the characterisation above for the separation point. However, so far, that this is indeed its locale has not been rigorously justified. In practice the geometry of the body surface and the character of the boundary-layer flow, whether it is laminar or turbulent, respectively whether the Reynolds number is relatively low or large, heavily influences the locale of the point of separation. For example, for a flow past a blunt body, such as that indicated in Figure 6.6, there will be a steep adverse pressure gradient $\partial P_{\mathrm{E}}/\partial x > 0$ relatively early in the flow around the body and thus separation occurs relatively early on. On the other hand, if the body is more streamlined with a thin trailing body surface, then strong adverse pressure gradients can be avoided and separation may not occur. Since separation increases drag, this streamlined design is more desirable – separation changes the effective shape of the body and modifies the bulk Euler flow. In experiments, the contrast between laminar and turbulent boundary-layer flows on the influence of the separation point is readily illustrated by flows past a circular cylinder as follows. Flow experiments show that for lower Reynolds numbers when the boundary-layer flow is laminar, the separation point is insensitive to the Reynolds number. However, for larger Reynolds numbers when the boundary-layer

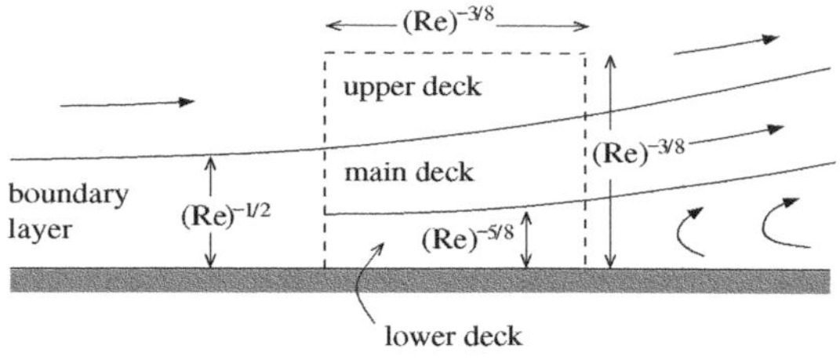

Figure 6.7 We illustrate the triple-deck structure. This is a reproduction of Fig. 2 from Stewartson, K. (1981) D'Alembert's paradox, *SIAM Review* **23**(3), pp. 308–343. The Prandtl boundary layer which scales like $(\text{Re})^{-1/2}$ in the non-dimensional y'-coordinate, where $y' = y/\delta$ and $\delta = L/(\text{Re})^{1/2}$, is split into a lower and main deck as shown. The matching between the three layers takes place in the $(\text{Re})^{-3/8} \times (\text{Re})^{-3/8}$ region shown in non-dimensional coordinates (x', y'). The thinner lower deck with scale $(\text{Re})^{-5/8}$ includes the separation and reverse flow. The main deck is the continuation of the upstream Prandtl boundary layer which is displaced away from the body surface. The upper deck denotes the bulk flow region affected by the boundary-layer effects from the other two decks.

flow is turbulent, then the separation point is delayed further downstream. See Kundu and Cohen (2008, pp. 366–70) or Panton (2013, pp. 570–71) for more details.

To deal with separation, so-called *triple-deck theory* was introduced in the 1950s and further developed in the 1960s and 1970s; see Stewartson (1981) for a review. To account for separation, the boundary layer is divided into two parts known as the lower and main decks. The third deck is the region of the bulk flow close to the main deck most affected by the boundary layer. Recall at the end of Section 6.1 that the boundary layer imposes correction effects on the bulk flow, which is particularly significant when separation occurs. See Figure 6.7 for the triple-deck structure. In the lower layer Prandtl's boundary-layer equations apply, though in a narrower region of non-dimensional scale $(\text{Re})^{-5/8}$. In this region viscous stress is important and after separation, the flow reverses direction. The main deck is the continuation of the upstream Prandtl boundary layer and so has a thickness of scale $(\text{Re})^{-1/2}$. It experiences a displacement perpendicular to the body surface and a slip velocity at its lower edge. The pressure gradient in the main deck is modified locally, affecting the lower layer. The main deck also interacts with the nearby bulk flow in the upper deck. For more details, see e.g. Stewartson (1969, 1981) or Smith (1977).

6.4 Notes

Some further details are as follows.

(1) *History of aerodynamics by Anderson (1998).* This a comprehensive tour of the early history of aerodynamics.

(2) *Ludwig Prandtl's boundary layer by Anderson (2005).* This is a succinct account of Prandtl's work and the very beginnings of boundary-layer theory.

(3) *Separation and triple-deck theory references.* There is much more to boundary-layer theory, separation and triple-deck theory than the extremely brief notes we provided above. For more details and references on early work on triple-deck theory, see Stewartson (1981) and Smith (1977). However, additional introductory accounts can be found in e.g. Acheson, (1990), Childress (2009), Kundu and Cohen (2008), Ockendon and Ockendon (1995) and Panton (2013).

(4) *Images and videos.* Besides the comprehensive set of images on boundary-layer phenomena found in Van Dyke (1982), there is a plethora of videos available online, even for example the highly recommended videos by Prof. Ascher H. Shapiro (MIT) from the 1960s.

(5) *Von Kármán vortex street.* For a flow past a cylinder with the Reynolds number between about 40 and 200, the wake behind the cylinder becomes unstable, developing a slow oscillation in time and increasing in amplitude downstream. The oscillating wake rolls up into staggered and alternately oppositely rotating vortices downstream, known as a von Kármán vortex street. See Van Dyke (1982) or go online for some striking images of such wakes.

(6) *Well-posedness of Prandtl boundary-layer equations.* There are two natural questions with regards to Prandtl's boundary-layer equations in Definition 6.1, matching the bulk flow field U_E, that require close scrutiny. The first is the well-posedness of the Prandtl equations, i.e. the existence, uniqueness and continuous dependence on the data of solutions. The second concerns whether, for the same initial data, a solution to the incompressible Navier–Stokes equations converges in the large Reynolds-number limit, with suitable scaling, to a solution of the incompressible Euler equations outside the boundary layer and to a corresponding solution of the Prandtl equations within the boundary layer. Such a result gives a justification for the expansion and boundary-layer equations. Indeed such a convergence result was proved locally in time for analytic initial data by Sammartino and Caflisch (1998b), on a half-space in two or three dimensions. With regards to the former well-posedness result, a useful summary can be found in E (2000).

In the case of the steady Prandtl equations (see Remark 6.5) briefly, the following results are known. Suppose the steady boundary-layer flow occurs in the domain $(x', y') \in [0, x'_*] \times [0, \infty)$ for an $x'_* > 0$ we specify presently. Assume the bulk Euler flow horizontal velocity at $y = 0$ is non-negative, i.e. $U_E := U_E(x, 0) \geqslant 0$, that u'_0 is non-negative and monotonically increasing in y' and $u' > 0$ for all $y > 0$ holds. Using the von Mises transformation, see Exercise 6.3, Oleinik (1963) proved the following. Suppose u'_0 and its first and second derivatives are bounded and Hölder continuous and u'_0 satisfies the compatibility condition $\partial^2 u'_0/\partial y'^2 - \partial P'_E/\partial x' \leqslant K y'^2$ at $x' = 0$, as $y' \to 0$ for some positive constant K. Then there exists an $x'_* > 0$ such that the boundary layer equations have a unique strong solution on the domain above, with u' and both its partial derivatives, as well as its second partial derivative with respect to y', bounded and continuous. If $\partial P_E/\partial x \geqslant 0$ then x'_* can be extended to infinity. Note that this is the favourable pressure gradient that occurs upstream of the curved body surface we considered in Section 6.3, and therefore precludes any possible separation. For more details and further results, see E (2000).

In the case of the unsteady Prandtl equations, we have the following. Suppose the boundary-layer flow occurs in the domain $(x', y') \in \mathbb{R} \times [0, \infty)$. Using the Crocco transformation, see Exercise 6.4, and assuming the initial data $u'_0 = u'_0(x', y')$ is monotonic in y', for example increasing in y', then Oleinik (1967) proved local in time well-posedness for the Prandtl boundary-layer equations. Under the additional assumption that the bulk flow pressure gradient $\partial P_E/\partial x \leqslant 0$, i.e. everywhere always favourable, Xin and Zhang (2004) established global in time well-posedness. Sammartino and Caflisch (1998a) established local in time well-posedness for analytic initial data in two or three dimensions, which was improved to only the assumption of analyticity with respect to x' for the initial data by Lombardo *et al.* (2003). However, without either the monotinicity or analyticity assumptions on the initial data, instabilities develop in solutions to the boundary-layer equations, for example leading to separation. These make well-posedness, at least globally in time, unlikely. Indeed, for example, E and Engquist (1997) showed that under fairly general technical conditions, global smooth solutions do not exist. Also see Grenier (2000). Further, Gérard-Varet and Dormy (2010) and Gérard-Varet and Nguyen (2012) proved the Prandtl equations are ill-posed in Sobolev spaces, respectively, linearly and nonlinearly. See Chapter 7 for more details on Sobolev spaces. We also remark that Gérard-Varet and Masmoudi (2013) have established well-posedness for Prandtl's equations in the intermediate setting between analytic and Sobolev data, namely the Gevrey class G^m with respect to the x' variable for the initial data with $m = 7/4$. Here G^m is the class of functions f such that, for some positive constants C and τ, we know the jth derivative of f is bounded in absolute value by $C\tau^{-j}(j!)^m$ for all $j \in \mathbb{N}$ and x' in a periodic domain.

(7) *Wind turbines and ocean/tidal energy.* Viscous fluid mechanics and boundary-layer theory is naturally fundamental to the optimal design of turbine blades for wind turbines or turbines for harnessing ocean/tidal/river energy. For more details, see e.g. Jamieson (2018) for wind turbines and Zeiner-Gundersen (2015) for ocean/tidal/river turbines.

(8) *The Magnus effect.* In contrast to the Kutta–Joukowski dynamic lift on an aerofoil – see Section 3.3 – the effect of flow past a rotating object is to induce an effective vortex (dependent on rotation speed) and a resulting side force. In a sporting context this is crucial in all ball sports – particularly cricket, football, lawn tennis, table tennis, baseball, golf, etc. This was described by Magnus in 1852, but was apparently noted by Newton for tennis in 1672, and by others in later applications to ballistics. A more recent application is to ships with 'rotor sails' – that is to say one (or two) vertical rotating cylinders giving propulsion as a result of passage through wind flow. Designed by Anton Flettner in 1924, the 'Buckau' was considered to be somewhat inefficient, but environmental concerns have led to a revival of interest in the concept and in new constructs.

6.5 Projects

Some suggested extended projects from this chapter are as follows.

(1) Falkner–Skan similarity solutions (analytical and numerical). Complete Exercise 6.7, which treats the case of a boundary-layer flow over a flat stationary rigid plate for the case when the bulk incompressible Euler flow $U_E = U_E(x)$ is x-dependent. Assume the constant K in part (d) of the exercise is zero. What constraint does this assumption impose? In this case show that $\delta^2 = \nu(2\alpha - \beta)x/U_E(x)$. Choose $2\alpha - \beta$ to be 1 or -1, distinguishing the cases when U_E and x have the same or opposite signs. In general, $2\alpha - \beta \neq 0$ can have any value. See Panton (2013, Sec. 20.5). Consequently, show, that $U_E = Cx^m$, where m equals β or $-\beta$ depending on whether U_E and x have the same or opposite signs, respectively. Show that the Falkner–Skan equation has the form

$$f''' + \tfrac{1}{2}(m+1)ff'' + m(1 - (f')^2) = 0.$$

Using the literature as a guide, see e.g. Panton (2013, Sec. 20.5) and then Rosenhead (1963), investigate the different values of m for which a solution exists and the solution character in different m-value regimes. Numerically plot the solutions, see Figure 6.4, in the different solution-characteristic regimes, including the regime $m < 0$. What are the different solution velocity profiles in the different regimes and what interpretation do they have?

(2) Boundary-layer separation survey. The theory of boundary-layer separation has a very rich history. In addition to the references listed in Section 6.4(3) above, there are detailed reviews in Stewartson (1981) and more recently in Cassel and Conlisk (2014) and Veldman (2017). Starting with these references, conduct a quantitative historical survey of the development of boundary-layer theory and in particular boundary-layer separation and triple-deck theory.

(3) Aerodynamics in sport. Aerodynamics plays a crucial role in many sports: running, cycling, skiing, ball sports, and so forth. Consider for example the careful unorthodox positioning of the support runners surrounding Eliud Kipchoge on his recent record-breaking marathon run under two hours, on 12th October 2019.

Or as another example, in the Tour de France, how is it that the peloton nearly always manages to catch breakaway groups in the final few kilometres of a day's race? This is because the larger group of riders in the peloton are shielded by a relatively smaller subdivision of riders working harder at the front, and who are continuously replaced by fresh riders coming up to take temporary frontline duties. How does riding in echelon formation help when there are strong cross-winds? Boundary layers, and whether they are laminar or turbulent, play a particularly important role in cricket and tennis for example, such as the 'swing' of a cricket ball. See the Magnus effect in Section 6.4(8) above. Conduct a survey of the role and subtle effects of aerodynamics, in particular of laminar and turbulent boundary layers and separation in sport. You may wish to focus on running or cycling or ball sports more specifically. You could start with Kundu and Cohen (2008, Ch. 10, Sec. 11) and then look more generally at Nørstrud (2008). For ball sports in particular, there are many articles by Mehta; you might like to start with Mehta (1985) or Nørstrud (2008, p. 229).

Exercises

6.1 Boundary layers (general): Porous circular cylinder. Consider a circular cylinder of radius a that rotates with a constant angular velocity Ω about its axis. The wall of the cylinder is porous and incompressible fluid is sucked into the cylinder with constant speed V. In other words, in cylindrical polar coordinates (r, θ, z) where the velocity field is $\boldsymbol{u} = u_r\,\hat{\boldsymbol{r}} + u_\theta\,\hat{\boldsymbol{\theta}} + u_z\,\hat{\boldsymbol{z}}$, the boundary condition at $r = a$ is $u_r = -V$, $u_\theta = \Omega a$ and $u_z = 0$. Indeed, hereafter assume the fluid flow is uniform in z and $u_z \equiv 0$.

(a) Assume the circulation κ of the fluid about the cylinder remains finite at an infinite distance from the cylinder. Further assume the flow is steady and such that $u_r = u_r(r)$ and $u_\theta = u_\theta(r)$ only, i.e. the flow is axisymmetric. The Reynolds number for this flow is $\text{Re} := Va/\nu$.

(i) Using the continuity equation in cylindrical polar coordinates, explain why $u_r = -Va/r$.

(ii) Explain why the vorticity field is given by $\boldsymbol{\omega} = \omega_z\,\hat{\boldsymbol{z}}$ and show, from the vorticity formulation of the Navier–Stokes equations in cylindrical polar coordinates (see the formulae in Section A.2 of the Appendix), $\omega_z = \omega_z(r)$ satisfies

$$u_r \frac{\partial \omega_z}{\partial r} = \frac{\nu}{r}\frac{\partial}{\partial r}\left(r\frac{\partial \omega_z}{\partial r}\right).$$

(iii) Using part (i) show that the solution to the differential equation for ω_z in part (ii) can be expressed in the form $\omega_z = \omega_0 + \mathrm{A}r^{-Va/\nu}$, where ω_0 and A are arbitrary constants.

(iv) Explain why $\omega_z = r^{-1}(\partial/\partial r)(ru_\theta)$ and combine this result with the result of part (iii) just above to show

$$u_\theta = \tfrac{1}{2}\omega_0 r + \frac{Ar^{1-Va/\nu}}{2 - Va/\nu} + \frac{B}{r},$$

where B is an arbitrary constant. Compute the circulation $\kappa = \int_0^{2\pi} u_\theta\, r\mathrm{d}\theta$.

(v) If Re < 2, explain why $u_\theta = \Omega a^2/r$ and $\kappa = 2\pi\,\Omega a^2$. If Re $= 2$, recompute u_θ and the circulation κ from stage (iv) and show that we still have $\kappa = 2\pi\,\Omega a^2$. If Re > 2 and the motion is started from rest, so the circulation at an infinite distance from the cylinder is zero, show that $u_\theta = \Omega a\,(a/r)^{\mathrm{Re}-1}$.

(b) Now consider a similar flow where the flow outside the cylinder $r = a$ is constrained so that it lies within a larger concentric stationary cylinder of radius $b > a$. In other words, the flow no longer extends to the whole of $\mathbb{R}^3$, but lies inside the larger concentric stationary cylinder. Assuming the same Reynolds number as in part (a), show that the circulations in this scenario match those in the cases Re < 2 and Re > 2 in part (a) above if $b \gg a$, as follows.

(i) Use the boundary condition $u_\theta = 0$ at $r = b$ to show

$$u_\theta = \Omega a \frac{\left(b/r - (b/r)^{\mathrm{Re}-1}\right)}{\left(b/a - (b/a)^{\mathrm{Re}-1}\right)}.$$

(ii) If Re < 2 show $u_\theta \to \Omega a^2/r$ as $b/a \to \infty$. If Re > 2 show $u_\theta \to \Omega a\,(a/r)^{\mathrm{Re}-1}$ as $b/a \to \infty$.

(iii) In the case Re > 2, by writing $r = a\,(1 + r'/(\mathrm{Re}))$, show $u_\theta \to \Omega a\,\mathrm{e}^{-r'}$ as Re $\to \infty$ and hence deduce that there exists a boundary layer at $r = a$ whose thickness scales like $(\mathrm{Re})^{-1}$.

6.2 Boundary layers (general): Ekman boundary layer A circular disc of very large radius rotates steadily about its axis, with angular velocity Ω, and bounds the semi-infinite region of fluid $z \geqslant 0$. Suppose the axis of rotation of the circular disc is the z-axis. Also suppose that far from the disc for $z \gg 1$ the fluid is in steady rigid-body rotation about the same axis with angular velocity $\gamma\Omega$, where $\gamma \geqslant 0$ is a constant. We assume the flow is axisymmetric. The steady axisymmetric incompressible Navier–Stokes equations in cylindrical polar coordinates (r, θ, z) with the velocity field $\boldsymbol{u} = u_r\,\hat{\boldsymbol{r}} + u_\theta\,\hat{\boldsymbol{\theta}} + u_z\,\hat{\boldsymbol{z}}$, with pressure field p, have the form

$$(\boldsymbol{u}\cdot\nabla)\,u_r - \frac{u_\theta^2}{r} = -\frac{1}{\rho}\frac{\partial p}{\partial r} + \nu\left(\Delta u_r - \frac{u_r}{r^2}\right),$$

$$(\boldsymbol{u}\cdot\nabla)\,u_\theta + \frac{u_r u_\theta}{r} = \nu\left(\Delta u_\theta - \frac{u_\theta}{r^2}\right),$$

$$(\boldsymbol{u}\cdot\nabla)\,u_z = -\frac{1}{\rho}\frac{\partial p}{\partial z} + \nu\Delta u_z,$$

$$\frac{1}{r}\frac{\partial}{\partial r}(r u_r) + \frac{\partial u_z}{\partial z} = 0.$$

The last equation is the incompressibility condition and we have

$$\boldsymbol{u}\cdot\nabla = u_r\frac{\partial}{\partial r} + u_z\frac{\partial}{\partial z} \quad \text{and} \quad \Delta = \frac{1}{r}\frac{\partial}{\partial r}\left(r\frac{\partial}{\partial r}\right) + \frac{\partial^2}{\partial z^2}.$$

Consider a similarity solution of the following form:

$$u_r = \Omega r f(\zeta), \quad u_\theta = \Omega r g(\zeta) \quad \text{and} \quad u_z = (\Omega\nu)^{1/2} h(\zeta),$$

where $\zeta := z\,(\Omega/\nu)^{1/2}$ and the functions $f = f(\zeta)$, $g = g(\zeta)$ and $h = h(\zeta)$ are to be determined.

(a) By substituting the similarity solution into the steady axisymmetric incompressible Navier–Stokes equations above, show that the functions f, g and h satisfy the system of ordinary differential equations

$$\begin{aligned} f^2 + hf' - g^2 &= f'' - \frac{1}{\rho r\Omega^2}\frac{\partial p}{\partial r}, \\ 2fg + hg' &= g'', \\ hh' &= h'' - \frac{1}{\rho(\nu\Omega^3)^{1/2}}\frac{\partial p}{\partial z}, \\ 2f + h' &= 0. \end{aligned}$$

(b) Set $\tilde{p}(\zeta, r) := p(z, r)$. Using the third equation from part (a), show

$$\tilde{p}(\zeta, r) = P(r) + \rho\nu\Omega \int_0^{\zeta} (h'' - hh')(\tilde{\zeta})\,\mathrm{d}\tilde{\zeta},$$

where $P = P(r)$ is an arbitrary function. Since as $z \to \infty$ we know $u_r \to 0$ and $u_\theta \to \gamma\Omega r$, i.e. $f \to 0$ and $g \to \gamma$ as $\zeta \to \infty$, deduce

$$P = C + \tfrac{1}{2}\rho\,(\gamma\Omega r)^2,$$

where C is an arbitrary constant. Hence show further that the first equation in part (a) can be expressed in the form

$$f^2 + hf' - g^2 = f'' - \gamma^2.$$

Verify that the no-slip boundary condition at $z = 0$ implies $f(0) = h(0) = 0$ and $g(0) = 1$.

(c) Suppose $\gamma = 1 - \epsilon$, where ϵ is small, and $f = \epsilon F$, $g = 1 + \epsilon G$ and $h = \epsilon H$ for corresponding functions $F = F(\zeta)$, $G = G(\zeta)$ and $H = H(\zeta)$. Substituting these forms into the second equation from part (a) and the equation for f in part (b), and neglecting terms of order ϵ^2, show $G'' = 2F$ and $F'' = -2G - 2$. Then by setting $U := F + \mathrm{i}G$, find the second-order ordinary differential equation satisfied by U and solve it to show

$$f = \epsilon\, \mathrm{e}^{-\zeta} \sin\zeta \quad \text{and} \quad g = 1 + \epsilon\,(\mathrm{e}^{-\zeta}\cos\zeta - 1).$$

(d) Now imagine what the flow system, specified by the similarity form and the explicit forms for f and g from part (c), looks like in the neighbourhood of a point on the disc when viewed relative to the disc, i.e. consider the flow prescribed by u_r and $v := u_\theta - \Omega r$. Show that the direction of the flow, i.e. the angle $\tilde{\theta}$ between u_r and v, satisfies

$$\tan\tilde{\theta} = \frac{\mathrm{e}^{-\zeta}\cos\zeta - 1}{\mathrm{e}^{-\zeta}\sin\zeta}.$$

Show that trajectories of such a relative velocity field are in fact spirals; these are known as *Ekman spirals*.

6.3 Prandtl equations: von Mises transformation. The von Mises transformation was used by Oleinik (1963) to prove the existence and uniqueness of strong solutions to the steady Prandtl boundary-layer equations, assuming monotonicity and other technical conditions on the data; see Section 6.4(6) just above. The goal of this question is to implement the von Mises transformation and derive the corresponding resulting degenerate parabolic equation. Recall the Prandtl equations from Definition 6.1. Assuming that the pressure is independent of y' has already been asserted, that $\partial p'/\partial x'$ has already been replaced by $\partial P'_{\mathrm{E}}/\partial x'$ where $P'_{\mathrm{E}} = P'_{\mathrm{E}}(x', 0, t')$, then dropping primes, the equations are

$$u\frac{\partial u}{\partial x} + v\frac{\partial u}{\partial y} = \frac{\partial^2 u}{\partial y^2} - \frac{\partial P_{\mathrm{E}}}{\partial x},$$
$$\frac{\partial u}{\partial x} + \frac{\partial v}{\partial y} = 0,$$

together with no-slip boundary conditions at $y = 0$. Note that there exists a stream function $\psi = \psi(x, y)$ such that

$$u = \frac{\partial \psi}{\partial y} \qquad \text{and} \qquad v = -\frac{\partial \psi}{\partial x}.$$

The von Mises transformation is defined as follows. We make a transformation of independent coordinates from (x, y) to (X, Y) where

$$X := x \qquad \text{and} \qquad Y := \psi(x, y),$$

and we set $U(X, Y) := u(x, y)$ and $V(X, Y) := v(x, y)$. (*Note:* Here U and V are new variables, not characteristic velocity magnitude scalings.)

(a) Using the chain rule, for example that

$$\frac{\partial u}{\partial y} = \frac{\partial U}{\partial X}\frac{\partial X}{\partial y} + \frac{\partial U}{\partial Y}\frac{\partial Y}{\partial y},$$

and the definitions for Y, U and V just above, show that

$$\frac{\partial u}{\partial y} = U\frac{\partial U}{\partial Y} \qquad \text{and} \qquad \frac{\partial u}{\partial x} = \frac{\partial U}{\partial X} - V\frac{\partial U}{\partial Y},$$

as well as

$$\frac{\partial^2 u}{\partial y^2} = U\left(U\frac{\partial^2 U}{\partial Y^2} + \left(\frac{\partial U}{\partial Y}\right)^2\right).$$

(b) Substitute the expressions for u and its partial derivatives in terms of U and its partial derivatives from part (a), to show after a cancellation of two terms that U satisfies the partial differential equation

$$U\frac{\partial U}{\partial X} = U\left(U\frac{\partial^2 U}{\partial Y^2} + \left(\frac{\partial U}{\partial Y}\right)^2\right) - \frac{\partial P_{\mathrm{E}}}{\partial X}.$$

(c) Show that the function $W = W(X, Y)$ defined by $W := U^2$ satisfies the degenerate nonlinear parabolic partial differential equation

$$\frac{\partial W}{\partial X} = \sqrt{W}\frac{\partial^2 W}{\partial Y^2} - 2\frac{\partial P_{\mathrm{E}}}{\partial X}.$$

6.4 Prandtl equations: Crocco transformation. Oleinik (1967) used another transformation, the Crocco transformation, to establish local in time well-posedness for the unsteady Prandtl boundary-layer equations, under the assumption that the initial data is monotonic in the y' variable. See Section 6.4(6) just above. The goal of this question is to derive the corresponding time-evolutionary degenerate parabolic partial differential equation arising from the Crocco transformation. As in Exercise 6.3 just above, recall the Prandtl equations from Definition 6.1, though here we consider the full unsteady equations. Since the pressure is independent of y' so that $\partial p'/\partial x'$ can be replaced by $\partial P'_{\mathrm{E}}/\partial x'$ where $P'_{\mathrm{E}} = P'_{\mathrm{E}}(x', 0, t')$, then dropping primes, the unsteady equations are

$$\frac{\partial u}{\partial t} + u\frac{\partial u}{\partial x} + v\frac{\partial u}{\partial y} = \frac{\partial^2 u}{\partial y^2} - \frac{\partial P_{\mathrm{E}}}{\partial x},$$

$$\frac{\partial u}{\partial x} + \frac{\partial v}{\partial y} = 0,$$

together with no-slip boundary conditions at $y = 0$. We also assume the matching condition $u \to U_{\mathrm{E}}(x, 0, t)$ as $y \to \infty$; translating the corresponding scaled and thus primed statement from Section 6.1.

(a) The vorticity field in the case of the boundary-layer equations is given by $\omega := -\partial u/\partial y$. By taking the partial derivative with respect to y of the evolution equation for u above, show, using the incompressibility condition, that the vorticity field satisfies the evolution equation

$$\frac{\partial \omega}{\partial t} + u\frac{\partial \omega}{\partial x} + v\frac{\partial \omega}{\partial y} = \frac{\partial^2 \omega}{\partial y^2}.$$

Further, using the evolution equation for u and no-slip boundary conditions on $y = 0$, show that ω satisfies the boundary condition $\partial\omega/\partial y = \partial P_{\mathrm{E}}/\partial x$ on $y = 0$. In addition, show that the far-field matching condition becomes $\partial\omega/\partial y \to 0$ as $y \to \infty$.

(b) Now consider the Crocco transformation. We make a transformation of independent coordinates from (t, x, y) to (T, X, Y), where

$$T := t, \qquad X := x \qquad \text{and} \qquad Y := u(x, y, t),$$

and we set $W(X, Y, T) := \omega(x, y, t)$. (*Note:* Again, here W is a new variable, and not a characteristic velocity magnitude scaling.) Use the chain rule, for example that,

$$\frac{\partial \omega}{\partial y} = \frac{\partial \omega}{\partial X}\frac{\partial X}{\partial y} + \frac{\partial \omega}{\partial Y}\frac{\partial Y}{\partial y} + \frac{\partial \omega}{\partial T}\frac{\partial T}{\partial y},$$

and the definitions for T, X, Y and W just above, to show that

$$\frac{\partial \omega}{\partial t} = \frac{\partial W}{\partial T} + \frac{\partial W}{\partial Y}\frac{\partial u}{\partial t},$$

$$\frac{\partial \omega}{\partial x} = \frac{\partial W}{\partial X} + \frac{\partial W}{\partial Y}\frac{\partial u}{\partial x},$$

$$\frac{\partial \omega}{\partial y} = \frac{\partial W}{\partial Y}\frac{\partial u}{\partial y},$$

$$\frac{\partial^2 \omega}{\partial y^2} = W^2\frac{\partial^2 W}{\partial Y^2} + \frac{\partial W}{\partial Y}\frac{\partial^2 u}{\partial y^2}.$$

(c) Substituting the terms shown in part (b) into the evolution equation for the vorticity ω derived in part (a), and utilising the evolution equation for the velocity field u from Prandtl's equations above, show that W satisfies the degenerate nonlinear parabolic partial differential equation

$$\frac{\partial W}{\partial T} + Y\frac{\partial W}{\partial X} = W^2\frac{\partial^2 W}{\partial Y^2} + \frac{\partial W}{\partial Y}\frac{\partial P_{\mathrm{E}}}{\partial X}.$$

6.5 Prandtl equations: Boundary with curvature. Suppose the $\mathbb{R}^2$ plane is parameterised by the Cartesian coordinates x_1 and x_2. Suppose there is a smooth simple curve C in this plane. Suppose we parameterise the plane by the curvilinear coordinates x and y defined as follows. Let x denote arclength along C so that x is constant on each normal to C, while y is the signed distance from C. Let $\kappa = \kappa(x)$ denote the curvature of C at position x along it. Arclengths in the curvilinear coordinates (x, y) are given by $(\mathrm{d}s)^2 = h^2(\mathrm{d}x)^2 + (\mathrm{d}y)^2$, where $h = h(x, y)$ is given by $h = 1 + \kappa y$. In these curvilinear coordinates the incompressible Navier–Stokes equations for a steady plane flow are

$$\frac{u}{h}\frac{\partial u}{\partial x} + v\frac{\partial u}{\partial y} + \frac{\kappa uv}{h} = -\frac{1}{\rho h}\frac{\partial p}{\partial x} + \nu\frac{\partial}{\partial y}\left(\frac{1}{h}\frac{\partial}{\partial y}(hu) - \frac{1}{h}\frac{\partial v}{\partial x}\right),$$
$$\frac{u}{h}\frac{\partial v}{\partial x} + v\frac{\partial v}{\partial y} - \frac{\kappa u^2}{h} = -\frac{1}{\rho}\frac{\partial p}{\partial y} + \frac{\nu}{h}\frac{\partial}{\partial x}\left(\frac{1}{h}\frac{\partial v}{\partial x} - \frac{1}{h}\frac{\partial}{\partial y}(hu)\right),$$
$$\frac{\partial u}{\partial x} + \frac{\partial}{\partial y}(hv) = 0.$$

Suppose L is a characteristic length scale in the x-direction. Further suppose U and V are characteristic velocity magnitudes in the u and v directions, while the characteristic length scale in the y-direction, the thickness of the boundary layer, is δ. Hence we introduce the non-dimensional variables

$$x' := \frac{x}{L},\quad y' := \frac{y}{\delta},\quad u' := \frac{u}{U},\quad v' := \frac{v}{V},\quad p' := \frac{p}{P}\quad \text{and}\quad \kappa' := \kappa L,$$

where P is a characteristic pressure yet to be determined. Note that we have $h = 1 + (\delta/L)\kappa' y'$. Assuming a bulk Euler flow $U_{\mathrm{E}} = U_{\mathrm{E}}(x)$ outside the boundary layer of characteristic scale U, our goal is to derive the Prandtl boundary-layer equations in this context on the side $y \geqslant 0$. The Reynolds number here is $\mathrm{Re} = UL/\nu$.

(a) Show, ignoring a correction term of order $(\delta/L)^2$, that to maintain incompressibility we must have $V = (\delta/L)U$.

(b) Substitute the relations between the dimensional and non-dimensional variables above into the curvilinear equation for u to show

$$(\mathrm{Re})\frac{\delta^2}{L^2}\left(\frac{u'}{h}\frac{\partial u'}{\partial x'} + v'\frac{\partial u'}{\partial y'}\right) + (\mathrm{Re})\frac{\delta^3}{L^3}\left(\frac{\kappa' u' v'}{h}\right)$$
$$= -\frac{P\delta^2}{\rho U L \nu}\left(\frac{1}{h}\frac{\partial p'}{\partial x'}\right) + \frac{\partial}{\partial y'}\left(\left(\frac{1}{h}\frac{\partial}{\partial y'}(hu')\right) - \frac{\delta^2}{L^2}\left(\frac{1}{h}\frac{\partial v'}{\partial x'}\right)\right).$$

(c) Explain why, to maintain the inertia terms with the vertical diffusion term in the limit as $\mathrm{Re} \to \infty$, we require $\delta^2 = L^2/(\mathrm{Re})$. Hence the boundary layer has characteristic scale $\delta = L/(\mathrm{Re})^{1/2}$. Consequently explain why, as $\mathrm{Re} \to \infty$, we have $h^{-1} \sim 1$. Thus show, with a suitable choice for P, that the limiting equation as Re $\to \infty$ is the steady version of the Prandtl equation for u' in Definition 6.1. By similar arguments show that the curvilinear equation for v above reduces to $\partial p'/\partial y' = 0$.

(*Note:* Matching the bulk flow as $y' \to \infty$ shows that $\partial p'/\partial x' = \partial P'_{\mathrm{E}}/\partial x' = -U'_{\mathrm{E}}\,\partial U'_{\mathrm{E}}/\partial x'$.)

(d) Suppose we now revert to the dimensional form of the steady Prandtl boundary-layer equations mentioned in Remark 6.7. Suppose that $u(x, y) > 0$ for $y > 0$ and that $v = 0$ on C. Use the von Mises transformation, see Exercise 6.3, to transform the steady dimensional Prandtl boundary-layer equations as follows. Transform from (x, y) coordinates to (X, Y) coordinates where $X := x$ and $Y := \psi(x, y)$, where $\psi = \psi(x, y)$ is the stream function for the boundary-layer flow. Setting $U(X, Y) := u(x, y)$ and $V(X, Y) := v(x, y)$, show that $U = U(X, Y)$ satisfies the nonlinear partial differential equation

$$U\frac{\partial U}{\partial X} = \tfrac{1}{2}\nu U\frac{\partial^2}{\partial Y^2}(U^2) + U_{\mathrm{E}}\frac{\mathrm{d}U_{\mathrm{E}}}{\mathrm{d}X}.$$

(e) Now consider a viscous flow in the domain $\mathcal{D} := \{(x_1, x_2) \in \mathbb{R}^2 : x_1^2 + x_2^2 \leqslant a\}$ with boundary C, where $a > 0$ is a given constant radius. Assume the velocity on C is $U = U_0(X)$ and $V = 0$, with $U_0(X) \geqslant 0$. Further assume inside $\mathcal{D}$ and outside the boundary layer close to C, the flow has the form of a rigid-body flow with unknown angular velocity Ω.

(i) Explain why $U_{\mathrm{E}} \to \Omega a$ as $y \to 0$, and thus why $\mathrm{d}U_{\mathrm{E}}/\mathrm{d}x \equiv 0$.

(ii) Integrate the equation for $U = U(X, Y)$ in part (d), and use periodicity in X, to show (here $A = A(X)$ and $B = B(X)$ are arbitrary functions)

$$\int_0^{2\pi a} U^2(X, Y)\,\mathrm{d}X = A(X) + B(X)Y.$$

(iii) Use that as $Y \to \infty$, $U \to U_{\mathrm{E}} = \Omega a$, to show for all $Y = \psi$:

$$\int_0^{2\pi a} U^2(X, Y)\,\mathrm{d}X = 2\pi\Omega^2 a^3.$$

(iv) Explain why $|\Omega|$ is determined by $U_0(X)$, and explain what determines the sign of Ω.

6.6 Prandtl equations: Three dimensions/yawed cylinder. Consider a long rigid cylinder placed in a steady uniform stream of fluid as follows. Suppose the cylinder is reflectively symmetric in a given plane, say the (x, z)-plane, and the uniform stream of velocity U is parallel to that plane. Suppose the cylinder is inclined at an angle α to the stream, see Figure 6.8. Further suppose the cross-section of the cylinder is streamlined so that no separation of the boundary layer occurs.

(a) Explain why the following system of equations represents, in dimensional variables, a three-dimensional system of steady boundary-layer equations – assume the variable y parameterises displacement across the boundary layer:

$$u\frac{\partial u}{\partial x} + v\frac{\partial u}{\partial y} + w\frac{\partial u}{\partial z} = -\frac{\partial p}{\partial x} + \nu\frac{\partial^2 u}{\partial y^2},$$

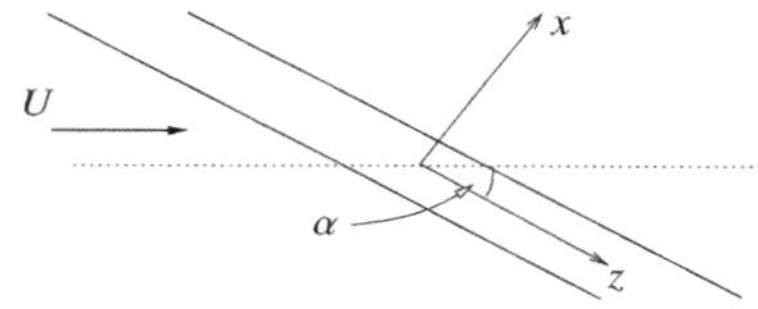

Figure 6.8 A long rigid cylinder is placed in a uniform stream of velocity U, both parallel to the (x, z)-plane, though yawed at the angle α as shown.

$$u\frac{\partial w}{\partial x} + v\frac{\partial w}{\partial y} + w\frac{\partial w}{\partial z} = -\frac{\partial p}{\partial z} + \nu\frac{\partial^2 w}{\partial y^2},$$
$$\frac{\partial p}{\partial y} = 0,$$
$$\frac{\partial u}{\partial x} + \frac{\partial v}{\partial y} + \frac{\partial w}{\partial z} = 0,$$

where u, v and w are the velocity components in the x-, y- and z-directions, respectively, $p = p(x, y, z)$ is the pressure field and ν is the constant kinematic viscosity. No-slip boundary conditions are assumed at $y = 0$.

(b) Explain why, for the given background flow, we also can deduce for the boundary-layer equations that $\partial p/\partial x = \partial p/\partial z = 0$. Further, note from Figure 6.8 that z is aligned along the cylinder. Explain why natural matching velocities as $y \to \infty$ are $u \to U \sin\alpha$ and $w \to U\cos\alpha$. Further explain why we know the whole system is independent of z.

(c) Using the identifications

$$\tilde{y} := y\,(\sin\alpha)^{1/2}, \qquad \tilde{u} := \frac{u}{\sin\alpha} \qquad \text{and} \qquad \tilde{v} := \frac{v}{(\sin\alpha)^{1/2}},$$

and parts (a) and (b), show that $\tilde{u}$ and $\tilde{v}$ satisfy the system of equations

$$\tilde{u}\frac{\partial \tilde{u}}{\partial x} + \tilde{v}\frac{\partial \tilde{u}}{\partial \tilde{y}} = \nu\frac{\partial^2 \tilde{u}}{\partial \tilde{y}^2},$$
$$\frac{\partial \tilde{u}}{\partial x} + \frac{\partial \tilde{v}}{\partial \tilde{y}} = 0,$$

together with the boundary conditions $\tilde{u} = 0$ and $\tilde{v} = 0$ on $\tilde{y} = 0$, and $\tilde{u} \to U$ as $\tilde{y} \to \infty$. Hence explain why we can deduce that the velocity component u has a distribution of the same form for all values of α, except those near 0 or π.

(d) Lastly, setting $\tilde{w} := w/\cos\alpha$, show that $\tilde{w}$ satisfies

$$\tilde{u}\frac{\partial \tilde{w}}{\partial x} + \tilde{v}\frac{\partial \tilde{w}}{\partial \tilde{y}} = \nu\frac{\partial^2 \tilde{w}}{\partial \tilde{y}^2},$$

and the boundary conditions $\tilde{w} = 0$ on $\tilde{y} = 0$ and $\tilde{w} \to U$ as $\tilde{y} \to \infty$.

6.7 Blasius problem: Falkner–Skan similarity solutions. Consider the following boundary-layer problem. In the upper half of the (x, y)-plane, where x represents the horizontal coordinate and y represents the vertical coordinate, we suppose there is a flow directed along a flat stationary rigid plate at $y = 0$. Close to the flat plate, between $0 \leqslant y \leqslant \delta(x)$, there is boundary-layer flow governed by the system of equations

$$\frac{\partial u}{\partial t} + u\frac{\partial u}{\partial x} + v\frac{\partial u}{\partial y} = -\frac{\partial p}{\partial x} + \nu\frac{\partial^2 u}{\partial y^2},$$
$$\frac{\partial p}{\partial y} = 0,$$
$$\frac{\partial u}{\partial x} + \frac{\partial v}{\partial y} = 0,$$

where u and v are the velocity components in the x- and y-directions, respectively, and $p = p(x, y, t)$ is the pressure field. The constant quantity ν is the kinematic viscosity. No-slip boundary conditions are assumed at $y = 0$. Above the boundary layer in the region $\delta(x) \leqslant y < \infty$, there is a *known* one-dimensional flow field $U_{\mathrm{E}} = U_{\mathrm{E}}(x, t)$ for which

$$-\frac{\partial p}{\partial x} = \frac{\partial U_{\mathrm{E}}}{\partial t} + U_{\mathrm{E}}\frac{\partial U_{\mathrm{E}}}{\partial x}.$$

Assume hereafter that the flow within, as well as above, the boundary layer is steady.

(a) Explain why the third boundary-layer equation above implies that there exists a function $\psi = \psi(x, y)$ such that the first boundary-layer equation above can be written in the form

$$\frac{\partial \psi}{\partial y}\frac{\partial^2 \psi}{\partial x \partial y} - \frac{\partial \psi}{\partial x}\frac{\partial^2 \psi}{\partial y^2} = -\frac{\partial p}{\partial x} + \nu\frac{\partial^3 \psi}{\partial y^3}.$$

(b) Use the second boundary-layer equation and the equation for the pressure for the region above the boundary layer to show that the boundary-layer equation has the form

$$\frac{\partial \psi}{\partial y}\frac{\partial^2 \psi}{\partial x \partial y} - \frac{\partial \psi}{\partial x}\frac{\partial^2 \psi}{\partial y^2} = U_{\mathrm{E}}\frac{\partial U_{\mathrm{E}}}{\partial x} + \nu\frac{\partial^3 \psi}{\partial y^3}.$$

(c) By looking for a similarity solution of the form $\psi = U_{\mathrm{E}}(x)\delta(x)f(\eta)$ where $\eta = y/\delta(x)$ to the boundary-layer equation in part (b), show that f satisfies the differential equation

$$f''' + \alpha f f'' + \beta(1 - (f')^2) = 0,$$

where

$$\alpha := \frac{\delta}{\nu}\frac{\partial}{\partial x}(\delta U_{\mathrm{E}}) \qquad \text{and} \qquad \beta := \frac{\delta^2}{\nu}\frac{\partial U_{\mathrm{E}}}{\partial x}.$$

Explain why α and β must be constants.

(d) Using the definitions for α and β from part (c) above, show that

$$\frac{\partial}{\partial x}(\delta^2 U_{\mathrm{E}}) = \nu(2\alpha - \beta).$$

Solve this differential equation and show that $\delta^2 U_{\mathrm{E}} = K + \nu(2\alpha - \beta)x$, where K is an arbitrary constant. Then, assuming $2\alpha \neq \beta$, use the definition of β from part (c) above to show that $U_{\mathrm{E}} = U_{\mathrm{E}}(x)$ is given by

$$U_{\mathrm{E}} = C(K + (2\alpha - \beta)\nu x)^{\beta/(2\alpha-\beta)},$$

where C is another arbitrary constant.

(e) Explain why the boundary conditions for the differential equation for f in part (c) are $f(0) = 0$, $f'(0) = 0$ and $f' \to 1$ as $\eta \to \infty$.

6.8 Blasius problem: Porous flat plate. Suppose that far from a semi-infinite flat plate at $y = 0$ and $x \geqslant 0$ there is uniform stream (bulk flow) prescribed by the constant horizontal flow field U_{E}. At the plate there is a normal flow blowing through the surface with velocity

$$v(x,0) = \gamma\left(\frac{\nu U_{\mathrm{E}}}{2x}\right)^{1/2},$$

where γ is a constant. Our goal is to verify a solution to the dimensional Prandtl boundary-layer equations, see Remark 6.7, in this context of the following similarity form:

$$\psi = U_{\mathrm{E}}\delta(x)f(\eta),$$

where $\psi = \psi(x,\eta)$ is the stream function for the flow, $\eta := y/\delta$ is the rescaled vertical boundary-layer coordinate and $\delta = \delta(x)$ is the boundary-layer thickness. You may take as given that

$$\delta = \left(\frac{2\nu x}{U_{\mathrm{E}}}\right)^{1/2}.$$

(a) Show that $f = f(\eta)$ necessarily satisfies the nonlinear ordinary differential equation

$$f''' + ff'' = 0,$$

and the boundary conditions $f(0) = -\gamma$, $f'(0) = 0$ and $f' \to 1$ as $\eta \to \infty$.

(b) Show that if $f = f(\eta)$ solves the nonlinear ordinary differential equation in part (a), then so does $g(\eta) := \alpha f(\alpha\eta)$ where α is a constant.

(c) Suppose hereafter $\gamma = 0$ and $f(0) = f'(0) = 0$ while $f' \to \beta$ as $\eta \to \infty$, where $\beta > 0$. Show that if $g(\eta) = \alpha f(\alpha\eta)$ is a solution satisfying the boundary conditions prescribed in part (a), then necessarily $\alpha = \beta^{-1/2}$. (*Note:* Hence, given such an f, the corresponding g satisfies the boundary conditions from part (a).)

(d) Explain how part (c) can aid in the numerical computation of a solution g satisfying the boundary conditions from part (a) when $\gamma = 0$.

6.9 Blasius problem: Jet flow. Consider the following fluid flow in the unbounded two-dimensional half space $x \geqslant 0$. Suppose a two-dimensional jet of fluid emerges from a thin slot in a plane wall, which respectively coincides with the origin and $y = 0$. The ambient fluid is initially everywhere at rest, and remains at rest in the far field. See Figure 6.9.

(a) For the moment, assume Cauchy's equation of motion from Theorem 4.2 applies to the flow for the steady two-dimensional incompressible velocity field $\boldsymbol{u} = \boldsymbol{u}(\boldsymbol{x})$ with $\boldsymbol{x} \in [0,\infty) \times \mathbb{R}$:

$$\rho\,(\boldsymbol{u}\cdot\nabla)\,\boldsymbol{u} = -\nabla p + \nabla\cdot\hat{\sigma}.$$

Here ρ is the constant mass density, $p = p(\boldsymbol{x})$ is the pressure field and $\hat{\sigma} = \hat{\sigma}(\boldsymbol{x})$ is the deviatoric stress. Consider the arbitrary, large, rectangular region $\mathcal{V}$ symmetrically straddling $y = 0$ shown in Figure 6.9. Assume that the deviatoric

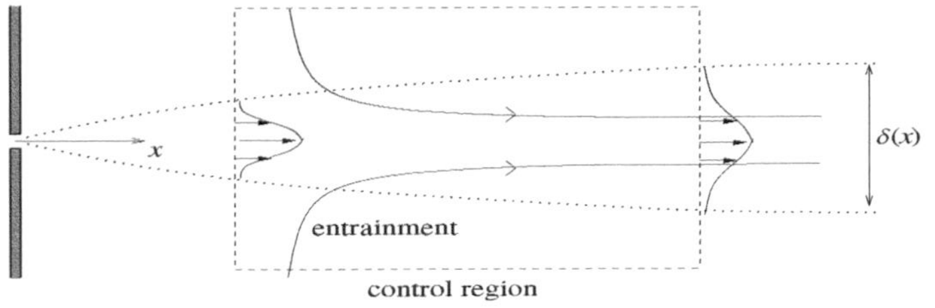

Figure 6.9 Flow associated with a jet emerging from a thin slot in a plane wall. Prandtl's boundary-layer equations apply in the central region indicated, of width $\delta = \delta(x)$. Fluid from the quiescent surrounding region is entrained into the jet flow as indicated. The control region $\mathcal{V}$ is also indicated.

stress and pressure field are negligible at the boundary $\partial\mathcal{V}$ of the region $\mathcal{V}$. By integrating Cauchy's equation of motion over the region $\mathcal{V}$ and then using the Divergence Theorem, show

$$\int_{\partial\mathcal{V}} \boldsymbol{u}\,(\boldsymbol{u}\cdot\boldsymbol{n})\,\mathrm{d}\boldsymbol{x} = 0,$$

where $\boldsymbol{n}$ is the outward normal to $\partial\mathcal{V}$. Explain why the total horizontal momentum (here u is the horizontal velocity field in the x-direction)

$$m := \rho \int_{-\infty}^{\infty} u^2\,\mathrm{d}y$$

is in fact independent of x.

(b) Explain why, sufficiently far from the slot, the boundary-layer equations apply to the central jet flow as follows. Set

$$x' := \frac{x}{X}, \quad y' := \frac{y}{\delta}, \quad u' := \frac{u}{U}, \quad v' := \frac{v}{V} \quad \text{and} \quad p' := \frac{p}{P}.$$

Here we assume: X is a characteristic horizontal scale; δ is the characteristic width of the jet; U and V are characteristic horizontal and vertical velocities, respectively; and P is a characteristic pressure to be determined presently. We naturally set U to be the velocity with which the fluid emerges from the thin slot. The Reynolds number here is $\mathrm{Re} = UX/\nu$.

(i) Show the incompressibility condition implies $V = (\delta/X)\,U$.

(ii) By substituting the non-dimensional variables above into the steady Navier–Stokes equation for the horizontal velocity field $u = u(x, y)$, show that the corresponding non-dimensional velocity field $u' = u'(x', y')$ satisfies

$$(\mathrm{Re})\frac{\delta^2}{X^2}\left(u'\frac{\partial u'}{\partial x'} + v'\frac{\partial u'}{\partial y'}\right) = \frac{\delta^2}{X^2}\left(\frac{\partial^2 u'}{\partial x'^2}\right) + \frac{\partial^2 u'}{\partial y'^2} - \frac{P\delta^2}{\mu X U}\left(\frac{\partial p'}{\partial x'}\right).$$

Explain why, for the inertia terms to balance the vertical diffusion term, we must choose $\delta = X/(\mathrm{Re})^{1/2}$ and to keep the pressure term we must choose $P = \mu X U/\delta^2$. Deduce that the horizontal diffusion term is consequently negligible in the limit $\mathrm{Re} \to \infty$.

(iii) By analogous arguments to those in part (ii) above, deduce that the equation for the non-dimensional vertical velocity field $v' = v'(x', y')$ is $\partial p'/\partial y' = 0$ in the limit $\mathrm{Re} \to \infty$. Use this fact and that the far-field ambient flow is quiescent to explain why we know $\partial p'/\partial x' = 0$ as well.

(iv) Suppose we set $m' := m/M$ for a characteristic momentum M. Explain why naturally $M = \rho U^2 \delta$. Hence show

$$\delta = \left(\frac{\rho \nu^2 X^2}{M}\right)^{1/3} \quad \text{and} \quad U = \left(\frac{M^2}{\rho^2 \nu X}\right)^{1/3}.$$

(c) Consider the boundary-layer equations derived in part (b) above.

(i) Reverting the boundary-layer equations to dimensional form, show, when expressed in terms of a stream function $\psi = \psi(x, y)$, that they become

$$\frac{\partial \psi}{\partial y}\frac{\partial^2 \psi}{\partial x \partial y} - \frac{\partial \psi}{\partial x}\frac{\partial^2 \psi}{\partial y^2} = \nu \frac{\partial^3 \psi}{\partial y^3}.$$

(ii) Look for a solution to the boundary-layer equations in part (c)(i) of the form $\psi = U(x)\,\delta(x) f(\eta)$, where $\eta := y/\delta$ and $\delta = \delta(x)$ and $U = U(x)$ have the form, i.e. x-dependence, shown in part (b)(iv). Show that the function $f = f(\eta)$ satisfies the nonlinear ordinary differential equation

$$3f''' + ff'' + (f')^2 = 0.$$

Explain why, assuming the flow remains symmetric about $y = 0$, suitable boundary conditions are $f(0) = 0$, $f''(0) = 0$ and as $\eta \to \infty$ we have $f' \to 0$. Further explain why in this case there is an additional (fourth!) constraint, namely

$$\int_{-\infty}^{\infty} (f'(\eta))^2 \, d\eta = 1.$$

(iii) Show that a solution to the nonlinear ordinary differential equation for f in part (ii) just above, satisfying the three stated boundary conditions, is

$$f = C \tanh(C\eta/6),$$

where C is an arbitrary constant. Show that the total momentum constraint implies $C = 6^{2/3}/2$.

(iv) Explain why the solution given in parts (ii) and (iii) just above is only valid for $X \gg \rho \nu^2/M$.

(d) Deduce the same conclusion as in part (a) by assuming the boundary-layer equations derived in part (b) govern the central part of the flow.

6.10 Blasius problem: Wake from a flat plate. Consider a shear flow $U_{\mathrm{E}} = \lambda\,|y|$ past a flat plate that coincides with $y = 0$, $x \leqslant 0$, where λ is a given constant. Suppose a wake develops behind the plate where $x > 0$ and the condition $\partial u/\partial y = 0$, i.e. zero shear, is maintained at $y = 0$ for $x > 0$. See Figure 6.10. Our goal in this question is to justify the assumption that the steady Prandtl boundary-layer equations apply in the region of the wake and look for a suitable similarity solution.

(a) Assuming for the moment a characteristic scale for U_{E} is U and X is a characteristic horizontal scale, use arguments exactly analogous to those in the

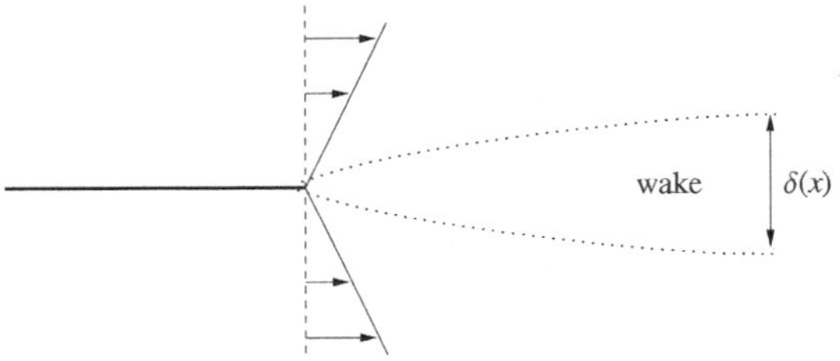

Figure 6.10 Wake flow beyond a long flat plate when the bulk Euler flow is a linear shear flow. Prandtl's boundary-layer equations apply in the central wake region indicated, of width $\delta = \delta(x)$.

last Exercise 6.9, parts (b)(i)–(iii), to explain why the steady boundary-layer equations apply in the region of the wake, which exists for $x > 0$ and has width δ. Explain why the Reynolds number is $\mathrm{Re} = UX/\nu$ and we must choose $\delta = X/(\mathrm{Re})^{1/2}$. Further explain why, since the ambient background velocity field U_{E} is x-independent, there is no external driving term in the boundary-layer flow.

(b) Now supposing $U_{\mathrm{E}} = \lambda\,|y|$, explain why

$$\delta = \left(\frac{\nu X}{\lambda}\right)^{1/3} \qquad \text{and} \qquad U = (\nu X \lambda^2)^{1/3}.$$

(c) Our results thus far imply that we can use the dimensional form for the boundary-layer equations in terms of a stream function $\psi = \psi(x, y)$ given in Exercise 6.9, part (c)(i). Look for a solution to this equation for the stream function of the form $\psi = U(x)\,\delta(x) f(\eta)$, where $\eta := y/\delta$ and $\delta = \delta(x)$ and $U = U(x)$ have the form, i.e. x-dependence, shown in part (b) just above. Show that the function $f = f(\eta)$ satisfies the nonlinear ordinary differential equation

$$3f''' + 2ff'' - (f')^2 = 0.$$

Explain why, assuming the flow remains symmetric about $y = 0$, suitable boundary conditions are $f(0) = 0$, $f''(0) = 0$ and $f' \to \eta$ as $\eta \to \infty$.

(d) Explain why the solution above is only valid for $X \gg (\nu/\lambda)^{1/2}$.

6.11 Blasius problem: Finite flat plate. Suppose a stream of uniform speed U at infinity flows past a finite plate aligned with the stream. A wake develops downstream of the plate. Let $u = u(x, y)$ denote the downstream velocity field with the x-direction parallel to the stream and the y-direction orthogonal to it in the standard juxtaposition.

(a) Using arguments analogous to those in Exercise 6.10 part (a) just above, explain why the steady boundary-layer equations are a suitable flow approximation in the wake. Explain why we must choose

$$\delta = \left(\frac{\nu X}{U}\right)^{1/2},$$

where X is a characteristic horizontal scale.

(b) Assuming the volume flux deficit

$$Q := \int_{-\infty}^{\infty} (U - u)\,\mathrm{d}y$$

in the wake is constant, explain why

$$U - u \propto Q\left(\frac{U}{\nu x}\right)^{1/2}.$$

(c) Linearise the steady dimensional boundary-layer equations as follows. Assume $u = U + \hat{u}$, where $|\hat{u}| \ll U$. By substituting this form into the boundary-layer equations, and assuming the velocity field v in the y-direction is such that $|v| \ll U$, show that they reduce to the heat equation

$$U\frac{\partial \hat{u}}{\partial x} = \nu\frac{\partial^2 \hat{u}}{\partial y^2}.$$

Explain why $\partial\hat{u}/\partial y = 0$ on $y = 0$ and $\hat{u} \to 0$ as $y \to \infty$ are suitable boundary conditions. Further, recall: (i) the solution of the heat equation is given by the convolution of the Gaussian density function of mean zero and variance ν/U with the data for $\hat{u}$ at $x = 0$, see Section 4.9(5); (ii) the discussion in Remark 6.5. With these in hand, explain why

$$u = U - Q\left(\frac{U}{4\nu\pi x}\right)^{1/2}\exp\left(-\frac{Uy^2}{4\nu x}\right) + \cdots.$$

(d) Consider a rectangular control region $\mathcal{V}$, as shown in Figure 6.11, which symmetrically straddles $y = 0$. Using the incompressibility condition to show that if all of $\partial\mathcal{V}$ is sufficiently far from the plate, and $\boldsymbol{n}$ is the unit outward normal, then

$$\int_{AB+CD} \boldsymbol{u}\cdot\boldsymbol{n}\,\mathrm{d}x = \int_{-h}^{h}(U-u)\,\mathrm{d}y.$$

(e) By considering Cauchy's equation of motion from Theorem 4.2 for the same control region $\mathcal{V}$ as $h \to \infty$, assuming the deviatoric stress is negligible on the resulting boundary $\partial\mathcal{V}_\infty$, show that the drag force F_{D} on the plate is given by

$$F_{\mathrm{D}} = -\rho\int_{\partial\mathcal{V}_\infty} \boldsymbol{u}\,(\boldsymbol{u}\cdot\boldsymbol{n})\,\mathrm{d}\boldsymbol{x}.$$

Hence using the result from part (d) applied in the limit $h \to \infty$, as well as part (b), show that the drag force $F_{\mathrm{D}} \approx \rho U Q$. See Panton (2013, pp. 579–81).

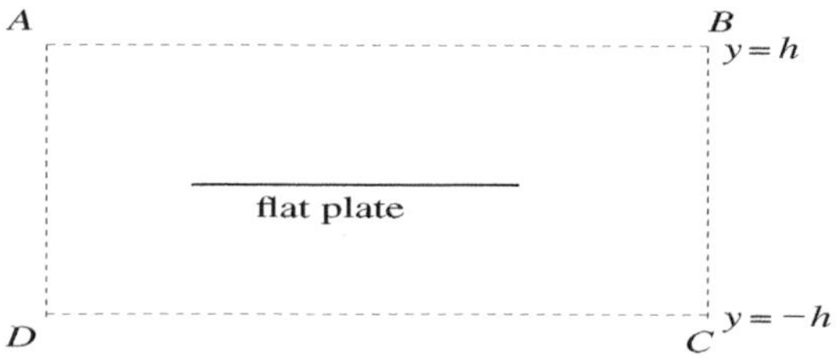

Figure 6.11 Wake flow beyond a finite flat plate when the bulk Euler flow is uniform. Prandtl's boundary layer equations apply in the wake. We also show a rectangular control region $\mathcal{V}$ around the finite flat plate.

7 Navier–Stokes Regularity

The well-posedness of smooth solutions to the three-dimensional incompressible Navier–Stokes equations globally in time is a major open mathematical problem. Our goal in this chapter is to provide a succinct though comprehensive introduction to the main well-posedness results that are known. In the three-dimensional case, we indicate how smooth solutions *may* develop singularities in finite time. At the same time we establish classical regularity assumptions/conditions that guarantee well-posedness globally in time, i.e. if we could prove three-dimensional incompressible Navier–Stokes solutions satisfied those assumptions/conditions, then we would establish global regularity. The determination of whether the three-dimensional incompressible Navier–Stokes equations develop a singularity in finite time (or not) is important, for if they do then 'our description of small scale flow must be missing some essential physics', quoting from Kerr and Oliver (2011). Many engineering and computational fluid dynamics (CFD) applications, for example in aerodynamics, rely on full three-dimensional incompressible Navier–Stokes simulation. These numerical simulations, and trust in them, relies on the fundamental assumption the equations are well-posed. The simulations do not resolve the finest scales and it is assumed the flow fields they reveal would persist in the limit of finest-scale resolution. If for classes of initial conditions the incompressible Navier–Stokes equations do develop a singularity in finite time, then such simulations would miss such a potentially catastrophic event. And we would need to take account of such possibilities in any CFD applications. Further, if solution flows do develop singularities in finite time, there would naturally be strong interest in classifying the initial conditions and mechanisms that cause them to develop. A succinct summary of classical regularity results can be found in, for example, Kerr and Oliver (2011) and/or Doering (2009). Possible routes to singularity, for example via the collapse of counter-rotating vortex tubes, and the role careful numerical simulations may play in establishing singularity formation, is discussed in detail in Kerr and Oliver (2011) – recall our discusson on vortex structures such as vortex rings and vortex tubes in Chapter 2. In this chapter we present the main results on Navier–Stokes regularity. These results narrow down the possible routes to singularity. Two recent results in particular stand out: the proof by Buckmaster and Vicol (2019) that a weak class of solutions (which do not necessarily satisfy the energy inequality) are not unique and the proof of singularity for an averaged Navier–Stokes equation by Tao (2016). See Remarks 7.26 and 7.35.

This chapter is structured as follows. We first introduce the essential background material in Section 7.1. We prove the global existence of weak solutions to the three-dimensional incompressible Navier–Stokes equations in Section 7.2. We prove all the results we need with two exceptions, the Aubin–Lions–Simon Lemma 7.25, whose proof can be found in Simon (1987), and the Sobolev–Gagliardo–Nirenberg inequality presented in Proposition 7.16(e). We discuss the existence of strong smooth solutions in Section 7.3 via classical a-priori energy estimates. These results on weak and strong solutions are classically found in, for example, Constantin and Foias (1988), Robinson *et al.* (2016) or Temam (1983). In Section 7.4 we outline a-priori estimates that indicate how local alignment or anti-alignment of vorticity supresses singularity formation. This establishes an important alternative assumption/condition that guarantees well-posedness of the three-dimensional incompressible Navier–Stokes equations, and is due to Constantin and Fefferman (1993). In Section 7.5 at the end of this chapter we point to further regularity results.

7.1 Lebesgue Integrability and Function Spaces

We outline principal results from functional analysis we require and notation we use. We state results herein without proof as they can be found in standard texts; for example Evans (1998). Throughout we assume $\mathcal{D} = \mathbb{T}^d$, the periodic domain of period 2π in dimension d, which is either 2 or 3. In other words, we assume the domain is the square or cube $[-\pi, \pi]^d$ with the boundary conditions

$$\boldsymbol{u}(\boldsymbol{x} + 2\pi\boldsymbol{h}, t) = \boldsymbol{u}(\boldsymbol{x}, t),$$

for all $\boldsymbol{x} \in \mathbb{R}^d$, $t \geqslant 0$ and for each of the cases $\boldsymbol{h} = \boldsymbol{i}$ and $\boldsymbol{h} = \boldsymbol{j}$ in two dimensions, and additionally $\boldsymbol{h} = \boldsymbol{k}$ in three dimensions. Quoting from Temam (1983): this boundary condition 'has no physical meaning', nevertheless 'the advantage of the boundary condition [above] is that it leads to a simpler functional setting, while many of the mathematical difficulties remain unchanged (except of course those related to the boundary layer difficulty, which vanish)'. In Section 7.4 the domain is $\mathcal{D} = \mathbb{R}^3$. We outline any adjustments we require as a result, therein.

We use $C^0(\mathbb{T}^d)$, or more succinctly $C(\mathbb{T}^d)$, to denote the space of continuous real-valued functions on $\mathbb{T}^d$. Similarly we use $C^1(\mathbb{T}^d)$ to denote the subspace of $C(\mathbb{T}^d)$ consisting of real-valued functions, all of whose first-order partial derivatives are also continuous. Naturally $C^\infty(\mathbb{T}^d)$ denotes the corresponding subset of functions whose partial derivatives to all orders are continuous. This is the space of *smooth functions* on $\mathbb{T}^d$. Let us now introduce the notion of *Lebesgue-integrable functions*. Thus far we have naturally interpreted any integrals in the Riemann sense. In other words, for a general domain $\mathcal{D}$, temporarily suspending that $\mathcal{D} = \mathbb{T}^d$ for the moment, we define the integral of a bounded real-valued function defined on $\mathcal{D}$ by partitioning $\mathcal{D}$ into suitable d-dimensional disjoint hyper-rectangles whose union is $\mathcal{D}$. We approximate the integral of the function over $\mathcal{D}$ using the partition in two ways. First by computing the supremum of the function in each partition and then considering the sum, across the partitions, of the product of the supremum in that hyper-rectangle and the area/volume (depending on whether d is 2 or 3) of the hyper-rectangle. The upper Riemann integral is the infimum of this sum over all such possible partitions. Second we can also compute the infimum of the function in each partition and consider the sum, across the partitions, of the product of the infimum in that hyper-rectangle and the area/volume of the hyper-rectangle. The lower Riemann integral is the supremum of this latter sum over all such possible partitions. If the upper and lower Riemann integrals coincide, the function is said to be Riemann integrable with the integral given by the coincidence value. To define the Lebesgue integral of a function we first need a notion of a Lebesgue measure function. And to introduce the latter we need a notion of a σ-algebra of $\mathcal{D}$. This is a collection of subsets Σ of $\mathcal{D}$ that includes $\mathcal{D}$ itself. The collection Σ is required to be closed under countable unions and also under the complement. These properties imply the collection Σ contains the empty set and is closed under countable intersections. We can endow the collection Σ with a measure, here the outer Lebesgue measure, as follows. Given any subset $\mathcal{V}$ of $\mathcal{D}$ in Σ, the outer Lebesgue measure $\mu(\mathcal{V})$ of $\mathcal{V}$ is the infimum of the sum, over all collections of hyper-rectangles whose union contains $\mathcal{V}$, of the volumes of the hyper-rectangles (areas in two dimensions). The Lebesgue measure satisfies the properties of a measure we would like: the measure of the empty set is zero; the measure of sets contained in other sets is smaller than the measure of the containing sets; and the measure of a union of subsets is less than the sum of the measure of the individual sets. A function $f\colon \mathcal{D} \to \mathbb{R}$ is Lebesgue measurable on $\mathcal{D}$ if the pre-image of every interval (u, ∞) lies in Σ endowed with the Lebesgue measure, i.e. if for all $u \in \mathbb{R}$:

$$\{\boldsymbol{x} \in \mathcal{D} \colon f(\boldsymbol{x}) > u\} \in \Sigma,$$

and is Lebesgue measurable. It remains to define the Lebesgue integral of such a function $f\colon \mathcal{D} \to \mathbb{R}$. The *characteristic function* $\chi_{\mathcal{V}}\colon \mathcal{D} \to \mathbb{R}$ of a subset $\mathcal{V}$ of $\mathcal{D}$ is such that $\chi_{\mathcal{V}}(\boldsymbol{x})$ is unity if $\boldsymbol{x} \in \mathcal{V}$ and zero otherwise. The function $\chi_{\mathcal{V}}$ is Lebesgue measurable if and only if $\mathcal{V}$ is a Lebesgue-measurable set. Suppose $s\colon \mathcal{D} \to [0, \infty)$ is a simple function of the form

$$s = \sum_{i=1}^{N} c_i\, \chi_{\mathcal{V}_i},$$

where the constants $c_i \geqslant 0$ and the subsets $\mathcal{V}_i \in \Sigma$. Then the integral of s with respect to μ is

$$\int s \,\mathrm{d}\mu = \sum_{i=1}^{N} c_i\, \mu(\mathcal{V}_i).$$

Definition 7.1 (Lebesgue integration) The Lebesgue integral of the positive, measurable, extended function $f\colon \mathcal{D} \to [0, \infty)$ is defined as

$$\int_{\mathcal{D}} f \,\mathrm{d}\mu := \sup\left\{\int s \,\mathrm{d}\mu \colon 0 \leqslant s \leqslant f\right\},$$

where the supremum is over all possible simple functions s satisfying the property indicated. For Lebesgue-measurable functions $f\colon \mathcal{D} \to \mathbb{R}$, we split the domain $\mathcal{D}$ into the subregion where f is positive and the subregion where f is negative. The

Lebesgue integral of f over the former subregion is defined as just above, while the Lebesgue integral of $-f$ over the latter region is similarly defined. The Lebesgue integral of f over $\mathcal{D}$ is then defined as the difference of these two integrals, i.e. the integral of f over the subregion where it is positive minus the integral of $-f$ over the subregion where f is negative.

For more details and a comprehensive account of Lebesgue measure and integration, see Hunter (2011). The Monotone Convergence Theorem and Dominated Convergence Theorem are particularly useful examples of results applicable for the Lebesgue integral. These imply the convergence of such integrals from the pointwise convergence of underlying sequences of integrands; see Hunter (2011, Sec. 4.5). Some further clarifications given therein are insightful here. The characteristic function on the rationals in [0, 1], i.e. the function $\chi_{\mathbb{Q}}\colon [0,1] \to \{0,1\}$, is not Riemann integrable – the lower and upper Riemann integrals are 0 and 1, respectively. However, $\chi_{\mathbb{Q}}$ is Lebesgue integrable with

$$\int \chi_{\mathbb{Q}} \, \mathrm{d}\mu = 1 \cdot \mu(\mathbb{Q} \cap [0,1]) = 0,$$

since the Lebesgue measure of $\mathbb{Q} \cap [0,1]$ is zero. On bounded domains, any Riemann-integrable function is Lebesgue measurable and thus Lebesgue integrable since it must be bounded. And of course the two integrals coincide. A Lebesgue-integrable function is not necessarily Riemann integrable. However, a Lebesgue-measurable function is Riemann integrable if and only if it is bounded and its set of discontinuities has Lebesgue measure zero. See Theorem 4.28 and Example 4.29 in Hunter (2004). One last caveat is that some improper Riemann integrals may exist while the corresponding Lebesgue integral does not; again see Hunter (2004, Example 4.12).

Remark 7.2 (Equivalence classes of almost everywhere equal functions) We identify Lebesgue-integrable functions that differ on a set of Lebesgue measure zero. In other words, if one Lebesgue-integrable function $f\colon \mathbb{T}^d \to \mathbb{R}$ differs from another on a set of Lebesgue measure zero in $\mathbb{T}^d$, then we consider them to be the same function. In particular, Lebesgue-integrable functions need only be defined *almost everywhere* on $\mathbb{T}^d$, meaning everywhere except on a set of Lebesgue measure zero. Consequently we thus consider *equivalence classes* of Lebesgue-integrable functions. One function can be used to represent all those that differ from it on a set of Lebesgue measure zero; they are all equivalenced.

Definition 7.3 (Lebesgue space) For any real number $p > 0$, the *Lebesgue space* $L^p(\mathbb{T}^d)$ is the space of equivalence classes of functions on $\mathbb{T}^d$ for which $|f|^p$ is Lebesgue integrable, i.e. it is the set of all Lebesgue-measurable functions $f\colon \mathbb{T}^d \to \mathbb{R}$ for which the following functional is finite:

$$\|f\|_{L^p(\mathbb{T}^d)} := \left(\int_{\mathbb{T}^d} |f|^p \, \mathrm{d}\mu \right)^{1/p}.$$

Remark 7.4 (Vector-valued functions) If the case of functions $f\colon \mathbb{T}^d \to \mathbb{R}^n$ we use $|f|^p$ in the functional in the definition above, where $|f|$ denotes the Euclidean length of $f \in \mathbb{R}^n$. We denote the corresponding space of functions $L^p(\mathbb{T}^d;\mathbb{R}^n)$, or simply $L^p(\mathbb{T}^d)$ if the image space is obvious from the context.

Remark 7.5 (Approximation by smooth functions) For $1 \leqslant p < \infty$, the spaces $C(\mathbb{T}^d)$ and $C^\infty(\mathbb{T}^d)$ are both dense in $L^p(\mathbb{T}^d)$. Hence it is always possible to find a smooth function arbitrarily close to an L^p function in the L^p-norm.

Definition 7.6 (Bounded functions) A measurable function $f\colon \mathbb{T}^d \to \mathbb{R}$ is *essentially bounded* if there exists a constant K such that $|f(\boldsymbol{x})| \leqslant K$ almost everywhere on $\mathbb{T}^d$. The greatest lower bound of such constants is called the essential supremum of $|f|$ on $\mathbb{T}^d$, which we write as $\operatorname{ess\,sup}_{\boldsymbol{x}\in\mathbb{T}^d} |f(\boldsymbol{x})|$. Then $L^\infty(\mathbb{T}^d)$ is the vector space consisting of all measurable functions f for which the norm $\|f\|_{L^\infty(\mathbb{T}^d)} := \operatorname{ess\,sup}_{\boldsymbol{x}\in\mathbb{T}^d} |f(\boldsymbol{x})|$ is finite.

Definition 7.7 (Inner product) On the space of square-integrable functions $L^2(\mathbb{T}^d;\mathbb{C}^n)$ we define the *inner product*

$$\langle \boldsymbol{f}, \boldsymbol{g}\rangle := \int_{\mathbb{T}^d} \boldsymbol{f}^\dagger \boldsymbol{g}\, \mathrm{d}\mu,$$

where $\boldsymbol{f}^\dagger$ denotes the complex conjugate transpose of $\boldsymbol{f}$. Equivalently we have $\boldsymbol{f}^\dagger\boldsymbol{g} = \boldsymbol{f}^*\cdot\boldsymbol{g}$, where $\boldsymbol{f}^*$ is the complex conjugate (only) of $\boldsymbol{f}$. The norm on $L^2(\mathbb{T}^d;\mathbb{C}^n)$ is naturally given by $\|\boldsymbol{f}\|^2_{L^2(\mathbb{T}^d;\mathbb{R}^n)} \equiv \langle \boldsymbol{f}, \boldsymbol{f}\rangle$.

Since the underlying domain is $\mathcal{D} = \mathbb{T}^d$, any square-integrable functions $\boldsymbol{f}\colon \mathbb{T}^d \to \mathbb{R}^n$ have a *Fourier series* representation of the form

$$\boldsymbol{f}(\boldsymbol{x}) = \sum_{\boldsymbol{k}\in\mathbb{Z}^d} F_{\boldsymbol{k}}\, \mathrm{e}^{\mathrm{i}\boldsymbol{k}\cdot\boldsymbol{x}},$$

for almost every $\boldsymbol{x} \in \mathbb{T}^d$. The $F_{\boldsymbol{k}}$ are the Fourier coefficients of $\boldsymbol{f}$, given by

$$F_{\boldsymbol{k}} = \frac{1}{(2\pi)^d}\int_{\mathbb{T}^d} \boldsymbol{f}(\boldsymbol{x})\, \mathrm{e}^{-\mathrm{i}\boldsymbol{k}\cdot\boldsymbol{x}}\, \mathrm{d}\boldsymbol{x}.$$

The inner product space $L^2(\mathbb{T}^d;\mathbb{C}^n)$ has an equivalent prescription as the space of all sequences $F\colon \mathbb{Z}^d \to \mathbb{C}^n$, where $F := \{F_{\boldsymbol{k}}\}_{\boldsymbol{k}\in\mathbb{Z}^d}$, which are square-summable. In particular $L^2(\mathbb{T}^d;\mathbb{C}^n)$ and the square-summable sequence space $\ell^2(\mathbb{Z}^d;\mathbb{C}^n)$ are naturally identified via equality of the inner product:

$$\langle \boldsymbol{f}, \boldsymbol{g}\rangle \equiv (2\pi)^d \sum_{\boldsymbol{k}\in\mathbb{Z}^d} F_{\boldsymbol{k}}^\dagger G_{\boldsymbol{k}},$$

which is Parseval's identity. Indeed $L^2(\mathbb{T}^d;\mathbb{C}^n)$, which is thus isomorphic to $\ell^2(\mathbb{Z}^d;\mathbb{C}^n)$, is a separable space. Fourier representation is useful in Section 7.2 as a natural context for a weak formulation of the incompressible Navier–Stokes equations.

We also need to define the *Sobolev space* $H^1(\mathbb{T}^d;\mathbb{C}^n)$, which is a natural subspace of $L^2(\mathbb{T}^d;\mathbb{C}^n)$. This subspace consists of those square-integrable functions whose 'derivatives' are also square-integrable in the sense that

$$\sum_{\boldsymbol{k}\in\mathbb{Z}^d} |F_{\boldsymbol{k}}|^2 + \sum_{\boldsymbol{k}\in\mathbb{Z}^d} |\boldsymbol{k}|^2 |F_{\boldsymbol{k}}|^2 < \infty.$$

Note that taking the gradient of the Fourier series expansion for f generates

$$\nabla f(x) = \mathrm{i} \sum_{k\in\mathbb{Z}^d} F_k\, k^{\mathrm{T}}\, \mathrm{e}^{\mathrm{i}k\cdot x}.$$

Hence the Fourier coefficient of the matrix ∇f is the matrix $\mathrm{i}F_k\, k^{\mathrm{T}}$. This is because the components of f are distributed across the rows of ∇f and partial derivatives are distributed across columns. The matrix $\mathrm{i}F_k\, k^{\mathrm{T}}$ reflects this structure in Fourier space. Hence if we consider the subspace of $L^2(\mathbb{T}^d;\mathbb{C}^n)$ whose 'derivatives' are also square-integrable, indeed component by component, then in addition to square-integrability of the functions f we should demand that the following quantity is finite:

$$\begin{aligned}
\int_{\mathbb{T}^d} \mathrm{tr}\,((\nabla f)^{\dagger}(\nabla f))\,\mathrm{d}x &= \mathrm{tr} \sum_{k,l\in\mathbb{Z}^d} (F_k\, k^{\mathrm{T}})^{\dagger} F_l\, l^{\mathrm{T}} \int_{\mathbb{T}^d} \mathrm{e}^{2\mathrm{i}(l-k)\cdot x}\,\mathrm{d}x \\
&= (2\pi)^d\, \mathrm{tr} \sum_{k\in\mathbb{Z}^d} k^* F_k^{\dagger} F_k\, k^{\mathrm{T}} \\
&= (2\pi)^d \sum_{k\in\mathbb{Z}^d} (\mathrm{tr}\, k^*\, k^{\mathrm{T}})\, |F_k|^2 \\
&= (2\pi)^d \sum_{k\in\mathbb{Z}^d} |k|^2 |F_k|^2,
\end{aligned}$$

where we have used that the integral of $\mathrm{e}^{2\mathrm{i}(l-k)\cdot x}$ over $\mathbb{T}^d$ is $(2\pi)^d \delta_{k,l}$, where $\delta_{k,l}$ is the Kronecker delta function. This representation in Fourier space is straightforwardly extended to a natural inner product.

Definition 7.8 (Sobolev Hilbert spaces) The Sobolev space $H^1(\mathbb{T}^d;\mathbb{C}^n)$ is a separable Hilbert space with inner product

$$\langle f, g\rangle_{H^1(\mathbb{T}^d;\mathbb{C}^n)} := \langle f, g\rangle_{L^2(\mathbb{T}^d;\mathbb{C}^n)} + \langle \nabla f, \nabla g\rangle_{L^2(\mathbb{T}^d;\mathbb{C}^{n\times n})},$$

or equivalently

$$\langle f, g\rangle_{H^1(\mathbb{T}^d;\mathbb{C}^n)} = (2\pi)^d \sum_{k\in\mathbb{Z}^d} (1+|k|^2)\, F_k^{\dagger} G_k.$$

The norm on $H^1(\mathbb{T}^d;\mathbb{C}^n)$ is given by $\|f\|^2_{H^1(\mathbb{T}^d;\mathbb{C}^d)} \equiv \langle f, f\rangle_{H^1(\mathbb{T}^d;\mathbb{C}^n)}$.

Remark 7.9 We define general (i.e. abstract) Hilbert spaces momentarily.

Definition 7.10 (Sequence space $h^1(\mathbb{Z}^d;\mathbb{C}^n)$) The sequence space $h^1(\mathbb{Z}^d;\mathbb{C}^n) \subset \ell^2(\mathbb{Z}^d;\mathbb{C}^n)$ represents the Fourier sequence space isomorphic to $H^1(\mathbb{T}^d;\mathbb{C}^n)$. Thus $h^1(\mathbb{Z}^d;\mathbb{C}^n)$ is the Hilbert space with inner product defined by

$$\langle F, G\rangle_{h^1(\mathbb{Z}^d;\mathbb{C}^n)} := (2\pi)^d \sum_{k\in\mathbb{Z}^d} (1+|k|^2) F_k^{\dagger} G_k,$$

for any two sequences $F, G\colon \mathbb{Z}^d \to \mathbb{C}^n$ and such that for every $F \in h^1(\mathbb{Z}^d;\mathbb{C}^n)$ the norm $\|F\|_{h^1(\mathbb{Z}^d;\mathbb{C}^n)} := (\langle F, F\rangle_{h^1(\mathbb{Z}^d;\mathbb{C}^n)})^{1/2}$ is finite.

Remark 7.11 Strictly, the 'derivatives' referred to above in our discussion preceding Definition 7.8, as well as the gradients in Definition 7.8 itself, are 'weak derivatives'. The definition of $H^1(\mathbb{T}^d;\mathbb{C}^n)$ does not require that its elements f are continuously differentiable. It requires that there exists a $v \in L^2(\mathbb{T}^d;\mathbb{C}^{n\times d})$, known as the weak gradient of f, which satisfies

$$\int_{\mathbb{T}^d} f^{\mathrm{T}}(\nabla w)\,\mathrm{d}x = -\int_{\mathbb{T}^d} w^{\mathrm{T}} v\,\mathrm{d}x,$$

for all test functions $w \in C^\infty(\mathbb{T}^d;\mathbb{C}^n)$. For more details, see e.g. Evans (1998, pp. 242–3). In our proof of the global existence of weak solutions to the incompressible Navier–Stokes equations in Section 7.2, we use the Fourier sequence representation for the solution and thus consider sequence spaces such as $\ell^2(\mathbb{Z}^d;\mathbb{C}^n)$ or $h^1(\mathbb{Z}^d;\mathbb{C}^n)$, as well as similar related sequence spaces. We can always view the square-integrability of the weak derivatives in Definition 7.8 of $H^1(\mathbb{T}^d;\mathbb{C}^n)$ via the respective weighted convergence of the Fourier sequences in the description of $h^1(\mathbb{Z}^d;\mathbb{C}^n)$ in Definition 7.10.

We denote a generic Banach space by $\mathbb{B}$ with norm $\|\cdot\|_{\mathbb{B}}$. Recall that a Banach space $\mathbb{B}$ is a linear space equipped with a norm which is complete, i.e. every Cauchy sequence in $\mathbb{B}$ converges within $\mathbb{B}$; see Evans (1998, p. 241). We denote a generic Hilbert space by $\mathbb{H}$. Recall that a Hilbert space is a Banach space endowed with an inner product $\langle\cdot,\cdot\rangle_{\mathbb{H}}\colon \mathbb{H}\times\mathbb{H}\to\mathbb{R}$ which generates the norm; see Evans (1998, p. 636). Examples we have already seen are $L^2(\mathbb{T}^d;\mathbb{C}^n)$, $\ell^2(\mathbb{Z}^d;\mathbb{C}^n)$ or $H^1(\mathbb{T}^d;\mathbb{C}^n)$. A *bounded linear functional* $\Lambda\colon \mathbb{B}\to\mathbb{R}$ on $\mathbb{B}$ is a real-valued linear operator for which $\sup_{F\in\mathbb{B}}\{\|\Lambda F\|_{\mathbb{R}}\colon \|F\|_{\mathbb{B}}\leqslant 1\}$ is finite. We denote the set of all bounded linear functionals on $\mathbb{B}$ by $\mathbb{B}^*$, which we call the *dual space* of $\mathbb{B}$. We define the following norm on $\mathbb{B}^*$:

$$\|\Lambda\|_{\mathbb{B}^*} := \sup_{F\in\mathbb{B}}\{|\Lambda F|\colon \|F\|_{\mathbb{B}}\leqslant 1\}.$$

See e.g. Reed and Simon (1975, pp. 43, 72) or Evans (1998, pp. 637–9). For Hilbert spaces we have the following.

Theorem 7.12 (Riesz Representation Theorem) *For each $\Lambda\in\mathbb{H}^*$, there exists a unique element $V\in\mathbb{H}$ such that for all $F\in\mathbb{H}$:*

$$\Lambda F = \langle V, F\rangle_{\mathbb{H}}.$$

The mapping $\Lambda\mapsto V$ is a linear isomorphism of $\mathbb{H}^*$ onto $\mathbb{H}$ which is isometric as $\|\Lambda\|_{\mathbb{H}^*} = \|V\|_{\mathbb{H}}$; the two spaces can be canonically identified.

Hence we observe that, for the dual of a Hilbert space, it is natural to set

$$\|\Lambda\|_{\mathbb{H}^*} := \sup_{F\in\mathbb{H}}\left\{|\langle V, F\rangle_{\mathbb{H}}|\colon \|F\|_{\mathbb{H}}\leqslant 1\right\}.$$

The notion of *compactness* will play an important role in Section 7.2, where we establish sequences of functions or sequences of infinite sets of Fourier coefficients, i.e. sequences of sequences, which are uniformly bounded. When we deal with sequences of elements in a finite-dimensional vector space, then for such uniformly bounded sequences we would invoke the Bolzano–Weierstrass Theorem and deduce that there exists a convergent subsequence whose limit lies within the finite-dimensional vector space. However for a sequence of functions in $L^2(\mathbb{T}^d;\mathbb{C}^n)$ or sequences of elements/sequences in $\ell^2(\mathbb{Z}^d;\mathbb{C}^n)$, for example, which are infinite-dimensional vector spaces, the direct corresponding result is that there only exists a subsequence that *weakly converges* in the infinite-dimensional vector space.

Definition 7.13 (Weak convergence) A sequence $F^K := \{F_{\boldsymbol{k}}^K\}_{\boldsymbol{k}\in\mathbb{Z}^d} \in \mathbb{B}$ converges weakly to $F \in \mathbb{B}$ if $\Lambda F^K \to \Lambda F$ for every $\Lambda \in \mathbb{B}^*$. In the case of a Hilbert space $\mathbb{H}$ an equivalent condition is $\langle V, F^K\rangle \to \langle V, F\rangle_{\mathbb{H}}$ for every $V \in \mathbb{H}$.

As already suggested, if F^K is a uniformly bounded sequence in $\mathbb{B}$, then there exists a subsequence F^{K_j} and some $F \in \mathbb{B}$ such that $\Lambda F^{K_j} \to \Lambda F$ for every $\Lambda \in \mathbb{B}^*$. This follows from the Bolzano–Weierstrass Theorem as for each $\Lambda \in \mathbb{B}^*$ the quantities ΛF^{K_j} represent a sequence of real numbers. For an example of a weakly convergent function sequence that is not strongly convergent, see Exercise 7.1. Naturally, by *strong convergence* we mean as $K \to \infty$,

$$\|F^K - F\|_{\mathbb{B}} \to 0.$$

If, however, the Banach space $\mathbb{B}$ is compactly embedded into a Banach space $\mathbb{B}'$, then we can conclude that for any uniformly bounded sequence in $\mathbb{B}$, there exists a subsequence that strongly converges in $\mathbb{B}'$. See Evans (1998, pp. 271–2).

Definition 7.14 (Continuous and compact embeddings) Let $\mathbb{B}$ and $\mathbb{B}'$ be Banach spaces. We say $\mathbb{B}$ is *continuously embedded* into $\mathbb{B}'$ and write $\mathbb{B} \hookrightarrow \mathbb{B}'$ if for every $F \in \mathbb{B}$ we have

$$\|F\|_{\mathbb{B}'} \leqslant c\,\|F\|_{\mathbb{B}},$$

for some constant $c > 0$. We say $\mathbb{B}$ is *compactly embedded* into $\mathbb{B}'$ and write

$$\mathbb{B} \hookrightarrow\!\!\!\!\rightarrow \mathbb{B}',$$

if $\mathbb{B}$ is continuously embedded into $\mathbb{B}'$ and each bounded sequence in $\mathbb{B}$ is precompact in $\mathbb{B}'$, i.e. there exists a subsequence that converges in $\mathbb{B}'$.

Example 7.15 We demonstrate that $H^1(\mathbb{T}^d;\mathbb{C}^n)$ is compactly embedded into $L^2(\mathbb{T}^d;\mathbb{C}^n)$, or equivalently $h^1(\mathbb{Z}^d;\mathbb{C}^n)$ is compactly embedded into $\ell^2(\mathbb{Z}^d;\mathbb{C}^n)$. Suppose $F^K\colon \mathbb{Z}^d \to \mathbb{C}^n$ for $K \in \mathbb{N}$ is a sequence $F^K := \{F_{\boldsymbol{k}}^K\}_{\boldsymbol{k}\in\mathbb{Z}^d}$ that is uniformly bounded in $h^1(\mathbb{Z}^d;\mathbb{C}^n)$, i.e. there exists a constant $c > 0$, independent of K, such that in particular

$$\sum_{\boldsymbol{k}\in\mathbb{Z}^d} |\boldsymbol{k}|^2 |F_{\boldsymbol{k}}^K|^2 \leqslant c.$$

This implies there exists a subsequence F^{K_j} that weakly converges to $F := \{F_{\boldsymbol{k}}\}_{\boldsymbol{k}\in\mathbb{Z}^d}$ in $h^1(\mathbb{Z}^d;\mathbb{C}^n)$. In other words, for every $V \in h^1(\mathbb{Z}^d;\mathbb{C}^n)$ we know $\langle V, F^{K_j}\rangle_{h^1(\mathbb{Z}^d;\mathbb{C}^n)} \to \langle V, F\rangle_{h^1(\mathbb{Z}^d;\mathbb{C}^n)}$. We prove F^{K_j} strongly converges to F in $\ell^2(\mathbb{Z}^d;\mathbb{C}^n)$. See Temam

(1983, p. 10). An immediate consequence of the uniform boundedness condition above is, for each $K_j \in \mathbb{N}$ and some constant $C > 0$:

$$\sum_{k \in \mathbb{Z}^d} |k|^2 |F_k^{K_j} - F_k|^2 \leqslant C.$$

Note that we can choose $V = E_k^n$, the sequence of vectors all of whose entries are zero vectors apart from the kth one, all of whose components are zero apart from the nth one, which is 1. For example, this means that the weak convergence result implies $F_k^{K_j} \to F_k$ as $K_j \to \infty$, for each $k \in \mathbb{Z}^d$. In other words we have pointwise, i.e. for each $k \in \mathbb{Z}^d$, convergence in sequence space. However our goal is to prove

$$\sum_{k \in \mathbb{Z}^d} |F_k^{K_j} - F_k|^2 \to 0.$$

To this end we observe that for every $\kappa > 0$ we have

$$\begin{aligned}
\|F_k^{K_j} - F_k\|^2_{\ell^2(\mathbb{Z}^d;\mathbb{C}^n)} &= \sum_{k \in \mathbb{Z}^d} |F_k^{K_j} - F_k|^2 \\
&\leqslant \sum_{|k| \leqslant \kappa} |F_k^{K_j} - F_k|^2 + \frac{1}{\kappa^2} \sum_{|k| > \kappa} |k|^2 |F_k^{K_j} - F_k|^2 \\
&\leqslant \sum_{|k| \leqslant \kappa} |F_k^{K_j} - F_k|^2 + \frac{C}{\kappa^2},
\end{aligned}$$

where we used $|k| > \kappa$ if and only if $1 < |k|^2/\kappa^2$. If we now take the limit $K_j \to \infty$ then we observe

$$\limsup_{K_j \to \infty} \|F_k^{K_j} - F_k\|^2_{\ell^2(\mathbb{Z}^d;\mathbb{C}^n)} \leqslant \frac{C}{\kappa^2}.$$

Since the upper bound can be chosen arbitrarily small, we conclude that $F^{K_j} \to F$ as $K_j \to \infty$. Hence we deduce $h^1(\mathbb{Z}^d;\mathbb{C}^n) \hookrightarrow\!\!\!\!\rightarrow \ell^2(\mathbb{Z}^d;\mathbb{C}^n)$.

Let us now outline some inequalities we need in Sections 7.2–7.4.

Proposition 7.16 (Useful inequalities) *The following inequalities hold:*

(a) ***Young's inequality.*** *For any $a \geqslant 0$, $b \geqslant 0$, $\epsilon > 0$ and pair of real numbers p and q with $p \in (1, \infty)$ satisfying $1/p + 1/q = 1$, we have*

$$ab \leqslant \frac{1}{p}(a\epsilon)^p + \frac{1}{q}\left(\frac{b}{\epsilon}\right)^q.$$

(b) ***Hölder's inequality.*** *Suppose the real numbers p and q with $p \in [1, \infty]$ satisfy $1/p + 1/q = 1$. If $f \in L^p(\mathbb{T}^d;\mathbb{R})$ and $g \in L^q(\mathbb{T}^d;\mathbb{R})$, then $f^* g \in L^1(\mathbb{T}^d;\mathbb{R})$ and we have*

$$\int_{\mathbb{T}^d} |f^*(x)\, g(x)|\, \mathrm{d}x \leqslant \|f\|_{L^p(\mathbb{T}^d;\mathbb{R})} \|g\|_{L^q(\mathbb{T}^d;\mathbb{R})}.$$

Equivalently, if $F \in \ell^p(\mathbb{Z}^d;\mathbb{C})$ and $G \in \ell^q(\mathbb{Z}^d;\mathbb{C})$, then $F^\dagger G \in \ell^1(\mathbb{Z}^d;\mathbb{C})$, where we interpret $F^\dagger G$ as a term-by-term product of the sequences, and we have

$$\sum_{k \in \mathbb{Z}^d} |F_k^* G_k| \leqslant \|F\|_{\ell^p(\mathbb{Z}^d;\mathbb{C})} \|G\|_{\ell^q(\mathbb{Z}^d;\mathbb{C})}.$$

(c) ***Young's convolution inequality.*** *Suppose the triplet of real numbers* $p, q, r \in [1, \infty)$ *satisfy* $1/p + 1/q + 1/r = 2$. *If* $F \in \ell^p(\mathbb{Z}^d; \mathbb{C}^n)$, $G \in \ell^q(\mathbb{Z}^d; \mathbb{C}^{n\times n})$ *and* $H \in \ell^r(\mathbb{Z}^d; \mathbb{C}^n)$, *then we have (the sum is over* $\boldsymbol{k}, \boldsymbol{l} \in \mathbb{Z}^d$*)*

$$\sum_{\boldsymbol{k},\boldsymbol{l}} |F_{\boldsymbol{k}}^\dagger G_{\boldsymbol{l}} H_{\boldsymbol{k}-\boldsymbol{l}}| \leqslant \|F\|_{\ell^p(\mathbb{Z}^d;\mathbb{C}^n)} \|G\|_{\ell^q(\mathbb{Z}^d;\mathbb{C}^{n\times n})} \|H\|_{\ell^r(\mathbb{Z}^d;\mathbb{C}^n)}.$$

(d) ***Poincaré's inequality.*** *Suppose* $\boldsymbol{f} \in H^1(\mathbb{T}^d; \mathbb{R}^n)$ *has zero mean over* $\mathbb{T}^d$, *i.e. the integral of* $\boldsymbol{f}$ *over* $\mathbb{T}^d$ *is zero. Then for some constant* $c > 0$ *we have*

$$\|\boldsymbol{f}\|_{L^2(\mathbb{T}^d;\mathbb{R}^n)} \leqslant c\, \|\nabla \boldsymbol{f}\|_{L^2(\mathbb{T}^d;\mathbb{R}^{n\times n})}.$$

(e) ***Sobolev–Gagliardo–Nirenberg inequality.*** *Suppose that* $\boldsymbol{f} \in H^1(\mathbb{T}^d; \mathbb{R}^n)$, $p \in [2, 2d/(d-2)]$ *and* $a = d(p-2)/2p$. *Then for some constant* $c > 0$ *we have*

$$\|\boldsymbol{f}\|_{L^p(\mathbb{T}^d;\mathbb{R}^n)} \leqslant c\, \|\nabla \boldsymbol{f}\|^a_{L^2(\mathbb{T}^d;\mathbb{R}^{n\times n})} \|\boldsymbol{f}\|^{1-a}_{L^2(\mathbb{T}^d;\mathbb{R}^n)}.$$

Remark 7.17 For the inequality statements in Proposition 7.16, inequalities (a), (b) and (c) are proved in the respective Exercises 7.2, 7.3 and 7.4. The Poincaré inequality (d) in our context here follows directly from the inequality

$$\sum_{\boldsymbol{k}\in\mathbb{Z}^d\setminus\{\boldsymbol{0}\}} F_{\boldsymbol{k}}^\dagger F_{\boldsymbol{k}} \leqslant \sum_{\boldsymbol{k}\in\mathbb{Z}^d} |\boldsymbol{k}|^2 F_{\boldsymbol{k}}^\dagger F_{\boldsymbol{k}},$$

since $\boldsymbol{f}$ has mean zero over $\mathbb{T}^d$ and thus the first Fourier coefficient $F_0 = \boldsymbol{0}$. If $\boldsymbol{f}$ doesn't have zero mean over $\mathbb{T}^d$, the Poincaré inequality still applies if $\boldsymbol{f}$ is replaced by the difference of $\boldsymbol{f}$ and its mean over $\mathbb{T}^d$ on the left. Further, Poincaré's inequality generalises for example to connected domains $\mathcal{D}$ with continuously differentiable boundaries, and for L^p-norms; see Evans (1998, p. 275). For the Sobolev–Gagliardo–Nirenberg inequality (e), see Nirenberg (1959). Further details and references can be found in Malham (1993).

In Sections 7.2–7.4 we consider functions of space $\boldsymbol{x} \in \mathbb{T}^d$ and time $t \in [0, T]$ for some $T > 0$. And we need to separate the time and spatial regularity of the functions concerned. We have already seen the notation $L^2(\mathbb{T}^d; \mathbb{R}^n)$ which denotes the space of vector-valued functions, each of whose n components are themselves square-integrable. For functions $\boldsymbol{f}\colon \mathbb{T}^d \times [0, T] \to \mathbb{R}^n$ we can distinguish those which lie in $C([0, T]; L^2(\mathbb{T}^d; \mathbb{R}^n))$. This is the space of functions $\boldsymbol{f} = \boldsymbol{f}(\boldsymbol{x}, t)$ on $\mathbb{T}^d \times [0, T]$ that are square-integrable over $\mathbb{T}^d$ and continuous in time on $[0, T]$. In other words, for such functions the quantity

$$\|\boldsymbol{f}(\cdot, t)\|_{L^2(\mathbb{T}^d;\mathbb{R}^n)}$$

is continuous in time. As another example, for some $p \in (0, \infty)$, consider the functions $\boldsymbol{f} \in L^p(0, T; H^1(\mathbb{T}^d; \mathbb{R}^n))$. Such functions satisfy the condition

$$\int_0^T \|\boldsymbol{f}(\cdot, t)\|^p_{H^1(\mathbb{T}^d;\mathbb{R}^n)}\, \mathrm{d}t < \infty.$$

This is an example of a *Bochner integral*.

7.2 Weak Solutions

We consider herein the three-dimensional incompressible Navier–Stokes equations in the domain $\mathcal{D} = \mathbb{T}^3$ and in the time interval $[0, T]$. We are concerned with global-in-time solutions and so want to prove the existence of suitable solutions on $[0, T]$ for arbitrary $T > 0$. In three dimensions, only the existence of finite-energy weak solutions global in time can be proved, and we establish this herein. To begin, let us *assume* the initial data $\boldsymbol{u}_0(\boldsymbol{x})$ is smooth and the solution $\boldsymbol{u} = \boldsymbol{u}(\boldsymbol{x}, t)$ to the three-dimensional incompressible Navier–Stokes equations

$$\frac{\partial \boldsymbol{u}}{\partial t} + (\boldsymbol{u} \cdot \nabla)\, \boldsymbol{u} = \nu\, \Delta \boldsymbol{u} - \nabla p,$$
$$\nabla \cdot \boldsymbol{u} = 0,$$

is *smooth* for all $t \in [0, T]$. Let us establish an a-priori, i.e. under this smoothness assumption, estimate for the energy. We directly compute $(\mathrm{d}/\mathrm{d}t)\frac{1}{2}\|\boldsymbol{u}\|_{L^2}^2$ as follows:

$$\begin{aligned}
\frac{\mathrm{d}}{\mathrm{d}t}\frac{1}{2}\|\boldsymbol{u}\|_{L^2}^2 &= \frac{1}{2}\int_{\mathbb{T}^d} \partial_t(\boldsymbol{u} \cdot \boldsymbol{u})\,\mathrm{d}\boldsymbol{x} \\
&= \frac{1}{2}\int_{\mathbb{T}^d} 2\,\boldsymbol{u} \cdot (\partial_t \boldsymbol{u})\,\mathrm{d}\boldsymbol{x} \\
&= \int_{\mathbb{T}^d} \boldsymbol{u} \cdot (-(\boldsymbol{u} \cdot \nabla)\,\boldsymbol{u} + \nu\,\Delta \boldsymbol{u} - \nabla p)\,\mathrm{d}\boldsymbol{x} \\
&= \int_{\mathbb{T}^d} \Big(-(\boldsymbol{u} \cdot \nabla)\,(\tfrac{1}{2}\boldsymbol{u} \cdot \boldsymbol{u}) + \nu\,\boldsymbol{u} \cdot \Delta \boldsymbol{u} - \boldsymbol{u} \cdot \nabla p\Big)\,\mathrm{d}\boldsymbol{x} \\
&= \int_{\mathbb{T}^d} \Big(-(\boldsymbol{u} \cdot \nabla)\,(p + \tfrac{1}{2}|\boldsymbol{u}|^2) + \nu\,\boldsymbol{u} \cdot \Delta \boldsymbol{u}\Big)\,\mathrm{d}\boldsymbol{x} \\
&= -\,\nu \int_{\mathbb{T}^d} |\nabla \boldsymbol{u}|^2\,\mathrm{d}\boldsymbol{x}.
\end{aligned}$$

In the last step we used the following identities. First we used that $\nabla \cdot (\phi \boldsymbol{u}) \equiv (\boldsymbol{u} \cdot \nabla)\phi + (\nabla \cdot \boldsymbol{u})\phi$ with $\phi := p + \frac{1}{2}|\boldsymbol{u}|^2$, which is identity #8 from Section A.1 of the Appendix. Second, as a consequence we used that

$$\int_{\mathbb{T}^d} \boldsymbol{u} \cdot \nabla(p + \tfrac{1}{2}|\boldsymbol{u}|^2)\,\mathrm{d}\boldsymbol{x} = \int_{\mathbb{T}^d} \nabla \cdot \Big(\boldsymbol{u}\,(p + \tfrac{1}{2}|\boldsymbol{u}|^2)\Big)\,\mathrm{d}\boldsymbol{x} - \int_{\mathbb{T}^d} (\nabla \cdot \boldsymbol{u})(p + \tfrac{1}{2}|\boldsymbol{u}|^2)\,\mathrm{d}\boldsymbol{x}.$$

Third, since $\boldsymbol{u} \cdot \Delta \boldsymbol{u} \equiv \boldsymbol{u}^{\mathrm{T}} \Delta \boldsymbol{u}$ we also have

$$\int_{\mathbb{T}^d} \boldsymbol{u} \cdot (\Delta \boldsymbol{u})\,\mathrm{d}\boldsymbol{x} = \int_{\mathbb{T}^d} \nabla \cdot (\boldsymbol{u}^{\mathrm{T}}(\nabla \boldsymbol{u}))\,\mathrm{d}\boldsymbol{x} - \int_{\mathbb{T}^d} |\nabla \boldsymbol{u}|^2\,\mathrm{d}\boldsymbol{x},$$

where $|\nabla \boldsymbol{u}|^2$ is the square of the Frobenius norm of $\nabla \boldsymbol{u}$, which is the sum of the squares of all the components of the matrix $\nabla \boldsymbol{u}$. Fourth, we used that $\boldsymbol{u}$ is divergence-free. Fifth, and last, we used that by the divergence theorem the integrals of the divergence terms on the right generate integrals which, due to the periodic boundary conditions assumed, are zero; see Remark 7.20(iii) below. We have thus established that, a priori, we have

$$\frac{\mathrm{d}}{\mathrm{d}t}\frac{1}{2}\|\boldsymbol{u}\|_{L^2(\mathbb{T}^3;\mathbb{R}^3)^2} + \nu\|\nabla \boldsymbol{u}\|_{L^2(\mathbb{T}^3;\mathbb{R}^{3\times 3})}^2 = 0$$

$$\Leftrightarrow \qquad \tfrac{1}{2}\|\boldsymbol{u}(t)\|^2_{L^2(\mathbb{T}^3;\mathbb{R}^3)} + \nu \int_0^t \|\nabla \boldsymbol{u}(\tau)\|^2_{L^2(\mathbb{T}^3;\mathbb{R}^{3\times 3})}\,\mathrm{d}\tau = \tfrac{1}{2}\|\boldsymbol{u}_0\|^2_{L^2(\mathbb{T}^3;\mathbb{R}^3)},$$

for any $t \in [0,T]$. Hence we expect solutions to have finite energy for all time and in particular we expect $\|\boldsymbol{u}(t)\|_{L^2(\mathbb{T}^3;\mathbb{R}^3)}$ to be uniformly bounded in time, i.e. $\boldsymbol{u} \in L^\infty(0,T;L^2(\mathbb{T}^3;\mathbb{R}^3))$, and $\|\nabla\boldsymbol{u}(t)\|_{L^2(\mathbb{T}^3;\mathbb{R}^{3\times 3})}$ to be square-integrable in time, so that we expect $\boldsymbol{u} \in L^2(0,T;H^1(\mathbb{T}^3;\mathbb{R}^3))$. In the proof of existence of weak solutions below, we are only able to establish the corresponding time-integrated version of the energy equality just above as an inequality.

Remark 7.18 (Energy equality) Under what regularity constraints the incompressible Navier–Stokes equations actually obey the energy equality above is a very active area of research, see e.g. Shinbrot (1974), Galdi (2019) and Berselli and Chiodaroli (2020).

To proceed to prove the existence of finite-energy global solutions requires the notion of a weak solution. We consider the dot product of the Navier–Stokes momentum equation with a smooth divergence-free function $\boldsymbol{v}\colon \mathbb{T}^3 \to \mathbb{R}^3$, integrate over $\mathbb{T}^3$ and then integrate in time over $[0,T]$. We use the identity $\nabla\cdot(\boldsymbol{v}^{\mathrm{T}}(\nabla\boldsymbol{u})) \equiv \operatorname{tr}((\nabla\boldsymbol{v})^{\mathrm{T}}(\nabla\boldsymbol{u})) + \boldsymbol{v}^{\mathrm{T}}\Delta\boldsymbol{u}$ to 'integrate by parts' the diffusion term. The divergence theorem and periodic boundary conditions imply the divergence/boundary term is zero. Then using density arguments, that there exist smooth divergence-free test functions $\boldsymbol{v}\colon \mathbb{T}^3 \to \mathbb{R}^3$ arbitrarily close to any functions in $H^1(\mathbb{T}^3;\mathbb{R}^3)$, we observe that the same identity holds for any divergence-free test functions $\boldsymbol{v} \in H^1(\mathbb{T}^3;\mathbb{R}^3)$. Indeed, we observe that we only require $\boldsymbol{v} \in H^1(\mathbb{T}^3;\mathbb{R}^3)$ for the identity to make sense. We thus arrive at the following classical formulation.

Definition 7.19 (Weak solution) A weak solution of the incompressible Navier–Stokes equations on $[0,T]$ with divergence-free initial data $\boldsymbol{u}_0 \in L^2(\mathbb{T}^3;\mathbb{R}^3)$ is a divergence-free function

$$\boldsymbol{u} \in L^\infty(0,T;L^2(\mathbb{T}^3;\mathbb{R}^3)) \cap L^2(0,T;H^1(\mathbb{T}^3;\mathbb{R}^3)),$$

satisfying the equation

$$\langle \boldsymbol{v},\boldsymbol{u}(t)\rangle - \langle \boldsymbol{v},\boldsymbol{u}_0\rangle = -\int_0^t \Big(\nu\langle\nabla\boldsymbol{v},\nabla\boldsymbol{u}(\tau)\rangle + \langle \boldsymbol{v},((\boldsymbol{u}\cdot\nabla)\,\boldsymbol{u})(\tau)\rangle\Big)\,\mathrm{d}\tau,$$

for every divergence-free $\boldsymbol{v} \in H^1(\mathbb{T}^3;\mathbb{R}^3)$ and every $t \in [0,T]$.

Remark 7.20 We observe:

(i) The first term on the right involves the inner product on $L^2(\mathbb{T}^3;\mathbb{R}^{3\times 3})$.

(ii) For the weak formulation of the equations above to make sense, notice that it involves the two terms on the left which are pointwise in time, the property of weak continuity in time for the solution is required. This can indeed be established; see Remark 7.27.

(iii) In the definition above we have used that $\langle \boldsymbol{v},\nabla p\rangle = 0$. This can be established by using identity #8 from Section A.1 of the Appendix as well as the divergence theorem and periodic boundary conditions. Note that when applying the divergence theorem we imagine six 'virtual faces' for $\mathbb{T}^3$, for example where $\boldsymbol{x}$ equals $(-\pi,y,z)^{\mathrm{T}}$,

$(\pi, y, z)^{\mathrm{T}}$, $(x, -\pi, z)^{\mathrm{T}}$, $(x, \pi, z)^{\mathrm{T}}$, $(x, y, -\pi)^{\mathrm{T}}$ and $(x, y, \pi)^{\mathrm{T}}$. The periodicity of the flow and that the outward normals of opposing faces point in opposite directions establish that divergence terms are zero.

Remark 7.21 (Leray projector) If we take the divergence of the Navier–Stokes momentum equation, we find

$$\Delta p = -\nabla \cdot ((\boldsymbol{u} \cdot \nabla)\, \boldsymbol{u}) \quad \Leftrightarrow \quad p = -\Delta^{-1}\big(\nabla \cdot ((\boldsymbol{u} \cdot \nabla)\, \boldsymbol{u})\big).$$

Subsituting this expression for the pressure p into the Navier–Stokes momentum equation, we can equivalently express the momentum equation in the form

$$\frac{\partial \boldsymbol{u}}{\partial t} + (I - \nabla(\Delta^{-1}\nabla \cdot\,))((\boldsymbol{u} \cdot \nabla)\, \boldsymbol{u}) = \nu\, \Delta \boldsymbol{u}.$$

The projection operator '$I - \nabla(\Delta^{-1}\nabla \cdot\,)$' is known as the *Leray projection operator.* Since any vector $\boldsymbol{u}$ can be decomposed into the form $\boldsymbol{u} = \nabla\phi + \nabla \times \boldsymbol{\psi}$ for some scalar potential function ϕ and some vector potential $\boldsymbol{\psi}$, the Leray projector projects $\boldsymbol{u}$ onto its solenoidal component $\nabla \times \boldsymbol{\psi}$ which is divergence-free.

Remark 7.22 (Why do we require a weak formulation?) The a-priori estimates above suggested that we seek global solutions $\boldsymbol{u}$ with finite energy and whose gradient $\nabla \boldsymbol{u}$ is square-integrable in space-time. In other words, we should at least seek to establish solutions $\boldsymbol{u}$ which for arbitrary $T > 0$, have finite L^2-norm at each time $t \in [0, T]$ and for which the L^2-norm of $\nabla \boldsymbol{u}$ is square-integrable in time for each $t \in [0, T]$. Establishing global solutions which are more regular, in particular smooth, in space and time would be much more desirable. However this has not been achieved in general in three dimensions so far. Indeed it may not be achievable. Whether it can or cannot be achieved in three dimensions is a very important problem, see Section 7.5(1). It can be achieved in two dimensions. See Section 7.3 where we explore results for such smoother solutions. In the three-dimensional case, as far as global solutions are concerned, we can at least establish the existence of such L^2-valued functions – weak solutions – which are in general fairly rough functions.

Summary (Proof of existence of weak solutions) For clarity and convenience we summarise the main components/ideas required to establish the global existence of weak solutions to the three-dimensional incompressible Navier–Stokes equations. We give a detailed proof in Steps 1–5 further below.

Approximate system by one with a smooth solution. The first step is to replace the incompressible Navier–Stokes equations with solution $\boldsymbol{u}$ by a system of equations for which we know a global smooth solution $\hat{\boldsymbol{u}}$ exists and for which, when the approximation is removed, we recover the Navier–Stokes equations. This is often achieved by projection of the Navier–Stokes equations onto a finite basis, say size $K \in \mathbb{N}$. Many approaches have been employed to achieve this, see Section 7.5 for some of the typical approaches. In our case, since we have periodic boundary conditions, we consider the Fourier expansion of the solution $\boldsymbol{u}$ and project the solution itself onto the first set of K modes, leaving us with a finite system of equations governing $\hat{\boldsymbol{u}}_K$. Standard ordinary differential equation existence theory establishes the global existence of unique solutions to this system. We outline this approximation process in detail as a preparatory step before starting the detailed proof.

Establish the energy inequality, uniform in K. We then establish estimates proving that $\hat{\boldsymbol{u}}_K$ has a finite L^2-norm at each time $t > 0$, and for which $\nabla\hat{\boldsymbol{u}}_K$ is square-integrable in time. This result helps to establish that the solutions to the finite system are indeed global (for details, see the main proof below). However, the other main goal in these estimates is to show that the upper bounds of the two norms involved, the L^2-norm of $\hat{\boldsymbol{u}}_K$ and the time integral of the square of the L^2-norm of $\nabla\hat{\boldsymbol{u}}_K$, are uniform with respect to K, i.e. independent of the basis size K. This is established in Step 1 in the detailed proof.

Establish basic convergence properties in the large K limit. We then consider the sequence of solutions generated by taking the limit of the basis size K to infinity to recover $\boldsymbol{u}$. If we were only dealing with a finite-dimensional vector space, then the Bolzano–Weierstrass Theorem would guarantee that there exists a subsequence that converges within the finite-dimensional vector space. However, the vector space of concern here is the infinite-dimensional vector space of functions L^2. For sequences of such functions we are only guaranteed that there exists a subsequence that *weakly converges* in L^2. Recall Definition 7.13 and the discussion thereafter in Section 7.1. Weak convergence corresponds to the property that there exists a subsequence $\hat{\boldsymbol{u}}_{K_j}$ such that for every $\boldsymbol{v} \in L^2$ we have $\langle \boldsymbol{v}, \hat{\boldsymbol{u}}_{K_j}\rangle \to \langle \boldsymbol{v}, \hat{\boldsymbol{u}}\rangle$. This is guaranteed by the Bolzano–Weierstrass Theorem as for each $\boldsymbol{v} \in L^2$ the quantities $\langle \boldsymbol{v}, \hat{\boldsymbol{u}}_K\rangle$ represent a sequence of real numbers. In finite-dimensional vector spaces the notions of weak and standard/strong convergence trivially coincide. The results of Step 1 are enough to establish the weak convergence of $\hat{\boldsymbol{u}}_{K_j}$ to $\boldsymbol{u}$ in the space of functions for which the L^2-norm is bounded in time and the L^2-norm of the gradient is square-integrable in time. These weak convergence results are sufficient to establish the convergence, for a corresponding weak formulation of the approximate system, of the linear terms to the corresponding terms in the weak formulation in Definition 7.19 globally in time. The specific terms included here are the two terms on the left and the first term on the right in the weak formulation. The computations in Step 1 used to establish these results do not involve the nonlinear term. It is projected out in the explicit energy estimates, essentially due to the incompressibility constraint.

The nonlinear term and recovering the Navier–Stokes equations. The most difficult task is to establish the weak convergence of the nonlinear term from the corresponding weak formulation of the approximate system, to the nonlinear term in the weak formulation in Definition 7.19, globally in time. In the main proof, Step 2 establishes suitable bounds, uniform in K, on the nonlinear term as well as for the time derivative of the solution. In Step 3 we invoke the abstract Aubin–Lions–Simon Lemma to establish, from the results of Step 2, the strong convergence of $\hat{\boldsymbol{u}}_{K_j}$ in the space of functions whose L^2-norm is square-integrable in time. This strong convergence result is then used in Step 4 to establish the suitable weak convergence of the nonlinear term in the weak formulation in Definition 7.19. Finally, Step 5 collects these results together, demonstrating the convergence of all the terms from the weak formulation of the approximate system to the weak formulation in Definition 7.19.

Instead of the classical formulation above, to keep the arguments as elementary as possible, we consider herein the Fourier formulation of the incompressible Navier–Stokes equations. We realise the plan for the proof of existence of global weak solutions outlined in the summary just above in this context as follows. Suppose the Fourier series expansion for $\boldsymbol{u} = \boldsymbol{u}(\boldsymbol{x}, t)$ is

$$\boldsymbol{u}(\boldsymbol{x}, t) = \sum_{\boldsymbol{k} \in \mathbb{Z}^3} U_{\boldsymbol{k}}(t)\, \mathrm{e}^{\mathrm{i}\boldsymbol{k}\cdot\boldsymbol{x}},$$

where $U_{\boldsymbol{k}} = U_{\boldsymbol{k}}(t) \in \mathbb{C}^3$ are the Fourier coefficients. Since at all times all three components of $\boldsymbol{u}$ are real, i.e. $\boldsymbol{u}^* = \boldsymbol{u}$, we must have for all $\boldsymbol{k} \in \mathbb{Z}^3$:

$$U_{\boldsymbol{k}}^* = U_{-\boldsymbol{k}}.$$

We now substitute the Fourier series expansion for $\boldsymbol{u}$ into the incompressible Navier–Stokes equations, and use the form for the momentum equation with the Leray projector above. After collecting together all the coefficients of $\mathrm{e}^{\mathrm{i}\boldsymbol{k}\cdot\boldsymbol{x}}$, we see that the incompressible Navier–Stokes equations are equivalent to the following infinite system of ordinary differential equations for the Fourier coefficients $U_{\boldsymbol{k}}$ together with the incompressibility constraint:

$$\frac{\mathrm{d}U_{\boldsymbol{k}}}{\mathrm{d}t} + \mathrm{i}\left(I - \frac{\boldsymbol{k}\boldsymbol{k}^{\mathrm{T}}}{|\boldsymbol{k}|^2}\right) \sum_{\boldsymbol{l} \in \mathbb{Z}^3} U_{\boldsymbol{l}}\, \boldsymbol{l}^{\mathrm{T}} U_{\boldsymbol{k}-\boldsymbol{l}} = -\nu |\boldsymbol{k}|^2 U_{\boldsymbol{k}},$$

$$\boldsymbol{k}^{\mathrm{T}} U_{\boldsymbol{k}} = 0,$$

for all $\boldsymbol{k} \in \mathbb{Z}^3$. Note that $\boldsymbol{k}^{\mathrm{T}} U_{\boldsymbol{k}} = U_{\boldsymbol{k}}^{\mathrm{T}} \boldsymbol{k}$ for all $\boldsymbol{k} \in \mathbb{Z}^3$. The linear terms above are straightforwardly identifiable from the classical formulation of the Navier–Stokes equations. The nonlinear advection term $(\boldsymbol{u} \cdot \nabla)\, \boldsymbol{u}$ can be expressed in the form $(\nabla \boldsymbol{u})\, \boldsymbol{u}$ involving matrix multiplication. The Fourier coefficients of the matrix $(\nabla \boldsymbol{u})$ are the matrices $U_{\boldsymbol{k}}\, \boldsymbol{k}^{\mathrm{T}}$. Hence if we substitute the Fourier series for $\nabla \boldsymbol{u}$ and $\boldsymbol{u}$ into $(\nabla \boldsymbol{u})\, \boldsymbol{u}$, we see

$$(\nabla \boldsymbol{u})\, \boldsymbol{u} = \sum_{\boldsymbol{l}, \boldsymbol{l}'} U_{\boldsymbol{l}}\, \boldsymbol{l}^{\mathrm{T}} U_{\boldsymbol{l}'}\, \mathrm{e}^{\mathrm{i}(\boldsymbol{l}+\boldsymbol{l}')\cdot\boldsymbol{x}} = \sum_{\boldsymbol{k}} \Big(\sum_{\boldsymbol{l}} U_{\boldsymbol{l}}\, \boldsymbol{l}^{\mathrm{T}} U_{\boldsymbol{k}-\boldsymbol{l}}\Big)\, \mathrm{e}^{\mathrm{i}\boldsymbol{k}\cdot\boldsymbol{x}},$$

where we set $\boldsymbol{l} + \boldsymbol{l}' = \boldsymbol{k}$. Hence the coefficient of $\mathrm{e}^{\mathrm{i}\boldsymbol{k}\cdot\boldsymbol{x}}$ in the advection term is the convolution sum involving the term $U_{\boldsymbol{l}}\, \boldsymbol{l}^{\mathrm{T}} U_{\boldsymbol{k}-\boldsymbol{l}}$. In Fourier space the action of the Leray projector is such that $V_{\boldsymbol{k}} \mapsto (I - |\boldsymbol{k}|^{-2} \boldsymbol{k}\boldsymbol{k}^{\mathrm{T}})\, V_{\boldsymbol{k}}$. This can be established by successively considering the actions of the divergence, Laplacian and gradient operations on the Fourier expansion of a function $\boldsymbol{v} = \boldsymbol{v}(\boldsymbol{x})$ with Fourier coefficients $V_{\boldsymbol{k}}$ with $\boldsymbol{k} \in \mathbb{Z}^3$. Note that if the pressure field $p = p(\boldsymbol{x}, t)$ has scalar Fourier coefficients $P_{\boldsymbol{k}} = P_{\boldsymbol{k}}(t)$, since $p = -\Delta^{-1} \nabla \cdot ((\boldsymbol{u} \cdot \nabla)\, \boldsymbol{u})$ we have

$$P_{\boldsymbol{k}} = -\mathrm{i}\, \frac{\boldsymbol{k}^{\mathrm{T}}}{|\boldsymbol{k}|^2} \sum_{\boldsymbol{l} \in \mathbb{Z}^3} U_{\boldsymbol{l}}\, \boldsymbol{l}^{\mathrm{T}} U_{\boldsymbol{k}-\boldsymbol{l}}.$$

The Fourier coefficients of ∇p are $\boldsymbol{k}\, P_{\boldsymbol{k}}$.

We note here, if the solution to the incompressible Navier–Stokes equations has zero mean initially, then its mean is zero thereafter. This corresponds to the conservation of total momentum. To see this, set $\boldsymbol{k} = \boldsymbol{0}$ in the Navier–Stokes momentum equation. Note that, replacing $\boldsymbol{l}$ by $-\boldsymbol{l}$, we have

$$\mathrm{i} \sum_{\boldsymbol{l}\in\mathbb{Z}^3} U_{\boldsymbol{l}}\, \boldsymbol{l}^{\mathrm{T}} U_{-\boldsymbol{l}} = -\mathrm{i} \sum_{\boldsymbol{l}\in\mathbb{Z}^3} U_{-\boldsymbol{l}}\, \boldsymbol{l}^{\mathrm{T}} U_{\boldsymbol{l}} = 0,$$

since $\boldsymbol{l}^{\mathrm{T}} U_{\boldsymbol{l}} = 0$ for all $\boldsymbol{l} \in \mathbb{Z}^3$. Hence we see

$$\frac{\mathrm{d}U_{\boldsymbol{0}}}{\mathrm{d}t} = 0 \quad \Leftrightarrow \quad U_{\boldsymbol{0}}(t) = U_{\boldsymbol{0}}(0).$$

In other words, since $U_{\boldsymbol{0}}$ represents the mean of $\boldsymbol{u}$ over $\mathbb{T}^3$, if the mean of $\boldsymbol{u}$ is zero initially, it remains zero thereafter. We assume this henceforth and thus in our Fourier series expansion for $\boldsymbol{u}$ we need only sum over $\boldsymbol{k} \in \mathbb{Z}^3\backslash\{\boldsymbol{0}\}$. For convenience we set $\mathbb{Z}^3_0 := \mathbb{Z}^3\backslash\{\boldsymbol{0}\}$.

The weak formulation of the incompressible Navier–Stokes equations has the following form in the present context. Note that we use $h^1(\mathbb{Z}^3_0;\mathbb{C}^3)$ to denote the class of sequences which are square-summable with respect to the weight $|\boldsymbol{k}|$, i.e. sequences $U := \{V_{\boldsymbol{k}}\}_{\boldsymbol{k}\in\mathbb{Z}^3_0}$ such that

$$\sum_{\boldsymbol{k}\in\mathbb{Z}^3_0} |\boldsymbol{k}|^2 V_{\boldsymbol{k}}^{\dagger} V_{\boldsymbol{k}} < \infty.$$

Since $|\boldsymbol{k}| \geqslant 1$ for all $\boldsymbol{k} \in \mathbb{Z}^3_0$, such sequences are also square-summable. Thus $h^1(\mathbb{Z}^3_0;\mathbb{C}^3)$ is isomorphic to the subspace of $H^1(\mathbb{T}^d;\mathbb{R}^3)$ consisting of all functions in $H^1(\mathbb{T}^d;\mathbb{R}^3)$ which have zero mean.

Definition 7.23 (Weak Fourier solution) A weak solution of the Fourier formulation of the incompressible Navier–Stokes equations on $[0,T]$ with initial data $U(0) \in \ell^2(\mathbb{Z}^3_0;\mathbb{C}^n)$, with coefficients satisfying $\boldsymbol{k}^{\mathrm{T}} U_{\boldsymbol{k}}(0) = 0$ for all $\boldsymbol{k} \in \mathbb{Z}^3_0$, is a sequence $U := \{U_{\boldsymbol{k}}\}_{\boldsymbol{k}\in\mathbb{Z}^3_0}$ such that

$$U \in L^\infty(0,T;\ell^2(\mathbb{Z}^3_0;\mathbb{C}^3)) \cap L^2(0,T;h^1(\mathbb{Z}^3_0;\mathbb{C}^3)),$$

whose coefficients satisfy $\boldsymbol{k}^{\mathrm{T}} U_{\boldsymbol{k}}(t) = 0$ for all $\boldsymbol{k} \in \mathbb{Z}^3_0$ and the equation

$$\sum_{\boldsymbol{k}} \left(V_{\boldsymbol{k}}^{\dagger} U_{\boldsymbol{k}}(t) - V_{\boldsymbol{k}}^{\dagger} U_{\boldsymbol{k}}(0) + \int_0^t \left(\nu |\boldsymbol{k}|^2 V_{\boldsymbol{k}}^{\dagger} U_{\boldsymbol{k}} + \mathrm{i} V_{\boldsymbol{k}}^{\dagger} \sum_{\boldsymbol{l}} U_{\boldsymbol{l}}\, \boldsymbol{l}^{\mathrm{T}} U_{\boldsymbol{k}-\boldsymbol{l}} \right) \mathrm{d}\tau \right) = 0,$$

for every sequence $V \in h^1(\mathbb{Z}^3_0;\mathbb{C}^3)$ and every $t \in [0,T]$, where the sums shown are over $\boldsymbol{k}, \boldsymbol{l} \in \mathbb{Z}^3_0$.

Remark 7.24 As previously, we note that the first two terms in the weak Fourier formulation of the equations in Definition 7.23 occur pointwise in time. For the equations to make sense we require the solution to be weakly continuous in time. This is established in Remark 7.27.

We now establish suitable global-in-time solutions to the weak Fourier formulation of the incompressible Navier–Stokes equations above. To achieve this, we begin by considering a truncated version of the equations, generated by setting all the coefficients in the sequence $U := \{U_k\}_{k\in\mathbb{Z}_0^3}$ for which $|k| > K$ for some $K \in \mathbb{N}$, equal to zero. Indeed, set

$$\hat{U}^K := \left\{\{U_k\}_{k\in\mathbb{Z}_0^3} : U_k = \mathbf{0} \text{ if } |k| > K\right\}.$$

We denote the components of $\hat{U}^K$ by $\hat{U}_k^K$ and often abbreviate $\hat{U}^K$ by $\hat{U}$ and $\hat{U}_k^K$ by $\hat{U}_k$ with the dependence on $K \in \mathbb{N}$ understood. If we replace U in the Fourier formulation of the incompressible Navier–Stokes equations above by $\hat{U}$, we obtain the following finite system of ordinary differential equations:

$$\frac{\mathrm{d}\hat{U}_k}{\mathrm{d}t} + \mathrm{i}\left(I - \frac{k\,k^{\mathrm{T}}}{|k|^2}\right)\sum_{|l|\leqslant K}\hat{U}_l\, l^{\mathrm{T}}\hat{U}_{k-l} = -\nu|k|^2\hat{U}_k,$$
$$k^{\mathrm{T}}\hat{U}_k = 0,$$

for all $k \in \mathbb{Z}_0^3$ with $|k| \leqslant K$. Note that in the nonlinear term, the terms with $|k - l| > K$ are also zero. The equations just above represent a finite autonomous system of ordinary differential equations with a quadratic nonlinearity that is Lipschitz continuous. Standard Picard–Lindelöf theory establishes the existence of a unique solution local in time. If we can show that all the components $\hat{U}_k$ remain bounded for all $t > 0$, then we can confirm global-in-time existence of the unique solution. We can establish the latter by demonstrating that $\|\hat{U}(t)\|_{\ell^2(\mathbb{Z}_0^3;\mathbb{C}^3)}$ is finite for all $t > 0$. This is established in Step 1 coming presently, which essentially follows the steps in the a-priori energy estimate we established at the very beginning of this section, though now in a rigorous setting and in more detail. The subsequent steps break down the proof of existence of global weak solutions to the weak Fourier formulation of the incompressible Navier–Stokes equations.

Step 1: The energy inequality. Our goal here is to establish a uniform upper bound for $\|\hat{U}\|_{\ell^2(\mathbb{Z}_0^3;\mathbb{C}^3)}$ and the time integral of $\|\hat{U}\|^2_{h^1(\mathbb{Z}_0^3;\mathbb{C}^3)}$ over $[0, t]$ for any $t \in [0, T]$ and $T > 0$. To achieve this it is expedient in the finite system of ordinary differential equations governing $\hat{U}$ to utilise that the Fourier coefficients of the associated approximate pressure field are given by

$$\hat{P}_k = -\mathrm{i}\frac{k^{\mathrm{T}}}{|k|^2}\sum_{|l|\leqslant K} U_l\, l^{\mathrm{T}} U_{k-l},$$

for $|k| \leqslant K$ and otherwise zero. The Fourier coefficients of the gradient of the associated pressure field are $k\,\hat{P}_k$. To estimate $\|\hat{U}\|_{\ell^2(\mathbb{Z}_0^3;\mathbb{C}^3)}$ we pre-multiply the finite system of ordinary differential equations governing $\hat{U}$ above by $\hat{U}_k^\dagger$ and sum over $k \in \mathbb{Z}_0^3$ with $|k| \leqslant K$. This generates the *energy relation*

$$\sum_{|k|\leqslant K}\hat{U}_k^\dagger\frac{\mathrm{d}\hat{U}_k}{\mathrm{d}t} + \mathrm{i}\sum_{k,l}\hat{U}_k^\dagger\hat{U}_l\, l^{\mathrm{T}}\hat{U}_{k-l} = -\nu\sum_{|k|\leqslant K}|k|^2\hat{U}_k^\dagger\hat{U}_k + \sum_{|k|\leqslant K}\hat{U}_k^\dagger k\,\hat{P}_k$$
$$\Leftrightarrow\quad \frac{1}{2}\frac{\mathrm{d}}{\mathrm{d}t}\sum_{|k|\leqslant K}\hat{U}_k^\dagger\hat{U}_k + \mathrm{i}\sum_{k,l}\hat{U}_k^\dagger\hat{U}_l\, l^{\mathrm{T}}\hat{U}_{k-l} = -\nu\sum_{|k|\leqslant K}|k|^2\hat{U}_k^\dagger\hat{U}_k.$$

Here we used that

$$\sum_{k} \hat{U}_{k}^{\dagger} \boldsymbol{k} \, \hat{P}_{k} = \sum_{k} \hat{U}_{-k}^{\mathrm{T}} \boldsymbol{k} \, \hat{P}_{k} = -\sum_{k} \hat{U}_{k}^{\mathrm{T}} \boldsymbol{k} \, \hat{P}_{-k} = 0,$$

where we replaced $\boldsymbol{k}$ by $-\boldsymbol{k}$ in the sum and recall $\hat{U}_{k}^{\mathrm{T}} \boldsymbol{k} = \boldsymbol{k}^{\mathrm{T}} \hat{U}_{k} = 0$ for all $\boldsymbol{k} \in \mathbb{Z}_0^3$ with $|\boldsymbol{k}| \leqslant K$. The contribution from the advection term in the energy relation above is also zero, which can be demonstrated as follows:

$$\begin{aligned}
\sum_{k,l} \hat{U}_{k}^{\dagger} \hat{U}_{l} \, \boldsymbol{l}^{\mathrm{T}} \hat{U}_{k-l} &= \sum_{k,l} \hat{U}_{-k}^{\mathrm{T}} \hat{U}_{k-l} \, \boldsymbol{k}^{\mathrm{T}} \hat{U}_{l} - \sum_{k,l} \hat{U}_{-k}^{\mathrm{T}} \hat{U}_{k-l} \, \boldsymbol{l}^{\mathrm{T}} \hat{U}_{l} \\
&= \sum_{k,l} \hat{U}_{-k}^{\mathrm{T}} \hat{U}_{l} \, \boldsymbol{k}^{\mathrm{T}} \hat{U}_{k-l} \\
&= \sum_{l,k} \hat{U}_{-l}^{\mathrm{T}} \hat{U}_{k} \, \boldsymbol{l}^{\mathrm{T}} \hat{U}_{l-k} \\
&= -\sum_{l,k} \hat{U}_{l}^{\mathrm{T}} \hat{U}_{-k} \, \boldsymbol{l}^{\mathrm{T}} \hat{U}_{k-l} \\
&= -\sum_{l,k} \hat{U}_{-k}^{\mathrm{T}} \hat{U}_{l} \, \boldsymbol{l}^{\mathrm{T}} \hat{U}_{k-l} \\
&= -\sum_{k,l} \hat{U}_{k}^{\dagger} \hat{U}_{l} \, \boldsymbol{l}^{\mathrm{T}} \hat{U}_{k-l}.
\end{aligned}$$

Hence this term is zero. In this calculation we: (i) swapped the roles of $\boldsymbol{l}$ and $\boldsymbol{k} - \boldsymbol{l}$ in the convolution sum and then utilised that $\boldsymbol{l}^{\mathrm{T}} \hat{U}_{l} = 0$; (ii) swapped the roles of $\boldsymbol{l}$ and $\boldsymbol{k} - \boldsymbol{l}$ back; (iii) relabelled $\boldsymbol{l}$ and $\boldsymbol{k}$; (iv) changed $\boldsymbol{k}$ to $-\boldsymbol{k}$ and similarly for $\boldsymbol{l}$; (v) utilised the symmetry in the scalar product $\hat{U}_{l}^{\mathrm{T}} \hat{U}_{-k}$; and (vi) used that $\hat{U}_{k}^{*} = \hat{U}_{-k}$. The initial energy for the approximations $\hat{U}(0)$ is bounded by the energy associated with the initial data for the full system, i.e. we have

$$\|\hat{U}(0)\|_{\ell^2(\mathbb{Z}_0^3;\mathbb{C}^3)} \leqslant \|U(0)\|_{\ell^2(\mathbb{Z}_0^3;\mathbb{C}^3)} = (2\pi)^{-d} \|\boldsymbol{u}_0\|_{L^2(\mathbb{T}^d;\mathbb{R}^3)}.$$

Consequently we deduce, after integrating the energy relation above in time and utilising the latter inequality, that the energy relation generates the *energy inequality* for the approximate system:

$$\tfrac{1}{2} \|\hat{U}(t)\|_{\ell^2(\mathbb{Z}_0^3;\mathbb{C}^3)}^2 + \nu \int_0^t \|\hat{U}(\tau)\|_{h^1(\mathbb{Z}_0^3;\mathbb{C}^3)}^2 \, \mathrm{d}\tau \leqslant \tfrac{1}{2} \|U(0)\|_{\ell^2(\mathbb{Z}_0^3;\mathbb{C}^3)}^2.$$

Since T and thus t are arbitrary, this establishes global existence of solutions to the finite system of ordinary differential equations which constitute the approximate system. Further, the upper bound is independent of K and thus uniform in K, and we deduce that there exists a subsequence of $\hat{U} = \hat{U}^K$ which converges weakly in $L^\infty(0, T; \ell^2(\mathbb{Z}_0^3; \mathbb{C}^3)) \cap L^2(0, T; h^1(\mathbb{Z}_0^3; \mathbb{C}^3))$.

Step 2: Bounding the nonlinear term. We now focus on the nonlinear term in the finite system of ordinary differential equations representing the approximation to the Fourier formulation of the incompressible Navier–Stokes equations. Consider the inner product in $\ell^2(\mathbb{Z}_0^3;\mathbb{C}^3)$ of the nonlinear term with $V := \{V_{\boldsymbol{k}}\}_{\boldsymbol{k}\in\mathbb{Z}_0^3}$ in $(h^1(\mathbb{Z}_0^3;\mathbb{C}^3))^*$, the dual space to $h^1(\mathbb{Z}_0^3;\mathbb{C}^3)$, as follows:

$$\sum_{\boldsymbol{k},\boldsymbol{l}} V_{\boldsymbol{k}}^{\dagger}\hat{U}_{\boldsymbol{l}}\,\boldsymbol{l}^{\mathrm{T}}\hat{U}_{\boldsymbol{k}-\boldsymbol{l}}.$$

The sum is over all $\boldsymbol{k},\boldsymbol{l} \in \mathbb{Z}_0^3$. Our goal herein is to derive a suitable bound for this double sum to help establish the convergence of solutions, to the finite approximate system, to the weak Fourier formulation of the incompressible Navier–Stokes equations in Definition 7.23. To begin, we use *Young's convolution inequality* from Proposition 7.16(c) as follows. We apply the inequality for $d = n = 3$ and $\boldsymbol{k},\boldsymbol{l} \in \mathbb{Z}_0^3$. If $p = r = 4/3$ then $q = 2$ and we have

$$\sum_{\boldsymbol{k},\boldsymbol{l}} |F_{\boldsymbol{k}}^{\dagger} G_{\boldsymbol{l}}\, H_{\boldsymbol{k}-\boldsymbol{l}}| \leqslant \|F\|_{\ell^{4/3}(\mathbb{Z}_0^3;\mathbb{C}^3)} \|G\|_{\ell^2(\mathbb{Z}_0^3;\mathbb{C}^{3\times 3})} \|H\|_{\ell^{4/3}(\mathbb{Z}_0^3;\mathbb{C}^3)}.$$

Then we use the Hölder inequality from Proposition 7.16(b) as follows with $p = 3/2$ and $q = 3$:

$$\sum_{\boldsymbol{k}} |F_{\boldsymbol{k}}|^{4/3} = \sum_{\boldsymbol{k}} |\boldsymbol{k}|^{4/3}|F_{\boldsymbol{k}}|^{4/3}|\boldsymbol{k}|^{-4/3} \leqslant \Big(\sum_{\boldsymbol{k}} |\boldsymbol{k}|^2|F_{\boldsymbol{k}}|^2\Big)^{2/3}\Big(\sum_{\boldsymbol{k}} |\boldsymbol{k}|^{-4}\Big)^{1/3},$$

where all sums are over $\boldsymbol{k} \in \mathbb{Z}_0^3$. The final factor is thus bounded and after taking the $3/4$ power across the inequality we have for some constant $c > 0$:

$$\|F\|_{\ell^{4/3}(\mathbb{Z}_0^3;\mathbb{C}^3)} \leqslant c\,\|F\|_{h^1(\mathbb{Z}_0^3;\mathbb{C}^3)}.$$

If we use the same inequality for $\|H\|_{\ell^{4/3}(\mathbb{Z}_0^3;\mathbb{C}^3)}$ and set $F_{\boldsymbol{k}} := V_{\boldsymbol{k}}$, $G_{\boldsymbol{k}} := \hat{U}_{\boldsymbol{k}}\,\boldsymbol{k}^{\mathrm{T}}$ and $H_{\boldsymbol{k}} := \hat{U}_{\boldsymbol{k}}$, we find for some finite constant $c > 0$:

$$\sum_{\boldsymbol{k},\boldsymbol{l}} |V_{\boldsymbol{k}}^{\dagger}\hat{U}_{\boldsymbol{l}}\,\boldsymbol{l}^{\mathrm{T}}\hat{U}_{\boldsymbol{k}-\boldsymbol{l}}| \leqslant c\,\|V\|_{h^1(\mathbb{Z}_0^3;\mathbb{C}^3)}\|\hat{U}\|^2_{h^1(\mathbb{Z}_0^3;\mathbb{C}^3)}.$$

Recall from Section 7.1 the characterisation of the dual space $\mathbb{H}^*$ to any Hilbert space $\mathbb{H}$ and the associated norm on $\mathbb{H}^*$. If $h^{-1}(\mathbb{Z}_0^3;\mathbb{C}^3)$ represents the dual space $(h^1(\mathbb{Z}_0^3;\mathbb{C}^3))^*$ to $h^1(\mathbb{Z}_0^3;\mathbb{C}^3)$, then we observe that the above inequality establishes

$$\sup\Big|\sum_{\boldsymbol{k},\boldsymbol{l}\in\mathbb{Z}_0^3} V_{\boldsymbol{k}}^{\dagger}\hat{U}_{\boldsymbol{l}}\,\boldsymbol{l}^{\mathrm{T}}\hat{U}_{\boldsymbol{k}-\boldsymbol{l}}\Big| \leqslant c\,\|\hat{U}\|^2_{h^1(\mathbb{Z}_0^3;\mathbb{C}^3)},$$

where the supremum is over all $V \in h^1(\mathbb{Z}_0^3;\mathbb{C}^3)$ with $\|V\|_{h^1(\mathbb{Z}_0^3;\mathbb{C}^3)} \leqslant 1$. The left-hand side is thus the norm on $h^{-1}(\mathbb{Z}_0^3;\mathbb{C}^3)$. Since $\hat{U} = \hat{U}^K$ is uniformly bounded with respect to K in $L^2(0,T;h^1(\mathbb{Z}_0^3;\mathbb{C}^3))$ from Step 1, we deduce the nonlinear term is uniformly bounded with respect to K in $L^1(0,T;h^{-1}(\mathbb{Z}_0^3;\mathbb{C}^3))$. We can identify $h^{-1}(\mathbb{Z}_0^3;\mathbb{C}^3)$ as the space of sequences $F := \{F_{\boldsymbol{k}}\}_{\boldsymbol{k}\in\mathbb{Z}_0^3}$ such that

$$\sum_{\boldsymbol{k}\in\mathbb{Z}_0^3} |\boldsymbol{k}|^{-2}|F_{\boldsymbol{k}}|^2 < \infty.$$

Then since $\hat{U} \in L^2(0,T;h^1(\mathbb{Z}_0^3;\mathbb{C}^3))$ uniformly with respect to K, the integral over $[0,T]$ of the sum of the terms $|\boldsymbol{k}|^2|\hat{U}_{\boldsymbol{k}}|^2$ over $\boldsymbol{k} \in \mathbb{Z}_0^3$ is uniformly bounded with respect to K. Hence we observe

$$\int_0^T \sum_{\boldsymbol{k}\in\mathbb{Z}_0^3} |\boldsymbol{k}|^{-2}||\boldsymbol{k}|^2\hat{U}_{\boldsymbol{k}}(\tau)|^2 \,\mathrm{d}\tau = \int_0^T \sum_{\boldsymbol{k}\in\mathbb{Z}_0^3} |\boldsymbol{k}|^2|\hat{U}_{\boldsymbol{k}}(\tau)|^2 \,\mathrm{d}\tau$$

is uniformly bounded with respect to K. Hence the diffusion term is uniformly bounded in $L^2(0,T;h^{-1}(\mathbb{Z}_0^3;\mathbb{C}^3))$, and by the natural ordering of L^p-norms on the bounded domain $[0,T]$, it is uniformly bounded with respect to K in $L^1(0,T;h^{-1}(\mathbb{Z}_0^3;\mathbb{C}^3))$. Hence, directly from the finite system of ordinary differential equations governing the evolution of $\hat{U}$ we deduce, uniformly with respect to K:

$$\frac{\mathrm{d}\hat{U}}{\mathrm{d}t} \in L^1(0,T;h^{-1}(\mathbb{Z}_0^3;\mathbb{C}^3)).$$

Step 3: Aubin–Lions–Simon Lemma. We now use the Aubin–Lions Lemma, which was extended by Simon (1987), to establish the strong convergence of a subsequence of $\hat{U}$ in $L^2(0,T;\ell^2(\mathbb{Z}_0^3;\mathbb{C}^3))$.

Lemma 7.25 (Aubin–Lions–Simon Lemma) *Suppose $\mathbb{B}_1$, $\mathbb{B}_0$ and $\mathbb{B}_{-1}$ are three Banach spaces such that $\mathbb{B}_1 \hookrightarrow\!\!\!\!\rightarrow \mathbb{B}_0 \hookrightarrow \mathbb{B}_{-1}$. Suppose a sequence F^K is uniformly bounded in $L^p(0,T;\mathbb{B}_1)$ and in addition $\mathrm{d}F^K/\mathrm{d}t$ is uniformly bounded in $L^q(0,T;\mathbb{B}_{-1})$, for $1 < p < \infty$ and $1 \leqslant q < \infty$. Then there exists a subsequence F^{K_j} that converges strongly in $L^p(0,T;\mathbb{B}_0)$.*

The extension by Simon (1987) is more general than stated here; we have restricted the lemma to the cases we need. We apply the lemma with $p = 2$, $q = 1$, $\mathbb{B}_1 := h^1(\mathbb{Z}_0^3;\mathbb{C}^3)$, $\mathbb{B}_0 := \ell^2(\mathbb{Z}_0^3;\mathbb{C}^3)$, $\mathbb{B}_{-1} := h^{-1}(\mathbb{Z}_0^3;\mathbb{C}^3)$ and $F^K := \hat{U}^K$. Note we established in Example 7.15 that $h^1(\mathbb{Z}_0^3;\mathbb{C}^3) \hookrightarrow\!\!\!\!\rightarrow \ell^2(\mathbb{Z}_0^3;\mathbb{C}^3)$. That $\ell^2(\mathbb{Z}_0^3;\mathbb{C}^3) \hookrightarrow h^{-1}(\mathbb{Z}_0^3;\mathbb{C}^3)$ follows from the fact that any sequence whose terms $F_{\boldsymbol{k}}^K$ with $\boldsymbol{k} \in \mathbb{Z}_0^3$ are square-summable will also be square-summable with respect to the weight $|\boldsymbol{k}|^{-2}$. With these identifications and the results from Steps 1 and 2, from the Aubin–Lions–Simon Lemma 7.25, we deduce that there exists a subsequence $\hat{U}^{K_j}$ and a $U \in L^2(0,T;\ell^2(\mathbb{Z}_0^3;\mathbb{C}^3))$ such that:

$$\int_0^T \|\hat{U}^{K_j} - U\|^2_{\ell^2(\mathbb{Z}_0^3;\mathbb{C}^3)} \,\mathrm{d}\tau \to 0,$$

as $K_j \to \infty$, i.e. the subsequence converges strongly in $L^2(0,T;\ell^2(\mathbb{Z}_0^3;\mathbb{C}^3))$.

Step 4: Convergence of the nonlinear term. Using the strong convergence of a subsequence $\hat{U}^{K_j}$ to U in $L^2(0,T;\ell^2(\mathbb{Z}_0^3;\mathbb{C}^3))$, our main goal in this step is to establish the convergence of the nonlinear term in the finite approximate system to the nonlinear term in the weak Fourier formulation. Indeed our goal is to show that for every $V \in h^1(\mathbb{Z}_0^3;\mathbb{C}^3)$:

$$\left|\int_0^T \sum_{\boldsymbol{k},\boldsymbol{l}} V_{\boldsymbol{k}}^{\dagger}\big(\hat{U}_{\boldsymbol{l}}\,\boldsymbol{l}^{\mathrm{T}}\hat{U}_{\boldsymbol{k}-\boldsymbol{l}} - U_{\boldsymbol{l}}\,\boldsymbol{l}^{\mathrm{T}}U_{\boldsymbol{k}-\boldsymbol{l}}\big)\,\mathrm{d}\tau\right| \to 0,$$

as $K_j \to \infty$, where now $\hat{U} := \hat{U}^{K_j}$. Note that for all $\boldsymbol{k},\boldsymbol{l} \in \mathbb{Z}_0^3$ we have

$$V_{\boldsymbol{k}}^{\dagger}\big(\hat{U}_{\boldsymbol{l}}\,\boldsymbol{l}^{\mathrm{T}}\hat{U}_{\boldsymbol{k}-\boldsymbol{l}} - U_{\boldsymbol{l}}\,\boldsymbol{l}^{\mathrm{T}}U_{\boldsymbol{k}-\boldsymbol{l}}\big) = V_{\boldsymbol{k}}^{\dagger}\big((\hat{U}_{\boldsymbol{l}} - U_{\boldsymbol{l}})\,\boldsymbol{l}^{\mathrm{T}}\hat{U}_{\boldsymbol{k}-\boldsymbol{l}} + U_{\boldsymbol{l}}\,\boldsymbol{l}^{\mathrm{T}}(\hat{U}_{\boldsymbol{k}-\boldsymbol{l}} - U_{\boldsymbol{k}-\boldsymbol{l}})\big).$$

For any four $\mathbb{C}^n$-valued vectors $\boldsymbol{a}$, $\boldsymbol{b}$, $\boldsymbol{c}$ and $\boldsymbol{d}$, we have $\boldsymbol{a}^{\mathrm{T}}\boldsymbol{b}\boldsymbol{c}^{\mathrm{T}}\boldsymbol{d} = \mathrm{tr}\,(\boldsymbol{a}\boldsymbol{c}^{\mathrm{T}}\boldsymbol{d}\boldsymbol{b}^{\mathrm{T}})$. Recall $\boldsymbol{k}^{\mathrm{T}}\hat{U}_{\boldsymbol{k}} = 0$ and $\boldsymbol{k}^{\mathrm{T}}U_{\boldsymbol{k}} = 0$ for all $\boldsymbol{k} \in \mathbb{Z}_0^3$. Thus we observe, exchanging $\boldsymbol{l}$ and $\boldsymbol{k}-\boldsymbol{l}$ in the convolution sum, for the term on the right above we have

$$
\begin{aligned}
&\sum_{\boldsymbol{k},\boldsymbol{l}} V_{\boldsymbol{k}}^{\dagger}\Big((\hat{U}_{\boldsymbol{l}} - U_{\boldsymbol{l}})\,\boldsymbol{l}^{\mathrm{T}}\hat{U}_{\boldsymbol{k}-\boldsymbol{l}} + U_{\boldsymbol{l}}\,\boldsymbol{l}^{\mathrm{T}}(\hat{U}_{\boldsymbol{k}-\boldsymbol{l}} - U_{\boldsymbol{k}-\boldsymbol{l}})\Big)\\
&\quad= \sum_{\boldsymbol{k},\boldsymbol{l}} V_{-\boldsymbol{k}}^{\mathrm{T}}\Big((\hat{U}_{\boldsymbol{k}-\boldsymbol{l}} - U_{\boldsymbol{k}-\boldsymbol{l}})\,\boldsymbol{k}^{\mathrm{T}}\hat{U}_{\boldsymbol{l}} + U_{\boldsymbol{k}-\boldsymbol{l}}\,\boldsymbol{k}^{\mathrm{T}}(\hat{U}_{\boldsymbol{l}} - U_{\boldsymbol{l}})\Big)\\
&\quad= \sum_{\boldsymbol{k},\boldsymbol{l}}\Big(\mathrm{tr}\left(V_{-\boldsymbol{k}}\,\boldsymbol{k}^{\mathrm{T}}\hat{U}_{\boldsymbol{l}}\,(\hat{U}_{\boldsymbol{k}-\boldsymbol{l}} - U_{\boldsymbol{k}-\boldsymbol{l}})^{\mathrm{T}}\right) + \mathrm{tr}\left(V_{-\boldsymbol{k}}\,\boldsymbol{k}^{\mathrm{T}}(\hat{U}_{\boldsymbol{k}-\boldsymbol{l}} - U_{\boldsymbol{k}-\boldsymbol{l}})\hat{U}_{\boldsymbol{l}}^{\mathrm{T}}\right)\Big).
\end{aligned}
$$

We now apply a variant of Young's convolution inequality given in Proposition 7.16(c) as follows. We set $d = n = 3$, $\boldsymbol{k}, \boldsymbol{l} \in \mathbb{Z}_0^3$ and assign $F_{\boldsymbol{k}} := V_{-\boldsymbol{k}}\,\boldsymbol{k}^{\mathrm{T}}$ as a matrix. Then in one instance we set $G_{\boldsymbol{l}} := \hat{U}_{\boldsymbol{l}}$ and $H_{\boldsymbol{k}-\boldsymbol{l}} := (\hat{U}_{\boldsymbol{k}-\boldsymbol{l}} - U_{\boldsymbol{k}-\boldsymbol{l}})^{\mathrm{T}}$, while in the other instance we set $G_{\boldsymbol{l}} := (\hat{U}_{\boldsymbol{k}-\boldsymbol{l}} - U_{\boldsymbol{k}-\boldsymbol{l}})$ and $H_{\boldsymbol{k}-\boldsymbol{l}} := \hat{U}_{\boldsymbol{l}}^{\mathrm{T}}$. In both cases we set $p = 1$ and $q = r = 2$. Consequently we deduce

$$
\begin{aligned}
&\left|\sum_{\boldsymbol{k},\boldsymbol{l}}\Big(\mathrm{tr}\left(V_{-\boldsymbol{k}}\,\boldsymbol{k}^{\mathrm{T}}\hat{U}_{\boldsymbol{l}}\,(\hat{U}_{\boldsymbol{k}-\boldsymbol{l}} - U_{\boldsymbol{k}-\boldsymbol{l}})^{\mathrm{T}}\right) + \mathrm{tr}\left(V_{-\boldsymbol{k}}\,\boldsymbol{k}^{\mathrm{T}}(\hat{U}_{\boldsymbol{k}-\boldsymbol{l}} - U_{\boldsymbol{k}-\boldsymbol{l}})\hat{U}_{\boldsymbol{l}}^{\mathrm{T}}\right)\Big)\right|\\
&\quad\leqslant 2\,\|V\|_{h^1(\mathbb{Z}_0^3;\mathbb{C}^3)}\|\hat{U}\|_{\ell^2(\mathbb{Z}_0^3;\mathbb{C}^3)}\|\hat{U} - U\|_{\ell^2(\mathbb{Z}_0^3;\mathbb{C}^3)}.
\end{aligned}
$$

Thus we observe

$$
\begin{aligned}
&\left|\int_0^T \sum_{\boldsymbol{k},\boldsymbol{l}} V_{\boldsymbol{k}}^{\dagger}\Big(\hat{U}_{\boldsymbol{l}}\,\boldsymbol{l}^{\mathrm{T}}\hat{U}_{\boldsymbol{k}-\boldsymbol{l}} - U_{\boldsymbol{l}}\,\boldsymbol{l}^{\mathrm{T}}U_{\boldsymbol{k}-\boldsymbol{l}}\Big)\,\mathrm{d}\tau\right|\\
&\quad\leqslant 2\,\|V\|_{h^1(\mathbb{Z}_0^3;\mathbb{C}^3)}\left(\int_0^T \|\hat{U}(\tau)\|^2_{\ell^2(\mathbb{Z}_0^3;\mathbb{C}^3)}\,\mathrm{d}\tau\right)^{1/2}\left(\int_0^T \|(\hat{U} - U)(\tau)\|^2_{\ell^2(\mathbb{Z}_0^3;\mathbb{C}^3)}\,\mathrm{d}\tau\right)^{1/2}.
\end{aligned}
$$

Recall that $\hat{U} := \hat{U}^{K_j}$ is uniformly bounded in $L^\infty(0,T;\ell^2(\mathbb{Z}_0^3;\mathbb{C}^3))$ and thus also in $L^2(0,T;\ell^2(\mathbb{Z}_0^3;\mathbb{C}^3))$. Further, from Step 3 we know $\hat{U}$ converges strongly in $L^2(0,T;\ell^2(\mathbb{Z}_0^3;\mathbb{C}^3))$ to U as $K_j \to \infty$. Hence the upper bound above converges to zero as $K_j \to \infty$. Hence we deduce that the nonlinear term converges weakly in $L^2(0,T;\ell^2(\mathbb{Z}_0^3;\mathbb{C}^3))$ to the appropriate limit term.

Step 5: Convergence. Let us consider each term in the weak Fourier formulation of the incompressible Navier–Stokes equations in Definition 7.23. We consider the subsequence $\hat{U} = \hat{U}^{K_j}$ from Step 3, which is strongly convergent in $L^2(0,T;\ell^2(\mathbb{Z}_0^3;\mathbb{C}^3))$ to U as $K_j \to \infty$. This subsequence is also uniformly bounded in $L^2(0,T;h^1(\mathbb{Z}_0^3;\mathbb{C}^3))$, so by choosing a further subsequence if necessary and relabelling it as $\hat{U}^{K_j}$, we know the strongly convergent subsequence $\hat{U}^{K_j}$ in $L^2(0,T;\ell^2(\mathbb{Z}_0^3;\mathbb{C}^3))$ is also weakly convergent in $L^2(0,T;h^1(\mathbb{Z}_0^3;\mathbb{C}^3))$ to U as $K_j \to \infty$. In Step 4 we already established the weak convegence of the nonlinear term in the finite approximate system to the nonlinear term in the weak Fourier formulation in Definition 7.23 in $L^2(0,T;\ell^2(\mathbb{Z}_0^3;\mathbb{C}^3))$. The weak convergence of $\hat{U}^{K_j}$ to U in $L^2(0,T;h^1(\mathbb{Z}_0^3;\mathbb{C}^3))$ guarantees in particular the weak convergence (corresponding to the diffusion term)

$$\sum_{k\in\mathbb{Z}_0^3}\int_0^t |k|^2 V_k^\dagger \hat{U}_k^{K_j}(\tau)\,\mathrm{d}\tau \to \sum_{k\in\mathbb{Z}_0^3}\int_0^t |k|^2 V_k^\dagger U_k(\tau)\,\mathrm{d}\tau,$$

for every $V \in h^1(\mathbb{Z}_0^3;\mathbb{C}^3)$. The strong convergence of $\hat{U} = \hat{U}^{K_j}$ to U as $K_j \to \infty$ in $L^2(0,T;\ell^2(\mathbb{Z}_0^3;\mathbb{C}^3))$ guarantees the convergence of the first two terms in the time-integrated version of the finite approximate system to the corresponding terms in the weak Fourier formulation of the incompressible Navier–Stokes equations in Definition 7.23.

The five steps above have thus established the existence of a weak solution to the incompressible Navier–Stokes equations on $\mathbb{T}^3 \times [0,T]$ for any $T > 0$.

Remark 7.26 (Uniqueness of weak solutions) The class of weak solutions we have considered above, which satisfy the energy inequality, are known as *Leray–Hopf* weak solutions. It is not known whether these are unique or not. However, Buckmaster and Vicol (2019) have proved the non-uniqueness of a weaker class of solutions, in particular a class that do not necessarily satisfy the energy inequality.

Remark 7.27 (Weak continuity in time) The weak solutions above are in fact weakly continuous in time in the following sense. Consider the weak formulation of the incompressible Navier–Stokes equations in Definition 7.23 at two different arbitrary times t and t_0, and take the difference. This generates

$$\sum_{k\in\mathbb{Z}_0^3}\big(V_k^\dagger U_k(t) - V_k^\dagger U_k(t_0)\big) = -\sum_{k\in\mathbb{Z}_0^3}\int_{t_0}^t \Big(\nu\,|k|^2 V_k^\dagger U_k(\tau) + \mathrm{i}V_k^\dagger \sum_{l\in\mathbb{Z}_0^3} U_l\, l^{\mathrm{T}} U_{k-l}\Big)\,\mathrm{d}\tau,$$

for every sequence $V \in h^1(\mathbb{Z}_0^3;\mathbb{C}^3)$. From Step 4 the last term on the right is bounded, and since $U \in L^2(0,T;h^1(\mathbb{Z}_0^3;\mathbb{C}^3))$, so is the first term on the right. We thus deduce that as $t \to t_0$ we have

$$\sum_{k\in\mathbb{Z}_0^3} V_k^\dagger U_k(t) \to \sum_{k\in\mathbb{Z}_0^3} V_k^\dagger U_k(t_0),$$

for every sequence $V \in h^1(\mathbb{Z}_0^3;\mathbb{C}^3)$, thus guaranteeing *weak continuity* in time in $\ell^2(\mathbb{Z}_0^3;\mathbb{C}^3)$. The initial condition is also attained in this sense.

Remark 7.28 (External forcing) It is straightforward to incorporate an external forcing term $\boldsymbol{f} = \boldsymbol{f}(\boldsymbol{x},t)$ in our results above and thus reproduce the global existence of weak solutions when $\boldsymbol{f} \in L^2(0,T;h^{-1}(\mathbb{Z}_0^3;\mathbb{C}^3))$. See Exercise 7.6.

7.3 Strong Solutions

We established the existence of weak solutions to the three-dimensional incompressible Navier–Stokes equations globally in time in Section 7.2. Our goal herein is to establish the existence, generally only locally in time, of so-called *strong solutions*, in both two and three dimensions. Indeed the goal is to establish the interval of time within which a quantity called the *enstrophy*, which is $\|\omega(t)\|_{L^2}^2$, is bounded. Solutions to the incompressible Navier–Stokes equations for which the enstrophy is bounded are called strong solutions. The reason why they are called strong solutions is that if we know the enstrophy is bounded then we know the solutions are in fact smooth. We do not prove this latter important result here. That the enstrophy being bounded implies the solutions are smooth provided the data (initial data and external forcing) are sufficiently smooth can be found for example in Temam (1983, Lemma 4.2). In two dimensions it is possible to prove the enstrophy is globally bounded in time; we address this case herein. We take as the domain of the fluid flow the bounded domain $\mathbb{T}^d$ with periodic boundary conditions as we did in Section 7.2. Further, we only provide a-priori estimates herein, i.e. we assume the solutions are smooth. We examine their properties as such – a-priori estimates provide a basis for a more rigorous approach like that in Section 7.2. The idea, having established properties of the solutions using a-priori estimates, is then to: (i) revert to the finite approximate system of ordinary differential equations such as the one in Section 7.2 for $\hat{U} = \hat{U}^K$ or any such similar approximate system for which we know existence; (ii) rederive the same estimates established a priori in the context of the finite approximate system, proving upper bounds uniform in the approximation parameter (such as K); and (iii) take the limit of the approximation parameter that removes the approximation to establish upper bounds corresponding to the a-priori upper bounds. As a starting point for the a-priori bounds we derive herein, we use the vorticity formulation from Corollary 4.15 of the incompressible Navier–Stokes equations on $\mathbb{T}^d$ which has the form

$$\begin{aligned}\frac{\partial \omega}{\partial t} + (\nabla \omega)\, \boldsymbol{u} &= \nu\, \Delta \omega + (\nabla \boldsymbol{u})\, \omega,\\ \Delta \boldsymbol{u} &= -\nabla \times \omega,\end{aligned}$$

where note for convenience, we write $(\boldsymbol{u} \cdot \nabla)\, \omega \equiv (\nabla \omega)\, \boldsymbol{u}$ and $(\omega \cdot \nabla)\, \boldsymbol{u} \equiv (\nabla \boldsymbol{u})\, \omega$. Note that the divergence-free condition $\nabla \cdot \boldsymbol{u} = 0$ is implicitly encoded in the second equation above. Further note that, in the two-dimensional case when $d = 2$ the term $(\nabla \boldsymbol{u})\, \omega$ on the right-hand side above should not be present. One way to see this is to view the two-dimensional flow as a three-dimensional flow with no z-dependence and $\boldsymbol{u} = (u, v, 0)^{\mathrm{T}}$. In this context the only non-zero component of the vorticity ω is the third component, say ω, and we observe

$$(\nabla \boldsymbol{u})\, \omega = \begin{pmatrix} \partial u/\partial x & \partial u/\partial y & 0 \\ \partial v/\partial x & \partial v/\partial y & 0 \\ 0 & 0 & 0 \end{pmatrix} \begin{pmatrix} 0 \\ 0 \\ \omega \end{pmatrix} = \mathbf{0}.$$

In other words, there is no vorticity stretching term in the two-dimensional case. This fact underlies the reason why global strong solutions can be established when $d = 2$. Lastly we also note in two or three dimensions, since $\boldsymbol{u}$ is divergence-free, the following quantities are equivalent, see Exercise 7.7:

$$\|\nabla \boldsymbol{u}\|_{L^2(\mathbb{T}^d;\mathbb{R}^{d\times d})}^2 \equiv \|\omega\|_{L^2(\mathbb{T}^d;\mathbb{R}^d)}^2.$$

Thus if $\mathcal{D} = \mathbb{T}^d$ then we can think of either quantity as the enstrophy. Note though we only defined weak solutions for the three-dimensional case in Definition 7.19, we can straightforwardly extend all our results in Section 7.2 to the two-dimensional case.

Definition 7.29 (Strong solution) A strong solution of the incompressible Navier–Stokes equations on $[0, T]$ with divergence-free data $\boldsymbol{u}_0 \in H^1(\mathbb{T}^d; \mathbb{R}^d)$ is a weak solution which satisfies

$$\boldsymbol{u} \in L^\infty(0, T; H^1(\mathbb{T}^d; \mathbb{R}^d)) \cap L^2(0, T; H^2(\mathbb{T}^d; \mathbb{R}^d)).$$

The main result we establish via a-priori estimates is as follows.

Proposition 7.30 (Strong solutions) *Suppose the data $\boldsymbol{u}_0 \in H^1(\mathbb{T}^d; \mathbb{R}^d)$. Then a-priori estimates for the incompressible Navier–Stokes equations establish:*

(1) If $d = 2$, for any $T > 0$ a strong solution exists.

(2) If $d = 3$, a strong solution exists on $[0, T_*)$, where with $c > 0$ a finite constant:

$$T_* := \nu^3 / \left(c \, \|\omega_0\|^2_{L^2(\mathbb{T}^3;\mathbb{R}^3)}\right).$$

In other words, when $d = 2$, strong solutions exist globally in time, while when $d = 3$, for general initial data, strong solutions exist up to time T_*.

Remark 7.31 (Small initial data) Ladyzhenskaya (1969) showed that if the $H^1(\mathbb{T}^3; \mathbb{R}^3)$-norm of the initial data is 'sufficiently small' or the viscosity ν is sufficiently large, and there is no external forcing, then in three dimensions there exist strong solutions to the incompressible Navier–Stokes equations globally in time. See Exercise 7.9.

We prove Proposition 7.30 alongside the following lemma.

Lemma 7.32 (Enstrophy a-priori estimates) *A-priori estimates for the incompressible Navier–Stokes equations establish:*

(1) If $d = 2$, the enstrophy satisfies

$$\frac{\mathrm{d}}{\mathrm{d}t} \tfrac{1}{2} \|\omega\|^2_{L^2(\mathbb{T}^2;\mathbb{R}^2)} + \nu \|\nabla \omega\|^2_{L^2(\mathbb{T}^2;\mathbb{R}^{2\times 2})} = 0.$$

(2) If $d = 3$, the enstrophy satisfies (for some constant $c > 0$)

$$\frac{\mathrm{d}}{\mathrm{d}t} \tfrac{1}{2} \|\omega\|^2_{L^2(\mathbb{T}^3;\mathbb{R}^3)} + \frac{\nu}{4} \|\nabla \omega\|^2_{L^2(\mathbb{T}^3;\mathbb{R}^{3\times 3})} \leqslant \frac{c}{4\nu^3} \left(\|\omega\|^2_{L^2(\mathbb{T}^3;\mathbb{R}^3)}\right)^3.$$

Remark 7.33 For more details, see e.g. Temam (1983, Ch. 3).

Proof [Lemma 7.32 and Proposition 7.30] We consider the L^2-inner product of ω with the first equation in the vorticity formulation of the incompressible Navier–Stokes equations above. In other words, we pre-multiply the vorticity evolution equation by ω^{T} and integrate over $\mathbb{T}^d$. This generates the equality

$$\frac{\mathrm{d}}{\mathrm{d}t} \tfrac{1}{2} \|\omega\|^2_{L^2(\mathbb{T}^d;\mathbb{R}^d)} + \nu \|\nabla \omega\|^2_{L^2(\mathbb{T}^d;\mathbb{R}^{d\times d})} = \int_{\mathbb{T}^d} \omega^{\mathrm{T}} (\nabla \boldsymbol{u}) \, \omega \, \mathrm{d}\boldsymbol{x}.$$

Note first, we used that

$$\int_{\mathbb{T}^d} \omega^{\mathrm{T}} \frac{\partial \omega}{\partial t}\,\mathrm{d}\boldsymbol{x} = \int_{\mathbb{T}^d} \frac{\partial}{\partial t}\frac{1}{2}|\omega|^2\,\mathrm{d}\boldsymbol{x},$$

and then swapped over the time derivative and spatial integration as we assume the solutions are smooth for a-priori estimates. Note second, we used that

$$\int_{\mathbb{T}^d} \omega^{\mathrm{T}}(\nabla\omega)\,\boldsymbol{u}\,\mathrm{d}\boldsymbol{x} = \int_{\mathbb{T}^d} \nabla(\tfrac{1}{2}|\omega|^2)\cdot\boldsymbol{u}\,\mathrm{d}\boldsymbol{x} = \int_{\mathbb{T}^d} \nabla\cdot(\tfrac{1}{2}|\omega|^2\boldsymbol{u})\,\mathrm{d}\boldsymbol{x} = 0.$$

The final term just above is zero by the Divergence Theorem and the periodic boundary conditions. We also used identity #8 from Section A.1 in the Appendix, $\nabla\cdot(\phi\boldsymbol{u}) = \phi(\nabla\cdot\boldsymbol{u}) + \boldsymbol{u}\cdot\nabla\phi$, with $\phi := \frac{1}{2}|\omega|^2$. Note third, we used that

$$\int_{\mathbb{T}^d} \omega^{\mathrm{T}}\Delta\omega\,\mathrm{d}\boldsymbol{x} = \int_{\mathbb{T}^d} \nabla\cdot(\omega^{\mathrm{T}}(\nabla\omega))\,\mathrm{d}\boldsymbol{x} - \int_{\mathbb{T}^d} |\nabla\omega|^2\,\mathrm{d}\boldsymbol{x},$$

where $|\nabla\omega|$ is the sum of the square of all the components of the matrix $\nabla\omega$, and to which we can also apply the Divergence Theorem and use the periodic boundary conditions. At this point we now distinguish between the two- and three-dimensional cases. Recall that when $d = 2$ the term on the right is zero.

(1) Assume $d = 2$. Since the term on the right in the enstrophy equality above is zero, if we integrate in time between $t = 0$ and some $t \in [0, T]$ for any $T > 0$, we observe

$$\tfrac{1}{2}\|\omega(t)\|^2_{L^2(\mathbb{T}^2;\mathbb{R}^2)} + \nu\int_0^t \|\nabla\omega(\tau)\|^2_{L^2(\mathbb{T}^2;\mathbb{R}^{2\times2})}\,\mathrm{d}\tau = \tfrac{1}{2}\|\omega_0\|^2_{L^2(\mathbb{T}^2;\mathbb{R}^2)}.$$

Hence we have established the first result (1) in Proposition 7.30.

(2) Assume $d = 3$. Now the vorticity stretching term $(\nabla\boldsymbol{u})\,\omega$ is an important aspect of the flow. Directly estimating the term on the right in the enstrophy equality above, we find

$$\begin{aligned}
\int_{\mathbb{T}^d} |\omega^{\mathrm{T}}(\nabla\boldsymbol{u})\,\omega|\,\mathrm{d}\boldsymbol{x}. &\leqslant \|\nabla\boldsymbol{u}\|_{L^2(\mathbb{T}^3;\mathbb{R}^{3\times3})}\|\omega\|^2_{L^4(\mathbb{T}^3;\mathbb{R}^3)}\\
&\leqslant c\,\|\nabla\boldsymbol{u}\|_{L^2(\mathbb{T}^3;\mathbb{R}^{3\times3})}\Big(\|\nabla\omega\|^{3/4}_{L^2(\mathbb{T}^3;\mathbb{R}^{3\times3})}\|\omega\|^{1/4}_{L^2(\mathbb{T}^3;\mathbb{R}^3)}\Big)^2\\
&= c\,\|\nabla\omega\|^{3/2}_{L^2(\mathbb{T}^3;\mathbb{R}^{3\times3})}\|\omega\|^{3/2}_{L^2(\mathbb{T}^3;\mathbb{R}^3)}\\
&\leqslant \frac{3}{4}\nu\,\|\nabla\omega\|^2_{L^2(\mathbb{T}^3;\mathbb{R}^{3\times3})} + \frac{c^4}{4\nu^3}\Big(\|\omega\|^2_{L^2(\mathbb{T}^3;\mathbb{R}^3)}\Big)^3.
\end{aligned}$$

Above in turn, from Proposition 7.16, we have used the: (i) Hölder inequality with $p = q = 2$; (ii) Sobolev–Gagliardo–Nirenberg inequality in the form

$$\|\omega\|_{L^4(\mathbb{T}^3;\mathbb{R}^3)} \leqslant c\,\|\nabla\omega\|^{3/4}_{L^2(\mathbb{T}^3;\mathbb{R}^{3\times3})}\|\omega\|^{1/4}_{L^2(\mathbb{T}^3;\mathbb{R}^3)},$$

i.e. with $d = 3$ and $p = 4$ so $a = 3/4$; and (iii) Young's inequality in the last step with $p = 4/3$, $q = 4$ and $\epsilon = \nu^{3/4}$. Using this bound for the term on the right in the enstrophy estimate above generates the enstrophy inequality shown in Lemma 7.32 after we replace the generic constant c^4 by c.

The estimate for T_* is established as follows. In the enstrophy inequality shown in Lemma 7.32 when $d = 3$, if we ignore the viscous term on the left, then using the Gronwall Lemma from Exercise 7.10 to integrate the cubic evolution inequality for $\|\omega(t)\|^2_{L^2(\mathbb{T}^3;\mathbb{R}^3)}$ in time, we find

$$\|\omega(t)\|^2_{L^2(\mathbb{T}^3;\mathbb{R}^3)} \leqslant \left(\|\omega_0\|^{-2}_{L^2(\mathbb{T}^3;\mathbb{R}^3)} - ct/\nu^3\right)^{-1/2}.$$

The estimate for T_* is generated by computing the time when the upper bound above blows up. Lastly we remark that if $\|\omega\|_{L^2(\mathbb{T}^d;\mathbb{R}^{d\times d})}$ is square-integrable in time then $\boldsymbol{u} \in L^2(0,T;H^2(\mathbb{T}^d;\mathbb{R}^d))$; see Exercise 7.7. We have thus completed the proof of both Lemma 7.32 and Proposition 7.30. □

Remark 7.34 There are many associated regularity results for the three-dimensional incompressible Navier–Stokes equations. These typically show the existence of strong solutions provided the solution is assumed to satisfy a particular property which as yet has not been estalished for solutions of the Navier–Stokes equations. We investigate in detail a geometric property assumption in the next Section 7.4. A brief summary of such results can be found in Section 7.5, and some of the results referred to therein are established in the exercises at the end of this chapter.

7.4 Vorticity Alignment/Anti-alignment

Among the conditional results for the global existence of strong solutions to the three-dimensional incompressible Navier–Stokes equations, the result of Constantin and Fefferman (1993) stands out due to the geometric nature of the required assumption. The domain of the flow is $\mathcal{D} = \mathbb{R}^3$.

Remark 7.35 (Blow-up for Tao's averaged Navier–Stokes equation) Tao (2016) proved finite-time blow-up for a modified version of the three-dimensional incompressible Navier–Stokes equations in $\mathbb{R}^3$ in which the nonlinear term is averaged over certain rotations and Fourier multipliers of order zero. One of the key conclusions from the result is that 'any proposed positive solution to the [three-dimensional incompressible Navier–Stokes] regularity problem which does not use the finer structure of the nonlinearity cannot possibly be successful'. For more details see Tao (2016, Sec. 1).

Our goal herein is to establish Constantin and Fefferman's result using a-priori estimates. These can be used as a basis to establish the result more rigorously by assuming a mollified advecting velocity field $\boldsymbol{u}_\delta$ so the nonlinear term is $(\boldsymbol{u}_\delta \cdot \nabla)\,\boldsymbol{u}$, and taking the limit $\delta \to 0$ to remove the mollification. See Section 7.5 for more details on the mollification procedure. The main result Constantin and Fefferman prove can be stated as follows; $T > 0$ is arbitrary.

Theorem 7.36 (Constantin and Fefferman, 1993) *Suppose there exist constants $\Omega > 0$ and $\rho > 0$ such that*

$$|\sin\theta| \leqslant \frac{|\boldsymbol{y}|}{\rho}$$

holds if both $|\omega(x,t)| > \Omega$ *and* $|\omega(x+y,t)| > \Omega$*, and* $0 \leqslant t \leqslant T$*. Here* $\omega = \omega(x,t)$ *is the vorticity field and* θ *is the angle between the vorticity vectors* $\omega(x,t)$ *and* $\omega(x+y,t)$*. Then the solution to the initial-value problem of the Navier–Stokes equations is strong and hence smooth on the time interval* $[0,T]$.

We prove this result via the following sequence of steps.

Step 1: The Biot–Savart Law. We use the vorticity formulation of the three-dimensional incompressible Navier–Stokes equations, see Corollary 4.15:

$$\frac{\partial \omega}{\partial t} + (\nabla \omega)\, u = \nu\, \Delta \omega + (\nabla u)\, \omega,$$
$$\Delta u = -\nabla \times \omega,$$

where we write $(u \cdot \nabla)\, \omega \equiv (\nabla \omega)\, u$ and $(\omega \cdot \nabla)\, u \equiv (\nabla u)\, \omega$. Recall, e.g. from Lemma 5.5, the divergence-free condition $\nabla \cdot u = 0$ implies there exists a vector potential ψ such that $u = \nabla \times \psi$. Hence taking the curl of u we observe $\omega = \nabla \times u = \nabla \times (\nabla \times \psi) = -\Delta \psi$ using identity #5 from Section A.1 of the Appendix. Hence $\psi = -\Delta^{-1} \omega$ and in $\mathbb{R}^3$ we can characterise the action of Δ^{-1} as follows, see e.g. Evans (1998, p. 23):

$$\psi(x) = \frac{1}{4\pi} \int_{\mathbb{R}^3} \frac{\omega(y)}{|x-y|}\, \mathrm{d}y.$$

We suppress the time dependence for the moment. Naturally we assume ω decays sufficiently rapidly to zero in the far-field for the integral to make sense. This is the Biot–Savart Law. By taking the curl of this expression, we find an expression for the velocity field u in terms of the vorticity field ω as follows:

$$\begin{aligned} u(x) &= -\nabla \times \left(\frac{1}{4\pi} \int_{\mathbb{R}^3} \frac{\omega(x-y)}{|y|}\, \mathrm{d}y \right) \\ &= -\frac{1}{4\pi} \int_{\mathbb{R}^3} \frac{\nabla_x \times \omega(x-y)}{|y|}\, \mathrm{d}y \\ &= \frac{1}{4\pi} \int_{\mathbb{R}^3} \frac{\nabla_y \times \omega(x-y)}{|y|}\, \mathrm{d}y \\ &= -\frac{1}{4\pi} \int_{\mathbb{R}^3} \nabla_y \left(\frac{1}{|y|} \right) \times \omega(x-y)\, \mathrm{d}y, \end{aligned}$$

where we: (i) made the change of variables $y' := x - y$ in the convolution integral and then relabelled y' as y; (ii) swapped over the integral and curl operations; (iii) used the chain rule to change the curl with respect to x to a curl with respect to y acting on $\omega = \omega(x-y)$ as shown; and (iv) used identity #10 from Section A.1 of the Appendix in the form $\nabla \times (\phi \omega) \equiv \phi \nabla \times \omega + \nabla \phi \times \omega$ with $\phi = |y|^{-1}$ and that ω decays sufficiently rapidly in the far field.

Step 2: The velocity gradient matrix. Recall from Section 1.9 that for any vector $h \in \mathbb{R}^3$, the action of the vorticity vector $\omega = \omega(x)$ on h is such that $\frac{1}{2}\omega \times h \equiv R\, h$, where $R = R(x)$ is the antisymmetric matrix

$$R := \frac{1}{2}\begin{pmatrix} 0 & -\omega_3 & \omega_2 \\ \omega_3 & 0 & -\omega_1 \\ -\omega_2 & \omega_1 & 0 \end{pmatrix}.$$

We thus observe that

$$\boldsymbol{u}(\boldsymbol{x}) = \frac{1}{2\pi}\int_{\mathbb{R}^3} R(\boldsymbol{x}-\boldsymbol{y})\,\nabla\Big(\frac{1}{|\boldsymbol{y}|}\Big)\,\mathrm{d}\boldsymbol{y}.$$

Using the chain rule, $\nabla(|\boldsymbol{y}|^{-1}) = -|\boldsymbol{y}|^{-3}\boldsymbol{y}$. Hence we observe

$$\begin{aligned}
\nabla\boldsymbol{u}(\boldsymbol{x}) &= -\frac{1}{2\pi}\nabla_{\boldsymbol{x}}\int_{\mathbb{R}^3} R(\boldsymbol{x}-\boldsymbol{y})\,\frac{\boldsymbol{y}}{|\boldsymbol{y}|^3}\,\mathrm{d}\boldsymbol{y}\\
&= \frac{1}{2\pi}\nabla_{\boldsymbol{x}}\int_{\mathbb{R}^3} R(\boldsymbol{y})\,\frac{(\boldsymbol{x}-\boldsymbol{y})}{|\boldsymbol{x}-\boldsymbol{y}|^3}\,\mathrm{d}\boldsymbol{y}\\
&= \frac{1}{2\pi}\int_{\mathbb{R}^3} R(\boldsymbol{y})\,\nabla_{\boldsymbol{x}}\Big(\frac{(\boldsymbol{x}-\boldsymbol{y})}{|\boldsymbol{x}-\boldsymbol{y}|^3}\Big)\,\mathrm{d}\boldsymbol{y}\\
&= \frac{1}{2\pi}\int_{\mathbb{R}^3} R(\boldsymbol{y})\,\nabla_{\boldsymbol{x}-\boldsymbol{y}}\Big(\frac{(\boldsymbol{x}-\boldsymbol{y})}{|\boldsymbol{x}-\boldsymbol{y}|^3}\Big)\,\mathrm{d}\boldsymbol{y}\\
&= -\frac{1}{2\pi}\int_{\mathbb{R}^3} R(\boldsymbol{x}-\boldsymbol{y})\,\nabla_{\boldsymbol{y}}\Big(\frac{\boldsymbol{y}}{|\boldsymbol{y}|^3}\Big)\,\mathrm{d}\boldsymbol{y},
\end{aligned}$$

where we: (i) made the change of variables $\boldsymbol{y}' := \boldsymbol{x}-\boldsymbol{y}$ in the convolution integral and then relabelled $\boldsymbol{y}'$ as $\boldsymbol{y}$; (ii) swapped over the integral and grad operations; (iii) used that $\nabla_{\boldsymbol{x}}$ is the same as $\nabla_{\boldsymbol{x}-\boldsymbol{y}}$ for the function $|\boldsymbol{x}-\boldsymbol{y}|^{-3}(\boldsymbol{x}-\boldsymbol{y})$; and (iv) again made the change of variables $\boldsymbol{y}' := \boldsymbol{x}-\boldsymbol{y}$ and then relabelled $\boldsymbol{y}'$ as $\boldsymbol{y}$. Recall of course the gradient of the vector $\boldsymbol{v}(\boldsymbol{y}) = |\boldsymbol{y}|^{-3}\boldsymbol{y}$ is the 3×3 matrix whose first column is the partial derivative of each of the components of $\boldsymbol{v}$ with respect to the first component of $\boldsymbol{y}$, the second column is the partial derivative of each of the components of $\boldsymbol{v}$ with respect to the second component of $\boldsymbol{y}$, etc. Hence if we take the y_j partial derivative of the ith component v_i of $\boldsymbol{v}(\boldsymbol{y}) = |\boldsymbol{y}|^{-3}\boldsymbol{y}$, we find the (i,j) component of the matrix $\nabla_{\boldsymbol{y}}(\boldsymbol{y}/|\boldsymbol{y}|^3)$ is

$$\frac{\partial}{\partial y_j}\Big(\frac{y_i}{|\boldsymbol{y}|^3}\Big) = \frac{\delta_{ij}}{|\boldsymbol{y}|^3} + y_i\frac{\partial}{\partial y_j}\Big(\frac{1}{|\boldsymbol{y}|^3}\Big) = \frac{1}{|\boldsymbol{y}|^3}\Big(\delta_{ij} - 3\frac{y_i y_j}{|\boldsymbol{y}|^2}\Big) = \frac{1}{|\boldsymbol{y}|^3}\big[(I - 3\hat{\boldsymbol{y}}\hat{\boldsymbol{y}}^{\mathrm{T}})\big]_{ij},$$

where δ_{ij} is the Kronecker delta function and we denote by $\hat{\boldsymbol{y}} := |\boldsymbol{y}|^{-1}\boldsymbol{y}$ the unit direction vector corresponding to $\boldsymbol{y}$. Further, $[A]_{ij}$ denotes the (i,j) component of the matrix A. Hence the velocity gradient matrix is given by

$$\nabla\boldsymbol{u}(\boldsymbol{x}) = -\frac{1}{2\pi}\int_{\mathbb{R}^3} R(\boldsymbol{x}-\boldsymbol{y})\,\frac{1}{|\boldsymbol{y}|^3}\big(I - 3\hat{\boldsymbol{y}}\hat{\boldsymbol{y}}^{\mathrm{T}}\big)\,\mathrm{d}\boldsymbol{y}.$$

Step 3: The vorticity stretching term. Now consider the vorticity stretching term $(\nabla\boldsymbol{u})\,\boldsymbol{\omega}$ in the vorticity formulation of the Navier–Stokes momentum equation. When we substitute the expression for the velocity gradient matrix above into $(\nabla \boldsymbol{u})\,\omega$, if we ignore the factor $|\boldsymbol{y}|^{-3}$ then inside the integral the term involving the identity matrix is

$$(R(\boldsymbol{x}-\boldsymbol{y})\,I)\,\omega(\boldsymbol{x}) = R(\boldsymbol{x}-\boldsymbol{y})\,\omega(\boldsymbol{x}) = \tfrac{1}{2}\omega(\boldsymbol{x}-\boldsymbol{y}) \times \omega(\boldsymbol{x}).$$

If we take the dot product of this term with $\omega = \omega(\boldsymbol{x})$, we generate a triple scalar product containing two identical terms so the result will be zero. Consequently, we see that if we take the dot product of the vorticity stretching term with $\omega = \omega(\boldsymbol{x})$ and use that $R\boldsymbol{h} \equiv \frac{1}{2}\omega \times \boldsymbol{h}$, we get

$$\begin{aligned}\big(\omega^{\mathrm{T}}((\nabla \boldsymbol{u})\,\omega)\big)(\boldsymbol{x}) &= \frac{3}{4\pi}\int_{\mathbb{R}^3} \omega^{\mathrm{T}}(\boldsymbol{x})\,(\omega(\boldsymbol{x}-\boldsymbol{y})\times\hat{\boldsymbol{y}})\,\frac{\hat{\boldsymbol{y}}^{\mathrm{T}}}{|\boldsymbol{y}|^3}\omega(\boldsymbol{x})\,\mathrm{d}\boldsymbol{y}\\ &= \frac{3}{4\pi}\int_{\mathbb{R}^3} \hat{\omega}^{\mathrm{T}}(\boldsymbol{x})\,(\hat{\omega}(\boldsymbol{x}-\boldsymbol{y})\times\hat{\boldsymbol{y}})\,\frac{\hat{\boldsymbol{y}}^{\mathrm{T}}}{|\boldsymbol{y}|^3}\omega(\boldsymbol{x})\,|\omega(\boldsymbol{x})||\omega(\boldsymbol{x}-\boldsymbol{y})|\mathrm{d}\boldsymbol{y}.\end{aligned}$$

Let us focus on the term

$$\hat{\omega}^{\mathrm{T}}(\boldsymbol{x})\,(\hat{\omega}(\boldsymbol{x}-\boldsymbol{y})\times\hat{\boldsymbol{y}}) = \hat{\omega}(\boldsymbol{x})\cdot(\hat{\omega}(\boldsymbol{x}-\boldsymbol{y})\times\hat{\boldsymbol{y}}) = -\hat{\boldsymbol{y}}\cdot(\hat{\omega}(\boldsymbol{x}-\boldsymbol{y})\times\hat{\omega}(\boldsymbol{x})),$$

where we have used that the triple scalar product is invariant to the cyclic rotation of the arguments and swapped the order of the terms in the cross product. Using the properties of the cross product we observe

$$|\hat{\boldsymbol{y}}\cdot(\hat{\omega}(\boldsymbol{x}-\boldsymbol{y})\times\hat{\omega}(\boldsymbol{x}))| \leqslant |\sin\theta\,|,$$

where θ is the angle between the vectors $\hat{\omega}(\boldsymbol{x}-\boldsymbol{y})$ and $\hat{\omega}(\boldsymbol{x})$.

Step 4: The enstrophy estimate. Now consider the L^2-inner product over $\mathbb{R}^3$ of ω with the vorticity formulation of the incompressible Navier–Stokes equations at the beginning of Step 1. Using the suitable decay of the solution in the far field instead of the periodic boundary conditions therein, we can exactly mirror the arguments at the beginning of Section 7.3 to generate the following identity for the evolution of the enstrophy $\|\omega\|_{L^2(\mathbb{R}^3;\mathbb{R}^3)} = \|\nabla \boldsymbol{u}\|_{L^2(\mathbb{R}^3;\mathbb{R}^{3\times 3})}$:

$$\frac{\mathrm{d}}{\mathrm{d}t}\tfrac{1}{2}\|\omega\|^2_{L^2(\mathbb{R}^3;\mathbb{R}^3)} + \nu\|\nabla\omega\|^2_{L^2(\mathbb{R}^3;\mathbb{R}^3)} = \int_{\mathbb{R}^3}\omega^{\mathrm{T}}((\nabla \boldsymbol{u})\,\omega)\,\mathrm{d}\boldsymbol{x}.$$

We now use the assumption stated in the theorem, i.e. in regions where the magnitude of the vorticity is large we suppose $|\sin\theta\,| \leqslant |\boldsymbol{y}|/\rho$ for some constant $\rho > 0$. Focusing on the term on the right above, which was generated by the vorticity stretching term, with the integrals with respect to $\boldsymbol{x}$ and $\boldsymbol{y}$ over the appropriate subregions of $\mathbb{R}^3$ where $|\omega(\boldsymbol{x})| > \Omega$ and $|\omega(\boldsymbol{x}-\boldsymbol{y})| > \Omega$, we observe

$$\begin{aligned}\int |\omega^{\mathrm{T}}(\boldsymbol{x})((\nabla \boldsymbol{u})\,\omega)(\boldsymbol{x})|\,\mathrm{d}\boldsymbol{x} &\leqslant \frac{3}{4\pi}\int |\omega(\boldsymbol{x})|^2 \int |\sin\theta\,|\,\frac{|\omega(\boldsymbol{x}-\boldsymbol{y})|}{|\boldsymbol{y}|^3}\,\mathrm{d}\boldsymbol{y}\,\mathrm{d}\boldsymbol{x}\\ &\leqslant \frac{C}{\rho}\|\omega\|^2_{L^4(\mathbb{R}^3;\mathbb{R}^3)}\left(\int\left(\int \frac{|\omega(\boldsymbol{x}-\boldsymbol{y})|}{|\boldsymbol{y}|^2}\,\mathrm{d}\boldsymbol{y}\right)^2\mathrm{d}\boldsymbol{x}\right)^{1/2},\end{aligned}$$

for some constant $C > 0$, where we used the Hölder inequality with $p = q = 2$. To bound the first factor in the upper bound on the right, we use the Sobolev–Gagliardo–Nirenberg inequality from Proposition 7.16(e), which also applies on $\mathbb{R}^d$. With $d = 3$ and $p = 4$ so $a = 3/4$, we find

$$\|\omega\|^2_{L^4(\mathbb{R}^3;\mathbb{R}^3)} \leqslant c\,\|\nabla\omega\|^{3/2}_{L^2(\mathbb{R}^3;\mathbb{R}^{3\times 3})}\|\omega\|^{1/2}_{L^2(\mathbb{R}^3;\mathbb{R}^3)}.$$

Step 5: The Hardy–Littlewood–Sobolev inequality. To bound the second factor in the upper bound on the right, we first need the Hardy–Littlewood–Sobolev inequality as follows.

Lemma 7.37 (Hardy–Littlewood–Sobolev inequality) *Suppose $p, q \in (1, \infty)$ and $r \in (0, d)$ satisfy $1/p + r/d = 1/q + 1$ with $p < q$. Set the integral I_r to be*

$$(I_r f)(x) := \int_{\mathbb{R}^d} \frac{f(x-y)}{|y|^r}\,\mathrm{d}y.$$

Then if $f \in L^p(\mathbb{T}^d;\mathbb{R}^n)$ for some constant $c > 0$, we have

$$\|I_r f\|_{L^q(\mathbb{R}^d;\mathbb{R}^n)} \leqslant c\,\|f\|_{L^p(\mathbb{R}^d;\mathbb{R}^n)}.$$

For a proof of Lemma 7.37, see Hardy *et al.* (1952) as well as Reed and Simon (1975, p. 30). With $d = 3$, $r = 2$ and $q = 2$ so $p = 6/5$, we see that

$$\|I_2\omega\|_{L^2(\mathbb{R}^3;\mathbb{R}^3)} \leqslant c\,\|\omega\|_{L^{6/5}(\mathbb{R}^3;\mathbb{R}^3)}.$$

Observe that

$$\|I_2\omega\|_{L^2(\mathbb{R}^3;\mathbb{R}^3)} = \left(\int\left(\int \frac{|\omega(x-y)|}{|y|^2}\,\mathrm{d}y\right)^2 \mathrm{d}x\right)^{1/2}$$

is the second factor in the upper bound on the right above. Second, we use the Hölder inequality as follows. With $\alpha + \beta = 1$ and $1/p + 1/q = 1$, we have

$$\begin{aligned}
\|\omega\|_{L^{6/5}(\mathbb{R}^3;\mathbb{R}^3)} &= \left(\int_{\mathbb{R}^3} |\omega|^{6/5}\,\mathrm{d}x\right)^{5/6}\\
&= \left(\int_{\mathbb{R}^3} |\omega|^{6\alpha/5}|\omega|^{6\beta/5}\,\mathrm{d}x\right)^{5/6}\\
&\leqslant \left(\int_{\mathbb{R}^3} |\omega|^{6\alpha p/5}\,\mathrm{d}x\right)^{5/6p}\left(\int_{\mathbb{R}^3} |\omega|^{6\beta q/5}\,\mathrm{d}x\right)^{5/6q}\\
&= \|\omega\|^{2/3}_{L^1(\mathbb{R}^3;\mathbb{R}^3)}\|\omega\|^{1/3}_{L^2(\mathbb{R}^3;\mathbb{R}^3)}
\end{aligned}$$

where in the final step we chose $\alpha = 2/3$, $\beta = 1/3$, $p = 5/4$ and $q = 5$.

Step 6: Young's inequality. Combining the results in Steps 4 and 5 together, we observe that the estimate for the enstrophy becomes

$$\frac{\mathrm{d}}{\mathrm{d}t}\frac{1}{2}\|\omega\|^2_{L^2(\mathbb{R}^3;\mathbb{R}^3)} + \nu\,\|\nabla\omega\|^2_{L^2(\mathbb{R}^3;\mathbb{R}^3)} \leqslant c\,\|\nabla\omega\|^{3/2}_{L^2(\mathbb{R}^3;\mathbb{R}^{3\times 3})}\|\omega\|^{5/6}_{L^2(\mathbb{R}^3;\mathbb{R}^3)}\|\omega\|^{2/3}_{L^1(\mathbb{R}^3;\mathbb{R}^3)},$$

for some constant $c > 0$. In Exercise 7.11 we show that $\|\omega\|_{L^1(\mathbb{R}^3;\mathbb{R}^3)}$ is bounded on any finite time interval. We absorb the upper bound for $\|\omega\|_{L^1(\mathbb{R}^3;\mathbb{R}^3)}$ on $[0, T]$ into the generic constant c shown. Then using Young's inequality from Proposition 7.16(a) with $p = 4/3$, $q = 4$ and $\epsilon = \nu^{3/4}$, we see that

$$c\,\|\nabla\omega\|^{3/2}_{L^2(\mathbb{R}^3;\mathbb{R}^{3\times3})}\,\|\omega\|^{5/6}_{L^2(\mathbb{R}^3;\mathbb{R}^3)} \leqslant \frac{3}{4}\nu\,\|\nabla\omega\|^2_{L^2(\mathbb{R}^3;\mathbb{R}^{3\times3})} + \frac{c^4}{4\nu^3}\|\omega\|^{10/3}_{L^2(\mathbb{R}^3;\mathbb{R}^3)}.$$

If we use these facts in the estimate for the enstrophy just above, we find

$$\frac{\mathrm{d}}{\mathrm{d}t}\frac{1}{2}\|\omega\|^2_{L^2(\mathbb{R}^3;\mathbb{R}^3)} + \frac{\nu}{4}\,\|\nabla\omega\|^2_{L^2(\mathbb{R}^3;\mathbb{R}^3)} \leqslant \frac{c^4}{4\nu^3}\|\omega(t)\|^{4/3}_{L^2(\mathbb{R}^3;\mathbb{R}^3)}\,\|\omega(t)\|^2_{L^2(\mathbb{R}^3;\mathbb{R}^3)},$$

where we have decomposed $\|\omega\|^{10/3}_{L^2(\mathbb{R}^3;\mathbb{R}^3)}$ into the two factors on the right shown. Using that the second term on the left above is non-negative, if we now use Gronwall's Lemma from Exercise 7.10, then for any $t \in [0, T]$ we find

$$\|\omega(t)\|^2_{L^2(\mathbb{R}^3;\mathbb{R}^3)} \leqslant \|\omega_0\|^2_{L^2(\mathbb{R}^3;\mathbb{R}^3)}\exp\left(\frac{c^4}{2\nu^3}\int_0^t \|\omega(\tau)\|^{4/3}_{L^2(\mathbb{R}^3;\mathbb{R}^3)}\,\mathrm{d}\tau\right).$$

Since we know for weak solutions $\boldsymbol{u} \in L^2(0, T; H^1(\mathbb{R}^3; \mathbb{R}^3))$ for any $T > 0$, the argument of the exponent is bounded for any $t \in [0, T]$ and any $T > 0$ as

$$\int_0^t \|\omega(\tau)\|^{4/3}_{L^2(\mathbb{R}^3;\mathbb{R}^3)}\,\mathrm{d}\tau \leqslant \left(\int_0^t \|\omega(\tau)\|^2_{L^2(\mathbb{R}^3;\mathbb{R}^3)}\,\mathrm{d}\tau\right)^{2/3} t^{1/3},$$

using Hölder's inequality with $p = 3/2$. Hence the enstrophy $\|\omega\|^2_{L^2(\mathbb{R}^3;\mathbb{R}^3)}$ is bounded for any $t \in [0, T]$ and any $T > 0$, i.e. globally in time. Finally, we recall from Section 7.3 that if the enstrophy is bounded in any time interval then solutions to the three-dimensional incompressible Navier–Stokes equations are strong and thus smooth solutions on that time interval.

Remark 7.38 (Weaker assumptions) It is possible to weaken the assumption of Constantin and Fefferman's result in Theorem 7.36. Indeed, the assumption $|\sin\theta| \leqslant |\boldsymbol{y}|^\gamma/\rho$ for any $\gamma \in [1/2, 1]$ suffices to reach the same conclusion; see Exercise 7.12 where these weaker assumptions are explored in detail.

7.5 Notes

Some further details are as follows.

(1) *Millennium Prize Problem.* A proof of whether three-dimensional incompressible Navier–Stokes solutions do exhibit a singularity in finite time or are smooth globally in time, in $\mathbb{T}^3$ or $\mathbb{R}^3$, constitutes one of the Millennium Prize Problems. See Fefferman (2000) for the problem criteria.

(2) *Finite domain and viscous boundary conditions.* The equivalent results to those proved in this chapter for the case of a finite domain and viscous boundary conditions can be found in Ladyzhenskaya (1969) or Constantin and Foias (1988).

(3) *Approximate systems.* To establish the existence of global-in-time weak solutions to the incompressible Navier–Stokes equations in Section 7.2, we considered an approximate Navier–Stokes system constructed by projecting the solution onto a finite number of Fourier modes. We were able to use standard analysis for finite systems of ordinary differential equations to establish global-in-time regularity for this finite approximate system of equations. We established upper bounds in appropriate norms of the solution to the approximate system which were uniform with respect to the variable K that parameterised the number of Fourier modes we projected onto. And then we passed to the limit $K \to \infty$. There are many other

analogous approaches to this procedure. The approach originally used by Leray (1934), and also for example by Constantin and Fefferman (1993), is to consider an approximate Navier–Stokes system in which the advection velocity field is mollified, i.e. smoothed. In other words, the nonlinear term $(\boldsymbol{u} \cdot \nabla)\boldsymbol{u}$ in the Navier–Stokes momentum equation is replaced by the term $(\boldsymbol{u}_\delta \cdot \nabla)\boldsymbol{u}$ where $\boldsymbol{u}_\delta$ is defined, for example in Constantin and Fefferman (1993), by

$$\boldsymbol{u}_\delta(\boldsymbol{x}, t) := \int_{\mathbb{R}^3} \delta^{-3}\phi(\delta^{-1}(\boldsymbol{x} - \boldsymbol{y}))\,\boldsymbol{u}(\boldsymbol{y}, t)\,\mathrm{d}\boldsymbol{y},$$

for $\delta > 0$, where ϕ is a smooth, compactly supported function whose integral over $\mathbb{R}^3$ equals one. Though the resulting approximate system is nonlinear, it is known it has global smooth solutions. Then, as described above, we proceed by establishing upper bounds in appropriate norms that are uniform with respect to the parameter δ. We then pass to the limit $\delta \to 0$ in which the full incompressible Navier–Stokes equations are recovered. Another approach in the case of bounded domains with viscous boundary conditions is Galerkin approximation. Here an approximate Navier–Stokes system is constructed by projection onto a finite number of eigenmodes of the underlying Stokes operator; see e.g. Constantin and Foias (1988). Other procedures include introducing hyperviscosity, or setting up an iterative system in which the advecting velocity $\boldsymbol{u}^{(n-1)}$ is one iteration behind the solution $\boldsymbol{u}^{(n)}$. The initial iterate is the initial data, i.e. $\boldsymbol{u}^{(0)} := \boldsymbol{u}_0$. There are other procedures, but the main ones are those just described.

(4) *Conditional regularity results (part I).* Many *conditional regularity results* for the three-dimensional incompressible Navier–Stokes equations have been established. These take the following form. If weak solutions to the three-dimensional incompressible Navier–Stokes equations satisfy a certain, as yet unproven, regularity criterion, then global-in-time strong solutions can be established. A classical result is due to Serrin (1962), see Exercise 7.8. This says that for any $T > 0$, if we *assume* $\boldsymbol{u} \in L^s(0, T; L^r(\mathbb{T}^3; \mathbb{R}^3))$ where $3/r + 2/s \leqslant 1$, then the solution to the incompressible Navier–Stokes equations is a strong solution on $[0, T]$.

(5) *Conditional regularity results (part II).* The significance of the conditional regularity result due to Constantin and Fefferman (1993) we considered in detail in Section 7.4 is that it provides a practical geometric constraint on the possible scenarios for blow-up of the three-dimensional incompressible Navier–Stokes equations in finite time. As mentioned in Remark 7.38, the result has since been strengthened in the sense that the same result can be established under weaker assumptions, as we consider in Exercise 7.12. See e.g. Beirão da Veiga and Berselli (2002) for these stronger results.

(6) *Partial regularity results.* We saw in Proposition 7.30 that in three dimensions, even with smooth initial data, we only know the solution to the incompressible Navier–Stokes equations is smooth for a finite time. It may be smooth thereafter or the solution may become singular. If the solution does become singular after a finite time then we can estimate the dimension of the singular set in space-time. This was first considered by Scheffer (1976, 1977). Caffarelli *et al.* (1982) proved that the one-dimensional parabolic Hausdorff measure of the singular set in space-time equals zero.

Exercises

7.1 Lebesgue integrability: Weak, without strong, convergence. We analyse an example of a sequence bounded in $L^2(\mathbb{T};\mathbb{C})$ that is weakly convergent to zero in $L^2(\mathbb{T};\mathbb{C})$ but not strongly convergent. Consider the sequence of functions $f_n(x) := \sin(nx)$ for $n \in \mathbb{N}$ on $\mathbb{T} = [-\pi, \pi]$.

(a) By direct computation show that

$$\int_{\mathbb{T}} |f_n(x)|^2\,\mathrm{d}x = \pi.$$

Hence we deduce $f_n \in L^2(\mathbb{T};\mathbb{C})$ and is uniformly bounded.

(b) Since $f_n \in L^2(\mathbb{T};\mathbb{C})$ from part (a), by the Riesz Representation Theorem 7.12 the criterion for weak convergence of f_n to zero in $L^2(\mathbb{T};\mathbb{C})$ as $n \to \infty$ is equivalent to establishing the condition

$$\int_{\mathbb{T}} v(x) f_n(x)\,\mathrm{d}x \to 0,$$

for all $v \in L^2(\mathbb{T};\mathbb{C})$. Suppose v has the Fourier representation

$$v(x) = \sum_{k\in\mathbb{Z}} V_k\, \mathrm{e}^{\mathrm{i}kx},$$

where the $V_k\colon \mathbb{Z} \to \mathbb{C}$ are the Fourier coefficients of v.

(i) Using Parseval's identity, explain why $V = \{V_k\}_{k\in\mathbb{Z}}$ is such that $V \in \ell^2(\mathbb{Z};\mathbb{C})$ and hence deduce $V_k \to 0$ as $k \to \infty$.

(ii) Again using Parseval's identity, show that

$$\int_{\mathbb{T}} v(x) f_n(x)\,\mathrm{d}x$$

is proportional to V_n and hence deduce the weak convergence of f_n to zero.

(c) Explain why the result from part (a) necessarily implies f_n does not strongly converge to zero in $L^2(\mathbb{T};\mathbb{C})$.

7.2 Lebesgue integrability: Young's inequality. The goal herein is to prove Young's inequality for products in Proposition 7.16(a). Hence assume $a > 0$ and $b > 0$ and further that $p, q \in (1,\infty)$ satisfy $1/p + 1/q = 1$. Let $f = f(x)$ be the monotonic function $f(x) = x^{p-1}$; note $f(0) = 0$.

(a) Show the inverse function $f^{-1}(x) = x^{q-1}$.

(b) Explain why we know

$$ab \leqslant \int_0^a f(x)\,\mathrm{d}x + \int_0^b f^{-1}(\xi)\,\mathrm{d}\xi.$$

(*Hint:* Identify all three areas present in the inequality in the plane.)

(c) By substituting the explicit forms for f and f^{-1} into the inequality in part (b) and performing the necessary integrations, show that Young's inquality for products holds.

7.3 Lebesgue integrability: Hölder's inequality. The goal of this question is to prove the Hölder inequality for sequences in Proposition 7.16(b). Suppose $p, q \in [1,\infty)$ satisfy $1/p + 1/q = 1$. Further suppose the sequences $F = \{F_{\boldsymbol{k}}\}_{\boldsymbol{k}\in\mathbb{Z}^d}$ and $G = \{G_{\boldsymbol{k}}\}_{\boldsymbol{k}\in\mathbb{Z}^d}$ are such that $F \in \ell^p(\mathbb{Z}^d;\mathbb{C})$ and $G \in \ell^q(\mathbb{Z}^d;\mathbb{C})$, respectively, for some $d \in \mathbb{N}$. If $\|F\|_{\ell^p(\mathbb{Z}^d;\mathbb{C})}$ and $\|G\|_{\ell^q(\mathbb{Z}^d;\mathbb{C})}$ denote the corresponding norms of F and G, then set

$$\hat{F} := F/\|F\|_{\ell^p(\mathbb{Z}^d;\mathbb{C})} \quad\text{and}\quad \hat{G} := G/\|G\|_{\ell^q(\mathbb{Z}^d;\mathbb{C})}.$$

Hence we know $\|\hat{F}\|_{\ell^p(\mathbb{Z}^d;\mathbb{C})} = 1$ and $\|\hat{G}\|_{\ell^q(\mathbb{Z}^d;\mathbb{C})} = 1$.

(a) Use Young's inequality for products, from Proposition 7.16(a) and Exercise 7.2, to show

$$|\hat{F}_{\boldsymbol{k}}^{\dagger}\hat{G}_{\boldsymbol{k}}| \leqslant \frac{1}{p}|\hat{F}_{\boldsymbol{k}}|^p + \frac{1}{q}|\hat{G}_{\boldsymbol{k}}|^q.$$

(b) By summing the terms on the left over $\boldsymbol{k} \in \mathbb{Z}^d$, show using the inequality in part (a) that

$$\sum_{\boldsymbol{k}\in\mathbb{Z}^d} |\hat{F}_{\boldsymbol{k}}^{\dagger}\hat{G}_{\boldsymbol{k}}| \leqslant 1.$$

(c) By using $F_{\boldsymbol{k}} = \hat{F}_{\boldsymbol{k}}\,\|F\|_{\ell^p(\mathbb{Z}^d;\mathbb{C})}$ and $G_{\boldsymbol{k}} = \hat{G}_{\boldsymbol{k}}\,\|G\|_{\ell^q(\mathbb{Z}^d;\mathbb{C})}$ show that

$$\sum_{\boldsymbol{k}\in\mathbb{Z}^d} |F_{\boldsymbol{k}}^{\dagger} G_{\boldsymbol{k}}| = \|F\|_{\ell^p(\mathbb{Z}^d;\mathbb{C})}\,\|G\|_{\ell^q(\mathbb{Z}^d;\mathbb{C})} \sum_{\boldsymbol{k}\in\mathbb{Z}^d} |\hat{F}_{\boldsymbol{k}}^{\dagger}\hat{G}_{\boldsymbol{k}}|,$$

and hence use part (b) to deduce Hölder's inequality.

7.4 Lebesgue integrability: Young's convolution inequality. The Young convolution inequality from Proposition 7.16(c) can be proved by an application of Hölder's inequality from Proposition 7.16(b) and Exercise 7.3 as follows. Suppose $p, q, r \in [1, \infty)$ satisfy $1/p + 1/q + 1/r = 2$ and the sequences $F = \{F_{\boldsymbol{k}}\}_{\boldsymbol{k}\in\mathbb{Z}^d}$, $G = \{G_{\boldsymbol{k}}\}_{\boldsymbol{k}\in\mathbb{Z}^d}$ and $H = \{H_{\boldsymbol{k}}\}_{\boldsymbol{k}\in\mathbb{Z}^d}$ are such that $F \in \ell^p(\mathbb{Z}^d;\mathbb{C})$, $G \in \ell^q(\mathbb{Z}^d;\mathbb{C})$ and $H \in \ell^r(\mathbb{Z}^d;\mathbb{C})$, respectively, for some $d \in \mathbb{N}$.

(a) Using the stated relation between p, q and r, show that

$$F_{\boldsymbol{k}}^* G_{\boldsymbol{l}} H_{\boldsymbol{k}-\boldsymbol{l}} \equiv \big((F_{\boldsymbol{k}}^*)^p (G_{\boldsymbol{l}})^q\big)^{\frac{r-1}{r}} \big((F_{\boldsymbol{k}}^*)^p (H_{\boldsymbol{k}-\boldsymbol{l}})^r\big)^{\frac{q-1}{q}} \big((G_{\boldsymbol{l}})^q (H_{\boldsymbol{k}-\boldsymbol{l}})^r\big)^{\frac{p-1}{p}}.$$

(b) Establish Hölder's inequality for the two-dimensional sequences $A = \{A_{\boldsymbol{k},\boldsymbol{l}}\}_{\boldsymbol{k},\boldsymbol{l}\in\mathbb{Z}^d}$, $B = \{B_{\boldsymbol{k},\boldsymbol{l}}\}_{\boldsymbol{k},\boldsymbol{l}\in\mathbb{Z}^d}$ and $C = \{C_{\boldsymbol{k},\boldsymbol{l}}\}_{\boldsymbol{k},\boldsymbol{l}\in\mathbb{Z}^d}$ in the form

$$\sum_{\boldsymbol{k},\boldsymbol{l}} |A_{\boldsymbol{k},\boldsymbol{l}} B_{\boldsymbol{k},\boldsymbol{l}} C_{\boldsymbol{k},\boldsymbol{l}}| \leqslant \Big(\sum_{\boldsymbol{k},\boldsymbol{l}} |A_{\boldsymbol{k},\boldsymbol{l}}|^{p'}\Big)^{1/p'} \Big(\sum_{\boldsymbol{k},\boldsymbol{l}} |B_{\boldsymbol{k},\boldsymbol{l}}|^{q'}\Big)^{1/q'} \Big(\sum_{\boldsymbol{k},\boldsymbol{l}} |C_{\boldsymbol{k},\boldsymbol{l}}|^{r'}\Big)^{1/r'},$$

where all the sums are over $\boldsymbol{k}, \boldsymbol{l} \in \mathbb{Z}^d$ and $1/p' + 1/q' + 1/r' = 1$, as follows. Successively apply Hölder's inequality from Proposition 7.16(b), for example, in the following manner. First, group $B_{\boldsymbol{k},\boldsymbol{l}}$ and $C_{\boldsymbol{k},\boldsymbol{l}}$ together and apply Hölder's equality. Second, apply Hölder's inequality to the resulting term involving $B_{\boldsymbol{k},\boldsymbol{l}} C_{\boldsymbol{k},\boldsymbol{l}}$ on the right.

(c) Use Hölder's inequality from part (b) with $p' := r/(r-1)$, $q' := q/(q-1)$, $r' := r/(r-1)$ and

$$A_{\boldsymbol{k},\boldsymbol{l}} := \big((F_{\boldsymbol{k}}^*)^p (G_{\boldsymbol{l}})^q\big)^{\frac{r-1}{r}},$$
$$B_{\boldsymbol{k},\boldsymbol{l}} := \big((F_{\boldsymbol{k}}^*)^p (H_{\boldsymbol{k}-\boldsymbol{l}})^r\big)^{\frac{q-1}{q}},$$
$$C_{\boldsymbol{k},\boldsymbol{l}} := \big((G_{\boldsymbol{l}})^q (H_{\boldsymbol{k}-\boldsymbol{l}})^r\big)^{\frac{p-1}{p}},$$

to establish Young's convolution inequality; note the double sums over $\boldsymbol{k}, \boldsymbol{l}$ for terms such as $(F_{\boldsymbol{k}}^*)^p (H_{\boldsymbol{k}-\boldsymbol{l}})^r$ separate, as the sum over $\boldsymbol{l} \in \mathbb{Z}^d$ of $(H_{\boldsymbol{k}-\boldsymbol{l}})^r$ is independent of $\boldsymbol{k} \in \mathbb{Z}^d$. See e.g. Yung (2017).

7.5 Weak solutions: Regularity in time. The standard result for the regularity of $\mathrm{d}\boldsymbol{u}/\mathrm{d}t$ in Step 2 of our proof of existence of global weak solutions in Section 7.2 is stronger than the regularity we establish. Indeed, the standard result establishes $\mathrm{d}\boldsymbol{u}/\mathrm{d}t \in L^{4/3}(0, T; H^{-1}(\mathbb{T}^3;\mathbb{R}^3))$; see Robinson (2011). In this question we reproduce this standard result, though in the physical space $\mathbb{T}^3$ as opposed to the corresponding Fourier space $\mathbb{Z}^3$. For any function $\boldsymbol{v} = \boldsymbol{v}(\boldsymbol{x})$ with $\boldsymbol{x} \in \mathbb{T}^3$, let $\hat{\boldsymbol{v}} = \hat{\boldsymbol{v}}_K(\boldsymbol{x})$ denote the approximation to $\boldsymbol{v}$ on $\mathbb{T}^3$ obtained by truncating the Fourier expansion of $\boldsymbol{v}$ to the first set of modes identified by $|\boldsymbol{k}| \leqslant K$. Let P_K denote the projection operator $P_K : \boldsymbol{v} \mapsto \hat{\boldsymbol{v}}$.

Now suppose $\hat{\boldsymbol{u}} = \hat{\boldsymbol{u}}_K(\boldsymbol{x}, t)$ is the solution to the finite system of equations obtained by applying the projector P_K to the incompressible Navier–Stokes equations. The resulting system is 'essentially' equivalent to the finite system of ordinary differential equations for the Fourier coefficients $\hat{U}_k$ established just before Step 1 in Section 7.2. Our initial goal is to seek a bound for the nonlinear term $P_K((\hat{\boldsymbol{u}} \cdot \nabla) \hat{\boldsymbol{u}})$, present in the projected system, in $H^{-1}(\mathbb{T}^3; \mathbb{R}^3)$.

(a) Explain why, for any $\boldsymbol{v} \in H^1(\mathbb{T}^3; \mathbb{R}^3)$, we can write

$$|\langle \boldsymbol{v}, P_K((\hat{\boldsymbol{u}} \cdot \nabla) \hat{\boldsymbol{u}})\rangle| = \left| \int_{\mathbb{T}^3} (P_K \boldsymbol{v}) \cdot ((\hat{\boldsymbol{u}} \cdot \nabla) \hat{\boldsymbol{u}}) \, \mathrm{d}\boldsymbol{x} \right|.$$

(b) Apply Hölder inequality to the result in part (a) to show that

$$|\langle \boldsymbol{v}, P_K((\hat{\boldsymbol{u}} \cdot \nabla) \hat{\boldsymbol{u}})\rangle| \leqslant \|\hat{\boldsymbol{v}}\|_{L^6(\mathbb{T}^3;\mathbb{R}^3)} \|\nabla \hat{\boldsymbol{u}}\|_{L^2(\mathbb{T}^3;\mathbb{R}^{3\times 3})} \|\hat{\boldsymbol{u}}\|_{L^3(\mathbb{T}^3;\mathbb{R}^3)}.$$

(c) Apply the Sobolev–Gagliardo–Nirenberg inequality to the result in part (b) twice: first with $p = 6$ to bound $\|\hat{\boldsymbol{v}}\|_{L^6(\mathbb{T}^3;\mathbb{R}^3)}$ by $\|\hat{\boldsymbol{v}}\|_{H^1(\mathbb{T}^3;\mathbb{R}^3)}$ and second with $p = 3$ to bound $\|\hat{\boldsymbol{u}}\|_{L^3(\mathbb{T}^3;\mathbb{R}^3)}$. Show, for some constant $c > 0$:

$$|\langle \boldsymbol{v}, P_K((\hat{\boldsymbol{u}} \cdot \nabla) \hat{\boldsymbol{u}})\rangle| \leqslant c \, \|\boldsymbol{v}\|_{H^1(\mathbb{T}^3;\mathbb{R}^3)} \left(\|\nabla \hat{\boldsymbol{u}}\|^{3/2}_{L^2(\mathbb{T}^3;\mathbb{R}^{3\times 3})} \|\hat{\boldsymbol{u}}\|^{1/2}_{L^2(\mathbb{T}^3;\mathbb{R}^3)} \right).$$

(d) Explain why we can deduce

$$\|P_K((\hat{\boldsymbol{u}} \cdot \nabla) \hat{\boldsymbol{u}}\|_{H^{-1}(\mathbb{T}^3;\mathbb{R}^3)} \leqslant c \, \|\nabla \hat{\boldsymbol{u}}\|^{3/2}_{L^2(\mathbb{T}^3;\mathbb{R}^{3\times 3})} \|\hat{\boldsymbol{u}}\|^{1/2}_{L^2(\mathbb{T}^3;\mathbb{R}^3)}.$$

(e) For weak solutions $\hat{\boldsymbol{u}} \in L^2(0, T; H^1(\mathbb{T}^3; \mathbb{R}^3))$ and so we know that $\Delta \hat{\boldsymbol{u}} \in L^2(0, T; H^{-1}(\mathbb{T}^3; \mathbb{R}^3))$. Combine this with the result from part (d) to deduce $\mathrm{d}\hat{\boldsymbol{u}}/\mathrm{d}t \in L^{4/3}(0, T; H^{-1}(\mathbb{T}^3; \mathbb{R}^3))$.

7.6 Weak solutions: Exponential bound for the energy. The goal of this question is to provide a more nuanced a-priori estimate for the energy to that at the beginning of Section 7.2, with the additional aspect of a non-zero external force $\boldsymbol{f} \colon \mathbb{T}^d \times [0, T] \to \mathbb{R}^d$.

(a) Use the identity $(\boldsymbol{u} \cdot \nabla) \boldsymbol{u} = \nabla(\frac{1}{2}|\boldsymbol{u}|^2) - \boldsymbol{u} \times (\nabla \times \boldsymbol{u})$ to show the Navier–Stokes momentum equation can be expressed in the form

$$\frac{\partial \boldsymbol{u}}{\partial t} + \boldsymbol{\omega} \times \boldsymbol{u} = \nu \, \Delta \boldsymbol{u} - \nabla(p + \tfrac{1}{2}|\boldsymbol{u}|^2) + \boldsymbol{f}.$$

(b) By direct computation, use part (a) to show that

$$\frac{\mathrm{d}}{\mathrm{d}t} \tfrac{1}{2} \|\boldsymbol{u}\|^2_{L^2(\mathbb{T}^d;\mathbb{R}^d)} = \int_{\mathbb{T}^d} \boldsymbol{u} \cdot \left(-\boldsymbol{\omega} \times \boldsymbol{u} + \nu \, \Delta \boldsymbol{u} - \nabla(p + \tfrac{1}{2}|\boldsymbol{u}|^2) + \boldsymbol{f} \right) \mathrm{d}\boldsymbol{x}.$$

(c) Using that $\boldsymbol{u} \cdot (\boldsymbol{\omega} \times \boldsymbol{u}) \equiv 0$, show using part (b) that

$$\frac{\mathrm{d}}{\mathrm{d}t} \tfrac{1}{2} \|\boldsymbol{u}\|^2_{L^2(\mathbb{T}^d;\mathbb{R}^d)} = -\nu \int_{\mathbb{T}^d} |\nabla \boldsymbol{u}|^2 \, \mathrm{d}\boldsymbol{x} + \int_{\mathbb{T}^d} \boldsymbol{u} \cdot \boldsymbol{f} \, \mathrm{d}\boldsymbol{x}.$$

(*Hint:* You may find some of the computations used for deriving the energy a-priori estimate at the beginning of Section 7.2 useful here.)

(d) Using the Hölder inequality and then Young's inequality for the last term on the right in part (c), show for some $\delta > 0$:

$$\frac{\mathrm{d}}{\mathrm{d}t} \tfrac{1}{2} \|\boldsymbol{u}\|^2_{L^2(\mathbb{T}^d;\mathbb{R}^d)} + \nu \, \|\nabla \boldsymbol{u}\|^2_{L^2(\mathbb{T}^d;\mathbb{R}^{d\times d})} \leqslant \frac{\delta}{2} \|\boldsymbol{u}\|^2_{L^2(\mathbb{T}^d;\mathbb{R}^d)} + \frac{1}{2\delta} \|\boldsymbol{f}\|^2_{L^2(\mathbb{T}^d;\mathbb{R}^d)}.$$

(e) Set $\delta = \nu$ and use Poincaré's inequality from Proposition 7.16(d), together with part (d), to show for some finite constant $c > 0$ that we have:

$$\frac{\mathrm{d}}{\mathrm{d}t}\|\boldsymbol{u}\|^2_{L^2(\mathbb{T}^d;\mathbb{R}^d)} + \frac{\nu}{c^2}\|\boldsymbol{u}\|^2_{L^2(\mathbb{T}^d;\mathbb{R}^d)} \leqslant \frac{1}{\nu}\|\boldsymbol{f}\|^2_{L^2(\mathbb{T}^d;\mathbb{R}^d)}.$$

(f) Using Gronwall's Lemma, see Exercise 7.10, show

$$\|\boldsymbol{u}(t)\|^2_{L^2(\mathbb{T}^d;\mathbb{R}^d)} \leqslant (\mathrm{e}^{-\nu t/c^2})\,\|\boldsymbol{u}_0\|^2_{L^2(\mathbb{T}^d;\mathbb{R}^d)} + \frac{c^2}{\nu^2}(1-\mathrm{e}^{-\nu t/c^2})\sup_{\tau\in[0,t]}\|\boldsymbol{f}(\tau)\|^2_{L^2(\mathbb{T}^d;\mathbb{R}^d)}.$$

This establishes a-priori for any $T > 0$ that we have $\boldsymbol{u} \in L^\infty(0,T;L^2(\mathbb{T}^d;\mathbb{R}^d))$, provided $\boldsymbol{f} \in L^\infty(0,T;L^2(\mathbb{T}^d;\mathbb{R}^d))$.

(g) Show that if we integrate the inequality in part (d) in time, we obtain

$$\|\boldsymbol{u}(t)\|^2_{L^2(\mathbb{T}^d;\mathbb{R}^d)} + \nu\int_0^T \|\nabla\boldsymbol{u}(\tau)\|^2_{L^2(\mathbb{T}^d;\mathbb{R}^{t\times d})}\,\mathrm{d}\tau$$
$$\leqslant \|\boldsymbol{u}_0\|^2_{L^2(\mathbb{T}^d;\mathbb{R}^d)} + \nu\int_0^T \|\boldsymbol{u}(\tau)\|^2_{L^2(\mathbb{T}^d;\mathbb{R}^d)}\,\mathrm{d}\tau + \frac{1}{\nu}\int_0^T \|\boldsymbol{f}(\tau)\|^2_{L^2(\mathbb{T}^d;\mathbb{R}^d)}\,\mathrm{d}\tau.$$

We can deduce for any $T > 0$, a-priori we have $\boldsymbol{u} \in L^2(0,T;H^1(\mathbb{T}^d;\mathbb{R}^d))$ provided $\boldsymbol{f} \in L^2(0,T;L^2(\mathbb{T}^d;\mathbb{R}^d))$. If we assume $\boldsymbol{f} \in L^\infty(0,T;L^2(\mathbb{T}^d;\mathbb{R}^d))$ as we do in part (f), then $\boldsymbol{f} \in L^2(0,T;L^2(\mathbb{T}^d;\mathbb{R}^d))$. Further note the middle term on the right above is bounded using the result from part (f).

7.7 Strong solutions: Enstrophy form. Recall from the beginning of Section 7.3 the claim that the enstrophy $\|\nabla\boldsymbol{u}\|_{L^2(\mathbb{T}^3;\mathbb{R}^{3\times3})}$ is equal to $\|\omega\|_{L^2(\mathbb{T}^3;\mathbb{R}^3)}$. Herein we establish this result. Suppose $\boldsymbol{u} = \boldsymbol{u}(\boldsymbol{x})$ and $\omega = \omega(\boldsymbol{x})$ have the respective Fourier representations:

$$\boldsymbol{u}(\boldsymbol{x}) = \sum_{\boldsymbol{k}\in\mathbb{Z}^3} U_{\boldsymbol{k}}\,\mathrm{e}^{\mathrm{i}\boldsymbol{k}\cdot\boldsymbol{x}} \quad\text{and}\quad \omega(\boldsymbol{x}) = \sum_{\boldsymbol{k}\in\mathbb{Z}^3} \Omega_{\boldsymbol{k}}\,\mathrm{e}^{\mathrm{i}\boldsymbol{k}\cdot\boldsymbol{x}},$$

with $U_{\boldsymbol{k}} \in \mathbb{C}^3$ and $\Omega_{\boldsymbol{k}} \in \mathbb{C}^3$ the respective Fourier coefficients. Note that we assume $\boldsymbol{u}$ is divergence-free, or equivalently, for every $\boldsymbol{k} \in \mathbb{Z}^3$, we know $\boldsymbol{k}^{\mathrm{T}}U_{\boldsymbol{k}} = 0$. Further, by definition $\omega = \nabla\times\boldsymbol{u}$.

(a) Explain why for each $\boldsymbol{k} \in \mathbb{Z}^3$ we know $\Omega_{\boldsymbol{k}} = \mathrm{i}\boldsymbol{k}\times U_{\boldsymbol{k}}$.

(b) If $\boldsymbol{k} = (k_1,k_2,k_3)^{\mathrm{T}}$, let $K_{\boldsymbol{k}}$ denote the 3×3 matrix

$$K_{\boldsymbol{k}} := \begin{pmatrix} 0 & -k_3 & k_2 \\ k_3 & 0 & -k_1 \\ -k_2 & k_1 & 0 \end{pmatrix}.$$

Explain why $\Omega_{\boldsymbol{k}} = \mathrm{i}K_{\boldsymbol{k}}U_{\boldsymbol{k}}$. (*Hint:* See Lemma 1.28 in Section 1.9.)

(c) Note the squared-norm $\|\omega\|^2_{L^2(\mathbb{T}^3;\mathbb{R}^3)}$ in terms of the Fourier coefficients of ω, by Parseval's identity, can be expressed in the form

$$\|\omega\|^2_{L^2(\mathbb{T}^3;\mathbb{R}^3)} = (2\pi)^3\sum_{\boldsymbol{k}\in\mathbb{Z}^3}|\Omega_{\boldsymbol{k}}|^2,$$

where $|\Omega_{\boldsymbol{k}}|^2 = \Omega_{\boldsymbol{k}}^\dagger\Omega_{\boldsymbol{k}}$. By direct computation, show that for each $\boldsymbol{k} \in \mathbb{Z}^3$:

$$|\Omega_{\boldsymbol{k}}|^2 = U_{\boldsymbol{k}}^\dagger(K_{\boldsymbol{k}}^{\mathrm{T}}K_{\boldsymbol{k}})\,U_{\boldsymbol{k}} = |\boldsymbol{k}|^2|U_{\boldsymbol{k}}|^2 - |\boldsymbol{k}^{\mathrm{T}}U_{\boldsymbol{k}}|^2.$$

(d) For each $\boldsymbol{k} \in \mathbb{Z}^3$ we have $\boldsymbol{k}^{\mathrm{T}} U_{\boldsymbol{k}} = 0$, so the last term in the expression for $|\Omega_{\boldsymbol{k}}|^2$ in part (c) is zero. Deduce $\|\nabla \boldsymbol{u}\|_{L^2(\mathbb{T}^3;\mathbb{R}^{3\times 3})} = \|\omega\|_{L^2(\mathbb{T}^3;\mathbb{R}^3)}$.

(e) Show that if $\|\nabla \omega\|_{L^2(\mathbb{T}^3;\mathbb{R}^{3\times 3})}$ is bounded then $\boldsymbol{u} \in H^2(\mathbb{T}^3;\mathbb{R}^3)$, by noting

$$\|\nabla \omega\|^2_{L^2(\mathbb{T}^3;\mathbb{R}^{3\times 3})} = (2\pi)^3 \sum_{\boldsymbol{k}\in\mathbb{Z}^3} |\boldsymbol{k}|^2 |\Omega_{\boldsymbol{k}}|^2,$$

and using the expression for $|\Omega_{\boldsymbol{k}}|^2$ in part (c).

7.8 Strong solutions: Serrin's result. There is a classical conditional global regularity result for the incompressible Navier–Stokes equations due to Serrin (1962). We reproduce Serrin's result here using a-priori estimates. Assume the domain of the flow is $\mathcal{D} = \mathbb{T}^3$. Recall the formulation of the Navier–Stokes momentum equation from Exercise 7.6(a):

$$\frac{\partial \boldsymbol{u}}{\partial t} + \omega \times \boldsymbol{u} = \nu \Delta \boldsymbol{u} - \nabla(p + \tfrac{1}{2}|\boldsymbol{u}|^2) + \boldsymbol{f}.$$

(a) By pre-multiplying the equation above by '$-\Delta \boldsymbol{u}$' and integrating over the domain $\mathbb{T}^3$, show that the solution to the incompressible Navier–Stokes equations satisfies

$$\frac{\mathrm{d}}{\mathrm{d}t}\tfrac{1}{2}\|\nabla \boldsymbol{u}\|^2_{L^2(\mathbb{T}^3;\mathbb{R}^{3\times 3})} + \nu\,\|\Delta \boldsymbol{u}\|^2_{L^2(\mathbb{T}^3;\mathbb{R}^3)} = \int_{\mathbb{T}^3} (\Delta \boldsymbol{u})\cdot(\omega\times\boldsymbol{u}) - (\Delta \boldsymbol{u})\cdot \boldsymbol{f}\,\mathrm{d}\boldsymbol{x}.$$

(*Hint:* A product rule, the Divergence Theorem and the periodic boundary conditions are required to generate the time-derivative term shown.)

(b) Use the Hölder inequality with $p = q = 2$ followed by Young's inequality from Proposition 7.16 to show that the term involving the external forcing from part (a) can be bounded as follows:

$$\int_{\mathbb{T}^3} |(\Delta \boldsymbol{u})\cdot \boldsymbol{f}|\,\mathrm{d}\boldsymbol{x} \leqslant \frac{\nu}{4}\|\Delta \boldsymbol{u}\|^2_{L^2(\mathbb{T}^3;\mathbb{R}^3)} + \frac{1}{\nu}\|\boldsymbol{f}\|^2_{L^2(\mathbb{T}^3;\mathbb{R}^3)}.$$

(c) Now we focus on the first term on the right in part (a).

(i) Use Hölder's inequality to show, with $1/p + 1/r = 1/2$, that we have

$$\int_{\mathbb{T}^3} |(\Delta \boldsymbol{u})\cdot(\omega\times\boldsymbol{u})|\,\mathrm{d}\boldsymbol{x} \leqslant \|\Delta \boldsymbol{u}\|_{L^2(\mathbb{T}^3;\mathbb{R}^3)} \|\nabla \boldsymbol{u}\|_{L^p(\mathbb{T}^3;\mathbb{R}^{3\times 3})} \|\boldsymbol{u}\|_{L^r(\mathbb{T}^3;\mathbb{R}^3)}.$$

(ii) Use the Sobolev–Gagliardo–Nirenberg inequality from Proposition 7.16(e) to show that for some constant $c > 0$ we have

$$\|\nabla \boldsymbol{u}\|_{L^p(\mathbb{T}^3;\mathbb{R}^{3\times 3})} \leqslant c\,\|\Delta \boldsymbol{u}\|^a_{L^2(\mathbb{T}^3;\mathbb{R}^3)} \|\nabla \boldsymbol{u}\|^{1-a}_{L^2(\mathbb{T}^3;\mathbb{R}^{3\times 3})},$$

where $a = 3(p-2)/(2p) = 3/r$. Hence deduce

$$\int_{\mathbb{T}^3} |(\Delta \boldsymbol{u})\cdot(\omega\times\boldsymbol{u})|\,\mathrm{d}\boldsymbol{x} \leqslant c\,\|\Delta \boldsymbol{u}\|^{1+3/r}_{L^2(\mathbb{T}^3;\mathbb{R}^3)} \|\nabla \boldsymbol{u}\|^{1-3/r}_{L^2(\mathbb{T}^3;\mathbb{R}^{3\times 3})} \|\boldsymbol{u}\|_{L^r(\mathbb{T}^3;\mathbb{R}^3)}.$$

(iii) Use Young's inequality with $p' := 2r/(r+3)$ and $q' := 2r/(r-3)$ to show, for some constant $C > 0$, that we have

$$\int_{\mathbb{T}^3} |(\Delta \boldsymbol{u})\cdot(\omega\times\boldsymbol{u})|\,\mathrm{d}\boldsymbol{x} \leqslant \frac{\nu}{2}\|\Delta \boldsymbol{u}\|^2_{L^2(\mathbb{T}^3;\mathbb{R}^3)} + \frac{C}{\nu^{\frac{r+3}{r-3}}}\|\nabla \boldsymbol{u}\|^2_{L^2(\mathbb{T}^3;\mathbb{R}^{3\times 3})} \|\boldsymbol{u}\|^{2r/(r-3)}_{L^r(\mathbb{T}^3;\mathbb{R}^3)}.$$

(*Hint:* It is useful to know in (i) above that the norms $\|\Delta \boldsymbol{u}\|_{L^2(\mathbb{T}^3;\mathbb{R}^3)}$ and $\|\nabla\nabla^{\mathrm{T}} \boldsymbol{u}\|_{L^2(\mathbb{T}^3;\mathbb{R}^{3\times 3\times 3})}$ are equivalent norms, and the same is true for the pair

$\|\nabla \boldsymbol{u}\|_{L^p(\mathbb{T}^3;\mathbb{R}^{3\times3})}$ and $\|\omega\|_{L^p(\mathbb{T}^3;\mathbb{R}^3)}$. That they are 'equivalent norms' means one can be bounded by a finite constant times the other, and vice versa.)

(d) Combine parts (b) and (c)(iii) to show

$$\frac{\mathrm{d}}{\mathrm{d}t}\|\nabla \boldsymbol{u}\|^2_{L^2(\mathbb{T}^3;\mathbb{R}^{3\times3})} \leqslant \frac{2C}{\nu^{\frac{r+3}{r-3}}}\|\nabla \boldsymbol{u}\|^2_{L^2(\mathbb{T}^3;\mathbb{R}^{3\times3})}\|\boldsymbol{u}\|^{2r/(r-3)}_{L^r(\mathbb{T}^3;\mathbb{R}^3)} + \frac{2}{\nu}\|\boldsymbol{f}\|^2_{L^2(\mathbb{T}^3;\mathbb{R}^3)}.$$

By integrating in time using the Gronwall Lemma (see Exercise 7.10), explain why if we *assume* $\boldsymbol{u} \in L^s(0,T;L^r(\mathbb{T}^3;\mathbb{R}^3))$ where

$$s \geqslant \frac{2r}{r-3} \quad \Leftrightarrow \quad \frac{3}{r} + \frac{2}{s} \leqslant 1,$$

then we know $\boldsymbol{u} \in L^\infty(0,T;H^1(\mathbb{T}^3;\mathbb{R}^3))$ for any $T > 0$ and the solution is a global-in-time strong solution.

7.9 Strong solutions: Global regularity for small energy initial data. There is a classical small initial data, or equivalently large viscosity, result due to Ladyzhenskaya (1969). This result establishes the regularity of the three-dimensional Navier–Stokes equations globally in time for any given viscosity if the initial enstrophy is sufficiently small. The upper bound criterion for the initial enstrophy is proportional to the viscosity coefficient. We reproduce this result here using a-priori estimates. Recall the inequality we proved for the enstrophy in Lemma 7.32(ii):

$$\frac{\mathrm{d}}{\mathrm{d}t}\frac{1}{2}\|\omega\|^2_{L^2(\mathbb{T}^3;\mathbb{R}^3)} + \frac{\nu}{4}\|\nabla\omega\|^2_{L^2(\mathbb{T}^3;\mathbb{R}^{3\times3})} \leqslant \frac{c}{\nu^3}\left(\|\omega\|^2_{L^2(\mathbb{T}^3;\mathbb{R}^3)}\right)^3,$$

for some constant $c > 0$.

(a) Use the Poincaré inequality to show, for some constant $C > 0$:

$$\frac{\mathrm{d}}{\mathrm{d}t}\frac{1}{2}\|\omega\|^2_{L^2(\mathbb{T}^3;\mathbb{R}^3)} + \frac{\nu}{4C}\|\omega\|^2_{L^2(\mathbb{T}^3;\mathbb{R}^{3\times3})} \leqslant \frac{c}{\nu^3}\left(\|\omega\|^2_{L^2(\mathbb{T}^3;\mathbb{R}^3)}\right)^3.$$

(b) Show that if

$$\frac{c}{\nu^3}\left(\|\omega\|^2_{L^2(\mathbb{T}^3;\mathbb{R}^3)}\right)^3 \leqslant \frac{\nu}{4C}\|\omega\|^2_{L^2(\mathbb{T}^3;\mathbb{R}^{3\times3})},$$

then $\|\omega\|^2_{L^2(\mathbb{T}^3;\mathbb{R}^3)}$ is non-increasing in time. Show that this last condition is equivalent to the criterion

$$\|\omega\|^2_{L^2(\mathbb{T}^3;\mathbb{R}^3)} \leqslant \frac{\nu^2}{2\sqrt{Cc}}.$$

(c) Explain why, if we assume the condition in part (b) holds at $t = 0$, then the enstrophy is uniformly bounded for all $t > 0$ and there exists a global strong solution to the three-dimensional Navier–Stokes equations.

7.10 Strong solutions: Gronwall's Lemma. Gronwall's Lemma is a very useful result providing an upper bound estimate for functions satisfying a linear differential inequality; see Evans (1998, p. 624). The statement of Gronwall's Lemma is as follows. Suppose $f = f(t)$ is a non-negative, absolutely continuous function on an interval $[0,T]$ which satisfies the linear differential inequality

$$\frac{\mathrm{d}}{\mathrm{d}t}f(t) \leqslant a(t)f(t) + b(t),$$

where $a = a(t)$ and $b = b(t)$ are integrable functions on $[0, T]$. Then for all $t \in [0, T]$ the function $f = f(t)$ has the bound

$$f(t) \leqslant e^{\int_0^t a(s)\,ds}\left(f(0) + \int_0^t b(s)\,ds\right).$$

(a) Using the linear differential inequality, show that for $\tau \in [0, T]$:

$$\frac{d}{d\tau}\left(f(\tau)\,e^{-\int_0^\tau a(s)\,ds}\right) \leqslant e^{-\int_0^\tau a(s)\,ds}\,b(\tau).$$

(b) By integrating the inequality in part (a), show that for all $t \in [0, T]$ we have

$$f(t)\,e^{-\int_0^t a(s)\,ds} \leqslant f(0) + \int_0^t b(s)\,ds.$$

7.11 Vorticity alignment/anti-alignment: Absolute integrability. In Step 6 of the proof of the Constantin and Fefferman result Theorem 7.36, we use that the L^1-norm of the vorticity in $\mathbb{R}^3$ is uniformly bounded in time. The goal of this question is to establish this result via a-priori estimates. More details can be found in Constantin and Fefferman (1993, p. 781). We begin with the vorticity formulation of the incompressible Navier–Stokes equations in $\mathbb{R}^3$ as given in Step 1 of the proof of Theorem 7.36:

$$\frac{\partial \omega}{\partial t} + (\nabla \omega)\,\boldsymbol{u} = \nu\,\Delta\omega + (\nabla \boldsymbol{u})\,\omega,$$
$$\Delta \boldsymbol{u} = -\nabla \times \omega.$$

We set $\hat{\omega} := \omega/|\omega|$, which is defined where $\omega \neq 0$. Taking the scalar product of the vorticity equation above with $\hat{\omega}$ generates the relation

$$\hat{\omega}^{\mathrm{T}}\frac{\partial \omega}{\partial t} + \hat{\omega}^{\mathrm{T}}(\nabla\omega)\,\boldsymbol{u} = \nu\,\hat{\omega}^{\mathrm{T}}(\Delta\omega) + \hat{\omega}^{\mathrm{T}}(\nabla\boldsymbol{u})\,\omega.$$

(a) By direct computation, show

$$\frac{\partial}{\partial t}|\omega| = \hat{\omega}^{\mathrm{T}}\frac{\partial\omega}{\partial t} \qquad \text{and} \qquad \nabla(|\omega|) = \hat{\omega}^{\mathrm{T}}(\nabla\omega).$$

(b) Again by direct computation, show

$$\Delta(|\omega|) = \hat{\omega}^{\mathrm{T}}\Delta\omega - \frac{|\nabla(|\omega|)|^2}{|\omega|} + \frac{|\nabla\omega|^2}{|\omega|},$$

where $|\nabla\omega|^2 = \sum_{j=1}^{3}(\nabla\omega_i)\cdot(\nabla\omega_i)$. Further, by direct computation show

$$|\nabla\hat{\omega}|^2 = \frac{|\nabla\omega|^2}{|\omega|^2} - \frac{|\nabla(|\omega|)|^2}{|\omega|^2}.$$

Combine these last two results to deduce

$$\hat{\omega}^{\mathrm{T}}(\Delta\omega) = \Delta(|\omega|) - |\omega|\,|\nabla\hat{\omega}|^2.$$

(c) Combine the results in parts (a) and (b) above to show, with $\boldsymbol{u}$ given by $\Delta \boldsymbol{u} = -\nabla \times \boldsymbol{\omega}$, that the quantity $|\boldsymbol{\omega}|$ satisfies the equation

$$\frac{\partial}{\partial t}|\boldsymbol{\omega}| + \nabla(|\boldsymbol{\omega}|)\,\boldsymbol{u} = \nu\Delta(|\boldsymbol{\omega}|) - \nu|\boldsymbol{\omega}|\,|\nabla\hat{\boldsymbol{\omega}}|^2 + \hat{\boldsymbol{\omega}}^{\mathrm{T}}(\nabla\boldsymbol{u})\,\boldsymbol{\omega}.$$

(d) Let f be a real-valued, twice-continuously differentiable function on $\mathbb{R}$ with $f(0) = f'(0) = 0$. Explain why the quantities

$$\int_{\mathbb{R}^3} \nabla\cdot(f(|\boldsymbol{\omega}|)\,\boldsymbol{u})\,\mathrm{d}\boldsymbol{x} \quad \text{and} \quad \int_{\mathbb{R}^3} \nabla\cdot(f'(|\boldsymbol{\omega}|)\,\nabla(|\boldsymbol{\omega}|))\,\mathrm{d}\boldsymbol{x}$$

are both zero. Use these facts to deduce that

$$\frac{\mathrm{d}}{\mathrm{d}t}\int_{\mathbb{R}^3} f(|\boldsymbol{\omega}|)\,\mathrm{d}\boldsymbol{x} + \nu\int_{\mathbb{R}^3} f''(|\boldsymbol{\omega}|)\,|\nabla(|\boldsymbol{\omega}|)|^2\,\mathrm{d}\boldsymbol{x} + \nu\int_{\mathbb{R}^3} f'(|\boldsymbol{\omega}|)\,|\boldsymbol{\omega}|\,|\nabla\hat{\boldsymbol{\omega}}|^2\,\mathrm{d}\boldsymbol{x}$$
$$= \int_{\mathbb{R}^3} f'(|\boldsymbol{\omega}|)\,\hat{\boldsymbol{\omega}}^{\mathrm{T}}(\nabla\boldsymbol{u})\,\boldsymbol{\omega}\,\mathrm{d}\boldsymbol{x}.$$

(e) Suppose the function f from part (d) has the form

$$f(x) := \int_0^x (x-y)\,\psi(y)\,\mathrm{d}y,$$

where ψ is a non-negative function that vanishes near the origin and also for all values $y > K_0$ for some parameter $K_0 \geqslant 0$. Suppose further that

$$\int_0^{K_0} \psi(y)\,\mathrm{d}y = 1.$$

Show, in addition to the properties $f(0) = f'(0) = 0$, that the function f here satisfies $0 \leqslant f'(x) \leqslant 1$, $f''(x) \geqslant 0$, $f''(0) = 0$ and $xf'(x) = x$ for $x \geqslant K_0$.

(f) Using the properties of the function f given in part (e), integrate the final relation in part (d) in time to show

$$\int_{\mathbb{R}^3} f(|\boldsymbol{\omega}(\boldsymbol{x},t)|)\,\mathrm{d}\boldsymbol{x} + \nu\int_0^t \int_{|\boldsymbol{\omega}(\boldsymbol{x},\tau)|>K_0} |\boldsymbol{\omega}(\boldsymbol{x},\tau)|\,|\nabla\hat{\boldsymbol{\omega}}(\boldsymbol{x},\tau)|^2\,\mathrm{d}\boldsymbol{x}\,\mathrm{d}\tau$$
$$\leqslant \int_{\mathbb{R}^3} |\boldsymbol{\omega}_0(\boldsymbol{x})|\,\mathrm{d}\boldsymbol{x} + \int_0^t \int_{\mathbb{R}^3} |\boldsymbol{\omega}(\boldsymbol{x},\tau)|^2\,\mathrm{d}\boldsymbol{x}\,\mathrm{d}\tau.$$

(*Hint:* You need to use the Hölder inequality and the relation between the L^2-norms of $\nabla\boldsymbol{u}$ and $\boldsymbol{\omega}$ from Exercise 7.7(d) which apply on $\mathbb{R}^3$.)

(g) Explain why, for weak solutions, the second term on the right in part (f) is uniformly bounded in time. Then, taking the limit $K_0 \to 0$, show $f(x) \to x$ and deduce $\|\boldsymbol{\omega}(t)\|_{L^1(\mathbb{R}^3;\mathbb{R}^3)}$ is uniformly bounded in time.

(*Hint:* You need to assume or prove here the properties of weak solutions we established in Section 7.2 in the periodic domain case carry over to $\mathbb{R}^3$. See Constantin and Fefferman (1993) for more details.)

7.12 Vorticity alignment/anti-alignment: Weaker assumptions. The assumption in Theorem 7.36 for Constantin and Fefferman's result can be weakened. The goal of this question is to explore this possibility. Indeed, once complete, the exercise reveals the assumption

$$|\sin\theta| \leqslant \frac{|y|^{\gamma}}{\rho},$$

for any $\gamma \in [1/2, 1]$, is sufficient to establish global regularity for the three-dimensional incompressible Navier–Stokes equations. Note that Theorem 7.36 deals with the case $\gamma = 1$. The proof carried through here uses the same first three steps as those outlined for the proof of Theorem 7.36. From Step 4 of that proof we also have the identity for the evolution of the enstrophy $\|\omega\|_{L^2(\mathbb{R}^3;\mathbb{R}^3)} = \|\nabla u\|_{L^2(\mathbb{R}^3;\mathbb{R}^{3\times 3})}$:

$$\frac{\mathrm{d}}{\mathrm{d}t}\frac{1}{2}\|\omega\|^2_{L^2(\mathbb{R}^3;\mathbb{R}^3)} + \nu\|\nabla\omega\|^2_{L^2(\mathbb{R}^3;\mathbb{R}^{3\times 3})} = \int_{\mathbb{R}^3} \omega^{\mathrm{T}}((\nabla u)\,\omega)\,\mathrm{d}x,$$

as well as the following estimate for the nonlinear term on the right-hand side of this enstrophy identity, that was established in Steps 1–3:

$$\int |\omega^{\mathrm{T}}(x)((\nabla u)\,\omega)(x)|\,\mathrm{d}x \leqslant \frac{3}{4\pi}\int |\omega(x)|^2 \int |\sin\theta|\frac{|\omega(x-y)|}{|y|^3}\,\mathrm{d}y\,\mathrm{d}x.$$

(a) By inserting the assumption on $|\sin\theta|$ stated at the very beginning, for a general value of $\gamma \in (0, 1]$, into the estimate of the nonlinear term just above, and using Hölder's inequality for general $p, q \in [0, \infty)$ satisfying $1/p + 1/q = 1$, show

$$\int |\omega^{\mathrm{T}}(x)((\nabla u)\,\omega)(x)|\,\mathrm{d}x \leqslant \frac{C}{\rho}\|\omega\|^2_{L^{2p}(\mathbb{R}^3;\mathbb{R}^3)}\|I_{3-\gamma}\,\omega\|_{L^q(\mathbb{R}^3;\mathbb{R}^3)},$$

for some constant $C > 0$. The integral $I_{3-\gamma}\,\omega$ is defined in the statement of the Hardy–Littlewood–Sobolev inequality Lemma 7.37.

(b) Use the Sobolev–Gagliardo–Nirenberg inequality from Proposition 7.16(e) on $\mathbb{R}^d$ with $d = 3$ to show, for $p \in [1, 3]$, $a = 3(p-1)/2p$ and some constant $c > 0$:

$$\|\omega\|^2_{L^{2p}(\mathbb{R}^3;\mathbb{R}^3)} \leqslant c\,\|\nabla\omega\|^{2a}_{L^2(\mathbb{R}^3;\mathbb{R}^{3\times 3})}\|\omega\|^{2(1-a)}_{L^2(\mathbb{R}^3;\mathbb{R}^3)}.$$

(c) Use the Hardy–Littlewood–Sobolev inequality Lemma 7.37 to show, for some constant $c > 0$ and with $1/p^* = 1/q + \gamma/3$:

$$\|I_{3-\gamma}\,\omega\|_{L^q(\mathbb{R}^3;\mathbb{R}^3)} \leqslant c\,\|\omega\|_{L^{p^*}(\mathbb{R}^3;\mathbb{R}^3)}.$$

Recall from part (a) that $1/p + 1/q = 1$. Explain why the condition $p^* < q$ in the Hardy–Littlewood–Sobolev inequality is naturally satisfied here.

(d) Recall the second part of Step 5 in the proof of Theorem 7.36. Using a similar sequence of steps, show that for parameters α, β, $\hat{p}$ and $\hat{q}$ satisfying $\alpha + \beta = 1$ and $1/\hat{p} + 1/\hat{q} = 1$, we have

$$\|\omega\|_{L^{p^*}(\mathbb{R}^3;\mathbb{R}^3)} \leqslant \|\omega\|^{\alpha}_{L^{\alpha\hat{p}p^*}(\mathbb{R}^3;\mathbb{R}^3)}\|\omega\|^{\beta}_{L^{\beta\hat{q}p^*}(\mathbb{R}^3;\mathbb{R}^3)}.$$

Making the choices $\alpha\hat{p}p^* = 1$ and $\beta\hat{q}p^* = 2$, show that $\alpha = (2-p^*)/p^*$ and determine β.

(e) Combine parts (a)–(d) to show that for $p \in [1, 3]$ and some constant $c > 0$:

$$\int |\omega^{\mathrm{T}}(x)((\nabla u)\,\omega)(x)|\,\mathrm{d}x \leqslant c\,\|\nabla\omega\|^{3(p-1)/p}_{L^2(\mathbb{R}^3;\mathbb{R}^{3\times 3})}\|\omega\|^{5/p-2\gamma/3-1}_{L^2(\mathbb{R}^3;\mathbb{R}^3)}\|\omega\|^{1-2/p+2\gamma/3}_{L^1(\mathbb{R}^3;\mathbb{R}^3)}.$$

(*Hint:* Note from part (c) that $1/p^* = 1 - 1/p + \gamma/3$.)

(f) Apply Young's inequality from Proposition 7.16(a) with parameters $\tilde{p}$ and $\tilde{q}$ satisfying $1/\tilde{p} + 1/\tilde{q} = 1$ to show that the right-hand side of the result in part (e) is bounded by

$$\frac{\nu}{2}\|\nabla\omega\|^{3\tilde{p}(p-1)/p}_{L^2(\mathbb{R}^3;\mathbb{R}^{3\times 3})} + \frac{c^{\tilde{q}}}{\tilde{q}}\left(\frac{2}{\nu\tilde{p}}\right)^{\tilde{q}/\tilde{p}} \|\omega\|^{\tilde{q}(5/p-2\gamma/3-1)}_{L^2(\mathbb{R}^3;\mathbb{R}^3)} \|\omega\|^{\tilde{q}(1-2/p+2\gamma/3)}_{L^1(\mathbb{R}^3;\mathbb{R}^3)}.$$

(g) It is natural to choose $\tilde{p}$ so that $3\tilde{p}(p-1)/p = 2$ in part (f). With this choice show that $\tilde{q} = 2p/(3-p)$ and use the estimate in part (f) in the enstrophy identity above to show

$$\frac{\mathrm{d}}{\mathrm{d}t}\|\omega\|^2_{L^2(\mathbb{R}^3;\mathbb{R}^3)} \leqslant 2\frac{c^{\tilde{q}}}{\tilde{q}}\left(\frac{2}{\nu\tilde{p}}\right)^{\tilde{q}/\tilde{p}} \|\omega\|^{\tilde{q}(5/p-2\gamma/3-1)}_{L^2(\mathbb{R}^3;\mathbb{R}^3)} \|\omega\|^{\tilde{q}(1-2/p+2\gamma/3)}_{L^1(\mathbb{R}^3;\mathbb{R}^3)}.$$

We know from Exercise 7.11 that $\|\omega\|_{L^1(\mathbb{R}^3;\mathbb{R}^3)}$ is uniformly bounded in time. Explain why we can only use this bound provided $p < 3$ and $1 - 2/p + 2\gamma/3 \geqslant 0$, or equivalently $p \geqslant 6/(3 + 2\gamma)$. Also note that the coefficients are bounded provided we further assume $p > 1$.

(h) Suppose the uniform bound in time on the coefficients and the factor involving $\|\omega\|_{L^1(\mathbb{R}^3;\mathbb{R}^3)}$ is K, so

$$\frac{\mathrm{d}}{\mathrm{d}t}\|\omega\|^2_{L^2(\mathbb{R}^3;\mathbb{R}^3)} \leqslant K\,\|\omega\|^{2p(5/p-2\gamma/3-1)/(3-p)}_{L^2(\mathbb{R}^3;\mathbb{R}^3)}.$$

Set $p' := 2p(5/p - 2\gamma/3 - 1)/(3 - p) - 2$ and use Gronwall's Lemma from Exercise 7.10 to show that if $\omega_0 = \omega_0(\boldsymbol{x})$ is the initial vorticity, then

$$\|\omega(t)\|^2_{L^2(\mathbb{R}^3;\mathbb{R}^3)} \leqslant \|\omega_0\|^2_{L^2(\mathbb{R}^3;\mathbb{R}^3)} \exp\left(K\int_0^t \|\omega(\tau)\|^{p'}_{L^2(\mathbb{R}^3;\mathbb{R}^3)}\,\mathrm{d}\tau\right).$$

Since for weak solutions $\omega \in L^2(0, T; L^2(\mathbb{R}^3; \mathbb{R}^3))$ for any $T > 0$, the term in the exponent is bounded provided $p' \leqslant 2$. Choose $p' = 2$ and show that this implies

$$p = \frac{3}{3 - 2\gamma}.$$

Recall the conditions on p we have accumulated thus far in part (g). Show that $1 < p < 3$ is equivalent to the condition $0 < \gamma < 1$. Show that the condition $p \geqslant 6/(3 + 2\gamma)$ implies

$$\gamma \geqslant \frac{1}{2}.$$

(*Note:* In the proof of Theorem 7.36 we make the choice $p' = 4/3$, which is equivalent to $\gamma = 1$.)

(i) As a reprise, with $p' = 2$, the special choice $\gamma = 1/2$ which is equivalent to $p = 3/2$ implies the power of the factor $\|\omega\|_{L^1(\mathbb{R}^3;\mathbb{R}^3)}$ in parts (f) and (g) is zero. In other words, in this case the L^1-estimate for ω does not play a role in the corresponding computations and is not required. Indeed, assume from the outset

$\gamma = 1/2$. First show the Sobolev–Gagliardo–Nirenberg inequality in part (b), in this case, has the form

$$\|\omega\|^2_{L^3(\mathbb{R}^3;\mathbb{R}^3)} \leqslant c\, \|\nabla\omega\|_{L^2(\mathbb{R}^3;\mathbb{R}^{3\times 3})} \|\omega\|_{L^2(\mathbb{R}^3;\mathbb{R}^3)}.$$

Second show the Hardy–Littlewood–Sobolev inequality in part (c) has the form

$$\|I_{5/2}\, \omega\|_{L^3(\mathbb{R}^3;\mathbb{R}^3)} \leqslant c\, \|\omega\|_{L^2(\mathbb{R}^3;\mathbb{R}^3)},$$

for some constant $c > 0$. Third combine these last two results together with part (a) to show, for some constant $C > 0$:

$$\int |\omega^{\mathrm{T}}(\boldsymbol{x})((\nabla \boldsymbol{u})\, \omega)(\boldsymbol{x})|\, \mathrm{d}\boldsymbol{x} \leqslant C\, \|\nabla\omega\|_{L^2(\mathbb{R}^3;\mathbb{R}^{3\times 3})} \|\omega\|^2_{L^2(\mathbb{R}^3;\mathbb{R}^3)}.$$

Fourth use Young's inequality from Proposition 7.16(a) with $p = q = 2$ to show

$$\|\omega(t)\|^2_{L^2(\mathbb{R}^3;\mathbb{R}^3)} \leqslant \|\omega_0\|^2_{L^2(\mathbb{R}^3;\mathbb{R}^3)} \exp\left(\frac{C^2}{\nu} \int_0^t \|\omega(\tau)\|^{p'}_{L^2(\mathbb{R}^3;\mathbb{R}^3)}\, \mathrm{d}\tau\right).$$

Hence deduce the corresponding result for this special case.

Appendix: Formulae

A.1 Multivariable Calculus Identities

We provide here some useful multivariable calculus identities. Here ϕ and ψ are generic scalars, and $\boldsymbol{u}$ and $\boldsymbol{v}$ are generic vectors.

1. $\nabla \times \boldsymbol{u} = \det \begin{pmatrix} \boldsymbol{i} & \boldsymbol{j} & \boldsymbol{k} \\ \partial/\partial x & \partial/\partial y & \partial/\partial z \\ u & v & w \end{pmatrix} = \begin{pmatrix} \partial w/\partial y - \partial v/\partial z \\ \partial u/\partial z - \partial w/\partial x \\ \partial v/\partial x - \partial u/\partial y \end{pmatrix}.$
2. $\nabla \cdot (\nabla \phi) = \nabla^2 \phi = \Delta \phi = \dfrac{\partial^2 \phi}{\partial x^2} + \dfrac{\partial^2 \phi}{\partial y^2} + \dfrac{\partial^2 \phi}{\partial z^2}.$
3. $\nabla \times (\nabla \phi) \equiv \boldsymbol{0}.$
4. $\nabla \cdot (\nabla \times \boldsymbol{u}) \equiv 0.$
5. $\nabla \times (\nabla \times \boldsymbol{u}) = \nabla(\nabla \cdot \boldsymbol{u}) - \nabla^2 \boldsymbol{u}.$
6. $\nabla(\phi\psi) = \phi\nabla\psi + \psi\nabla\phi.$
7. $\nabla(\boldsymbol{u} \cdot \boldsymbol{v}) = (\boldsymbol{u} \cdot \nabla)\boldsymbol{v} + (\boldsymbol{v} \cdot \nabla)\boldsymbol{u} + \boldsymbol{u} \times (\nabla \times \boldsymbol{v}) + \boldsymbol{v} \times (\nabla \times \boldsymbol{u}).$
8. $\nabla \cdot (\phi \boldsymbol{u}) = \phi(\nabla \cdot \boldsymbol{u}) + \boldsymbol{u} \cdot \nabla \phi.$
9. $\nabla \cdot (\boldsymbol{u} \times \boldsymbol{v}) = \boldsymbol{v} \cdot (\nabla \times \boldsymbol{u}) - \boldsymbol{u} \cdot (\nabla \times \boldsymbol{v}).$
10. $\nabla \times (\phi \boldsymbol{u}) = \phi \nabla \times \boldsymbol{u} + \nabla \phi \times \boldsymbol{u}.$
11. $\nabla \times (\boldsymbol{u} \times \boldsymbol{v}) = \boldsymbol{u}\,(\nabla \cdot \boldsymbol{v}) - \boldsymbol{v}\,(\nabla \cdot \boldsymbol{u}) + (\boldsymbol{v} \cdot \nabla)\boldsymbol{u} - (\boldsymbol{u} \cdot \nabla)\boldsymbol{v}.$

A.2 Navier–Stokes Equations in Cylindrical Polar Coordinates

The incompressible Navier–Stokes equations in *cylindrical polar coordinates* (r, θ, z) have the following form in terms of the velocity field $\boldsymbol{u} = u_r\, \hat{\boldsymbol{r}} + u_\theta\, \hat{\boldsymbol{\theta}} + u_z\, \hat{\boldsymbol{z}}$ (for the incompressible Euler equations set $\nu = 0$):

$$
\begin{aligned}
\frac{\partial u_r}{\partial t} + (\boldsymbol{u} \cdot \nabla)\, u_r - \frac{u_\theta^2}{r} &= -\frac{1}{\rho}\frac{\partial p}{\partial r} + \nu\left(\Delta u_r - \frac{u_r}{r^2} - \frac{2}{r^2}\frac{\partial u_\theta}{\partial \theta}\right) + f_r, \\
\frac{\partial u_\theta}{\partial t} + (\boldsymbol{u} \cdot \nabla)\, u_\theta + \frac{u_r u_\theta}{r} &= -\frac{1}{\rho r}\frac{\partial p}{\partial \theta} + \nu\left(\Delta u_\theta + \frac{2}{r^2}\frac{\partial u_r}{\partial \theta} - \frac{u_\theta}{r^2}\right) + f_\theta, \\
\frac{\partial u_z}{\partial t} + (\boldsymbol{u} \cdot \nabla)\, u_z &= -\frac{1}{\rho}\frac{\partial p}{\partial z} + \nu \Delta u_z + f_z,
\end{aligned}
$$

where p is the pressure, ρ is the mass density and $\boldsymbol{f} = f_r\,\hat{\boldsymbol{r}} + f_\theta\,\hat{\boldsymbol{\theta}} + f_z\,\hat{\boldsymbol{z}}$ is the body force per unit mass. Here we also have

$$\boldsymbol{u}\cdot\nabla = u_r\frac{\partial}{\partial r} + \frac{u_\theta}{r}\frac{\partial}{\partial\theta} + u_z\frac{\partial}{\partial z} \quad\text{and}\quad \Delta = \frac{1}{r}\frac{\partial}{\partial r}\left(r\frac{\partial}{\partial r}\right) + \frac{1}{r^2}\frac{\partial^2}{\partial\theta^2} + \frac{\partial^2}{\partial z^2}.$$

Further, the gradient operator and the divergence of a vector field $\boldsymbol{u}$ are given in cylindrical coordinates, respectively, by

$$\nabla = \hat{\boldsymbol{r}}\frac{\partial}{\partial r} + \hat{\boldsymbol{\theta}}\frac{1}{r}\frac{\partial}{\partial\theta} + \hat{\boldsymbol{z}}\frac{\partial}{\partial z} \quad\text{and}\quad \nabla\cdot\boldsymbol{u} = \frac{1}{r}\frac{\partial}{\partial r}(ru_r) + \frac{1}{r}\frac{\partial u_\theta}{\partial\theta} + \frac{\partial u_z}{\partial z}.$$

In cylindrical coordinates $\nabla\times\boldsymbol{u}$ is given by

$$\nabla\times\boldsymbol{u} = \left(\frac{1}{r}\frac{\partial u_z}{\partial\theta} - \frac{\partial u_\theta}{\partial z}\right)\hat{\boldsymbol{r}} + \left(\frac{\partial u_r}{\partial z} - \frac{\partial u_z}{\partial r}\right)\hat{\boldsymbol{\theta}} + \left(\frac{1}{r}\frac{\partial}{\partial r}(ru_\theta) - \frac{1}{r}\frac{\partial u_r}{\partial\theta}\right)\hat{\boldsymbol{z}}.$$

Lastly the diagonal components of the deformation matrix D are

$$D_{rr} = \frac{\partial u_r}{\partial r}, \qquad D_{\theta\theta} = \frac{1}{r}\frac{\partial u_\theta}{\partial\theta} + \frac{u_r}{r} \qquad\text{and}\qquad D_{zz} = \frac{\partial u_z}{\partial z},$$

while the off-diagonal components are

$$2D_{r\theta} = r\frac{\partial}{\partial r}\left(\frac{u_\theta}{r}\right) + \frac{1}{r}\frac{\partial u_r}{\partial\theta}, \quad 2D_{rz} = \frac{\partial u_r}{\partial z} + \frac{\partial u_z}{\partial r} \quad\text{and}\quad 2D_{\theta z} = \frac{1}{r}\frac{\partial u_z}{\partial\theta} + \frac{\partial u_\theta}{\partial z}.$$

A.3 Navier–Stokes Equations in Spherical Polar Coordinates

The incompressible Navier–Stokes equations in *spherical polar coordinates* (r,θ,φ), with θ the angle to the south–north pole axis and φ the azimuthal angle, have the following form in terms of the velocity field $\boldsymbol{u} = u_r\,\hat{\boldsymbol{r}} + u_\theta\,\hat{\boldsymbol{\theta}} + u_\varphi\,\hat{\boldsymbol{\varphi}}$ (for the incompressible Euler equations set $\nu = 0$):

$$\begin{aligned}
\frac{\partial u_r}{\partial t} + (\boldsymbol{u}\cdot\nabla)\,u_r - \frac{u_\theta^2}{r} - \frac{u_\varphi^2}{r} &= -\frac{1}{\rho}\frac{\partial p}{\partial r} \\
&\quad + \nu\left(\Delta u_r - 2\frac{u_r}{r^2} - \frac{2}{r^2\sin\theta}\frac{\partial}{\partial\theta}(u_\theta\sin\theta) - \frac{2}{r^2\sin\theta}\frac{\partial u_\varphi}{\partial\varphi}\right) + f_r, \\
\frac{\partial u_\theta}{\partial t} + (\boldsymbol{u}\cdot\nabla)\,u_\theta + \frac{u_r u_\theta}{r} - \frac{u_\varphi^2\cos\theta}{r\sin\theta} &= -\frac{1}{\rho r}\frac{\partial p}{\partial\theta} \\
&\quad + \nu\left(\Delta u_\theta + \frac{2}{r^2}\frac{\partial u_r}{\partial\theta} - \frac{u_\theta}{r^2\sin^2\theta} - 2\frac{\cos\theta}{r^2\sin^2\theta}\frac{\partial u_\varphi}{\partial\varphi}\right) + f_\theta, \\
\frac{\partial u_\varphi}{\partial t} + (\boldsymbol{u}\cdot\nabla)\,u_\varphi + \frac{u_r u_\varphi}{r} + \frac{u_\theta u_\varphi\cos\theta}{r\sin\theta} &= -\frac{1}{\rho r\sin\theta}\frac{\partial p}{\partial\varphi} \\
&\quad + \nu\left(\Delta u_\varphi + \frac{2}{r^2\sin\theta}\frac{\partial u_r}{\partial\varphi} + \frac{2\cos\theta}{r^2\sin^2\theta}\frac{\partial u_\theta}{\partial\varphi} - \frac{u_\varphi}{r^2\sin^2\theta}\right) + f_z,
\end{aligned}$$

where p is the pressure, ρ is the mass density and $\boldsymbol{f} = f_r\,\hat{\boldsymbol{r}} + f_\theta\,\hat{\boldsymbol{\theta}} + f_\varphi\,\hat{\boldsymbol{\varphi}}$ is the body force per unit mass. Here we also have

$$\boldsymbol{u}\cdot\nabla = u_r\frac{\partial}{\partial r} + \frac{u_\theta}{r}\frac{\partial}{\partial\theta} + \frac{u_\varphi}{r\sin\theta}\frac{\partial}{\partial\varphi}$$

and

$$\Delta = \frac{1}{r^2}\frac{\partial}{\partial r}\left(r^2\frac{\partial}{\partial r}\right) + \frac{1}{r^2\sin\theta}\frac{\partial}{\partial\theta}\left(\sin\theta\frac{\partial}{\partial\theta}\right) + \frac{1}{r^2\sin^2\theta}\frac{\partial^2}{\partial\varphi^2}.$$

Further, the gradient operator and the divergence of a vector field $\boldsymbol{u}$ are given in spherical coordinates, respectively, by

$$\nabla = \hat{\boldsymbol{r}}\frac{\partial}{\partial r} + \hat{\boldsymbol{\theta}}\frac{1}{r}\frac{\partial}{\partial\theta} + \hat{\boldsymbol{\varphi}}\frac{1}{r\sin\theta}\frac{\partial}{\partial\varphi}$$

and

$$\nabla\cdot\boldsymbol{u} = \frac{1}{r^2}\frac{\partial}{\partial r}(r^2u_r) + \frac{1}{r\sin\theta}\frac{\partial}{\partial\theta}(\sin\theta\,u_\theta) + \frac{1}{r\sin\theta}\frac{\partial u_\varphi}{\partial\varphi}.$$

In spherical coordinates $\nabla\times\boldsymbol{u}$ is given by

$$\nabla\times\boldsymbol{u} = \frac{1}{r\sin\theta}\left(\frac{\partial}{\partial\theta}(\sin\theta\,u_\varphi) - \frac{\partial u_\theta}{\partial\varphi}\right)\hat{\boldsymbol{r}} + \left(\frac{1}{r\sin\theta}\frac{\partial u_r}{\partial\varphi} - \frac{1}{r}\frac{\partial}{\partial r}(ru_\varphi)\right)\hat{\boldsymbol{\theta}}$$
$$+\left(\frac{1}{r}\frac{\partial}{\partial r}(ru_\theta) - \frac{1}{r}\frac{\partial u_r}{\partial\theta}\right)\hat{\boldsymbol{\varphi}}.$$

Lastly the diagonal components of the deformation matrix D are

$$D_{rr} = \frac{\partial u_r}{\partial r},\qquad D_{\theta\theta} = \frac{1}{r}\frac{\partial u_\theta}{\partial\theta} + \frac{u_r}{r}\quad\text{and}\quad D_{\varphi\varphi} = \frac{1}{r\sin\theta}\frac{\partial u_\varphi}{\partial\varphi} + \frac{u_r}{r} + \frac{u_\theta\cot\theta}{r},$$

while the off-diagonal components are given by

$$2D_{r\theta} = r\frac{\partial}{\partial r}\left(\frac{u_\theta}{r}\right) + \frac{1}{r}\frac{\partial u_r}{\partial\theta},\qquad 2D_{r\varphi} = \frac{1}{r\sin\theta}\frac{\partial u_r}{\partial\varphi} + r\frac{\partial}{\partial r}\left(\frac{u_\varphi}{r}\right)$$

and

$$2D_{\theta\varphi} = \frac{\sin\theta}{r}\frac{\partial}{\partial\theta}\left(\frac{u_\varphi}{\sin\theta}\right) + \frac{1}{r\sin\theta}\frac{\partial u_\theta}{\partial\varphi}.$$

References

Acheson, D. J. 1990. *Elementary Fluid Dynamics*, Oxford Applied Mathematics and Computing Science Series. Oxford University Press, Oxford.

Akhmetov, D. G. 2009. *Vortex Rings*. Springer, New York.

Ambrosio, L. and Figalli, A. 2008. On the regularity of the pressure field of Brenier's weak solutions to incompressible Euler equations. *Calc. Var.* **31**, pp. 497–509.

Ambrosio, L. and Figalli, A. 2010. *Lecture Notes on Variational Models for Incompressible Euler Equations*. HAL Archives-Overtes: hal-00518481.

Anderson, J. D. Jr. 1998. *A History of Aerodynamics*. Cambridge University Press, New York.

Anderson, J. D. Jr. 2005. Ludwig Prandtl's boundary layer. *Physics Today* **Dec**, pp. 42–8.

Anderson, J. 2017. *Fundamentals of Aerodynamics*, 6th ed. McGraw–Hill-Education, New York.

Arnold, V. I. and Keshin, B. 1998. *Topological Methods in Hydrodynamics*. Springer, New York.

Baines, P. G. 1987. Upstream blocking and airflow over mountains. *Ann. Rev. Fluid Mech.* **19**, pp. 75–97.

Batchelor, G. K. 1967. *An Introduction to Fluid Dynamics*. Cambridge University Press, New York.

Bayada, G. and Váquez, C. 2007. A survey on mathematical aspects of lubrication problems. *Bol. Soc. Esp. Mat. Apl.* **39**, pp. 31–74.

Beale, J. T., Kato, T. and Majda, A. 1984. Remarks on the breakdown of smooth solutions for the 3-D Euler equations. *Commun. Math. Phys.* **94**, pp. 61–6.

Benjamin, T. B. 1975. The alliance of practical and analytical insights into the nonlinear problems of fluid mechanics. In Germain, P. and Nayroles, B. (eds), *Applications of Methods of Functional Analysis to Problems in Mechanics*. Lecture Notes in Mathematics, vol. **503**, pp. 8–29. Springer, Berlin.

Beirão da Veiga, H. and Berselli, L. C. 2002. On the regularizing effect of the vorticity direction in incompressible viscous flows. *Diff. Int. Eq.* **15**(3), pp. 345–56.

Berkshire, F. H. and Warren, F. W. G. 1970. Some aspects of linear Lee wave theory for the stratosphere. *Q. J. Roy. Met. Soc.* **96**(407), pp. 50–66.

Berselli, L. C. and Chiodaroli, E. 2020. On the energy equality for the 3D Navier–Stokes equations. *Nonlinear Anal.* **192**, 111704.

Besse, N. 2020. Regularity of the geodesic flow of the incompressible Euler equations on a manifold. *Commun. Math. Phys.* doi.org/10.1007/s00220-019-03656-5

Bianchi, L. A. and Flandoli, F. 2020. Stochastic Navier–Stokes equations and related models. *Milan J. Math.* **88**, pp. 225–46.

Blasius, P. R. H. 1908. *Z. Math. Phys.* **56**, p. 1.

Buckmaster, T. and Vicol, V. 2019. Nonuniqueness of weak solutions to the Navier–Stokes equation. *Ann. Math.* **189**(1), pp. 101–44.

Buckmaster, T. and Vicol, V. 2019. *Convex integration and phenomenologies in turbulence*, arXiv:1901.09023v2.

Burkill, J. C. and Burkill, H. 1970. *A Second Course in Mathematical Analysis*. Cambridge University Press, Cambridge.

Busnello, B., Flandoli, F. and Romito, M. 2005. A probabilistic representation for the vorticity of a three-dimensional viscous fluid and for general systems of parabolic equations. *Proc. Edin. Math. Soc.* **48**(2), pp. 295–336.

Caffarelli, L., Kohn, R. and Nirenberg, L. 1982. Partial regularity of suitable weak solutions of the Navier–Stokes equations. *Commun. Pure Appl. Math.* **XXXV**, pp. 771–831.

Cassel, K. W. and Conlisk, A. T. 2014. Unsteady separation in vortex-induced boundary layers. *Phil. Trans. R. Soc. A* **372**: 20130348.

Childress, S. 2009. *An Introduction to Theoretical Fluid Dynamics*, Courant Lecture Notes 19. American Mathematical Society, Providence, RI.

Chorin, A. J. and Marsden, J. E. 1990. *A Mathematical Introducton to Fluid Mechanics*, 3rd ed. Springer-Verlag, New York.

Constantin, P. 2007. On the Euler equations of incompressible fluids. *Bull. Amer. Math. Soc.* **44**(4), pp. 603–21.

Constantin, P. and Fefferman, C. 1993. Direction of vorticity and the problem of global regularity for the Navier–Stokes equations. *Indiana Univ. Math. J.* **42**(3), pp. 775–89.

Constantin, P., Fefferman, C. and Majda, A. 1996. Geometric constraints on potentially singular solutions for the 3-D Euler equations. *Commun. Partial Diff. Eq.* **21**, pp. 559–71.

Constantin, P. and Foias, C. 1988. *Navier–Stokes Equations*, Chicago Lectures in Mathematics. University of Chicago Press, Chicago, IL.

Constantin, P. and Iyer, G. 2008. A stochastic Lagrangian representation of the three-dimensional incompressible Navier–Stokes equations. *Commun. Pure Appl. Math.* **61**, pp. 330–45.

Crowdy, D. G. 2004. Stuart vortices on a sphere, *J. Fluid Mech.* **498**, pp. 381–402.

De Lellis, C. and Székelyhidi, L. Jr. 2014. Dissipative Euler flows and Onsager's conjecture. *J. Eur. Math. Soc.* **16**(7), pp. 1467–505.

Doering, C. R. 2009. The 3D Navier–Stokes Problem. *Ann. Rev. Fluid Mech.* **41**, pp. 109–28.

Durran, D. R. 2015. Mountain meteorology, in: *Encyclopedia of Atmospheric Sciences*, 2nd ed. Elsevier, Amsterdam, pp. 69–74.

E, W. 2000. Boundary layer theory and the zero viscosity limit of the Navier–Stokes equation. *Acta Math. Sin.* **16**, pp. 207–18.

E, W. and Engquist, B. 1997. Blowup of solutions of the unsteady Prandtl's equation. *Commun. Pure Appl. Math.* **50**(12), pp. 1287–93.

Emden, R. 1907. *Gaskugeln: Anwendungen der Mechanischen Wärmetheorie auf Kosmologische und Meteorologische Probleme*. Teubner, Berlin.

Evans, L. C. 1998. *Partial Differential Equations*, Graduate Studies in Mathematics 19. American Mathematical Society, Providence, RI.

Falkner, V. W. and Skan, S. W. 1931. Solutions of the boundary layer equations, *Phil. Mag. (Ser. 7)* **12**, pp. 865–96.

Fefferman, C. L. 2000. *Existence and smoothness of the Navier–Stokes equation*. Millenium Prize Problem.

Fraenkel, L. E. 1970. On steady vortex rings of small cross-section in an ideal fluid, *Proc. Roy. Soc. Lond. A* **316**, pp. 29–62.

Fraenkel, L. E. 1972. Examples of steady vortex rings of small cross-section in an ideal fluid, *J. Fluid Mech.* **51**(1), pp. 119–35.

Franosh, T., Grimm, M., Belushkin, M., Mor, F. M., Foffi, G., Forró, L. and Jeney, S. 2011. Resonances arising from hydrodynamic memory in Brownian motion, *Nature* **478**, pp. 85–8.

Galdi, G. P. 2019. On the energy equality for distributional solutions to Navier–Stokes equations. *Proc. Amer. Math. Soc.* **147**, pp. 785–92.

Gérard-Varet, D. and Dormy, E. 2010. On the ill-posedness of the Prandtl equation, *J. Amer. Math Soc.* **23**(2), pp. 591–609.

Gérard-Varet, D. and Nguyen, T. 2012. Remarks on the ill-posedness of the Prandtl equation, *Asymp. Anal.* **77**, pp. 71–88.

Gérard-Varet, D. and Masmoudi, N. 2013, Well-posedness issues for the Prandtl boundary layer equations. *Séminaire Laurent–Schwarz — EDP et applications*. Centre de mathématiques Laurent Schwarz, Exposé no. XV, pp. 1–10.

Grenier, E. 2000. On the nonlinear instability of Euler and Prandtl equations. *Commun. Pure Appl. Math.* **53**(9), pp. 1067–91.

Guillot, T. and Gautier, D. 2015. Giant planets, in *Treatise on Geophysics* (2nd ed.). Editor-in-Chief: Gerald Schubert, Vol. 10. Elsevier, Amsterdam, pp. 529–57.

Gurtin, M. E. 1981. *An Introduction to Continuum Mechanics*, Mathematics in Science and Engineering, Vol. 158. Academic Press, New York.

Hancock, C., Lewis, E. and Moffatt, H. K. 1981. Effects of inertia in forced corner flows. *J. Fluid Mech.* **112**, pp. 315–27.

Hardy, G. H., Littlewood, J. E. and Polya, G. 1952. *Inequalities*. Cambridge University Press, Cambridge.

Held, I. M., Pierrehumbert, R. T., Garner, S. T. and Swanson, K. L. 1995. Surface quasi-geostrophic dynamics. *J. Fluid Mech.* **282**, pp. 1–20.

Hicks, W. M. 1899. II. Researches in vortex motion. Part III. On spiral or gyrostatic vortex aggregates. *Phil. Trans. R. Soc. Lond. A* **192**, pp. 33–99.

Hieber, M. and Saal, J. 2018. The Stokes equation in the L^p-setting: Well-posedness and regularity properties. In *Handbook of Mathematical Analysis in Mechanics of Viscous Fluids*, Y. Giga and A. Novotný (eds). Springer, New York.

Hill, M. J. M 1894. On a spherical vortex. *Phil. Trans. R. Soc. Lond. A* **185**, pp. 213–45.

Holm, D.D., Marsden, J.E. and Ratiu, T. 1986. Nonlinear stability of the Kelvin–Stuart cat's eyes flow. In *Nonlinear Systems of Partial Differential Equations in Applied Mathematics Part 2*, Lectures in Applied Mathematics No. 23. American Mathematical Society, Providence, RI, pp. 171–86.

Hoskins, B. J. 1982. The mathematical theory of frontogenesis. *Ann. Rev. Fluid Mech.* **14**, pp. 131–51.

Hunt, J. C. R., Feng, Y., Linden, P. F., Greenslade, M. D. and Mobbs, S. D. 1997. Low-Froude-number stable flows past mountains. *Il Nuovo Cimento* **20**(3), pp. 261–72.

Hunter, J. K. 2004. Prandtl's boundary layer theory. *Applied Asymptotic Analysis Lecture Notes*, `www.math.ucdavis.edu/~hunter/m204/m204.html`

Hunter, J. K. 2011. Lecture notes on measure theory.
www.math.ucdavis.edu/~hunter/measure_theory/measure_theory.html

Jamieson, P. 2018. *Innovation in Wind Turbine Design*. Wiley, Chichester.

Johnson, F. T., Tinoco, E. N. and Jong Yu, N. 2005. Thirty years of development and application of CFD at Boeing Commercial Airplanes, Seattle. *Comput. Fluids* **34**, pp. 1115–51.

Kaplun, S. 1957. Low Reynolds number flow past a circular cylinder. *J. Math. Mech.* **6**, pp. 585–603.

Kelvin, Lord 1867. The translatory velocity of a circular vortex ring. *Phil. Mag.* **33**, pp. 511–12.

Kelvin, Lord 1880. On a disturbing infinity in Lord Rayleigh's solution for waves in a plane vortex stratum. *Nature* **23**, pp. 45–6.

Kerr, R. M. and Oliver, M. 2011. The ever-elusive blowup in the mathemtical description of fluids. In D. Schleicher and M. Lackmann (eds), *An Invitation to Mathematics*. Springer, Berlin.

King, A. C., Billingham, J. and Otto, S. R. 2003. *Differential Equations: Linear, Nonlinear, Ordinary, Partial*. Cambridge University Press, Cambridge.

Kundu, P. K. and Cohen, I. M. 2008. *Fluid Mechanics*, 4th ed. Academic Press as an imprint of Elsevier, Amsterdam.

Ladyzhenskaya, O. A. 1969. *The Mathematical Theory of Viscous Incompressible Flow*. Gordon & Breach, London.

Lane, J. H. 1870. On the theoretical temperature of the Sun under the hypothesis of a gaseous mass maintaining its internal heat and depending on the laws of gases known to terrestial experiment. *Amer. J. Sci. Arts* **50**, pp. 57–74.

Leray, J. 1934. Essai sur le mouvement d'un liquide visqueux emplissant l'espace. *Acta Math.* **63**, pp. 193–248.

Lighthill J. 1958. On displacement thickness, *J. Fluid Mech.* **4**(4), pp. 383–92.

Lighthill J. 1978. *Waves in Fluids*, Cambridge University Press, Cambridge.

Lighthill J. 1986. *An Informal Introduction to Theoretical Fluid Mechanics*. Clarendon Press, Oxford.

Lombardo, M. C., Cannone, M. and Sammartino, M. 2003. Well-posedness of the boundary layer equations. *SIAM J. Math. Anal.* **35**(4), pp. 987–1004.

Majda, A. J. and Bertozzi, A. L. 2002. *Vorticity and Incompressible Flow*, Cambridge Texts in Applied Mathematics. Cambridge University Press, Cambridge.

Malham, S. J. A. 1993. *Regularity assumptions and length scales for the Navier–Stokes equations*, PhD thesis, University of London.

Matthews, M. T. 2012. Complex mapping of aerofoils. *Int. J. Math. Ed. Sci. Technol.* **43**(1), pp. 43–65.

Mavroyiakoumou, C. and Berkshire, F. 2020. Collinear interaction of vortex pairs with different strengths – criteria for leapfrogging. *Phys. Fluids* **32**, 023603.

McCallum, W. G. 1997. *Multivariable Calculus*. Wiley, New York.

Mehta, R. D. 1985. Aerodynamics of sports balls. *Ann. Rev. Fluid Mech.* **17**, pp. 151–89.

Meyer, C. D. 2000. *Matrix Analysis and Applied Linear Algebra*. SIAM, Philadelphia, PA.

Milne-Thomson, L. M. 1952. *Theoretical Aerodynamics*, 2nd ed. Macmillan, New York.

Moffatt, H. K. 1964. Viscous and resistive eddies near a sharp corner. *J. Fluid Mech.* **18**(1), pp. 1–18.

Moffatt, H. K. and Moore, D. W. 1978. The response of Hill's spherical vortex to a small axisymmetric disturbance. *J. Fluid Mech.* **87**, pp. 749–60.

Nickelsen, D. 2017. Master equation for She–Leveque scaling and its classification in terms of other Markov models of developed turbulence. *J. Stat. Mech.: Theory Exp.* **2017**, 073209.

Nirenberg, L. 1959. On elliptic partial differential equations. *Annali della Scuola Norm. Sup. Di Pisa* **13**, pp. 115–62.

Norbury J. 1972. A steady vortex ring close to Hill's spherical vortex. *Proc. Camb. Phil. Soc.* **72**, pp. 253–84.

Norbury J. 1973. A family of steady vortex rings. *J. Fluid Mech.* **57**(3), pp. 417–31.

Nørstrud, H. 2008. *Sport Aerodynamics*, CISM Courses and Lectures 506. Springer, Wien.

Ockendon, H. and Ockendon, J. R. 1995. *Viscous Flow*, Cambridge Texts in Applied Mathematics. Cambridge University Press, Cambridge.

Oleinik, O. A. 1963. The Prandtl system of equations in boundary layer theory. *Soviet Math. Dokl.* **4**, pp. 583–6.

Oleinik, O. A. 1967. Construction of the solutions of a system of boundary layer equations by the method of straightlines. *Soviet Math. Dokl.* **8**, pp. 775–9.

Oseen, C. W. 1910. *Über die Stokes'sche Formel, unde über eine verwandt Aufgabe in der Hydrodynamik*, Ark Math. Astrom. Fys. **6**(29).

Panton, R. L. 2013. *Incompressible Flow*, 4th ed. Wiley, Chichester.

Paterson, A.R. 1983. *A First Course in Fluid Dynamics*. Cambridge University Press, Cambridge.

Pedlosky J. 1979. *Geophysical Fluid Dynamics*. Springer, New York.

Ponnusamy, S. and Silverman, H. 2006. *Complex Variables with Applications*. Birkhäuser, Boston.

Pozrikidis, C. 1986. The nonlinear instability of Hill's vortex. *J. Fluid Mech.* **168**, pp. 337–67.

Prandtl, L. 1904. In Verhandlungen des dritten internationalen Mathematiker-Kongresses in Heidelberg 1904 (A. Krazer, ed.), Teubner, Leipzig, p. 484 [English trans. in *Early Developments of Modern Aerodynamics*, J. A. K. Ackroyd, B. P. Axcell and A. I. Ruban (eds), Butterworth–Heinemann, Oxford, 2001, p. 77].

Protas, B. 2019. Linear instability of inviscid vortex rings to axisymmetric perturbations. *J. Fluid Mech.* **874**, pp. 1115–46.

Proudman, I. and Pearson, J. R. A. 1957. Expansions at small Reynolds numbers for the flow past a sphere and a circular cylinder. *J. Fluid Mech.* **2**, pp. 237–62.

Rayleigh, J.W.S. 1883. The soaring of birds. *Nature* **27**, pp. 534–5.

Reed. M. and Simon, B. 1975. *Methods of Modern Mathematical Physics: I Functional Analysis*. Academic Press, New York.

Reed. M. and Simon, B. 1980. *Methods of Modern Mathematical Physics: II Fourier Analysis, Self-Adjointness*. Academic Press, New York.

Reich, S. and Cotter, C. 2015. *Probabilistic Forecasting and Bayesian Data Assimilation*. Cambridge University Press, Cambridge.

Richardson, P. L. 2019. Leonardo da Vinci's discovery of the dynamic soaring by birds in wind shear. *Notes Rec.* **73**, pp. 285–301.

Robinson, J. C. 2011. *An introduction to the rigorous theory of weak solutions of the Navier–Stokes equations*. LMS–EPSRC Short Course on Mathematical Fluid Dynamics, Heriot-Watt University.

Robinson, J. C., Rodrigo, J. L. and Sadowski, W. 2016. *The Three-dimensional Navier–Stokes Equations: Classical Theory*, Cambridge Studies in Advanced Mathematics. Cambridge University Press, Cambridge.

Rosenhead, L. 1963. *Laminar Boundary Layers*. Oxford University Press, Oxford.

Rubenstein, D. A., Yin, W. and Frame, M. D. 2016. *Biofluid Mechanics: An Introduction to Fluid Mechanics, Macrocirculation and Microcirculation*, 2nd ed. Academic Press, New York.

Saffman, P. G. 1992. *Vortex Dynamics*, Cambridge Monographs in Mechanics and Applied Mathematics. Cambridge University Press, Cambridge.

Sammartino, M. and Caflisch, R. E. 1998a. Zero viscosity limit for analytic solutions of the Navier–Stokes equation on a half-space. I. Existence for the Euler and Prandtl equations. *Commun. Math. Phys.* **192**, pp. 433–61.

Sammartino, M. and Caflisch, R. E. 1998b. Zero viscosity limit for analytic solutions, of the Navier–Stokes equation on a half-space. II. Construction of the Navier–Stokes solution. *Commun. Math. Phys.* **192**, pp. 463–91.

Scheffer, V. 1976. Partial regularity of solutions to the Navier–Stokes equations, *Pacific J. Math.* **66**(2), pp. 535–52.

Scheffer, V. 1977. Hausdorff measure and the Navier–Stokes equations, *Comm. Math. Phys.* **55**(2), pp. 97–112.

Scorer, R. S. 1978. *Environmental Aerodynamics*. Wiley, Chichester.

Serrin, J. 1962. On the interior regularity of weak solutions of the Navier–Stokes equations. *Archiv. Rat. Mech. Anal.* **9**, pp. 187–195.

Shenoy, A. 2020. *Rheology of Drag Reducing Fluids*. Springer, New York.

Shinbrot, M. 1974. The energy equation for the Navier–Stokes system. *SIAM J. Math. Anal.* **5**(6), pp. 948–54.

Simon, J. 1987. Compact sets on the space $L^p(0, T; B)$. *Annali di Matematica Pura ed Applicata (IV)* **CXLVI**, pp. 65–96.

Smith, F. T. 1977. The laminar separation of an incompressible fluid streaming past a smooth surface. *Proc. R. Soc. Lond. A* **356**, pp. 443–63.

Smith, R. B. 2015. Smith, R. B. 2015. Dynamic Meteorolgy. In *Encyclopedia of Atmospheric Sciences*, 2nd ed. Elsevier, Amsterdam, pp. 332–3.

Stashchuk, N., Inall, M. and Vlasenko, V. 2007. Analysis of supercritical tidal flow in a Scottish Fjord. *J. Phys. Oceanogr.* **37**(7), pp. 1793–810.

Stewartson, K. 1969. On the flow near the trailing edge of a flat plate – II, *Mathematika* **16**, pp. 106–21.

Stewartson, K. 1981. D'Alembert's paradox. *SIAM Review* **23**(3), pp. 308–43.

Stuart, J. T. 1971. Stability problems in fluids. In *Mathematical Problems in Geophysical Sciences*, Lectures in Applied Mathematics 13. American Mathematical Society, Providence, RI, pp. 139–55.

Tao, T. 2016. Finite time blow up for an averaged three-dimensional Navier–Stokes equation. *J. Amer. Math. Soc.* **29**, pp. 601 74.

Taylor, G. I. 1962. On scraping viscous fluid from a plane surface. In *Miszellaneen der Angewandten Mechanik (Festschrift Walter Tollmien)*. Akademie Berlin, Berlin, pp. 313–15.

Temam, R. 1983. *Navier–Stokes Equations and Nonlinear Functional Analysis*. SIAM, Philadelphia, PA.

Van Dyke, M. 1982. *An Album of Fluid Motion*. Stanford University Press, Stanford, CA.

Veldman, A. E. P. 2017. Entrainment and boundary layer separation: a modeling history. *J. Eng. Math.* **107**, pp. 5–17.

Widnall, S. E., Yin Tsai, C. and Stuart, J. T. 1977. The instability of the thin vortex ring of constant velocity. *Phil. Trans. R. Soc. Lond. A* **287**(1344), pp. 273–305.

Xin, Z. and Zhang, L. 2004. On the global existence of solutions to Prandtl's system. *Adv. Math.* **181**(1), pp. 88–133.

Yung, P-L. 2017. *Topics in Harmonic Analysis*, Lecture Notes.
`www.math.cuhk.edu.hk/~plyung/math6081/index.html`

Zeiner-Gundersen, D. H. 2015. Turbine design and field development concepts for tidal, ocean, and river applications. *Energy Sci. Eng.* **3**(1), pp. 27–42.

Further Reading

Adams, R. A. 1975. *Sobolev Spaces.* Academic Press, New York.

Birkhoff, G. 1960. *Hydrodynamics: A Study in Logic, Fact and Similitude.* Princeton University Press, Princeton, NJ.

Cantwell, B. 2019. *Applied Aerodynamics Lectures.* web.stanford.edu/cantwell/

Currie, I. G. 2003. *Fundamental Mechanics of Fluids*, 3rd ed. Marcel Dekker, New York.

Doering, C. R. and Gibbon, J. D. 1995. *Applied Analysis of the Navier–Stokes Equations*, Cambridge Texts in Applied Mathematics. Cambridge University Press, Cambridge.

Drazin, P. G. and Reid, W. H. 1981. *Hydrodynamic Stability.* Cambridge University Press, Cambridge.

Faber, T. 1995. *Fluid Dynamics for Physicists.* Cambridge University Press, Cambridge.

Fulton, W. and Harris, J. 2004. *Representation Theory: A First Course*, Graduate Texts in Mathematics 129. Springer, New York.

Goldstein, S. 1965. *Modern Developments in Fluid Mechanics.* Dover, New York.

Kaper, H. and Engler, H. 2013. *Mathematics and Climate.* SIAM, Philadelphia, PA.

Keener, J. P. 2000. *Principles of Applied Mathematics: Transformation and Approximation.* Perseus Books: New York.

Lamb, H. 1932. *Hydrodynamics*, 6th ed. Cambridge University Press, Cambridge.

Landau, L. D. and Lifshitz, E. M. 1959. *Fluid Mechanics*, Vol. 6 of Course of Theoretical Physics. Pergamon Press, New York.

Lautrup, B. 2005. *Physics of Continuous Matter – Exotic and Everyday Phenomena in the Macroscopic World.* Institute of Physics Publishing, Bristol.

Marsden, J. E. and Ratiu, T. S. 1999. *Introduction to Mechanics and Symmetry*, 2nd ed. Springer, New York.

Messiter, A. F. 1970. Boundary-layer flow near the trailing edge of a flat plate, *SIAM J. Appl. Math.* **18**, pp. 241–57.

Pozrikidis, C. 2011. *Introduction to Theoretical and Computational Fluid Dynamics*, 2nd ed. Oxford University Press, Oxford.

Schlichting, H. 1979. *Boundary-Layer Theory*, 7th ed. McGraw-Hill, New York.

Stein, E. M. 1970. *Singular Integrals and Differentiability Properties of Functions*, Princeton Mathematical Series. Princeton University Press, Princeton, NJ.

Tritton, D. J. 1988. *Physical Fluid Dynamics*, 2nd ed. Oxford Science Publications, Oxford.

von Kármán, T. 1954. *Aerodynamics.* Cornell University Press, Ithaca, NY.

White, F. 2011. *Fluid Mechanics*, 7th ed. McGraw-Hill, New York.

Yih, C.-S. 1969. *Fluid Mechanics.* McGraw-Hill, New York.

Index